UNDERSTANDING
NUMBERS

MARIANNE FREIBERGER
& RACHEL THOMAS

UNDERSTANDING NUMBERS

SIMPLIFY LIFE'S
MATHEMATICS.
DECODE THE WORLD
AROUND YOU.

WHITE LION
PUBLISHING

Brimming with creative inspiration, how-to projects and useful information to enrich your everyday life, Quarto Knows is a favourite destination for those pursuing their interests and passions. Visit our site and dig deeper with our books into your area of interest: Quarto Creates, Quarto Cooks, Quarto Homes, Quarto Lives, Quarto Drives, Quarto Explores, Quarto Gifts or Quarto Kids.

First published in 2019 by White Lion Publishing
an imprint of The Quarto Group
The Old Brewery, 6 Blundell Street
London N7 9BH
United Kingdom

www.QuartoKnows.com

A catalogue record for this book is available from the British Library.

ISBN 978 1 78131 815 7
Ebook ISBN 978 1 78131 816 4
10 9 8 7 6 5 4 3 2 1
2023 2022 2021 2020 2019

Designed and illustrated by Stuart Tolley of Transmission Design

Printed in China

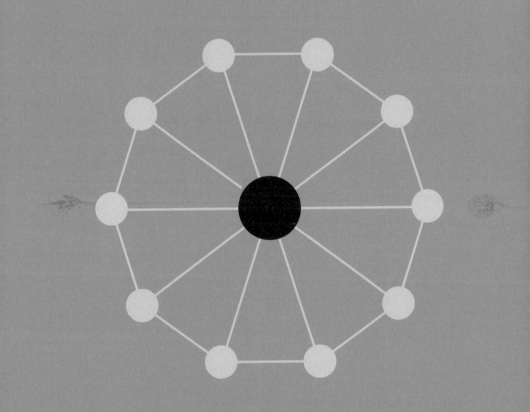

CONTENTS

INTRODUCTION

We have an affinity for patterns. We use this innate skill at spotting regularities to make sense of the world around us. Taking note of the passing of the seasons, the coming and going of the tides, the periodic motion of celestial bodies, and searching for their root causes helped our ancestors to understand the world around them. Monitoring what we produce – through hunting, farming and industry – has helped us to be more productive. Perceiving and studying the ebbs and flows of our bodies has kept us safe and healthy. Understanding the rhythms of the natural world has allowed us to shape it to serve our needs.

Mathematics is an indispensable tool in this context. It is the language of patterns and forms, whether they're found within numbers, physical shapes or processes that play out over time and space. Maths is the language of the sciences, and the key to unlocking the wealth of data that technology provides. Knowingly or not, you rely on its power every day.

Look at the world with a mathematical eye and you quickly find out how maths reveals the patterns in life, the universe and everything. And why these patterns make sense of things.

This book explores how looking at the world with a mathematical eye helps make sense of it. It is divided into five chapters: health, the environment, society, relationships and communication. Though it may not be obvious at first glance, each of these areas can be examined from a mathematical viewpoint, sometimes with surprising consequences.

This book will give you the language you need to understand some of the structures that underlie modern life, and allow you to make informed decisions that might change the way you live.

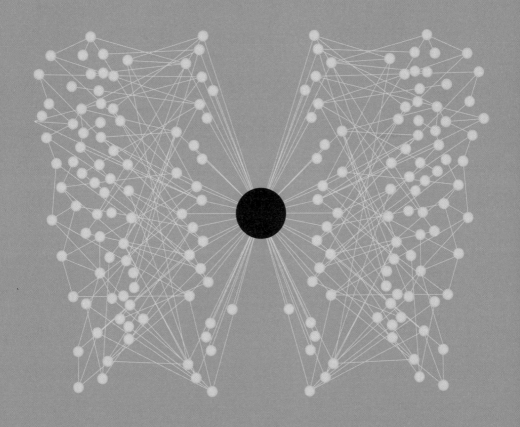

HOW TO USE THIS BOOK

This book is organized into five parts and 20 lessons covering the mathematical patterns at the heart of our modern world.

Each lesson introduces you to an important concept,

and explains how you can apply what you've learned to everyday life.

As you go through the book, TOOLKITS help you keep track of what you've learned so far.

At BUILD+BECOME we believe in building knowledge that helps you navigate your world. So, dip in, take it step-by-step, or digest it all in one go – however you choose to read this book, enjoy and get thinking.

Specially curated FURTHER LEARNING notes give you a nudge in the right direction for those things that most captured your imagination.

KNOWINGLY
YOU RELY O
MATHEMATI
EVERY DAY

OR NOT,

N
CS

HEALTH

LESSONS

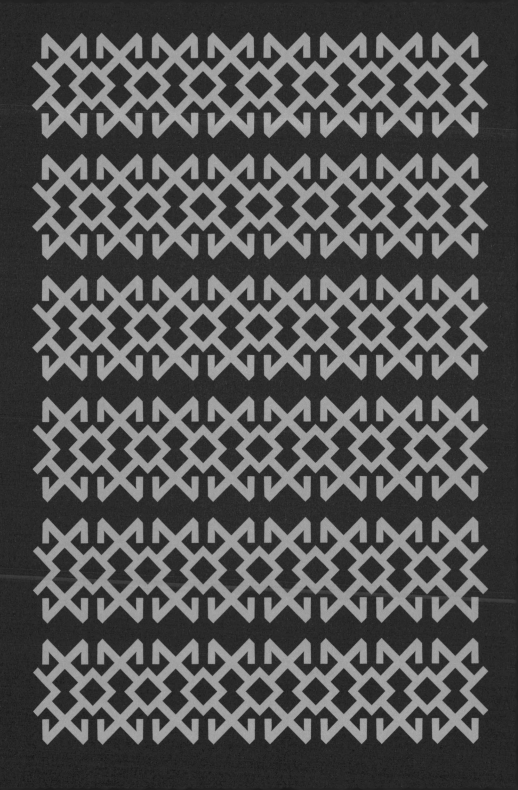

Today all reputable medical research uses the scientific method. That's good to know, but the resulting jungle of numbers and percentages can be hard to penetrate, let alone interpret.

Many people still instinctively think that bed rest is the best way to treat back pain. Indeed this was the treatment recommended by doctors until the mid-1990s. Then, in 1995, a team from Finland used a *randomized controlled trial* (RCT), the gold standard in evidence-based medicine, to turn this thinking on its head. The clear statistical arguments and rigorous mathematical design of RCTs provide the best known way to rule out bias and unknown factors to answer the question: what is the best treatment? Now patients with back pain are advised to stay active.

Today all reputable medical research uses the scientific method. That's good to know, but the resulting jungle of numbers and percentages can be hard to penetrate, let alone interpret. What does a 5% overall risk of a disease mean for you, personally? How do you know a treatment really works?

And should you spend money on costly alternative remedies that aren't supported by your healthcare system or insurance?

Newspapers and media outlets have an important, and not always positive, role to play in this context. Health scares and miracle cures make great headlines, and they are easy to come by if you're happy to massage the numbers to suit your purpose. Pharmaceutical companies also have an incentive to juggle the stats to suit their message. Even those in the medical profession don't appear always to see through the number jungle.

So what is the poor lay person to do? In this chapter we look at four situations in which very personal healthcare decisions depend on numbers and percentages. The results aren't always intuitive, so it's good to know where those numbers come from and what exactly they mean.

PREVENTING DISEASE

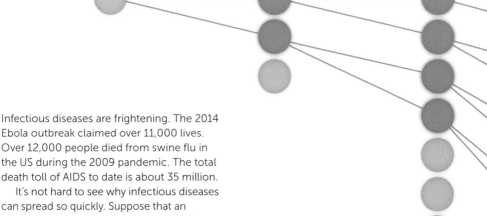

Infectious diseases are frightening. The 2014 Ebola outbreak claimed over 11,000 lives. Over 12,000 people died from swine flu in the US during the 2009 pandemic. The total death toll of AIDS to date is about 35 million.

It's not hard to see why infectious diseases can spread so quickly. Suppose that an infected person goes on to infect two other people during the course of their disease — not an unrealistic assumption if you consider the coughing and spluttering that goes on on public transport. A single infected person will infect another two people, giving a total of $1 + 2 = 3$ infected people. The two newly infected people will infect another two each, giving a total of $1 + 2 + 4 = 7$ infected people. The four newly infected people then infect another two each, giving a total of $1 + 2 + 4 + 8 = 15$ people, and so on.

Continuing in this vein, you see that the number of infected people grows very fast. In fact, it grows exponentially. If each infected person infects their two victims within the first day of catching the disease, then it will only take 26 days to infect a population larger than that of the UK. And that's starting with a single infected individual. (You might want to work out for yourself that the number of infected people after n days would be $2^0 + 2^1 + 2^2 + \ldots + 2^{n-1}$

Luckily, you don't need to tap this long sum into a calculator to get the result: a sum of this form is always equal to $2^n - 1$).

The number 2 obviously plays an important role in this example. If an infected person infected more than two people a day, then the disease would spread a lot quicker. And if an infected person infected fewer than two people a day, the disease would spread more slowly. In fact, it turns out that the number 1 is the watershed in this context. If an infected person infects, on average, more than one other person, the number of sick people will grow beyond any bound, as long as nothing bars the path of the disease. If, on the other hand, an infected person infects

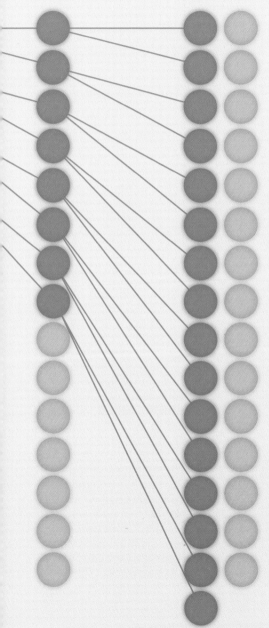

fewer than one other person on average, the spread will eventually come to a halt of its own accord.

Epidemiologists, those tasked with analyzing the spread of disease, have a name for the number of individuals that are, on average, infected by a person who has a particular disease, assuming that all the population is susceptible to catch the disease: it's called the *basic reproduction number* of the disease. Looking up basic reproduction numbers of common diseases gives you a good idea of how dangerous they are. The basic reproduction number of Ebola is between 1.5 and 2.5. For AIDS it lies somewhere between 2 and 5. For influenza (the 1918 epidemic strain) it's between 2 and 3. And for measles it's between 12 and 18!

What is to be done in the face of such ferocious exponential growth? Epidemiologists use complex mathematical models to see how a disease might spread. Importantly these models can be used to test the effect an intervention, such as vaccination or perhaps a travel ban, might have. The results don't always chime with intuition and can lead to outraged headlines. But rest assured: the advice epidemiologists come up with is based on thorough mathematical investigation.

WHY DOES VACCINATION WORK?

To give an idea of how even some basic maths can help, let's turn to the sometimes contentious subject of vaccination. The idea behind vaccination is to make people immune to a disease by injecting them with a pathogen, but it isn't without problems. It can be difficult and costly to get hold of everyone in a population; some people may be put at risk because of underlying health problems; and others may flatly refuse to be vaccinated. Luckily, though, you don't need to vaccinate everybody in a population to ensure the disease eventually fizzles out. Here's a short calculation to show why.

Suppose you have vaccinated a proportion p of the community, so these people are now immune to the disease. This means a proportion $1 - p$ is still susceptible to catching the disease. The basic reproduction number, call it R, gives the number of people a sick person infects, on average, in a totally susceptible population. Since after vaccinating only a proportion, $1 - p$ of the population, are still susceptible, the reproduction number is now only a proportion, $1 - p$, of what it was in a totally susceptible population: the basic reproduction number of R turns into an effective reproduction number of

$R \times (1 - p)$. In order for the disease to eventually fizzle out, we'd like the effective reproduction number to be less than 1, so

$$R \times (1 - p) < 1$$

A bit of rearranging will show that p, the proportion vaccinated, must therefore be at least $1 - 1/R$:

$$1 - 1/R < p$$

In other words, to ensure the disease dies out, you need to vaccinate a proportion of at least $1 - 1/R$ of the population.

For a basic reproduction number of 2, you only need to vaccinate $1 - 1/2 = 1/2$ of the population. If R is 3, the upper bound for influenza, you should vaccinate $1 - 1/3 = 2/3$ of the population. Importantly, our calculation shows that not only vaccinated people benefit from vaccination. People who haven't been vaccinated do as well, because their overall risk of catching the disease has decreased. In this way people who cannot have a vaccination for whatever reason can still be protected. Vaccination isn't just for you, it's for everyone!

These back-of-the-envelope calculations already provide quite a lot of insight into how diseases spread, but the models used by epidemiologists are more sophisticated. They take into account that not everyone in a population is at the same risk of catching a disease, that people may recover and become immune, and they try and reflect the complex contact patterns between people within a population. Whenever a new epidemic threatens, a race is on, not only to find a vaccine and effective drugs, but also to find good mathematical models to predict how the disease might spread.

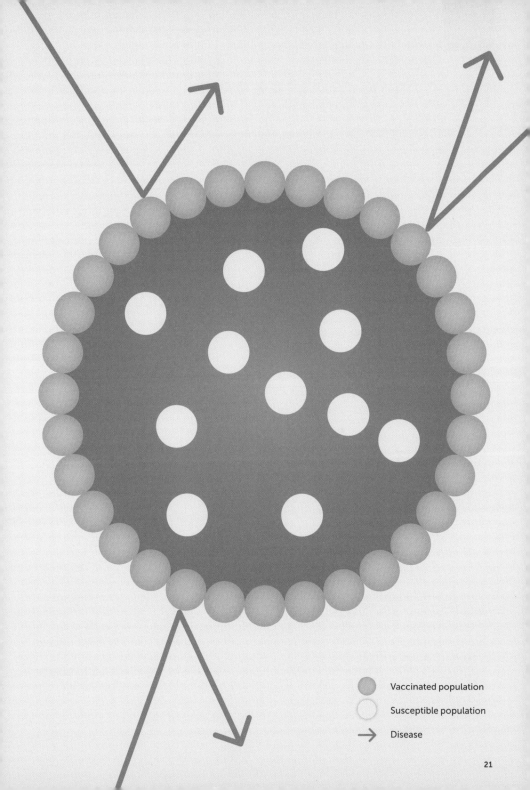

Vaccinated population

Susceptible population

→ Disease

DETECTING DISEASE

It seems obvious that knowledge is better than ignorance, that it is better to find out if you are likely to get a disease, rather than wait until you experience symptoms. But the example of screening for diseases illustrates that this isn't always the case. Screening programmes have both benefits (saving lives) and harms (which we will discuss below) – and these have to be balanced using careful statistical analysis. This is why screening programmes are so carefully researched before they are approved, and rigorously monitored to ensure the balance between benefit and harm is preserved.

The first thing to remember is that screening is not the same as diagnosis. Screening programmes check for well-understood markers that are a clear indication that a person is at a high risk of having a disease. In almost all cases the test is not the same as the diagnostic test used when a patient goes to see the doctor with symptoms of the disease (one exception is the screening test for HIV, hepatitis B and syphilis in pregnancy, which is the same as the diagnostic test). Screening is used to pick up these disease markers in the general population who, otherwise, have no symptoms of the disease.

10,000

people are tested for a disease that 1% of the population has at any one time.

So if your screening result shows that you do have the markers for a disease, usually called an *abnormal* result, then this does not mean that you have the disease. You will be called in for more tests, to confirm if you do or do not have the disease. And, in fact, the majority of people who get abnormal results in a screening programme do not have the disease in question.

This might seem surprising at first, but it stems from the fact that screening tests are not 100% accurate in their prediction as to whether a person will go on to develop the disease.

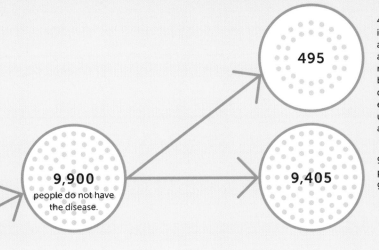

495 people will incorrectly receive an abnormal result, called a *false positive*. They do not have the disease, but will go through diagnostic tests and experience anxiety until they are given the all-clear.

9,405 (95% of 9,900) people will correctly be given a normal result.

An incorrect normal result — called a *false negative* — will be given to 5 people. This is one of the potential harms of screening — a relatively small number of people will be falsely reassured they're not at risk.

95 people (95% of 100) will correctly be given an abnormal result. It is worth noting that some of them may never have been harmed by the disease but may end up undergoing treatment unnecessarily, with all the risks that entails (known as overdiagnosis).

Suppose the screening test correctly predicts that someone has the disease 95% of the time. Out of the 10,000 people tested, 590 (the 95 with the disease and the 495 without) received abnormal results. The test has 95% accuracy so, if you were one of the 590 with an abnormal result, you might at first think that you have a 95% chance of having the disease. This is very high and would of course be very worrying.

But in fact, of the 590 people receiving positive tests, only just over 16% (95 out of 590) will have the disease. Even if you get a positive result it's actually much more likely (84%) that you still don't have the disease.

UNDERSTANDING SCREENING NUMBERS

Before you had the screening test you would have believed you had a 1% chance of having the disease, represented by the area of the blue circle (people having the disease) out of the white circle (the whole population) in the picture opposite. The screening test gave you new information (that you are in the green circle), which allows you to update your probability of having the disease – it is the area of the intersection of the green and blue circles (you have the disease and had a positive result) compared to the area of the green circle. This picture illustrates an incredibly useful result from probability theory, called *Bayes' theorem*. It allows you to update your beliefs about a particular state of affairs in the light of new evidence.

Working through this example illustrates how important it is for screening programmes to test for markers that reliably indicate that a person with an abnormal result is very likely to either have, or go on to develop, the disease in question. And this is the reason why medical experts are very careful in their design, and ongoing monitoring, of screening programmes. For example, they did not extend cervical cancer screening to women younger than 25, despite public pressure and media campaigns. The cervical cancer screening programme looks for changes in cells in the cervix, which is a strong indicator that a woman over 25 has, or will develop, cervical cancer. However it is not a good indicator for younger women, who are more likely to have changes in the cells in their cervix without going on to have cancer. Screening of this younger age group would have resulted in many more false positives, without guaranteeing to save any more lives.

Ultimately it is your decision whether or not to participate in a screening programme. Health organizations around the world are continuously improving their information about screening to help make that an informed decision. You need to understand the limitations of screening as outlined here and elsewhere, and weigh up the evidence and what it means for you.

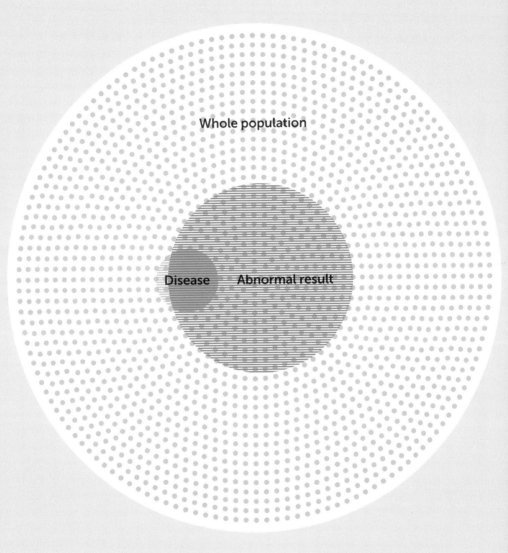

Whole population

Disease   Abnormal result

WHAT DO H STATISTICS FOR YOU?

EALTH
MEAN

TESTING TREATMENTS

01
Single-blinded trial

02
Double-blinded trial

03
Triple-blinded trial

How do you know that a new medicine or treatment works? It's tempting to think that if something worked for a friend, or a friend of a friend, or some blogger on the internet, that it will work for you. But there are many reasons why anecdotal evidence is not reliable as proof that a treatment or change of lifestyle will cure an illness.

The person in the story might have got better because of some other change in their life, or they might have got better regardless of any action they took. It's impossible to know what was responsible for someone's recovery from anecdotal evidence alone. Instead, the medical community requires treatments to be tested using *randomized controlled trials* (RCTs), and this takes some careful statistics.

How do we know?
RCTs have been developed over the last century to remove the possibility of intentionally or unintentionally biasing the results of testing treatments. They involve two groups of participants: the *control group* and the *study group*. The study group will be given the treatment in question. If the effectiveness of this treatment is being assessed against the treatment normally used for that particular condition, then the control group will be given the normal treatment. This allows researchers to assess the difference between the outcomes of these two types of treatment.

If there is no current treatment for the condition, however, the control group will be given a *placebo* – something that mimics treatment but has no physiological effect, such as a sugar pill. This allows someone

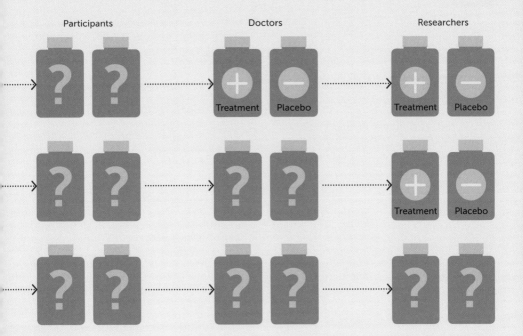

receiving treatment to be compared to someone not receiving treatment, but it also allows for something called the *placebo effect*: there is a great deal of research showing that the mere idea that they are receiving treatment, or medical attention, can make some people better. Even if the new treatment was not effective, then some people in the study group may still improve thanks to the placebo effect, but this should be comparable to any similar improvement in the control group.

In order to get an objective view of any difference between the study and control groups, it's important that people involved in the study don't know who is in the control group and who is in the study group — if, for example, people in the control group know they are getting the placebo treatment then there is less chance of any placebo effect

coming into play. To avoid this, RCTs are usually blinded. A single-blinded trial is one where the participants don't know which group they are in, and a double-blinded trial is one where both the doctors and participants don't know who is in the study group. A trial can also be triple-blinded if the researchers doing the final analysis of the results are also in the dark.

A blinded trial could still be influenced by the choice of who goes in which group. If the people in your study group were less sick, then you might get more favourable results for your treatment when compared to the sicker patients in the control group. The most powerful aspect of RCTs is that participants are *randomly allocated* to both groups. This balances out the healthier and sicker patients between the groups, and also evens out any other unknown factors that could affect the progression of a participant's illness.

HAS IT WORKED?

Given the parameters of an RCT, it would be surprising if we saw an improvement for those in the study group for a treatment that doesn't work (aside from the placebo effect). But rare and surprising things do happen, and maths provides the tools to quantifying these.

Suppose you have a pair of dice. If you rolled them 20 times you wouldn't be surprised if you rolled a double six once, or not at all. (The probability of rolling a double six is $\frac{1}{6} \times \frac{1}{6} = \frac{1}{36}$.) But if you rolled several double sixes out of the 20 rolls, you might start to doubt the fairness of the dice, and think they were weighted in some way. A similar approach is used to statistically understand the results of an RCT.

Significance level

RCTs are designed so that there is only a 5% probability that the difference in outcomes observed between the control group (taking a placebo) and the study group would occur by chance alone. To use the technical term, they are designed at a 5% *significance level*. That is, if the treatment were no more effective than a placebo, and you ran 20 trials, at most one trial would have shown

up this difference in observed outcomes by chance alone. Compare this to rolling our pair of dice: if the dice were both fair (which we can think of as our treatment being no more effective than a placebo) then we wouldn't be surprised if we rolled a double six out of 20 rolls. But we would be very surprised if we rolled several double sixes.

Point measure

The effect that is measured in one trial is called a *point measure*. This is the average of effects for that particular group of people at that particular time.

Confidence interval

What you are actually interested in is the *underlying real effect* of the treatment for the whole population who have that disease. This is why an RCT report refers to what is called a *confidence interval*. Usually, this is the point measure plus or minus an amount that depends on the variability in the trial's data.

The confidence interval is calculated so that researchers are 95% sure that the underlying real effect of the treatment being

Observed treatment effect

Point measure

Researchers are 95% sure that the underlying real effect of the treatment is in this confidence interval.

Confidence interval

tested will be somewhere in that confidence interval. The level of confidence, 95%, is equivalent to the significance level of 5%. The precise definition of a confidence interval means that if we repeated the trial 20 times for different groups taken from the general population of people with the disease, then 19 of the confidence intervals arising from these trials would contain the true underlying improvement in that population.

If the confidence interval given for the effect detected in an RCT does not include 0 (no difference in outcomes between the two groups), then that indicates the treatment's effect was *statistically significant*: we can be confident that the effect was really there. But statistical significance isn't the final hurdle – the effect must also be *clinically significant*. That is, it must make a marked difference on the outcome of the patients.

Individual anecdotes about the effects of treatments don't allow us to understand fully what works and what doesn't: careful design of trials and statistical analysis are crucial in finding the best treatments.

As we've noted, rare things do happen, so it is still possible to bias the reporting of RCTs by running lots of trials and only publishing results of those you like. One way to combat this is with greater transparency, with some countries requiring all trials to be registered in an easily accessible database, so that no unfavourable results can be hidden. The medical community continues its efforts to produce the best evidence for making medical decisions. And we, the general public and political decision-makers, must make an effort to understand this evidence in order to make the best decisions.

STATISTICALLY SPEAKING

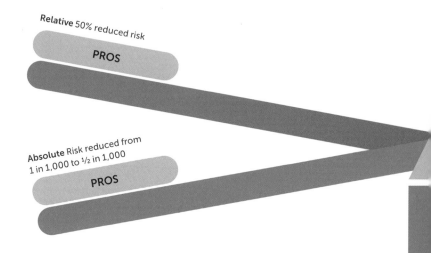

Relative 50% reduced risk

PROS

Absolute Risk reduced from 1 in 1,000 to ½ in 1,000

PROS

We are faced with statistics on a daily basis. And when it comes to matters of health, these can be daunting, misleading and, well, confusing. Suppose you're a menopausal woman considering hormone replacement therapy (HRT). You've heard that HRT can increase your risk of getting breast cancer, so you seek further information. Here's what you're told:

HRT increases the risk of breast cancer by 6 in 1,000: out of 1,000 women on HRT, six will develop breast cancer that otherwise wouldn't. However, HRT also reduces the risk of colon cancer by 50%.

What would you make of this? It's very tempting to conclude that the benefit outweighs the risk. A huge 50% risk reduction for the very nasty colon cancer, versus a tiny 6 in 1,000 increase for breast

cancer. Should you go for HRT, safe in the knowledge you've made an informed decision based on the evidence?

The answer is no, not yet. The information you have been given here is not enough to base a decision on. Let's start with the 50% reduction in the risk of colon cancer. That sounds great, it's a halving of the risk. The question is, however, how big was your risk to start with. For the sake of the argument, let's suppose that 80 in 100 women get colon cancer when they are at the age at which they might also take HRT (nowhere near as many women get colon cancer, but like we say, it's for the sake of the argument). The 50% reduction brings that down to 40 women in 100. In other words, your risk of colon cancer is reduced from 80% to 40% by HRT: a reduction that may well be worth

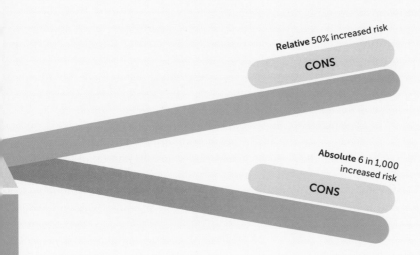

Relative 50% increased risk

CONS

Absolute 6 in 1,000
increased risk

CONS

putting up with other side effects, including an increased risk of breast cancer.

But now let's suppose that only about one woman in 1,000 gets colon cancer when they are at the age at which they might also take HRT. This means that your risk of getting colon cancer is 1 in 1,000. A 50% reduction would take this down to a risk of ½ in 1,000 (or 1 in 2,000 if you prefer whole numbers). In absolute terms, that's a tiny reduction in risk, which may well not be worth an increased risk of breast cancer.

The problem with the statement above is that it only gives you the *relative* change in risk due to HRT without giving you the baseline risk — the *absolute* risk — to start with.

Now let's look at the first part of the statement above. When taking HRT your risk of developing breast cancer increases by

6 in 1,000, which means that out of 1,000 women taking HRT, six will develop breast cancer as a result of the therapy. These six wouldn't have got breast cancer if they hadn't taken it. This risk increase does sound very small, but suppose, again for the sake of the argument and not reflecting real figures, that the baseline risk of breast cancer is 12 in 1,000. Then the supposedly tiny addition of 6 in 1,000 cases actually corresponds to a 50% increase in risk.

Confused? You might well be. The problem with the initial statement is that it gives the benefit of HRT, the reduction in the risk of colon cancer, in *relative terms*, and the harm of HRT, the increase in the risk of breast cancer, in *absolute terms* — and the relative number (50%) is a lot bigger than the absolute number (6 in 1,000).

MAKING SENSE OF HEALTH STATISTICS

The technique of presenting benefits and harms of a particular drug or treatment in different metrics, so that one appears bigger than the other, is called *mismatched framing* and it seems to be endemic in the health sector. A 2007 study of papers published in three prestigious medical journals, *The British Medical Journal, The Lancet* and *The Journal of the American Medical Association*, found that one in three articles used mismatched framing. Another 2007 study, which surveyed 150 GPs, found that around one in three GPs did not understand the difference between relative risk and absolute risk. It's clear why people interested in selling a drug or furthering their research might use mismatched framing. When it comes to GPs, we can only assume they suffer from the same haziness that afflicts most of us when confronted with statistics.

So what is a poor lay person to do? First, make sure you really understand the difference between relative and absolute quantities. The concepts themselves don't have anything to do with medicine, so let's phrase them in more familiar terms. What if your employer suggested scrapping your annual bonus in exchange for a 10% pay rise each year? You'd only consider this deal if 10% of your salary amounted to more than the bonus. You'd either need to convert the percentage into an actual sum of money or work out what percentage your bonus represents of your salary.

Once you have understood the difference, keep in mind the following checklist to be sure you don't get fooled:

01. Whenever you read that something has increased or decreased, check to see if the number that follows is a percentage (eg 'decreased by *x*%') or an absolute number (eg 'decreased by *x* dollars, *x* people, *x* grams', etc).

02. If it's a percentage, ask yourself 'x% of what?' An increase of 50% might not amount to much when the total number the percentage is taken from is very small.

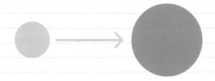

03. If it's an absolute number, check to see what the absolute number means in percentage terms. An increase of 10,000 dollars, people or grams doesn't mean much if it relates to a total of 10 billion dollars, people or grams.

04. When both a percentage and an absolute number are mentioned in the same breath, check to see if the percentage is calculated from the absolute number (eg x% of a total of y dollars, people or grams). That's fine, in fact, it's exactly what you want.

05. If that isn't the case, for example if the percentage measures an increase and the absolute number a decrease, then it's likely someone is trying to pull the wool over your eyes. That's mismatched framing.

TOOLKIT

01

Mathematics explains why infectious diseases can be so devastating. If an infected individual infects more than one other person on average, then the number of infected people will grow exponentially — unless the disease encounters a natural or artificial barrier. Vaccination can provide such a barrier by making people immune. And as the maths shows, not everyone needs to be vaccinated for the disease to die out.

02

If you get an abnormal screening result for a disease, it is still more likely that you don't have the disease. This is because the screening tests used are not diagnostic tests, and most abnormal results will be *false positives*.The actual chance you have a disease depends on the probabilities involved in the prevalence of the disease and the accuracy of the test, and can be calculated using *Bayes' theorem*.

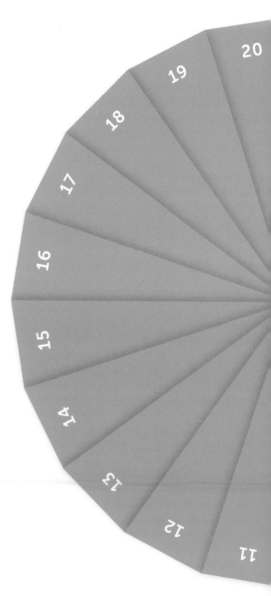

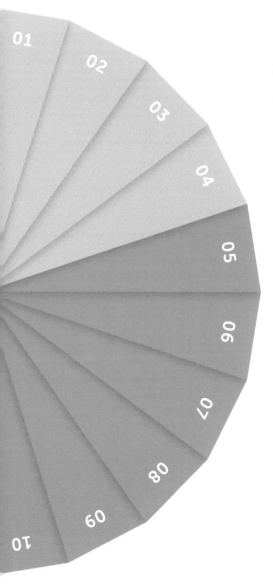

03

Although personal stories of recovery can resonate strongly with you if someone you know (or read about) has been unwell, this sort of anecdote is not evidence that the treatment or lifestyle change involved was the cause of that recovery. There are many factors that affect our health, and the best way we have of pinpointing what is effective is a randomized controlled trial.

04

When confronted with the risks and benefits of a treatment or lifestyle choice, always check whether risks and benefits are presented in relative and absolute terms, and ideally find out the baseline risks. Think of it this way: if your employer suggested scrapping your annual bonus, in exchange for a 10% pay rise each year, you wouldn't take the offer without checking that 10% of your salary amounted to more than the bonus, would you?

FURTHER LEARNING

READ

Reckoning with risk
Gerd Gigerenzer (Penguin, 2003)

The mathematics of diseases
Matthew Keeling (*Plus*, 2001)
www.plus.maths.org/content/mathematics-diseases

Protecting the nation
Marianne Freiberger (*Plus*, 2009)
www.plus.maths.org/content/os/latestnews/sep-dec09/vaccines/index

Making sense of screening: weighing up the benefits and harms of health screening programmes
Sense About Science (2015)
www.senseaboutscience.org/wp-content/uploads/2016/11/Makingsenseofscreening

NHS population screening explained
Public Health England (2013)
www.gov.uk/guidance/nhs-population-screening-explained

Evaluating a medical treatment – how do you know it works?
Sarah Garner and Rachel Thomas (*Plus*, 2010)
www.plus.maths.org/content/evaluating-medical-treatment-how-do-you-know-it-works

WATCH

Bacon sandwiches
Professor David Spiegelhalter explores the risk in this YouTube video

DO

The zombie outbreak puzzle
Adam Kucharski, *The Guardian*, October 26, 2015

VISIT

Broadwick Street pump and John Snow pub, London
This was the centre of 1854 cholera outbreak, which provides one of the first examples of evidence-based medicine.

ENVIRONMENT

LESSONS

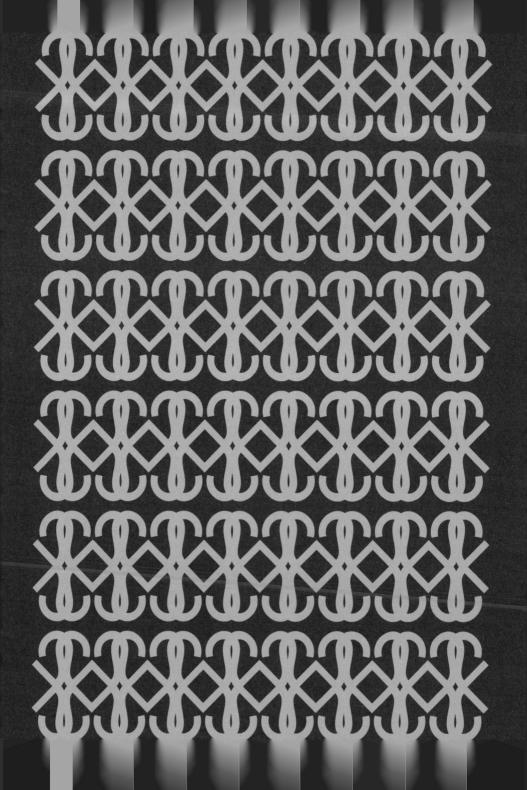

In modern physics, mathematical theories often guide the way to new discoveries, rather than the other way around.

As a population, we have always strived to understand the world around us. We are never satisfied to just exist in our environment – we want to know why it is like it is, and we famously (or infamously) are one of the few animals who adapt our environment to suit ourselves.

Mathematics provides us with the language to describe and understand our environment. Scientific rigour appears when the apparent laws of nature can be codified in mathematical equations and tested against experiments. Newton revolutionized science in this way, phrasing his laws of motion and gravitation in simple equations back in the 17th century. And at the beginning of the 20th century, Albert Einstein superseded Newton's theory of gravity with his supremely mathematical theory of relativity. In modern physics, mathematical theories often guide the way to new discoveries, rather than the other way around. Mathematics is pivotal in other sciences too, for example chemistry, medicine, biology, genetics, psychology and the social sciences. What has been described as the 'unreasonable effectiveness' of mathematics continues to be demonstrated.

Today, one of our greatest concerns is the effect our actions have on the environment and the effect environmental changes will have on us. These concepts can seem too big and complicated to grasp, passing before our eyes in a blur of figures and controversy in the press. So in this chapter we explore some of the maths used to understand our environment, the consequences of our adaptations to the world around us, and to make predictions about the future.

ARCHITECTURE

Mathematics has long been associated with architecture, from the tools and heuristics used by stonemasons over millennia to the fantastic forms that add to our skylines today. On 29 January 1957, when his sail-like sketches were announced as the winning design for the Sydney Opera House, Jørn Utzon had a problem – he didn't know exactly how he would build them. The problem still wasn't solved when construction began on 2 March 1959.

Inspired by the harbour location, the young Danish architect had envisioned a series of curved shells, but in order to build these, the shapes had to be described mathematically to accurately calculate the loads and stresses on the building. When asked by the engineers to specify the curves, Utzon bent a ruler to trace the curves he wanted. Over the next four years they tried various mathematical forms – ellipses, parabolas – to try to capture Utzon's design.

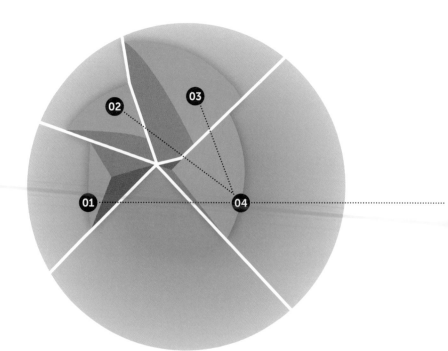

Finally, in October 1961, Utzon found the 'key to the shells': every sail was formed from a wedge cut from a single sphere (and its mirror image). This ingenious solution not only provided a mathematical description of the roof of his design, it also solved all the problems of constructing such a complex structure. Previously each shell appeared to be different, and it would have been almost impossible, in terms of both engineering and finance, to construct these huge bespoke parts. Instead, with the shells having a common spherical geometry (based on a sphere with a radius of 75 metres), they could be constructed using standard parts that could be mass-produced then assembled.

Describing a design mathematically made it possible to create one of the most iconic buildings in the world. The mathematics of the design also exemplified Utzon's artistic vision, providing 'full harmony between all the shapes in this fantastic complex'.

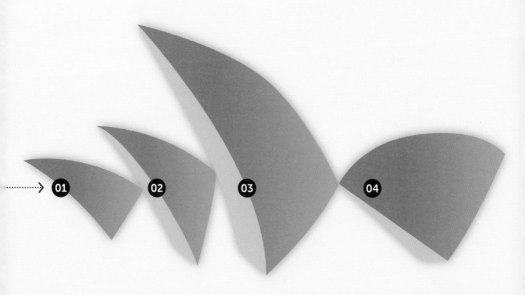

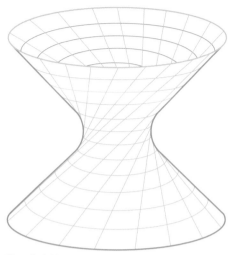

Hyperboloid

BUILDING THE IMPOSSIBLE

It is now standard practice for large buildings to be mathematically modelled in three-dimensions on a computer, so that the effect of any small changes in architectural design or practical construction (such as moving the position of piping or electrics) immediately cascades through the model of the building. This highlights any unforeseen clashes, and allows for the most economic use of materials and people during the construction process.

An understanding of mathematical forms has also allowed modern architects to create surprising architectural designs, such as the billowing forms of Frank Gehry's Guggenheim Museum in Bilbao or the Walt Disney Concert Hall in Los Angeles. His team used *parametric models* – computerized three-dimensional models of the structures that are defined by interacting mathematical rules that then define surfaces and forms, rather than specifying the exact size and shape of each element. These models allowed Gehry's team to specify how to build the structures by minimizing the number of panels required to build the surfaces, and making the shapes of these panels as simple as possible.

Parametric modelling is now commonly used by many architects and engineers, allowing them to design using functional rules, rather than being driven only by external forms.

Finding the answer

The design for the velodrome for the London 2012 Olympics began with the concerns of spectators and competitors. The aim was to ensure all spectators had good sight-lines onto the the steeply angled curved track in the centre of the space. This had to be balanced by access issues and building regulations that restricted the inclination of staircases. In addition, the steepest parts of the track are at the ends, and velodromes usually have no seating here. But feedback from cyclists was that this resulted in a huge difference between

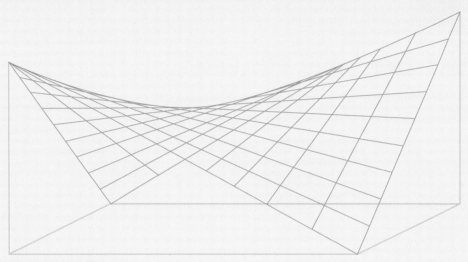

Hyperbolic paraboloid

the noise and energy from crowds in the straights to the deathly silence at the ends.

These issues were fed into the parametric model in a form of mathematical rules, along with the geometry of the track. As a result, the London velodrome has the bulk of the seating on the sides, but continues seating around both ends. And the finished shape features a doubly curved roof that mimics, but curves in the opposite way to, the rise and fall of the track inside.

Mathematics continues to provide the answers for our built environment. Mathematical modelling builds more efficient buildings in terms of materials, energy and construction costs. And maths also improves people's lives – the people using the buildings as well as those admiring their forms from the outside.

Antoni Gaudí's designs, including that for the Sagrada Família in Barcelona, were inspired by mathematical surfaces he saw in nature. These undulating and curving surfaces, such as *helicoids, hyperbolic paraboloids* and *hyperboloids*, can be created using straight lines and so are called 'ruled surfaces'. The nature of these surfaces, which can be built from repeating simple elements, meant that Gaudí could produce complex shapes that also provided great structural strength.

These shapes don't just produce beautiful structures. Hyperboloids, created by twisting a cylinder, are regularly used in constructing strong, lightweight towers. This elegant solution first appeared in a water tower designed by Vladimir Shukhov in 1896, which could be built with minimal materials and labour.

MODELLING TRAFFIC

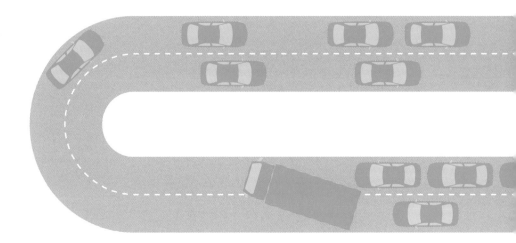

Mathematical models are used in almost every area of science and industry you can think of. They deliver predictions about the safety, usability and efficiency of all kinds of structures and devices, from the cars we drive and the bridges we cross, to the planes we fly and the financial products we hope will make us richer. Building a model is like stripping a system to its bare bones, and the X-ray vision provided by the maths allows us to discern phenomena that might otherwise be hard to see.

Take traffic jams. They're a pain, especially when they occur for no apparent reason. Even when a road is only moderately full and there are no obstacles, traffic can bunch up, slow down, and even come to a complete halt. Traffic researchers have long been interested in such phantom jams — and mathematical models can tell us why

they happen. Let's build a model based on two very simple assumptions:
- A number of cars travel in a single lane around a ring (that way we can assume that the number of cars stays the same, which makes things easier).
- Drivers vary their speed according to the distance between themselves and the vehicle in front. If the distance decreases because the car in front brakes, a driver will slow down. If the distance increases, a driver will speed up until they reach the speed they desire (or the speed dictated to them by a speed limit).

It's relatively straightforward to capture this situation using mathematical equations that express the acceleration of each car in terms of the headway in front of them and their current speed. Given an initial speed

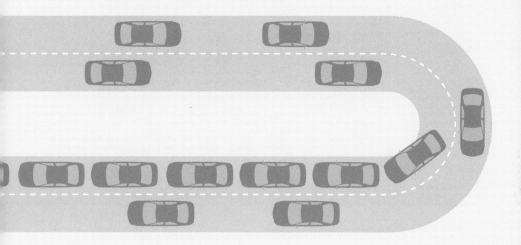

and distribution of cars, the solutions to the equations will tell you how fast the cars will be travelling at any given time.

This toy model of traffic is so simple, you'd expect the traffic to settle down into an even flow going round and round the ring. That's exactly the kind of situation we'd want in real life, but we all know that in reality things can play out very differently.

One crucial thing the model is missing is the fact that drivers aren't machines: they might be fiddling with the radio, talking to the kids in the back or arguing with the satnav. In short, drivers don't react immediately to a change in the distance to the car in front. And a little delay could make quite a difference to how hard they need to brake.

We can add this fact into the model by including a reaction-time-delay parameter.

When you do this, the system changes its overall nature: apart from the stable uniform flow behaviour, the equations now also admit a stop-and-go wave that can persistently travel down the road in the opposite direction to the traffic flow, even if the density of traffic isn't all that bad. The crucial ingredient here is the drivers' delayed reactions to even minor disruptions of traffic: it's those delays that can trigger the stop-and-go wave, the annoying phantom jam that seems to appear out of nowhere. In real life, the reactions that cause such delays might result from relatively minor incidents you wouldn't normally think could trigger a jam, such as a lorry ahead changing lanes.

Our mathematical model has thus provided a possible explanation for phantom jams, which previously wasn't obvious.

MATHS AND THE PERFECT CITY

Traffic jam waves spreading backwards through the traffic have been predicted by other mathematical traffic models too. They have even been observed in a real-life experiment which reconstructs the simple set-up we described: it had 22 cars travelling around a ring at a constant speed of roughly 30mph, and even though traffic started out flowing smoothly, jam waves were soon triggered by little variations in speed.

Crucially, such models don't just explain some of the mysterious phenomena we observe in real life, they also suggest solutions. One lesson to learn from the car-following model outlined is to pay attention to the traffic. If drivers don't react too violently, traffic can keep flowing smoothly. But there is also a lesson for the authorities: variable speed limits can help 'guide' traffic out of a jam and back into uniform flow, and restrictions on overtaking might prevent a jam from being triggered. Other traffic models might predict, for example, how planned new roads or changes to traffic regulations are likely to impact traffic flow.

Add the prospect of driverless cars into the picture and the power of maths conjures up visions of perfect cities. Traffic models can be used to determine a perfect road layout on which optimally programmed cars

travel without human interference. Guided by computer algorithms, the vehicles will act predictably and react optimally to whatever traffic situations they find in front of them. Traffic jams will become a thing of the past, as passengers glide along in their automated vehicles, passing their time engaged in more important, or pleasurable, activities than driving.

Whether this vision will really come into being remains to be seen. Designers of driverless cars still have to cope with the human element. Imagine a world full of perfectly functioning driverless cars programmed to never, ever run over a person in front of them. If you were a pedestrian in this world, in a hurry to get to work and needing to cross a busy road, would you wait at a red traffic light? Or would you cross the road anyway when there's a suitable gap in the traffic, knowing that the cars will all stop, heedless of the jam that might ensue? It's probably the latter. One way of avoiding this is to programme the cars so that they don't always stop for a person. If there is a chance that a car will react just a little too slowly, or perhaps not at all, this knowledge would be enough to stop hordes of pedestrians bringing traffic to a standstill. A little randomness might still be necessary after all.

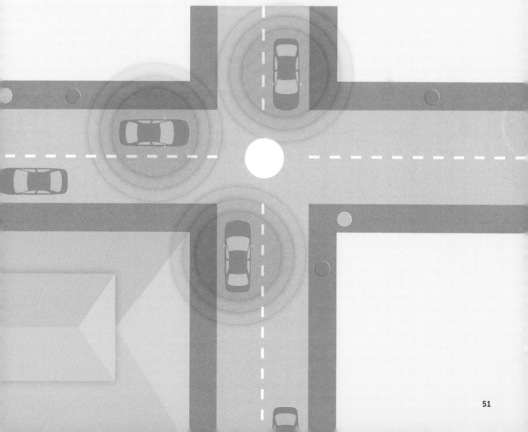

MATHEMATI

CAN, AND

PREDICT T

CIANS
DO,
HE 'FUTURE.

PREDICTING THE UNPREDICTABLE

Mathematicians can, and do, predict the future. We rely on their forecasts for economic and political decisions, as well as for needing to take an umbrella. These predictions are the results of mathematical models: sets of equations designed to describe the process in question, be it the evolution of the stock market or the upcoming weather.

Mathematical modelling comes with challenges. Any mathematical model will only ever be an approximation of reality, so it's necessary to extensively test any model against reality and all available data. The mathematical equations in a model may also be too complex to solve exactly, or too time-intensive to run repeatedly on a computer. In these cases, numerical methods are used to approximate the solutions.

But perhaps the most famous, and most fascinating, aspect of forecasting is *chaos*: the propensity of even seemingly simple dynamical systems to display complex and entirely unpredictable behaviour in the long run.

Chaos theory

When we think of chaotic behaviour, we might imagine something behaving randomly, making it impossible to predict what will happen next. Randomness is inherently impossible to predict. But chaotic behaviour doesn't have to be random, it can also be deterministic – if we had perfect knowledge about everything affecting that behaviour we could exactly predict what would happen next. But we rarely have perfect knowledge. The unpredictability in chaotic behaviour can come from a sensitivity to initial conditions – even a slight difference in the starting conditions, say a rounding down of your measurement rather than a rounding up – might result in a totally different outcome.

This phenomenon was poetically described as the butterfly effect in 1972 by mathematician Edward Lorenz when he was running computer simulations for forecasting the weather. The first time the simulation ran it started with an initial value of 0.506127, but the second time Lorenz typed in the number by hand and rounded the number to 0.506. This tiny difference, which he compared to the flap of a butterfly's wings, over time resulted in wildly different outcomes, a calm day versus a tornado.

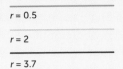

$r = 0.5$

$r = 2$

$r = 3.7$

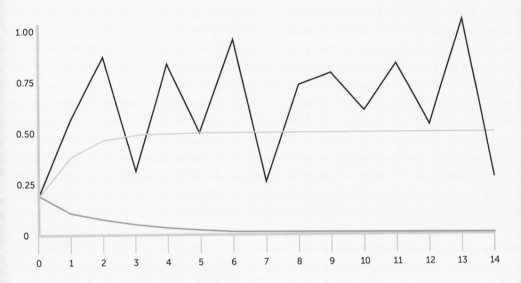

Even simple mathematical models can lead to chaotic behaviour. In 1976, when biologist Robert May was describing the change over time in a population of insects, he used this simple equation:

$$P_{\text{next year}} = r \times p_{\text{this year}} \times (1 - p_{\text{this year}})$$

Here $p_{\text{this year}}$ and $p_{\text{next year}}$ are the proportions of animals alive (this year or next respectively) out of some maximum number possible, and r controls how quickly the population changes size from year to year. If there are fewer animals there is enough food for everyone to eat and so the population will increase. But once the population has reached or gone over the maximum, animals start to starve and the population starts to decrease.

The surprising thing is that as you vary the value of r, the future of the population can vary wildly. If r is less than 1, the population will die out. If r is between 1 and 3, the population will eventually stabilize. Above this, strange things start to happen. For some values of r the population will oscillate between two or more values. And if r is bigger than 3.57, the population varies wildly from year to year, with just a small change in the population one year resulting in huge differences in the population the next. This system, despite being described by such a simple mathematical equation, is *chaotic*.

AND NOW THE WEATHER

Meteorologists today have a very good understanding of the weather and all the physical processes that are involved. The numerical simulations used by Lorenz and modern meteorologists are based on the interactions between the laws of thermodynamics and the *Navier–Stokes equations*, which describe the motion of fluids, whether that fluid is the atmosphere, the swirling of milk in your coffee or the flow of water down a pipe. These equations are fiendishly difficult to solve (we actually only know how to solve them exactly for a few simple cases), but solutions can be approximated using very fast computers, so extremely accurate simulations of the weather are possible.

The problem isn't that we don't understand the weather, and can't predict how it will develop over time. Instead the real problem with our mathematical weather simulations, as Lorenz first discovered, is that they can exhibit mathematical chaos due to their sensitivity to initial conditions. The problem in accurately predicting the weather comes from our ignorance about the starting conditions for any simulation.

It is impossible to continuously measure the temperature, air pressure, humidity and wind conditions at every point around the globe. (Although the amount of data used in weather simulations continues to increase,

particularly due to the increase in satellite weather observations.) And it would also be computationally impossible to run simulations on such a fine scale.

Instead, forecasters base their forecasts on simulations that run at points on a three-dimensional grid wrapped across the globe. Across the UK the points of the grid are spaced 2.2km apart across the surface and are layered in 70 levels upwards through the atmosphere. The simulation starts with initial values at each point in the grid and runs forward in time, predicting the weather for the next 54 hours (or up to a week ahead for global forecasts simulated on a slightly sparser grid).

As Lorenz discovered, a slight error in the starting conditions for one of the grid points could drastically change the outcome of the simulation. To account for this chaotic behaviour, meteorologists run *ensemble forecasts* – simulations run many times, each with slightly different starting conditions at each of the points on the grid predicting a possible future weather. Meteorologists can then use this ensemble of possible future weathers to predict the most likely forecast. If the forecast for your area says there's a 10% chance of rain today, then it was rainy today in 10% of the ensemble of future weathers. If the forecast says 90% chance of rain, you'd be wise to take your umbrella.

CHANGING CLIMATES

$$seT^4$$

$$(1 - a)S$$

Never before has the Earth's climate been more on our minds than it is today. To slow down climate change, or at least assess its impacts and brace against them, we need to be able to forecast what the climate is going to do over the next few decades and centuries. The climate is a complex system, however, involving the Sun, the oceans, vegetation, ice and topography, and of course human activity. The only hope we have of forecasting the climate of the future and dealing with the uncertainties involved is mathematics.

To see how mathematics can be used, let's have a look at a very basic climate model. Imagine coming inside from the cold. At first your body will absorb more heat from the warm air around you than it loses, so your body will heat up. But after a while it'll reach an equilibrium temperature where the heat it absorbs exactly equals the heat it loses. Hopefully that'll be a comfortable

temperature, so you can snuggle down with a good book.

The Earth, having spent a long time in the warm glow of the Sun, is also in a thermal equilibrium: the amount of energy it absorbs from the Sun equals the amount of energy it emits into space. The heat energy from the Sun can be written as:

$$(1 - a)S$$

Here S stands for the total power that comes from the Sun (around 342 Watts per square metre on average) and a measures the proportion of the solar energy that is immediately reflected back, for example by large areas of ice. This proportion of energy reflected back is called the *albedo* of the Earth and its current value is about 0.31.

The laws of thermodynamics also tell us how much energy the Earth radiates out into space, which depends on the Earth's average

a Albedo
e Emissivity
s Stefan-Boltzmann constant
S Power from the Sun
T Temperature

temperature. Writing T for that temperature, the amount of energy emitted is:

$$seT^4$$

The number s is a constant from thermodynamics (the *Stefan-Boltzmann constant*, with a value of $5.67 \times 10^{-8} Wm^{-2}K^{-4}$) and e is the Earth's *emissivity*, which measures how transparent the atmosphere is. The emissivity for Earth is currently around 0.6.

Since the Earth is in thermal equilibrium we get a surprisingly simple equation:

$$(1 - a)S = seT^4$$

And here is the beauty of this climate model: it's possible to solve this equation to find out the value of Earth's temperature T (measured on the Kelvin scale). Plugging the values of S, a, s and e into the formula and solving it for temperature, we get:

$$T = ((1 - 0.31)S/(5.67 \times 10^{-8} \times 0.6))^{0.25}$$

in which $T = 288$, or 15 degrees centigrade. This is almost exactly the current average temperature of the Earth as measured by NASA.

Our simple energy balance model also enables us to see what might happen if the parameters involved change. Imagine, for example, the albedo (that energy the Earth reflects back) becomes smaller because the ice caps melt. The equation tells us that in this case the temperature goes up and it also tells us the rate at which this will happen. Similarly, a lower emissivity, for example as a result of greenhouse gases in the atmosphere, will also increase the Earth's temperature. The model explains why, if we don't want the Earth to become warmer, we need to control both albedo and emissivity.

MODELLING THE FUTURE

The models used by real climate scientists are much more complex of course. They are based on Newton's laws of motion and the laws of thermodynamics and can take account of a variety of factors, from the Earth's rotation to the effect of vegetation on climate. Unlike the energy balance model above, these models contain a time parameter: once you have solved the equations you get a prediction for what the climate will look like in the future.

Are the predictions of these complex climate models 100% reliable? The answer is no, they can't be. Whenever you use a mathematical model to make predictions about reality you come up against three main sources of uncertainty:

01. Model uncertainty

The mathematical equations we use to model a process may not represent reality accurately.

02. Uncertainty from approximation

The equations used in modelling processes often can't be solved exactly. We have to rely on approximate solutions found by a computer, often at huge computational expense. Approximate solutions always come with a degree of error.

03. Input uncertainty

We have already met sensitive dependence on initial conditions in Lesson 7 on chaos theory. Uncertainty about a parameter that goes into the model (for example the albedo or emissivity) also leads to uncertainty in the predictions of a modelling process.

In climate modelling the dominant source of uncertainty is model uncertainty: scientists are not sure how best to represent the many factors that make the climate, and their interactions, mathematically. Uncertainty due to initial conditions comes a close second. Only uncertainty from approximation is under good control. There are mathematical methods that keep a lid on the errors involved and also quantify them, so we know by how much predictions are going to be off.

A necessary skill in mathematical modelling is to cope with the uncertainties inherent in the modelling process: identify them, reduce them, quantify them. There is a whole area of mathematics that concerns itself with uncertainty quantification and produces tools that can be used in many different modelling scenarios.

A highly effective method in climate modelling involves starting your model using initial data from some time in the past

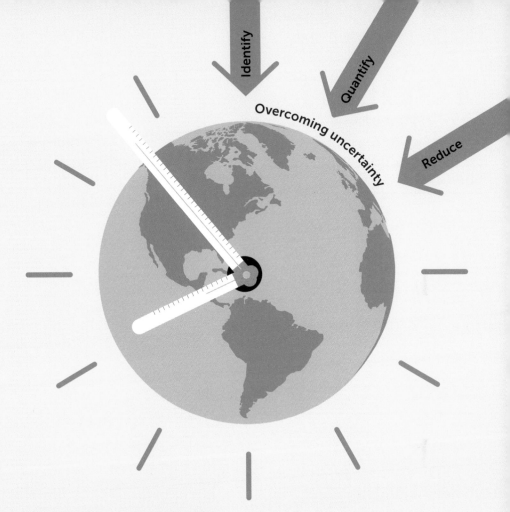

Identify

Quantify

Overcoming uncertainty

Reduce

(the beginning of the 20th century is often used as a starting point) and seeing if it correctly predicts today's climate. You can do this using different mathematical ways of representing things you are not sure about, such as the dependence of the climate on human-made agents like greenhouse gases. This not only tests the model, it also helps you to infer the effects of human activity on climate.

Current climate models perform well when compared to past and present climate data in this way. They also chime in their predictions, as far as they all predict a rise in global temperature: different models are being used by different climate centres but all predict that the Earth will warm up. Underlying these models is well-established science that has proved reliable in many other contexts. On balance, then, we should take the predictions of climate models seriously — very seriously given the impact a changing climate is likely to have on our planet.

BUILD +
BECOME

TOOLKIT

05

Efficiency – in terms of cost, energy and materials – has become very important in modern architecture. Mathematically modelling a building allows architects to explore the best solutions before a single brick has been laid. The next time you see a spectacular new building, remember that the geometrical forms used are not just there for looks: they also often serve the function and efficiency of a building.

06

Modelling traffic mathematically can explain puzzling phenomena and suggest solutions, and also allows us to predict how the layout of roads and interventions such as speed limits might impact on traffic flow. Driverless cars will also rely heavily on mathematics, in the shape of the algorithms that control them. But it's not all about perfect predictability: to take account of the human element we might have to inject a little randomness into the models.

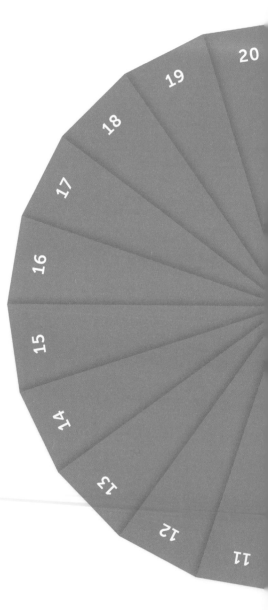

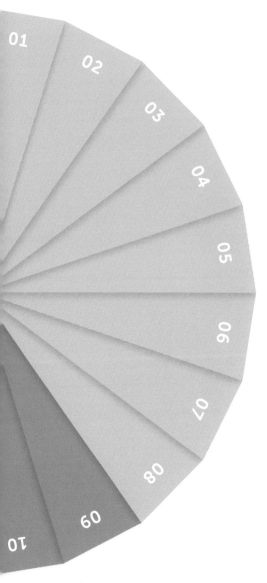

07

The reason the weather forecast is not always correct is because the weather displays something termed *mathematical chaos*. Meteorologists account for chaos by making many forecasts with slightly different initial conditions; a 30% chance of rain in the reported forecast means that it rained in 30% of these simulated forecasts.

08

Any mathematical model of any physical process comes with uncertainties. A key skill in mathematical modelling is to identify, quantify and if possible reduce these uncertainties, so that we can gauge how accurate a model's predictions are. In climate modelling, a range of checks and balances is used to assess the quality of model predictions. Different models all predict that the Earth will warm up, so we should take these predictions seriously.

FURTHER LEARNING

READ

Climate change: Does it all add up?
Chris Budd (*Plus*, 2016)
www.plus.maths.org/content/climate-change-does-it-all-add

Chaos
James Gleick (Penguin, 1997)

Perfect buildings: the maths of modern architecture
Marianne Freiberger (*Plus*, 2007)
www.plus.maths.org/content/perfect-buildings-maths-modern-architecture

How the velodrome found its form
Rachel Thomas (*Plus*, 2001)
www.plus.maths.org/content/how-velodrome-found-its-form

WATCH

The mathematics of climate change
Chris Budd, Gresham College Lecture
www.gresham.ac.uk/lectures-and-events/mathematics-climate-change

The emergent patterns of climate change
Gavin Smith
TED Talk

Shockwave traffic jam recreated for first time
New Scientist, March 4, 2008
www.newscientist.com/article/dn13402-shockwave-traffic-jam-recreated-for-first-time

DO

Play with the chaotic motion of a double pendulum
Either online (www.web.mit.edu/jorloff/www/chaosTalk/double-pendulum/double-pendulum-en.html) or build your own (www.youtube.com/watch?v=d2E5oojoXjk)

VISIT

Edge of Chaos
An installation by artists Vasilija Abramovic and Ruairi Glynn and scientist Bas Overwelde.

Maths in the City
You can visit some mathematical architecture on the self-led Maths in the City tours of London and Oxford.

SOCIETY

LESSONS

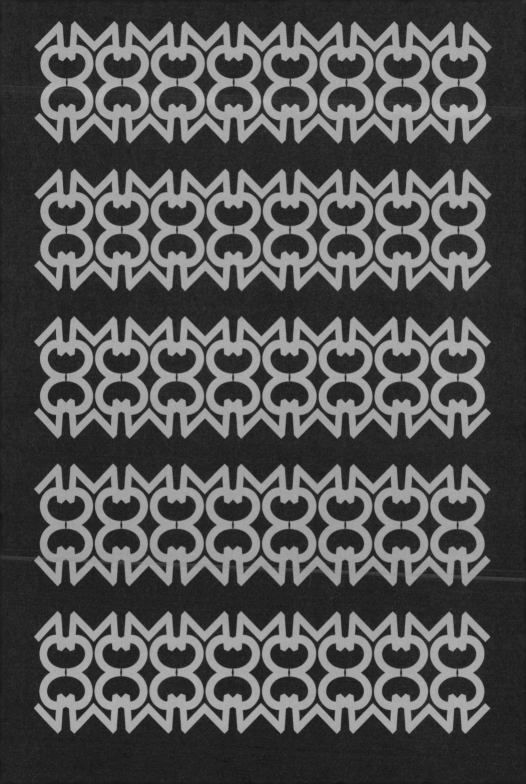

From a mathematical viewpoint, the boundary between the many and the few, the line at which individual agents merge to form a unified whole, is fruitful ground.

'There is no such thing as society', the former British Prime Minister Margaret Thatcher famously said. The idea being that it was not possible for many individuals to form a common whole, which can be held responsible for what is happening within it.

From a mathematical viewpoint, the boundary between the many and the few, the line at which individual agents merge to form a unified whole, is fruitful ground. The art of statistics is all about spotting patterns, trends and perhaps even fundamental underlying principles in the jumble of data collected from many individuals. Every democratic process requires the consolidation of many individual preferences into a single outcome that politicians like to call 'the will of the people', and poses a wealth of mathematical questions. Indeed, the idea that many individuals, be they humans, ants, birds or bees, can form a whole that is greater than the sum of its parts, sometimes with unintended consequences (think of the economy), is not news for mathematicians.

In this chapter we will look at some of the mathematics that affects society, either because it bridges that treacherous boundary between the individual and the collective, or because it affects every single one of us directly. We will look at the statistical notion of an average, and why it needs to be treated with caution. We will examine voting and voting systems. We will see how that most individual of our features, our DNA sequence, can be used to impose justice, but also lead to injustice if not treated carefully. And we explore something none of us can get away from: the mathematics of saving and borrowing.

EVERYDAY AVERAGES

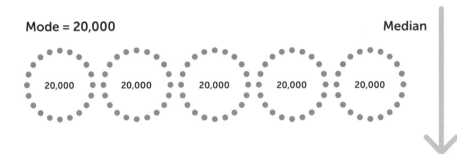

Mode = 20,000   Median

20,000   20,000   20,000   20,000   20,000

'The average man, indeed, is in a nation what the centre of gravity is in a body', wrote the Belgian statistician Adolphe Quetelet in his *Treatise on Man*, published in 1835. 'It is by having that central point in view that we arrive at the apprehension of all the phenomena of equilibrium and motion.'

In Quetelet's day, the idea that human characteristics, physical as well as psychological, should follow statistical patterns was still relatively new. Today, in spite of the rise of individualism, we take the idea for granted. When faced with a jumble of numbers describing one characteristic or another in a population we immediately feel compelled to work out the average: after all, it's quicker to communicate and may also tell us something profound about the population as a whole. But useful as averages are, they need to be treated with caution.

This fact is poignantly illustrated using the example of yearly incomes. Imagine a small village with 1,000 inhabitants of which 900 earn $20,000 a year and 100 earn $500,000 a year. To work out the average of a set of numerical values you add up all the values and then divide by the total number of values. In our example this gives:

$$(900 \times 20,000 + 100 \times 500,000)/1,000 = 68,000,000/1,000 = 68,000$$

The average yearly income per person is $68,000, more than three times what 90% of the people in the village earn. The example shows that the average is sensitive to *outliers* — unusual and extreme values. In the village, a small number of super-high earners skew the average to such an extent that it doesn't accurately reflect the true wealth per person. This sensitivity can lead to all sorts of bizarre

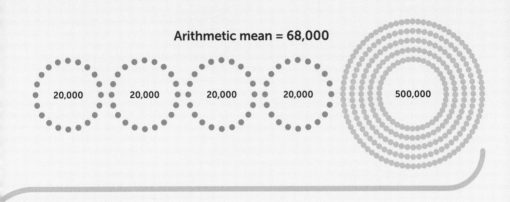

Arithmetic mean = 68,000

20,000   20,000   20,000   20,000   500,000

results, such as the fact that the vast majority of people have more than the average number of noses (as you might want to work out for yourself: it only takes one person to lose a nose to make the average less than 1).

Luckily the average we have just considered isn't the only one there is. Technically, it is known as the *arithmetic mean*. Another notion of average is called the *median*. Given a set of numerical values, such as number of noses or yearly incomes, list them all in order (including repeated numbers) and find the one that sits right in the middle of your list — that's the median. (In our noses example, the median would be 1.) If there isn't a middle value, because there are an even number of values, then the median is the number that's halfway between the two middle values of the list. Roughly half of a set of values lies below the median of the set and the rest lies above it.

In our example of yearly incomes the list would start with nine hundred 20,000s followed by one hundred 500,000s. The two numbers in the middle come in the 500th and 501st place on the list. They are both 20,000. Trivially, the number halfway between 20,000 and 20,000 is 20,000, so the median in this case is 20,000. That's much more representative of the yearly income of the majority of village dwellers than the arithmetic mean.

Another type of average you might compute is the mode. That's the number that appears most frequently in a list of numbers. In the example above the mode is the same as the median, $20,000. If our village contained only 500 people earning $20,000, with another 400 earning $30,000 and 100 earning $500,000 as before, then the median would be $25,000, the mode would be $20,000, and the arithmetic mean would be $72,000. Take your pick.

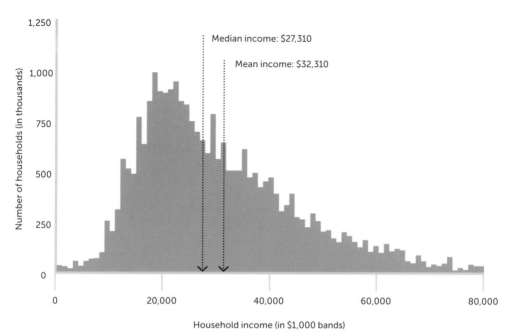

Median income: $27,310

Mean income: $32,310

KNOWING YOUR AVERAGES

All this means that, as an average consumer of the news, you need to be wary whenever you hear the word 'average'. As a rule of thumb, ask yourself the following question: would I care if I knew that the average quoted (eg the average income per capita) has been skewed by a few outliers (eg a few super rich people)? If the answer is yes, then you need more information. Dig up additional sources to see if the word 'average' refers to the arithmetic mean (it usually does), the median, or (rarely) the mode. Ideally, find out how the quantity in question (eg money) is distributed among the data points it's averaged over (eg people).

This kind of detective work isn't always as time-consuming as you might fear. As far as income is concerned, a quick internet search

will hopefully reveal just the kind of graphic you would want.

The figure above is an example of a frequency plot, showing how many households in a population have a given income. In this example it tells us that around 1.1 million households have an income in the $17,000 to $18,000 bracket. Because it corresponds to the peak of the plot, the $17,000 to $18,000 bracket also corresponds to the mode of the income.

The mean and median can't be guessed just by looking at the plot, but they are marked. The fact that they differ by around $5,000 is explained by the far right-hand tail of the income distribution: it indicates the presence of high earners who skew the average upwards. At a single glance, this

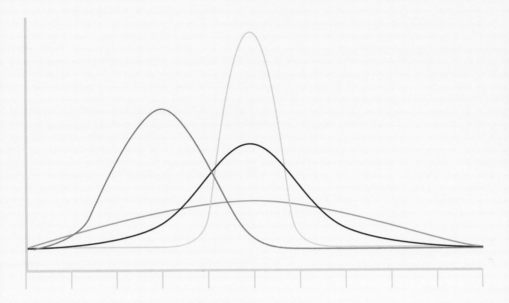

graphic gives you a lot more information than a single number will ever do.

Any data set can be summarized using such a frequency plot, and interestingly, you don't see a different shape every single time. In his desire to define the average (wo)man Quetelet himself discovered that many human characteristics – such as height or weight – give rise, not to a skewed shape as the one opposite, but to a lovely symmetric bell-shaped curve known as the *normal distribution* (shown above). The location of the peak of the bell and its height and width may be different depending on which phenomenon you are looking at, but the characteristic shape is the same and it can be described using a certain type of mathematical equation.

In any data set that is normally distributed, the arithmetic mean, the median and the mode are all equal: they correspond to the peak of the distribution around which non-average values are spread out symmetrically.

The normal distribution is especially ubiquitous, but there are also other distributions that come up a lot, each in its own particular circumstance. Any first course on statistics will introduce those distributions and teach students how to best analyze data sets that adhere to them. The average, be it arithmetic mean, median or mode, is just the first step towards such an analysis. Next time you hear it mentioned, remember it's just a peak of a highly sophisticated, and very useful, mathematical iceberg.

VOTING

Candidate B

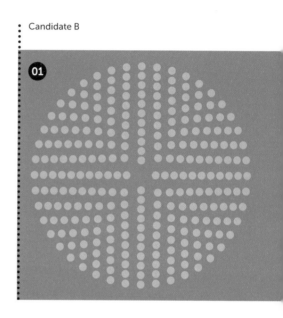

01

01. First past the post

In a first-past-the-post system, the winner takes all. The head of government of a country isn't voted directly by the people, but by a body composed of representatives of geographical regions (MPs in the UK and members of the Electoral College in the US). Those representatives are voted in by the people using majority vote: the party/candidate that gets the most votes in a region determines the region's representation.

That's simple enough (indeed simplicity is one of the advantages of a first-past-the-post-system), but a bit of maths reveals a major disadvantage: the winner of a

There are few things that stretch our numeracy skills quite as much as elections. Glued to our screens, we are bombarded with percentages, trends, pie charts and histograms. And while the final result soon turns to incontrovertible fact, the mathematical workings under the surface often remain obscure.

That's no surprise: the ins and outs of individual voting systems are complex. What we will look at here are not the nuts and bolts of elections, but two different approaches: a first-past-the-post approach, as used in the UK and the US, and proportional representation, as used by many European countries.

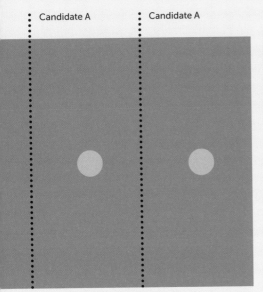

Candidate A   Candidate A

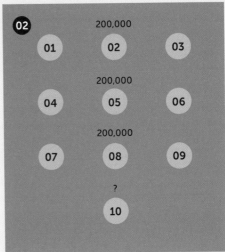

first-past-the-post election isn't always the candidate that got the larger share of the vote. In the 2016 US election, for example, Hillary Clinton won by over 2.8 million votes, receiving 48.2% of the popular vote as compared to Donald Trump's 46.1%.

To see how this anomaly comes about, here's a simple example. Imagine a nation made of three states, one state with 100,000 citizens and the other two states with one citizen each. If the two lone citizen states vote for candidate A and the other 100,000 citizens for candidate B, then A has taken the majority of states and becomes president — despite winning only 0.002% of the popular vote.

02. Proportional representation

To avoid such an anomaly you might opt for *proportional representation*. The idea is that a party that got *x*% of the vote should get *x*% of the seats in the body that is being elected (and goes on to form a government). The obvious advantage here is better representation of the people, but maths again picks out a disadvantage: percentages don't always translate into whole numbers. For example, if there are 600,000 voters in an election for 100 seats and three parties who each got 200,000 votes, then each party should get exactly one-third of the seats. Since one-third of 100 is 33.33 and politicians can't be carved up, that's impossible.

UNDERSTANDING VOTING SYSTEMS

It's easy to get caught up with the particulars of individual voting systems. So perhaps we should start from scratch, stipulate the minimal requirements of democracy and work from there? The economist Kenneth Arrow pursued this approach in the 1950s, stripping the concept of a voting system down to its bare bones. He assumed that in an election each voter has a preference ranking of candidates or parties, which he or she then communicates at the ballot box. The task of the voting system is to take all the preference rankings as input and return one ranking as output — that one final ranking decides who wins the election.

Arrow also decided that a democratic voting system should satisfy the following four conditions:

01. The system should reflect the wishes of more than just one individual (so there's no dictator).
02. If all voters prefer candidate x to candidate y, then x should come above y in the final result (this condition is called *unanimity*).
03. The voting system should always return exactly one clear result (this condition is called *universality*).
04. In the final result, whether one candidate is ranked above another, say x above y, should only depend on how individual voters ranked x compared to y.

It shouldn't depend on how they ranked either of the two compared to a third candidate, z (this condition is called *independence of irrelevant alternatives*).

The result Arrow came up with is surprising: he proved mathematically that no voting system involving three or more candidates can satisfy all four conditions. The result earned Arrow the 1972 Nobel Prize in Economics.

At first sight Arrow's so-called *impossibility theorem* appears to spell death for democracy, but things probably aren't that bad. Social choice scientists have examined Arrow's theoretical framework (admittedly loosely stated here) in detail and some have argued that it's more demanding of democracy than it appears to the untrained eye.

The general lesson to take from Arrow's theorem is that democracy is not an exact science. The mathematical approach can tell us how unintended anomalies can come about, and suggest how to avoid them. Indeed, there is a whole body of mathematical social choice and fair division literature, which examines different notions of fairness, applying to anything from politics to divorce, and how to achieve it. In the end, choosing a voting system is a balancing act between those mathematical considerations, politics and ethics.

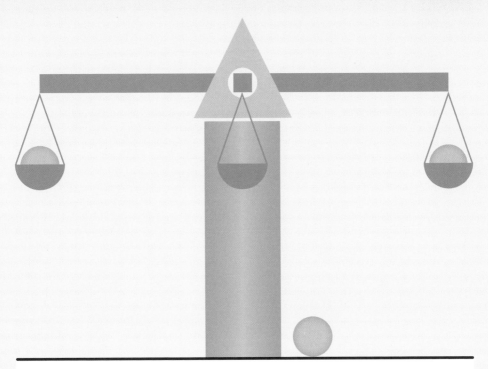

To deal with the proportional representation issue, we need an extra layer of complexity; a method to translate percentages into seats. One common way is the *d'Hondt* method. The idea is that one seat should 'cost' a certain number of votes. Each party should be able to buy as many seats (from the total number of seats) as its vote money allows, because if a party can't get all it can pay for, it's not fairly represented. Once each party has bought all the seats it can, no seats should be left over.

Setting the correct price per seat to achieve this seems tricky, but there's an iterative process that delivers the desired result. You start off by giving the party with the largest number of votes one seat. Then, for each party, you divide the number of votes it got by the number of seats it already has plus 1, to get a number N. The expression for this is $N = V/(s + 1)$, where V is the number of votes the party got in total and s the number of seats it already has (at the beginning of the process s is 0 for all but the largest party, for which it is 1). The second seat is given to the party with the highest N. You then work out each party's N again, with s increased as appropriate. The party with the highest N gets the third seat. And continue this way until all seats are gone.

Since proportional representation always involves rounding up or down the number of seats a party should get, no system designed to deliver it is perfect. The *d'Hondt* system, for example, tends to over-represent the largest party. Which system you choose depends on which of several possible problems you're happiest to live with.

AVERAGES.
PROBABILIT
STATISTICS
SPOTTING
AND TREND

IES.

. ARE YOU

PATTERNS

S?

EVIDENCE

We've all watched crime dramas on TV: once the cops have a DNA match between the crime scene and a suspect, the case is closed. But in real life things are a little more complicated. DNA matches used in forensics are not unique to an individual; instead, they offer only a likelihood that the sample found comes from the suspect in question.

DNA is the famous molecule found inside every one of our cells, shaped like a twisting ladder. Each strand of DNA is made up of four basic building blocks (bases) denoted by the letters A, G, T and C – the links between the As and the Ts, and between the Gs and the Cs provide the rungs of the ladder. These letters create a sequence containing many identifiable patterns called *genes*.

Your complete genome (the entire set of DNA within a cell) will almost always be unique to you, the exceptions being identical twins, triplets, and so on. But it's not possible to look at the full DNA sequence when comparing to a sample from a crime scene; instead we count the number of times patterns repeat in specific parts of the DNA, known as a DNA profile. A DNA profile looks at several specific sections on the ladder, known as *markers*. (These markers are in your 'spare code' – the non-coding parts of DNA that don't have a particular function at that particular location on the ladder.)

For example, let's look at the section called the THO1 marker. We know that THO1 has the pattern AATG repeated somewhere between 3 and 14 times. The exact number

of times AATG repeats varies across the population. It might even vary within your DNA: the DNA strand from your mum might have a different number of repeats for this marker to the DNA strand from your dad.

It is entirely possible that two unrelated people might have the same DNA profile, while having different DNA sequences. We rely on mathematics to work out the probability of this happening. For each marker in a DNA profile, we need to know the *frequency* of the different number of times a pattern repeats across a population. These frequencies are calculated from databases collected by forensic laboratories.

All the markers used within a DNA profile are specifically chosen so that we can assume the genetic code at one marker occurs independently from that at any other marker. By the rules of probability theory, we can multiply the frequencies to calculate the probability that particular variants of markers occur together. Multiplying the frequencies for all the markers in the DNA profile gives a *match probability* for that DNA profile: the probability that a randomly chosen person in the population has that DNA profile.

While the frequency for a single marker might be quite high, full DNA match probabilities tend to be small. It's similar to playing the lottery, where the chance of getting one ball right is 1 in 49, but the chance of getting six right is only around 1 in 14 million. DNA profiles consist of between 10 and 14 markers, and routinely come with match probabilities of 1 in a billion.

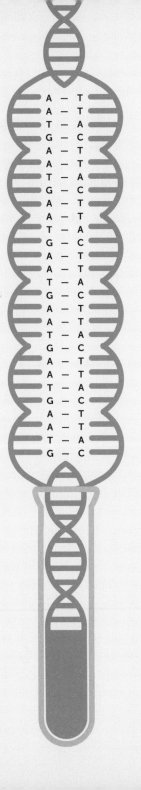

TH01 marker

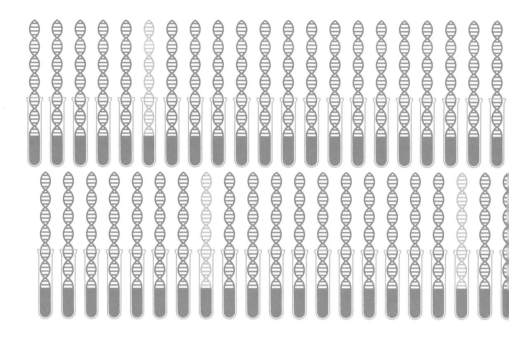

WHAT DOES A DNA MATCH TELL US?

Probabilities underpin many important decisions in and out of the courtroom, such as those regarding health, lifestyle and economics. DNA evidence is a good illustration of the two things you need to make an informed decision. One is accurate probabilities, the other a good understanding of how to use these probabilities to update your view in the face of new evidence.

Imagine you are a juror on a trial involving DNA evidence, with the match probability estimated to be 1 in 20 million. This is the probability that a random person matches the DNA profile. It is vital to understand this is not the same as the probability that the suspect is innocent. Mixing up these two probabilities is called the *prosecutor's*

fallacy: confusing the probability of the evidence assuming someone's innocence (the match probability) with the probability of someone's innocence given the evidence. This mistake is incredibly easy to make, and has been made many times in legal proceedings involving statistical evidence. These mistakes can be challenged on appeal, but the lawyers, judges and juries all need to be able to understand how to interpret the match probability correctly.

Assessing the evidence

The correct way to work with such probabilities is to use *Bayes' theorem*, which updates your beliefs (say, the probability someone is innocent) based on

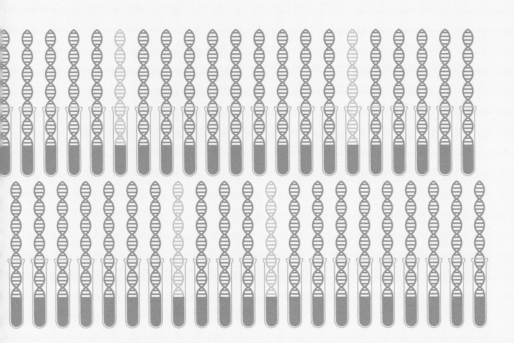

the evidence you've seen. This has been attempted in trials, coaching juries to assess probabilities of all the evidence involved and make calculations, but such a technical approach hasn't been successful as of yet.

Instead, juries are often guided by judges to understand match probabilities in the following way. If the match probability is 1 in 20 million, having the same DNA profile as the crime sample is unlikely but not impossible. In a country with a population of 60 million there are probably three people with the same profile, or even as many as six or seven. The effect of the DNA evidence is to narrow down the pool of people from everyone in the country to just those six or seven. On the basis of the DNA evidence alone, there is a 1 in 3 to a 1 in 7 chance that the defendant isn't the source of the sample from the crime scene.

Regardless of how aware juries are of the prosecutor's fallacy and how well they can apply *Bayes' theorem*, they won't be able to accurately take into account all the evidence in this way. This is because DNA evidence is one of the only types of evidence that comes with a clear probabilistic understanding of its reliability, and its fallibility. Evidence with less scientific basis, such as witness statements and fingerprint evidence, can incorrectly seem more definitive than DNA, exactly because it is impossible to assess the probabilities involved in these types of evidence.

DEBTS AND SAVINGS

Compound interest is a fact of life for almost everyone who has a bank account, credit card or loan. It is surprisingly simple to understand, but, despite this, it is easy to be caught unaware by how quickly debt can blow up. Although compound interest works in your favour on your savings, it acts against you with your debts. Banks are businesses, so it's best to assume that banking is like playing roulette against the house – the bank is going to make more money from people in debt than it will give people in interest on their savings. It's important you know the rules of this game.

Exponential growth

Let's start with the advantages of compound interest. Suppose you were lucky enough to have $1,000 and a bank account with an annual interest rate of 20%. If the interest is applied annually, then after four years you'd have doubled your money, with $2,073.60 in the bank. Every year the bank pays you interest on your original deposit of $1,000, but also pays interest on all the interest you've earned along the way. Every time the

interest is compounded, and the amount of money added to your account increases.

If you left your account untouched and at this rate of interest, after n years it would contain $1,000 \times (1.2)^n$. It's not just that the total amount grows, the interest added each year grows too – the interest added grows exponentially, increasing dramatically the longer this process carries on. Suppose you forgot all about your bank account, and you left it to sit there quietly growing for decades. After 38 years you'd discover that your original $1,000 has grown to over a million dollars – $1,020,674.70 to be exact – all thanks to the exponential growth of compound interest.

Exponential growth appears anywhere the future value of something is the present value raised to some power. It accounts for how a tiny initial population of microscopic algae (which reproduce by dividing, doubling the size of the population with each cycle) can bloom into a cloud of algae in the right conditions and smother a river or pond. The surprising power of exponential growth has caught many unawares.

$1,000 + $200 = $1,200

$1,200 + $240 = $1,440

$1,440 + $288 = $1,728

$1,728 + $345.60 = $2,073.60

$ 4692 9623 0571 0713   03.08.18

$ 4692 9623 0571 0713   03.08.18

$ 4692 9623 0571 0713   03.08.18

$ 4692 9623 0571 0713   03.08.22   03.08.18

BALANCING THE BOOKS

Interest compounds and grows exponentially whatever rate is used, whether 2% or 20%. But the bad news is that you're only likely to find an interest rate of the order of 20% applied to your debts, such as the money you owe on your credit card. (You're more likely to have an interest rate of around 2% for savings at the moment.) Turning the previous example around, if you owed $1,000 on your credit card at an interest rate of 20% a year, compounded annually, and didn't make any payments, as we calculated above you'd owe double that in four years.

Although interest rates are quoted as annual figures to make things easier to compare, banks actually compound the interest every month, usually using the annual rate divided by 12. For our example of a credit card balance starting out at $1,000 in debt, and an annual rate of 20%, the bank would use the rate of 1.67% per month (this is 20/12). This might sound like a lot less, but remember this will be applied every month. After one month the balance owed will be $1,016.70 (which is $1,000 \times 1.0167$); after two months it will be $1,033.68 (which is $1,016.70 \times 1.0167 = 1,000 \times (1.0167)^2$); and after 12 months it will be $1,219.87 (which is $1,000 \times (1.0167)^{12}$). So although this monthly percentage of 1.67% is much lower than the 20% annual rate, because it is compounded more frequently, you will end up owing $19.87 more than the $1,200 you'd owe if you were only compounding once a year. In the real world no bank is going to sit back and let you leave an unpaid debt to grow exponentially – you have to pay back what

you owe. And with compound interest, the quicker you pay debts off, the less interest you will pay overall.

Carrying on with our example, if you paid $200 a month off this debt, it would take six months to pay back everything you owe and the interest accrued. Paying off $200 a month means you would pay back a total of $1,053.24.

Month 1 $1,000 \times 1.0167 - 200 = 816.70$
Month 2 $816.70 \times 1.0167 - 200 = 630.34$
Month 3 $630.34 \times 1.0167 - 200 = 440.87$
Month 4 $440.87 \times 1.0167 - 200 = 248.23$
Month 5 $248.23 \times 1.0167 - 200 = 52.37$
Month 6 $52.37 \times 1.0167 = 53.24$ final payment

What if you upped your payment to $400 a month? Then you'd pay off the total debt in three months, paying back a total amount of $1,030.79. Paying off more means paying off your debt faster and less interest compounded, saving you $22.45.

Month 1 $1,000 \times 1.0167 - 400 = 616.70$
Month 2 $616.70 \times 1.0167 - 400 = 227.00$
Month 3 $227 \times 1.0167 = 230.79$ final payment

Equally, paying less means that it will take a lot longer to pay off your debt, and the compound interest added to your debt will grow larger. Paying off the debt at $50 a month would take just over two years and cost a total of $1,227.10. And pay off even less per month and the debt begins to rocket: at just $20 a month, it will take over four years and will cost $2,175.82 – more than twice the original amount you owed.

Suppose we're at the pub and you've run out of cash. I offer to lend you $20, you offer to repay it next month and buy me a beer ($5) to say thanks. This doesn't seem like such a big deal, but represents a horrific interest rate that you might expect from the worst loan sharks! Can you work out the annual interest rate this represents if you are compounding monthly?

BUILD +
BECOME

TOOLKIT

09

When you are confronted with an average, try to find out what average is being used and if it is likely to be representative of the information that is being transmitted. Sometimes a lot of information can be gleaned by comparing different averages. For example, during the 1980s the median income in the US rose slower than the arithmetic mean of the income. This tells us that the increase in wealth benefitted rich people more than it did average people.

10

Voting systems are complex and often come with anomalies, such as the person with the largest share of the poupular vote ending up as the loser. Designing a voting system comes down to choosing between various imperfect alternatives. Understanding a voting system often requires a good look at the small print, but is essential in knowing how your vote works.

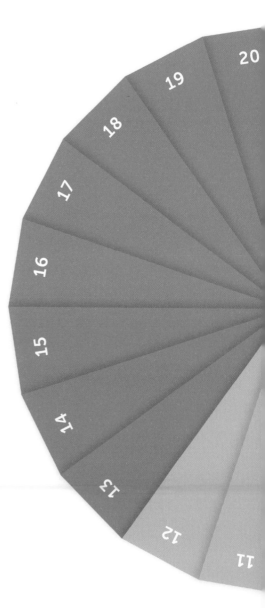

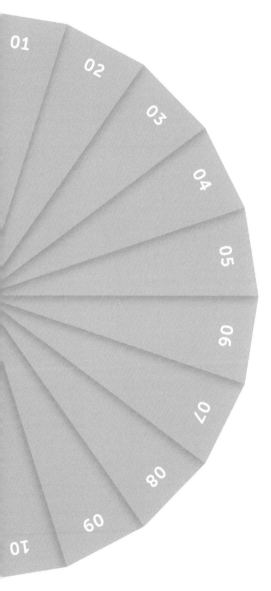

11

DNA profiles used as evidence are not unique, instead they are based on the frequencies of different variants of certain DNA markers in the population. These are used to calculate the *match probability* – the probability that a random person's DNA profile, such as the defendant's, will match the DNA profile in the evidence. It's vital to understand that this is not the same as the probability that the defendant is innocent.

12

To make things easier to compare, companies who lend money legally have to give the APR, or *annual percentage rate*, for their loans or credit cards. This is the figure it would cost you to borrow for a year, and should also take account of any fees or charges. Be aware if your bank tells you it is changing your monthly interest rate – even a small increase in the monthly rate used for calculating the monthly interest can have a huge effect on compounding your debt.

Another good way of judging a loan, or comparing two loans, is the total cost of borrowing at the end of the loan period.

FURTHER LEARNING

READ

The Rich, the Right, and the Facts: Deconstructing the Income Distribution Debate
Paul Krugman (*The American Prospect*, 2014)
www.prospect.org/article/rich-right-and-facts-deconstructing-inequality-debate

All about averages
Andrew Stickland (*Plus*, 2005)
www.plus.maths.org/content/all-about-averages

Mathematics and Democracy
Steven J. Brams (Princeton University Press, 2008)

Appealing Statistics
Peter Donnelly (*Significance*, February 2005)

It's a match!
Rachel Thomas (*Plus*, 2010)
www.plus.maths.org/content/os/issue55/features/dnacourt/index

WATCH

The Galton Board
IMA maths, YouTube

The best stats you've ever seen
Hans Gosling
TED Talk

The paradox of democracy: Arrow impossibility explained
University of Leeds, YouTube

DO

Check your compound interest calculations using a mortgage calculator, such as
www.citizensadvice.org.uk/housing/moving-and-improving-your-home/mortgage-calculator/

EXPLORE

Congressional exhibitions and briefings by the American Mathematical Society,
www.ams.org/government/dc-outreach

RELATIONSHIPS

LESSONS

A simple mathematical argument helps make the case that we should all be nicer to each other.

Understanding human relationships is a messy business. When it comes to the way we interact, the choices we make when we are in love and our sexual behaviour, it often seems as though logic has left the room and shut the door behind it. And don't get us started on how we interact with our families – just ask a fly on the wall at a family get-together!

But it turns out the cold, hard light of mathematical logic can give us a clear-headed view on all this, often revealing surprising results and stimulating new questions and understandings of how people interact. This chapter starts with a surprising mathematical explanation for the evolution of human kindness, even though being nice wouldn't appear to do any of us any individual good. Then we see how mathematics has revolutionized the way we find people to be particularly nice to, with the algorithms of online dating.

Of course, once you've found that special someone, chemistry hopefully takes hold – but maths can help us understand the way we think and have sex as well. And finally, the outcome of all that sex is all those relatives. And a simple mathematical argument helps make the case that we should all be nicer to each other because we are, within surprisingly few generations, one big family.

THE EVOLUTION OF HUMAN KINDNESS

Mathematics isn't usually associated with psychology. Human behaviour isn't based on axioms and it certainly doesn't follow logical rules. Yet an area of mathematics called *game theory* has proved surprisingly useful in understanding aspects of our own behaviour and also that of other animals.

Game theory is based on the idea that an interaction between two people, for example in closing a business deal, can play out much like a game. The 'players' contemplate their possible moves, anticipate counter-moves, and come up with a strategy they hope will benefit them in some way: perhaps by making them richer, or by optimizing some other quantity, such as happiness or lack of guilt.

The idea is to use mathematical games as proxies for such interactions. The games are designed to capture the essence of a particular type of interaction. Their rules are clear-cut, the possible moves well defined, and the benefit or harm of a particular move quantifiable. This means that the games can be systematically analyzed to see which strategies lead to which outcomes. Game theory is a major tool in economics, but also finds applications in psychology and the biological sciences.

A particularly interesting application of game theory is in the field of evolution. The idea is to model a population of individuals as agents that interact through mathematical games. Each agent has a particular approach to playing the game – a particular strategy – which doesn't necessarily have to be optimal and could even vary from game to game. The success of an individual in playing the game over and over against different partners is reflected in the number of offspring they produce: the higher your gain in playing the game, the more children you will have. The offspring inherit their parents' strategy (save for some random mutations perhaps), which means that, over the generations, strategies can become dominant, die out or come and go.

Using evolutionary game theory, scientists have been able to shed light on the evolution of a human (and animal) characteristic that appears to fly in the face of the central idea of Darwinian evolution: capacity for altruism.

	A CONFESSES	A DOESN'T CONFESS
B CONFESSES	Both get 8 years	B gets 2 years, A gets 10
B DOESN'T CONFESS	B gets 10 years, A gets 2	Both get 5 years

THE PRISONER'S DILEMMA

Imagine two criminals have been caught in a stolen car, who are also suspected of having robbed a bank. The police decide to interview the two separately and offer each the same deal, which involves a shorter prison sentence for confessing. The exact terms of the deal are shown in the table above.

In analyzing their options, each (selfish and rational) suspect will realize that, no matter what the other prisoner does, they are better off confessing to the robbery. If the other suspect doesn't confess, the first suspect gets away with a lenient two years, much better than the five years

that comes with both not confessing. If the other suspect does confess, the first suspect gets eight years, still better than the ten years that would have come with not confessing.

Thus, both prisoners will decide to confess and get eight years each. If they had only trusted each other and remained silent, they would have got five years. It's a situation in which a lack of trust leads to a worse outcome for both than cooperation would have done. The prisoner's dilemma epitomizes some of the impossible situations humans can get themselves into.

PLAYING THE GAME

The prisoner's dilemma is essentially a study of trust – or at least cooperation. To place the prisoner's dilemma in an evolutionary setting, imagine a population of individuals in which each individual plays all others, not just once, but repeatedly.

Individuals don't play exactly the same way each time. Instead, an individual A's decision of whether to cooperate with B depends on what B did in the last interaction: if B cooperated the last time around, then A cooperates this time with some probability p. If B didn't cooperate the last time around, then A cooperates this time with some probability q. This mimics real life, in which trust is built over repeated interactions. The exact values of an individual's p and q is therefore a measure of how cooperative and forgiving they are.

This iterated version of the prisoner's dilemma, where individuals play each other repeatedly, can then be turned into an evolutionary game that is played out over generations. You start with a population in which each individual's values of p and q are picked at random, and allow those who gained the highest pay-off in their lifetime (that is, minimized their time in jail) to have more offspring than others. Individuals pass their innate willingness to cooperate (their values of p and q) on to their offspring, so over generations it becomes clear which strategy — being cooperative or being mean — does better.

When game theorists put this approach into practice using computer simulations, they found a fascinating result: after only a few generations, a generous strategy emerged as dominant (**01**). In that strategy, an individual will always cooperate if their opponent cooperated last time (so $p = 1$), but even if their opponent didn't cooperate last time they will still cooperate with some probability that's not zero (so q is not 0).

What is more, the society would become more and more generous as generations went on, until it was dominated by individuals who always cooperated (**02**). Once that was the case, non-cooperators stood to gain again. A few mutations of strategies would be enough for these defectors to become dominant (**03**), and the whole cycle would start again from the beginning.

This evolutionary game may be very simple, but it suggests a mechanism by which altruistic behaviour may have evolved, not just in humans, but also in many animal species: in the long run, spanning generations, cooperation and trust can indeed pay.

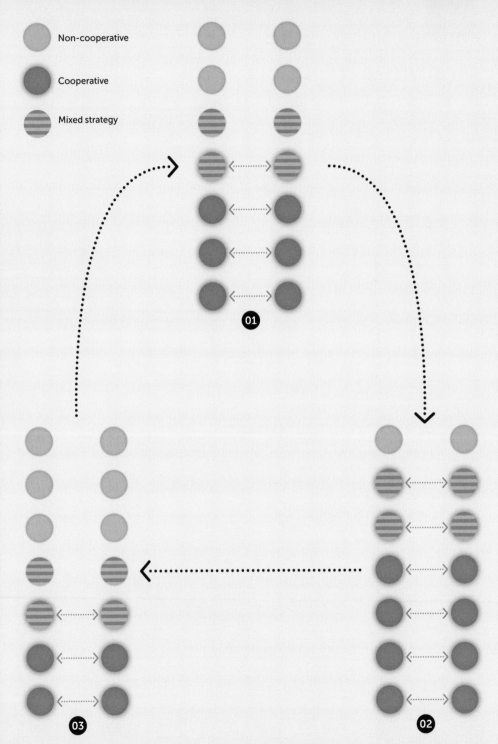

Non-cooperative

Cooperative

Mixed strategy

HOW WE SEARCH

Are you looking for love? Or perhaps you're looking for a gift for a friend? Or where to go on holiday and how to get there? Then you are almost certainly going to use a mathematical algorithm to help you find the perfect solution. The sheer volume of information and choices available to us means we need some mathematical help to match us to our desires.

An obvious example is the internet. There is so much information out there that it would take millions of years to read it all. But we are used to finding answers to our questions thanks to search engines such as Google. In 1998, Google unleashed its *PageRank* algorithm, which proved surprisingly good at sifting out the wheat from the chaff in our search for information online.

PageRank decides how important a page is by staging a popularity contest: the more times a web page is linked to by others on the web, the more important it is, and the higher it will appear in your search results. The algorithm constructs a huge grid of numbers (called a *matrix*), with each column in the grid representing each web page. The entries of the column corresponding to a page *P* indicate which other pages *P* links to. The page *P* passes on a proportion of its own importance to those pages it links

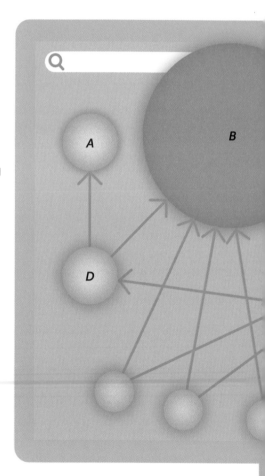

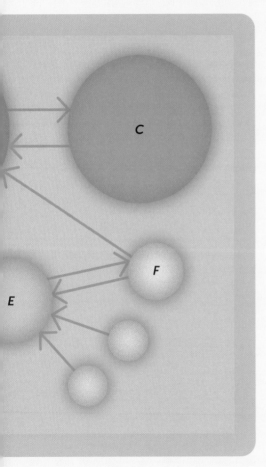

to, and its own importance is calculated by adding up all the contributions made by the other pages linking to it.

This all sounds pretty circular but, in fact, calculating the numbers in this grid is relatively straightforward, even if the matrix involved has billions of rows and columns, thanks to some established mathematics called *linear algebra*.

PageRank is just one of the mathematical algorithms contributing to Google's search results, and demonstrates the vital role maths plays in our daily lives. Aggregation is now the name of the game and multiple websites offer ways to match us with flights, hotels, clothes, obscure collectible figurines or our heart's desire. People and companies regularly try to game the results (just search for 'search engine optimization' to see the market for doing this in search results), but, as we'll see, it took a mathematician to try to game the results for online dating.

The PageRank algorithm is a bit like a popularity contest, where each web page votes for the pages it links to. But some votes, such as that for page *B*, are worth more than others if that page is itself highly ranked. This is why page *C* is ranked higher than page *E*, despite having just one vote.

LOOKING FOR LOVE

A particularly personal example of how search algorithms work can be seen on dating websites and apps, all of which rely on mathematics to link you up with potential dates – whether it's by shared interests based on surveys, compatibility based on personality tests, proximity or even automatic profiling produced from your use of social media.

OkCupid was started by four maths students from Harvard in 2004 and has a sales pitch that states: 'we use maths to find you dates'. Their matching algorithm relies on users answering multiple-choice questions that vary from 'Are carbohydrates something you think about?' to 'Do you want to have children?'. Users can answer as many questions as they want, and for each they also record how they'd like their potential match to answer. The user also indicates how important each question is to them personally: from irrelevant to mandatory. Mathematically speaking, the importance you give each question changes the *weight* of that question in the calculations: your rating of a question's importance decides how many points that question will contribute to your potential match's score, and vice versa.

The final step of the algorithm, calculating the 'match percentage', combines both you and your potential match's scores in a type of average. Rather than an *arithmetic mean*,

a *geometric mean* (taking the nth root of a product of n numbers) is used instead. Geometric means are particularly useful when the quantities you are averaging might be quite different, perhaps even measuring different properties.

The match percentage is based on the scores you gave each other from the n questions you both answered. These are multiplied together and the nth root is taken of this answer. If you have two questions in common, with scores of 98% for your potential match and 91% for you, the 'match percentage' is: $\sqrt{0.98 \times 0.91}$ = 94%. If your satisfaction scores were based on three questions, we would have instead have taken the cube root and your match percentage would have been: $\sqrt[3]{0.98 \times 0.91}$ = 96%. We'd have taken the fourth root if you'd had four questions in common: $\sqrt[4]{0.98 \times 0.91}$ = 97%.

Because of the properties of this mathematical operation, the more questions you answer – the higher the value of n in the nth root – the higher your match percentage will be.

The site then uses this information to calculate a 'match percentage' between you and other users and suggests potential dates to you based on these. The higher your match percentage with a person, the higher up your profile will appear in their results and their profile in yours.

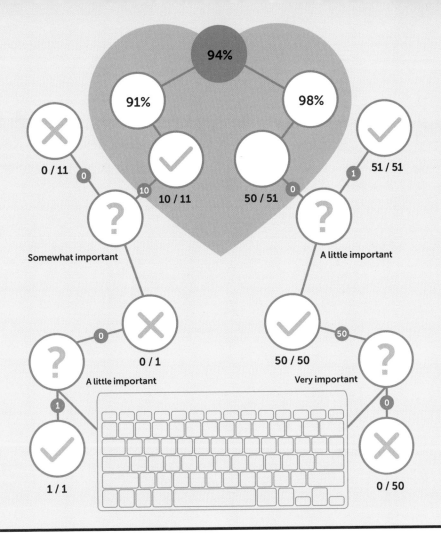

94%

91%   98%

0 / 11   51 / 51

10 / 11   50 / 51

Somewhat important   A little important

0 / 1   50 / 50

A little important   Very important

1 / 1   0 / 50

In 2012 a maths PhD student, Chris McKinlay, famously 'hacked' the OkCupid algorithm. He didn't actually hack into their machines; instead he applied his maths skills to the process of choosing which questions to answer (and he answered these honestly), and what importance to give them, to increase his match percentages dramatically. He went from just 100 matches over 90%, to being the top match for more than 30,000 women. This didn't automatically make these matches successful relationships, it just meant he got a chance to be higher in other people's results and so get that elusive first date. It took 88 of these first dates to meet his true love, and they were engaged by the time the story hit the press in 2014. Reader, he married them.

HUMANS M

ALWAYS FO

LOGIC. BU

CAN EXPLAI

BEHAVIOUR

GHT NOT
LLOW
LOGIC
N HUMAN

THE STATISTICS OF SEX

Sex isn't a mathematical activity, even mathematicians would agree, but trying to understand our collective sexual behaviour is. It's a statistical activity, to be precise. As in all fields of the social sciences (and all other sciences for that matter), to really understand what is going on, you need data. You need to know what people are doing, how much of it they are doing, why they are doing it, and what their attitudes towards their doings are.

As you might expect, people only started systematically collecting data about sexual attitudes and behaviours relatively recently. You might remember Shere Hite's studies on female sexuality published in the 1970s, or Alfred Kinsey's pioneering books published around 1950. The exceedingly private part of human life that those studies covered is fascinating in its own right, but there are plenty of other reasons for collecting information about sexual behaviour.

It became apparent during the HIV/AIDS epidemic in the 1980s that you can only understand and prevent the spread of an STD if you know about people's sexual contact patterns (see Lesson 1). Sex is inextricably linked to babies, and we can only predict future demographics if we know who has sex with whom, how often, and what their attitudes and access to contraception and abortion are. Sexual attitudes are also intimately linked to issues like gender equality and sexual crime. More fundamentally, sex is an important part of the human psyche and worth understanding for that reason alone.

Kinsey's and Hite's studies shocked the world and had important impacts at the time – in Hite's case on the women's liberation movement – but unfortunately they also provide examples of how not to do statistics. Hite, for instance, sent her first survey to women's organizations, including groups fighting for abortion rights and women's centres at universities, targeting a population of women that was hardly representative of the female US population as whole. Only 3% of women replied, so Hite ended up with a potentially unrepresentative sample of what was already an already unrepresentative sample. Response rates to later surveys didn't fare much better, and this laid her open to strong criticism. Her doubtlessly important message was undermined as a result. In his fascinating book *Sex by Numbers*, David Spiegelhalter labels Hite's statistics as 'inaccurate' and Kinsey's as 'numbers that could be out by a very long way'.

These days, statistical tools – such as significance levels and confidence intervals – are staples in most social scientists' toolboxes. What singles sexual statistics out from other statistics in the social sciences is the difficulty of coming by the numbers. Not everyone is willing to part with private information about sex, or part with it truthfully. Studies need to be set up carefully and a lot of detective work is needed when analyzing them. All this turns sexual statistics into a statistician's playground – or nightmare.

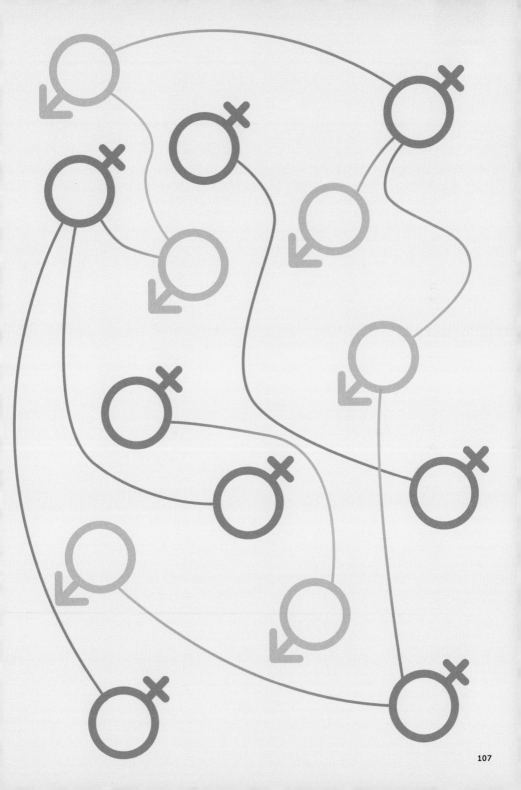

TRUTH AND LIES IN NUMBERS

An example of how a little mathematical awareness can go a long way in sexual statistics comes from the third National Survey of Sexual Attitudes and Lifestyles (Natsal), which interviewed 15,000 adults aged 16–74 years between 2010 and 2012. A result of the survey was that straight men have 14 sexual partners, on average, and straight women only seven. This might not sound surprising at first — until you realize that it's mathematically impossible. In a closed population with roughly as many men as women, the averages need to be equal.

This result isn't too hard to prove. Imagine you have n men and n women, some of whom have had sex with each other. Line them up to face each other, a row of women facing a row of men, and draw a line between a woman and a man if they have had sex. Write w for the total number of lines emanating from the row of women and m for the total number of lines emanating from the row of men. The average number of sexual partners per woman is then w/n. The average number of sexual partners per man is then m/n. However, since every line that emanates from a woman ends up at a man, and vice versa, the total numbers of sexual partners, w and m, are equal. But if the $w = m$ then $w/n = m/n$. In other words, the average are equal.

What, then, is behind the numbers of Natsal-3? An obvious possibility is that men tend to inflate their numbers, presumably to show off, and women to deflate them, perhaps for fear of stigma, even if their responses to the survey are treated confidentially. Perhaps they are trying to delude themselves, or please that imaginary moral authority we all fear is looking over our shoulders at all times.

But there are also more subtle explanations that illustrate the difficulty of conducting surveys into such intimate and subjective subjects as sex. For example, our mathematical proof assumed a closed population, that is, a population in which every individual only has sex with people who are also part of the population, not outsiders. This may be more or less true for the entire population of Britain, but it's not true for the people who took part in the survey. If a reasonable proportion of men had sex with women not included in the survey, then it is possible for the averages not to balance. These women could include, for example, those falling below the age limit of the survey, or sex workers, who also weren't questioned.

Another issue is what people might interpret as a sexual partner or a sexual experience. Perhaps men have a tendency to count experiences as sexual that women wouldn't. For example, if an experience involving sex was unpleasant for a woman, she might not count it in a survey that appears to refer to voluntary and pleasurable events. Scientists have tried to adjust for these, and other, explanations for the statistical discrepancy, but to our knowledge, the jury is still out. Whatever the answer, or answers, may turn out to be, without the Natsal-3 survey, this interesting discrepancy wouldn't have been detected.

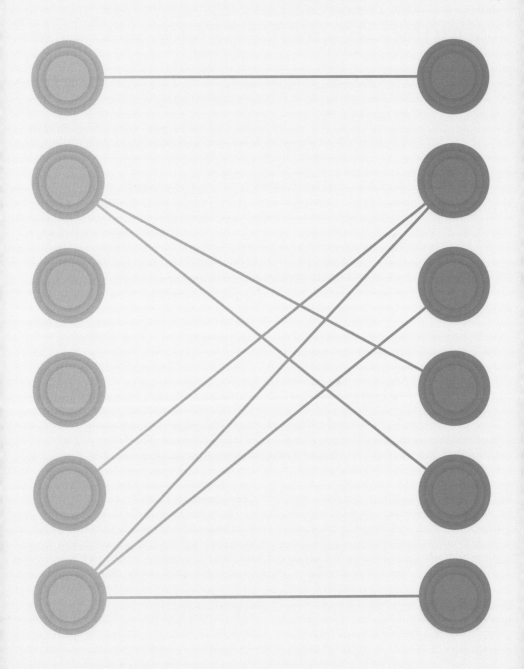

MAKING RELATIVE SENSE

Families are complicated. Sometimes you might even wonder how you can be related to your nearest and dearest. Luckily maths can make even the most complicated relationships crystal clear.

Understanding your direct ancestors or descendants is fairly straightforward. Your parents' parents are your grandparents, their parents are your great-grandparents, theirs are your great-great-grandparents... and so, too, your children's children are your grandchildren, their children great-grand-children, and so on. But as soon as you start considering anyone's siblings, things quickly get confusing without a bit of maths to set things straight.

If you have siblings, your children will be first cousins to your siblings' children. Your grandchildren will be second cousins to your siblings' grandchildren. Your great-grandchildren will be third cousins to your siblings' great-grandchildren. An easy way to remember this is that if two people are *n*th cousins, their respective children will be (*n* + 1)th cousins. And that *n*th cousins share two ancestors who are (*n* + 1) generations

back – first cousins share grandparents, second cousins share great-grandparents etc.

But how are the descendants of your cousins related to you? Your first cousin's children are your *first cousin once removed*. Their children are your *first cousin twice removed*, and so on. The number of times 'removed' refers to the number of generations between you.

This separation can work up the family tree (into the past) as well as down. A sibling of your grandparent is your great-uncle or great-aunt. Their children would be first-cousins of your parent. As they are in the previous generation to you, they are once-removed from you. They share two ancestors with you – your great-grandparents (three generations back from you) who were their grandparents (two generations back from them). They are your *first cousin once removed*. The number of times removed designates how many generations are between you, and the degree *n* of the cousin comes from the shortest distance back (*n* + 1 generations back) to your shared ancestors.

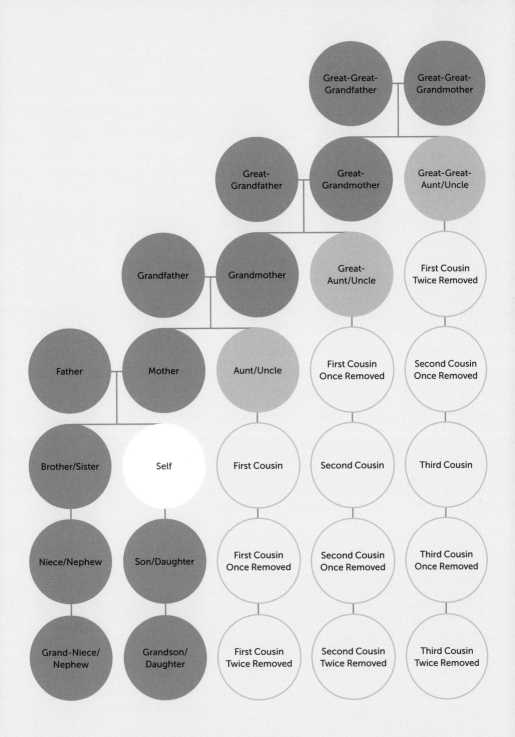

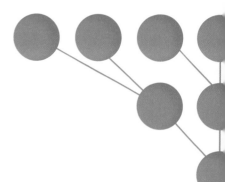

WE ARE FAMILY

Biologically, you are the product of two parents – your mother and your father. And they each had two parents – your maternal grandmother and grandfather, and your paternal grandmother and grandfather. Each of your ancestors had two parents, increasing the size of each previous generation of your ancestors by a factor of two.

The size of these previous generations blooms with exponential growth. Ten generations back, about 300 years ago (using the common assumption of a generation being about 30 years), that layer of your family tree consisted of $2^{10} = 1,024$ ancestors. Your 20th generation of ancestors, alive about 600 years ago, consisted of over a million people. And 900 years ago, your 30th generation of ancestors was over a billion people. But here's where we hit a

problem: the population of the entire world at that time was thought to be less than 400 million. That means that at some time in the past the branches of your family tree started to intertwine and overlap.

This same argument applies for every person alive today. There were not enough people living in the past to allow for all your potential ancestors to be distinct from all of my ancestors, or anyone else's. All of our family trees eventually intertwine: at some point in the past we all share a common ancestor. From that point our ancestral generations overlap more and more until there is a time in the past where everyone alive then was either an ancestor of everyone who is alive today, or no one who is alive today.

This mathematical argument has been explored in various ways, and incorporated

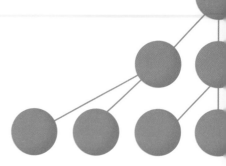

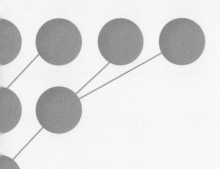

in various models of human migration and mating. Estimates of when the most recent shared common ancestor was alive are surprisingly close – definitely within the last few thousand years, and for people of European descent less than a thousand years ago – well within the timeframe of our historical records. A result of these models are convincing arguments that all people of European descent are descendants of King Charlemagne from the eighth century, or indeed descendants of anyone else alive at that time (say Charlemagne's lowliest servant, as long as they went on to have descendants alive today).

This mathematical model of shared ancestry is also backed up by genetic studies. We get half of our genes from our mother, and half from our father. We share DNA sequences with our ancestors, the longest with our parents, with these sequences getting shorter the further back our ancestors are from us. Geneticists have identified shared genetic sequences in populations that demonstrate that everyone who was alive in Europe a thousand years ago and who had descendants is an ancestor of everyone alive today of European ancestry.

Humanity has a tendency to group together, with one group differentiating from another on the basis of location, language or external differences. But this mathematical perspective emphasizes the fact that we each share a long history with every person on this planet. And the maths shows this shared ancestry is probably a lot more recent than you might at first imagine. You can't choose your family but you can embrace the fact that all humanity is related.

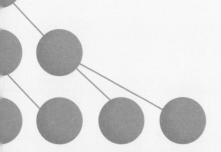

TOOLKIT

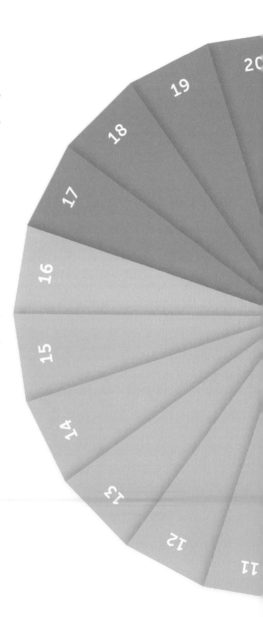

13

Next time you are wondering whether or not to be generous to a fellow human, remember that being generous may not just benefit the individual you are dealing with, but society as a whole. According to game models, sufficient numbers of generous individuals enable others to feel they can afford to be generous also, and benefit from it in the long run, so a happily generous community can evolve.

14

We all rely on algorithms to find what we need in this information age. One you probably use every day is PageRank - the maths behind Google's search engine. This mathematical algorithm is now used in surprising places: including studying molecules in chemistry, genes in biology, neural networks in neuroscience, and predicting traffic flow on road networks.

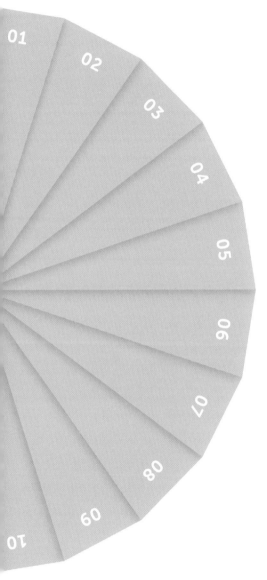

15

Understanding our collective sexual behaviour is not just interesting, but also useful. For example, it can help us understand how sexually transmitted diseases spread and how our population might grow or decline in the future. Sexual statistics can be hard to come by, however, because not everyone is willing to part with such private information. This means that surveys about sex need to be designed and interpreted carefully.

16

The number of our ancestors blooms with exponential growth as we go back each generation: two parents, four grandparents . . . 2^n great($^{(n-2)}$)–grandparents. As you go further back through the generations, due to the size of the world's population, the branches of your family tree start to intertwine and overlap with that of every other person alive today, until, at some point, we all share a common ancestor.

BUILD +
BECOME

FURTHER LEARNING

READ

Sex by Numbers
David Spiegelhalter
(Wellcome Collection, 2015)

Does it pay to be nice?
Rachel Thomas (*Plus*, 2012)
www.plus.maths.org/content/does-it-pay-be-nice-maths-altruism-part-i

SuperCooperators: Evolution, altruism and why we need each other to succeed
Martin Nowak with Roger Highfield
(Free Press, 2013)

On Chris McKinlay's hacking of OkCupid
Kevin Poulson (*Wired*, 2014)
www.wired.com/2014/01/how-to-hack-okcupid/

More on geometric mean
John D. Barrow (*Plus*, 2008)
www.plus.maths.org/content/outer-space-pretty-mean-prices

Modelling the recent common ancestry of all living humans
Rohde, D.L.T, Olson, S. & Chang, J.T., *Nature*, 431, 562–566 (2004)

WATCH

Inside OKCupid: The math of online dating
Christian Rudder, YouTube

Are we all related?
PBS, YouTube

DO

Explore **Sex by Numbers** with David Speigelhalter and the Wellcome Collection with this animation: www.sexbynumbers.wellcomecollection.org

COMMUNICATION

LESSONS

The digital world grew from mathematics, its technologies are built using mathematics, and it's mathematics that keeps our communications safe.

Over the last decades, our means of communication have expanded beyond recognition. Much of this is down to the digital revolution, which is firmly grounded in mathematics. If you have bought this book online, then this single transaction will have used maths in at least three ways: maths is behind the algorithms that enable the ordering process; maths has encrypted your credit card details; and it has also enabled the book to travel through the air to your device. The digital world grew from mathematics, its technologies are built using mathematics, and it's mathematics that keeps our communications safe.

What enables modern technologies is not just the mathematics of the 0s and 1s we associate with computers. Every time you use the satnav in your car, you evoke the classic geometry of ancient Greece and Einstein's rather esoteric theory of relativity. Our communications play out on networks whose features are best understood using the mathematics of network theory; under the sharp eye of mathematical abstraction, even the wildest networks tend to exhibit some order.

Some worry about the rise of artificial intelligence. With algorithms now capable of sifting through the large volume of data we leave in our digital trails and independently learning from what they see, will they at some point take over the world? This fear is real and a little suspicion is justified. In many cases, however, the fear is down to our lack of understanding of how our communication technologies really work. Perhaps it's time we all looked under the hood of the modern information age – mathematically speaking.

NETWORKS

How popular are you on social media? Not as popular as you might think, is the most likely answer. A study of 5.8 million Twitter users conducted in 2016 found that over 93% of users had fewer followers than the people they followed, on average. The only users who escaped the humiliation were those with over 155,000 followers. And since only 1% of users had over 460 followers, there weren't many of those.

This phenomenon had already been observed by the sociologist Scott L. Field in 1991 before social media even existed. Most people have fewer friends than their friends do, on average, but when you ask them, they tend to say they believe that they've got more friends than their friends. Because of this discrepancy, the phenomenon has become known as the *friendship paradox*.

This paradox is one of the many puzzling things the world of social media throws at us, but can be explained by some relatively simple maths. One possibility is that the paradox is just a statistical artefact. Imagine a friendship network, be it real or virtual, that contains a few people who are extremely popular. If you are friends with such a so-called *hub* and calculate the average of the number of friends your friends have, then the single highly popular hub will drive that average up.

This is best illustrated by drawing the network on a piece of paper. In the example

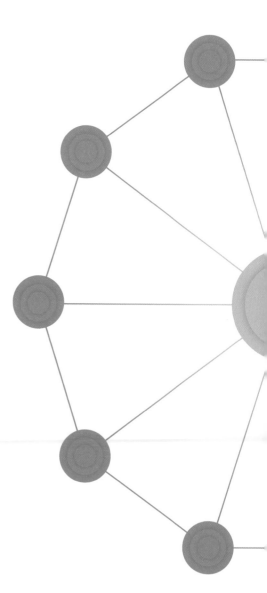

here, the hub in the middle has ten friends, while all the others have only three. The average number of friends of friends for the non-hubs is therefore $(3 + 3 + 10)/3 = 16/3 = 3.3$. Since this is greater than three, all the non-hub nodes suffer from the friendship paradox. But this doesn't mean they are particularly unpopular. They just happen to be less popular than the one superstar hub in the middle.

There is also a more interesting way in which the friendship paradox can arise. In a form of social climbing, people may prefer to only befriend people who are more popular than themselves. If this is the case, then a picture of the network won't contain the telltale stars that indicate the presence of highly popular hubs. Instead, it will contain hierarchies. In a Twitter network things are slightly different as the following relationship isn't generally reciprocal, but similar results apply. And hierarchies are exactly what the study of the Twitter users found: people prefer to 'follow up' rather than 'follow down', and only people with few followers are happy to 'follow across'. The trend was so strong that even users in the top 0.5% suffered from the friendship paradox: the average number of followers of the people they followed was larger than their own number of followers.

The friendship paradox isn't really a paradox, then, but a result of the way in which the network has grown.

GROWTH OF NETWORKS

Network science is important because networks are absolutely everywhere in modern life. We travel along road and transport networks. We depend on the power grid. Even our bodies rely on networks, formed, for example, by the neurons in our brains. Understanding networks is essential for improving the flow of things that we need – water, electricity, information – and disrupting the spread of things we fear – terrorism, fake news or disease.

Understanding all the different networks might seem an impossible task, but the god of mathematics has been kind: many networks, even if they arise in different contexts, show up similar features.

One of these features is called *scale-freeness*, a concept that burst onto the mathematical scene around the year 2000 in the work of Albert-László Barabási and Réka Albert, who claimed it existed in social networks, the internet, and protein regulatory networks, to name just a few examples.

Previously people had assumed that most complex networks are essentially random: you'd get the same kind of structures if you started with a set of nodes and decided the connections between them at random. Such networks tend to be 'democratic': most nodes have roughly the same number of links. That typical number of links tells us whether we are dealing with a 'large scale' network, where the typical number is high, or a 'small scale' one.

Scale-free networks, by contrast, don't have such a characteristic scale: most nodes have only a few links but there are also some that have a huge number of links. Mathematically, a scale-free network has a link distribution that follows a *power law*: the number y of nodes with a given number of links x turns out to be proportional to $1/x^k$, where k is a constant which is typically small, often between two and four.

To understand why so many networks show up this kind of structure, you need to let go of the context in which the networks arise and concentrate on the mathematical rules that govern a network's growth. Barabási and Albert discovered that scale-freeness is a result of the *preferential attachment* we saw in the Twitter study: if the new nodes that join a network prefer to attach themselves to nodes that already have many links (and we can see why that might happen) then scale-freeness arises naturally.

Knowing that a network is scale-free tells us how robust it is against disruption. If, in a scale-free transport network, a random node fails, for example due to bad weather, chances are that this won't affect the whole network too badly: even though there are many highly connected hubs, the vast majority still don't have too many connections, so one of them failing isn't too big a deal. On the other hand, though, scale-free networks are very vulnerable to targeted attacks: if a major airport is out of action, then the whole network is likely to fall apart.

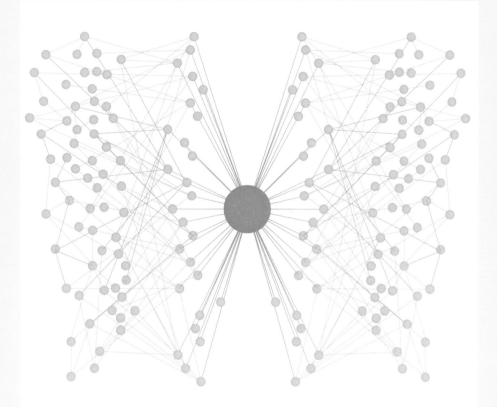

Because many modern networks are incredibly large, it can be hard to tell which ones really do exhibit scale-freeness. However, the assumption of scale-freeness can at least shed light on processes that are otherwise hard to understand. For example, in a 2017 study, Michael J. Spivey from the University of California simulated the spread of fake news in a scale-free network, assuming that nodes don't immediately believe a piece of news but try to verify it using another source. Assuming, as is the case in reality, that the supposedly independent second source is actually connected to the first, Spivey found that it's 'remarkably easy for a node in the network to be fooled into thinking it has received independent verification of a false rumor, when in fact that "second source" can be traced back to the original source'. Spivey's work is only a simulation, of course, but if we are going to prevent the spread of fake news, this kind of theoretical understanding is important in combatting its rise.

STAYING SECURE

Internet security is more important than ever before. Until not so long ago it was mostly just personal emails that needed to be protected from prying eyes, but today it's highly sensitive information of many kinds: financial transactions and bank details, medical records, access to social media accounts, to name just a few. All these pieces of information need to be protected – and mathematics is essential in making sure the protection is safe.

To see how, let's start with a basic type of protection we all come into contact with every day: internet passwords.

Internet passwords

We are often told to use truly random combinations of letters, numbers and symbols at least eight digits long, but it's much easier to use things like p@ssw@rd or Michael88. It only takes a tiny bit of maths, however, to see how important randomness really is.

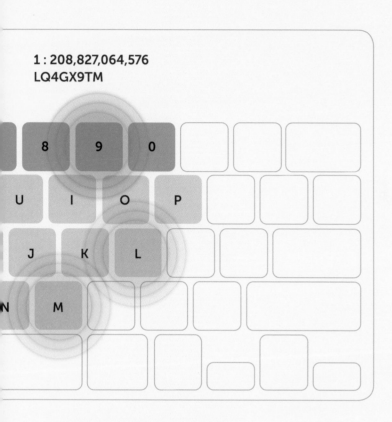

1 : 208,827,064,576
LQ4GX9TM

Imagine trying to hack a password that is eight characters long and made up only of the 26 lower-case letters of the alphabet – never mind any symbols or numbers, which would make it harder to crack. If the password is a truly random string of eight characters, then you need to be prepared to try over 200 billion possibilities before getting lucky: for each of the eight characters there are 26 possible letters, giving a total of $26^8 = 208,827,064,576$ combinations. By contrast, there are only around 80,000 eight-letter words in the English language. If the password is one of those, and you as a hacker suspect this, you would need at most 80,000 guesses. That's 0.04% of 200 million – easy work for a computer. And if you include personal information about your victim in your hacking attempt, focusing on passwords made up of their name and birth year, for example, your task becomes even easier.

CRYPTOGRAPHY

As we have just described it, a password acts like a lock to a box containing sensitive information. The trouble is, however, that when it comes to internet accounts, the box isn't in your possession. It lives on some other computer (for example your bank's), so your password still needs to be sent over the internet and stored elsewhere (in the bank's database). This brings us to the art of cryptography: to send and store sensitive information safely, you need to encrypt it using some secret code that is hard to crack. A good way of doing this involves mathematical problems that are easy to do but hard to undo.

A good example comes from two of the most basic mathematical operations, multiplication and division. Given two numbers it's relatively easy to multiply them, but given their product alone, it's hard to work out what its factors are. For example, it's relatively easy to work out that $7 \times 13 = 91$ in your head, but if you were given just the number 91 and asked to find its factors you'd probably have to resort to pen and paper.

When the numbers involved are large, this is true even for computers. We have algorithms that can find the factors of any given number, no matter how large, but the number of steps these algorithms need to perform to find the factors grows exponentially with the size of the number you give them as input. If the input number is large enough, then even the fastest supercomputer won't be able to factor it before the Universe comes to an end, at least not with known algorithms.

The RSA encryption scheme, which is widely used to encrypt messages sent over the internet, is based on just this hard-to-reverse nature of multiplication.

- To start with, the sender of a message notifies the intended receiver that they want to send them something.
- The receiver picks two large prime numbers and multiplies them together to get an even larger number N, which it sends back to the sender.
- The sender then uses a particular algorithm to encrypt their message, an algorithm that depends crucially on the number N.
- To decrypt the message, you need to know the factors of N. For the intended receiver, this is no problem, as it already knows the factors.
- An attacker who has intercepted the message, however, would first need to factorize N, and as we said above, if N is large, that's a hopeless endeavour.

Returning to the lock analogy, the RSA scheme acts like a padlock. The intended receiver sends the sender an open padlock (the number N). The sender puts the message in a box and snaps the padlock shut (encrypts the message using the number N), and sends the box to the receiver. The receiver then opens the padlock with their key (the factors of N).

So whenever you make a transaction on the internet, whether it's logging into your bank account or sending a private email, it's maths that keeps you safe.

$N = 1{,}103{,}477{,}490{,}486{,}037$

$N = 27{,}644{,}437 \times 39{,}916{,}801$

THE MODER
INFORMAT
IS BUILT B
MATHEMAT

N

ON AGE
Y
CS.

BIG DATA

Every minute we post over 300,000 photos on social media, upload hundreds of hours of video, send millions of tweets and emails, spend a million dollars shopping online, and ask the virtual PAs in our phones over 100,000 questions. Of all the data around today, 90% was created in just the last two years. We used to leave behind footprints, dead cells and the occasional bit of trash, but now we shed data as we move through our lives every minute of every day.

Data scientists regularly deal with measurements of thousands of quantities at once, without knowing which are responsible for the outcomes they are interested in. Microarray genetic data, for example, are the simultaneous measurements of thousands of genes. Trying to identify which out of these thousands of variables actually matter in what you're trying to understand can feel more like a fishing expedition than research. And while, in the old days, you might have been able to plot data points on a two-dimensional piece of paper, these variables are points in higher-dimensional spaces, each point representing hundreds or thousands of different pieces of information. Statisticians and data scientists now work on developing new techniques to make sense of such profoundly different data.

Machine learning

One way of understanding and using big data is to use machine learning. Rather than being explicitly told what to do, a machine-learning algorithm uses a set of data (the training data) to tune a mathematical model – a set of mathematical equations – so that it will automate some process.

As an example, suppose you want to automatically recognize pictures of cats. Thankfully the internet is teeming with pictures of cats, as well as slightly fewer images that aren't cats. You start with the training data: a set of pictures that are accurately labelled as either CAT or NO CAT.

One way of automating the process of recognizing a cat is with a mathematical model inspired by our own brain. Artificial neural networks are made up of many

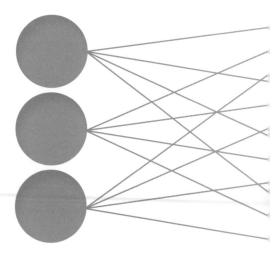

Input layer

mathematical equations, each acting as a neuron, linked together in a network. The network starts with a layer of inputs, which takes the digital representation of the image as a grid of pixels. These inputs are then passed through to successive layers of mathematical equations, until the equation that is acting as the output neuron gives a result of either CAT or NO CAT.

As you send the training data through the network, the contribution of each neuron to the final answer depends on the strength of the connections between it and the rest of the network – a weakly connected neuron doesn't play much of a role in the final answer. If the network gives the wrong answer for a particular image in the training data, the rules the machine uses to learn will increase or decrease the strength of the connections between the neurons in the network to make it as likely as possible that the network will give the correct result. The idea is that if you repeat this for enough training data, you'll end up with a network that will automatically and accurately recognize if an image contains a CAT.

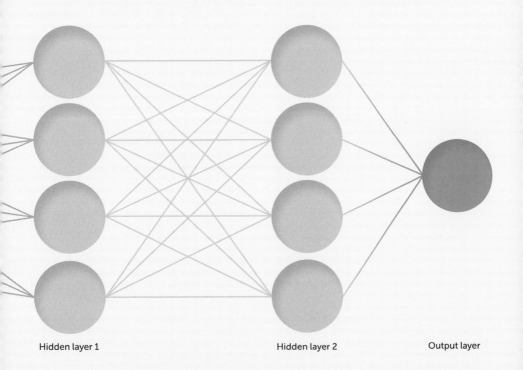

Hidden layer 1   Hidden layer 2   Output layer

ARTIFICIAL INTELLIGENCE

The above task sounds pretty banal (although we are always interested in finding more pictures of cats). But to many, a machine learning how to do something without our explicit help might seem a little alarming. Machine learning is a branch of artificial intelligence (AI), something often sensationalized in the media as the agent of a future apocalypse.

We have, however, been living with artificial intelligence, where machines emulate human behaviour, for some time. It is now common for social media sites to recognize images of people known to you, to encourage you to tag them in that photo. Online retailers suggest other products we might like to look at or buy, and ads are continuously served to us as we travel around the internet, specifically tailored to our past browsing history. And many of us now have artificial intelligences in our phones or in our homes that respond to spoken questions – understanding what has been asked and taking the appropriate action or providing the correct answer with impressive success.

Artificial intelligence is clearly benefiting humanity in slowing the tsunami of spam email, searching for new drugs, and detecting fraudulent uses of our credit cards. But it is also the powerhouse behind fake news and social media echo chambers. Many of the machine-learning algorithms in use in social media news websites are only interested in you clicking on links and spending more time on their websites. So they suggest similar connections based on our own views and online behaviours, and serve us stories that are like those we and our friends have read before, that can lead to us living in online bubbles where we only see and hear information that we are likely to agree with. None of these examples of artificial intelligence are sentient, and none have an evil master plan other than to achieve the (usually economic) goals they have been set.

Most researchers believe that the danger from AI won't be from some network achieving consciousness and deciding to get rid of the wetware (i.e. the humans), but from the unintended consequences of non-sentient AI changing the way we behave or the information we see.

AI IS ALREADY HERE

- Google maps uses data from people's phones to analyze traffic and suggest the fastest routes for travel.
- Mechanical AI has been used for over a century, such as the *Watt's governor* used in steam engines to keep a steady speed. A mechanical feedback mechanism automated the task of controlling the amount of fuel or speed that entered the engine.
- Rather than rely on simple rules that spammers can outrun, spam filters learn from the messages that come in, as well as from your decisions as to what is spam.
- Whether you are shopping online, or choosing a film to stream, the system you are using will recommend products based on what others with a similar purchase history have gone on to watch or buy.

KNOWING WHERE YOU ARE

The days of paper maps have long passed for most of us. We print maps off the internet, use satnav in our car or navigate with the maps on our smartphones. You will find a lot of us staring at a blue dot on our phones as we find our way through a city.

But how does our phone or car satnav know where we are? These devices rely on an ingenious piece of technology, GPS, that was developed by the military and is now a part of our everyday lives. And while it is very high tech, it also relies on maths that is thousands of years old, and some seemingly esoteric theoretical physics.

Ancient maths

As you might expect, mathematicians have always been interested in geometric objects like circles and spheres. These were first comprehensively studied over two millennia ago, by the Greek mathematicians Euclid and Archimedes, among others. They were particularly interested in the *conic sections* – those shapes made by slicing through a double cone with a flat plane.

You get a *parabola* if your plane is parallel to the slope of the cone. If you slice the double cone with a plane that's perpendicular to its vertical axis, you get a *circle*. If you tilt the plane a little you get an *ellipse*, and a *hyperbola* if the plane is vertical.

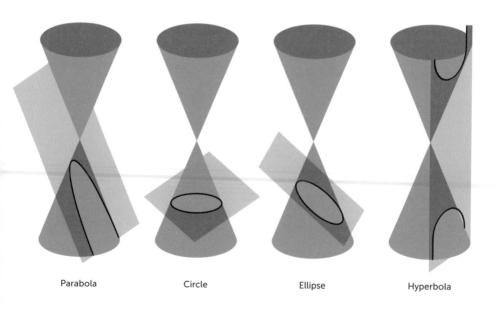

| Parabola | Circle | Ellipse | Hyperbola |

All of these shapes can be described geometrically. For example, a circle is the set of points that are the same distance, *r*, from a centre point. But you can also describe a circle using an equation. In a typical *Cartesian coordinate system*, which you might have encountered at school, a circle is the set of points with coordinates (*x*,*y*), such that:

$$x^2 + y^2 = r^2$$

Each of the conic sections has an equivalent geometric and algebraic description.

As we will see, the circle lies at the heart of GPS. But the other shapes are also useful in modern communication. Parabolas are used in all sorts of receivers, such as the satellite TV dishes that sprout from the sides of buildings across the city. The geometry of a parabola means that this shape focuses incoming rays at a focal point at the centre.

Another of the conic sections, the hyperbola, is used in *multilateration* – another way of determining location using incoming signals. This was used in World War I to locate enemy artillery ranges by the sound of their gunfire. Hyperbolas can still be used today to locate a target by silently listening for the signals it transmits.

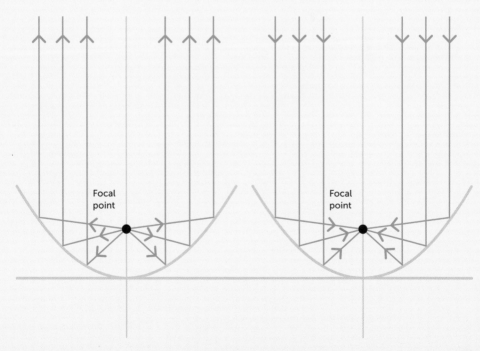

GPS

GPS stands for Global Positioning System and is the name given to a network of 31 satellites orbiting the Earth so that at least four are overhead at any time from any position on the Earth.

Each satellite is on a carefully planned orbit and so knows its own position relative to the Earth very precisely. Also on board is an incredibly accurate clock whose time is set to match that on Earth. These satellites constantly bellow out their position in messages sent as radio waves travelling at the speed of light, messages that also include the time the message was sent from the satellite.

A GPS-enabled device, such as your satnav, phone or even your camera, has a receiver that listens out for these signals. When it receives such a message it can easily calculate the exact distance between you and the satellite. The distance the message travelled was:

$$d = c \times t$$

where c is the speed of light and t is the time the message took to reach you (the difference between the time the message was sent and when it was received on your device, respective to GMT).

Your position is the point at which all three spheres intersect.

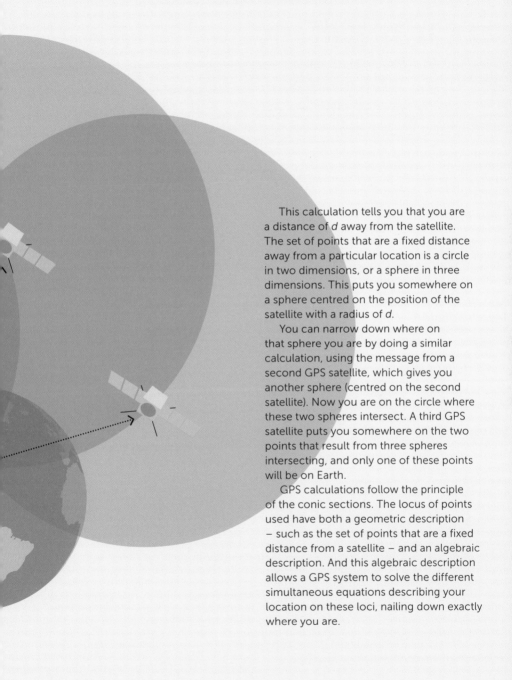

This calculation tells you that you are a distance of d away from the satellite. The set of points that are a fixed distance away from a particular location is a circle in two dimensions, or a sphere in three dimensions. This puts you somewhere on a sphere centred on the position of the satellite with a radius of d.

You can narrow down where on that sphere you are by doing a similar calculation, using the message from a second GPS satellite, which gives you another sphere (centred on the second satellite). Now you are on the circle where these two spheres intersect. A third GPS satellite puts you somewhere on the two points that result from three spheres intersecting, and only one of these points will be on Earth.

GPS calculations follow the principle of the conic sections. The locus of points used have both a geometric description – such as the set of points that are a fixed distance from a satellite – and an algebraic description. And this algebraic description allows a GPS system to solve the different simultaneous equations describing your location on these loci, nailing down exactly where you are.

TOOLKIT

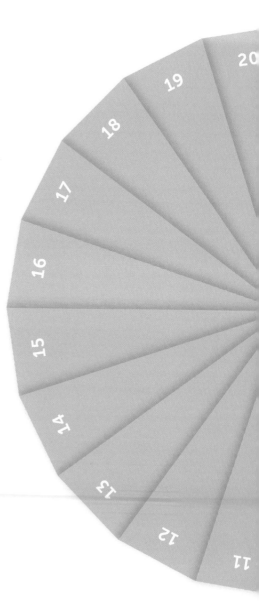

17

Networks are everywhere in modern life and arise in all sorts of different contexts. However, once you forget about the particular setting and only concentrate on the connectivity of the network, you often find similar structural features appear again and again. Studying these features sheds light on a huge range of real world phenomena, from human relationships to neuroscience.

18

Randomness is good when it comes to passwords, but truly random passwords are also hard to remember. One way of creating a memorable and reasonably secure password is to think of a memorable phrase and then pick the first letter from each word. The resulting string could then form the basis of your password, which you can further embellish with numbers and symbols.

19

Every minute of every day we are shedding data at an alarming rate – especially online and on social media. Artificial intelligence is used to understand these vast swathes of data, including machine learning based on mathematics inspired by the human brain. Neural networks are systems of linked equations that can be tuned using existing data to perform automatic processes, from recognizing pictures of cats to improving medical diagnoses.

20

Every time you use your phone or sat nav to help you find your way, you are relying on ancient mathematics – such as the geometric and algebraic descriptions of shapes like simple cones. GPS uses the 3D version of the geometric description – spheres, centred on GPS satellites. And the algebraic description means we can calculate where these spheres intersect and find your location.

**BUILD +
BECOME**

FURTHER
LEARNING

READ

Linked
Albert-László Barabási and Jennifer Frangos
(Perseus Books, 2003)

Safety in numbers
Rachel Thomas (*Plus*, 2002)
plus.maths.org/content/safety-numbers

The Code Book
Simon Singh (Fourth Estate, 2002)

Machine learning
The Royal Society (2017)
www.royalsociety.org/topics-policy/projects/
machine-learning/

Can AI be taught to explain itself?
Cliff Kuang (*New York Times*, 2017)
www.nytimes.com/2017/11/21/magazine/
can-ai-be-taught-to-explain-itself

Conic section hide and seek
Rachel Thomas (*Plus*, 2012)
www.plus.maths.org/content/conic-section-
hide-seek

Space segment
The official US government site on GPS:
www.gps.gov/systems/gps/space/

WATCH

What is machine learning?
The Royal Society, www.royalsociety.org/
topics-policy/projects/machine-learning/
videos-and-background-information/

Encryption and huge numbers
James Grime for Numberphile, YouTube

DO

Play **Six Degrees of Kevin Bacon**
www.oracleofbacon.org

Can machines really learn?
Test yourself against the machines in this
infographic from the Royal Society,
www.royalsociety.org/topics-policy/projects/
machine-learning/what-is-machine-
learning-infographic/

VISIT

Bletchley Park
Home of WWII code breakers

EPILOGUE

Galileo Galilei called mathematics 'the language of the Universe' and this has never been truer than it is today. Almost every aspect of our lives, from the gadgets we use to how we understand and explain the world around us, is touched by maths.

The reason for its power is abstraction: when you try to capture the essence of a process as succinctly as you can, you are inevitably led to mathematics. It is the language of patterns, forms and structures; a natural tool with which to separate the essential from the incidental.

We hope that this book illuminates the role of maths as a powerful tool used to build and understand the world we live in, and also to understand ourselves. The theory of networks (see Chapter 5) is perhaps the clearest example of how abstraction leads to insight: only when you forget the specific context of a particular network and focus on its connectivity can you spot the features that many networks have in common, explain these features and use what you find to improve networks you can control.

Physical processes, from traffic flow to weather patterns, can be described by mathematical models, which simulate these processes, predict how they will play out in the future and what might happen if parameters change. Even human interactions have been examined in this way. Modelling always comes with uncertainty, of course, and the people who make the models don't always get it right. But a good model based on solid science and the best data available is infinitely better than anecdotes or guesswork.

Almost every aspect of our lives, from the gadgets we use every day to understanding the world in which we live, is touched by maths.

This book also contains a fair amount of statistics. Numbers, whether they measure the efficacy of a medical treatment or the popularity of a political party, have the power to confuse. They even have the power to mislead, whether by accident or design. To distil a true meaning from them, to tell a blip from a trend, and a true effect from a fluke, statistical awareness is essential.

Its immense power as a tool is not the only thing we love about mathematics. First and foremost, we're enchanted by the beauty that comes from its elegance, clarity and (believe it or not) simplicity. We hope that this book conveys some of this beauty, and helps you to see the world through mathematical eyes.

BIBLIOGRAPHY

Barabási, Albert-László and Albert, Réka, 'Emergence of scaling in random networks', *Science*, 286, 5439 (1999)

Brams, Steven J. and Taylor, Alan D., *Fair Division* (Cambridge University Press, 1998)

Brooks-Pollock, Ellen and Eames, Ken, 'Pigs didn't fly but swine flu' in *50: Visions of Mathematics* (Oxford University Press, 2014)

Chang, Joseph T., 'Recent Common Ancestors of All Present-day Individuals', *Advances in Applied Probability*, 3, pp1002-1026 (1999), http://www.stat.yale.edu/~jtc5/papers/CommonAncestors/AAP_99_CommonAncestors_paper.pdf

Dizikes, Peter, 'When the Butterfly Effect Took Flight', *MIT Technology Review* (2011), https://www.technologyreview.com/s/422809/when-the-butterfly-effect-took-flight/

Freiberger, Marianne, 'Britain in love', *Plus* (2018) https://plus.maths.org/content/brits-love

Freiberger, Marianne, 'Solving the genome puzzle', *Plus* (2010), https://plus.maths.org/content/os/issue55/features/sequencing/index

Freiberger, Marianne, 'The graphs and network package', *Plus* (2007), https://plus.maths.org/content/graphs-and-networks

Freiberger, Marianne and Thomas, Rachel, *Numericon: A journey through the hidden lives of numbers* (Quercus, 2014)

Freiberger, Marianne and Thomas, Rachel, 'Spin doctors: The truth behind health scare headlines', *New Scientist* (2011), https://www.newscientist.com/article/mg20927991-700-spin-doctors-the-truth-behind-health-scare-headlines/

Gigerenzer, Gerd, Gaissmaier, Wolfgang, Kurtz-Milcke, Elke, Schwartz, L.M. & Woloshin, Steven, 'Helping Doctors and Patients Make Sense of Health Statistics', *Psychological Science in the Public Interest*, 8, 2 (2007)

Gigerenzer, Gerd, Wegwarth, Odette &
Feufel, Markus, 'Misleading communication
of risk', *British Medical Journal*, 342, 7777
(2012)

Gleich, David F., 'PageRank beyond the Web',
SIAM Review, 57(3), pp321-363 (2015), https://
arxiv.org/abs/1407.5107

Hordijk, Wim, 'The mathematics of kindness',
Plus (2016), https://plus.maths.org/content/
mathematics-kindness

Joyce, Helen, 'Beyond reasonable doubt',
Plus (2002), https://plus.maths.org/content/
beyond-reasonable-doubt

May, Robert M., 'Simple Mathematical
Models with Very Complicated Dynamics',
Nature, 261, pp459-467 (1976), https://www.
researchgate.net/publication/237005499_
Simple_Mathematical_Models_With_Very_
Complicated_Dynamics

Orosz, G. and Stépán, G., 'Subcritical Hopf
bifurcations in a car-following model with
reaction-time delay', *Proceedings of the
Royal Society A*, 462, 2073 (2006)

Pearson, Mike and Short, Ian,
'Understanding uncertainty: Visualising
probabilities', *Plus* (2011), https://plus.maths.
org/content/understanding-uncertainty-
visualising-probabilities

Sedrakyan, Artyom and Shih, Chuck,
'Improving Depiction of Benefits and Harms',
Medical Care, 45, 10, 2 (2007)

Spiegelhalter, David and Pearson,
Mike, '2845 ways of spinning risk', *Plus*
(2009), https://plus.maths.org/content/
understanding-uncertainty-2845-ways-
spinning-risk-0

Spivey, Michael J., 'Fake news and
false corroboration: Interactivity in
rumor networks', https://pdfs.
semanticscholar.org/7df1/c310
d79d9be69752b67bf660278965d9eea1.pdf

Watts, Duncan J. and Strogatz, Steven,
'Collective dynamics of 'small world'
networks', *Nature*, 393 (1998)

At BUILD+BECOME we believe in building knowledge that helps you navigate your world.

Our books help you make sense of the changing world around you by taking you from concept to real-life application through 20 accessible lessons designed to make you think. Create your library of knowledge.

BUILD +
BECOME
www.buildbecome.com
buildbecome@quarto.com

@buildbecome
@QuartoExplores

Using a unique, visual approach, Gerald Lynch explains the most important tech developments of the modern world — examining their impact on society and how, ultimately, we can use technology to achieve our full potential.

From the driverless transport systems hitting our roads to the nanobots and artificial intelligence pushing human capabilities to their limits, in 20 dip-in lessons this book introduces the most exciting and important technological concepts of our age, helping you to better understand the world around you today, tomorrow and in the decades to come.

Gerald Lynch is a technology and science journalist, and is currently Senior Editor of technology website TechRadar. Previously Editor of websites Gizmodo UK and Tech Digest, he has also written for publications such as *Kotaku* and *Lifehacker*, and is a regular technology pundit for the BBC. Gerald was on the judging panel for the James Dyson Award. He lives with his wife in London.

GERALD LYNCH

BUILD+ BECOME

GET TECHNOLOGY

BE IN THE KNOW. UPGRADE YOUR FUTURE.

KNOW TECHNOLOGY TODAY, TO EQUIP YOURSELF FOR TOMORROW.

Using a unique, visual approach to explore philosophical concepts, Adam Ferner shows how philosophy is one of our best tools for responding to the challenges of the modern world.

From philosophical 'people skills' to ethical and moral questions about our lifestyle choices, philosophy teaches us to ask the right questions, even if it doesn't necessarily hold all the answers. With 20 dip-in lessons from history's great philosophers alongside today's most pioneering thinkers, this book will guide you to think deeply and differently.

Adam Ferner has has worked in academic philosophy both in France and the UK, but much prefers working outside the academy in youth centres and other alternative learning spaces. He is the author of *Organisms and Personal Identity* (2016) and has published widely in philosophical and popular journals. He is an associate editor of the Forum's *Essays*, and a member of Changelings, a North London fiction collaboration.

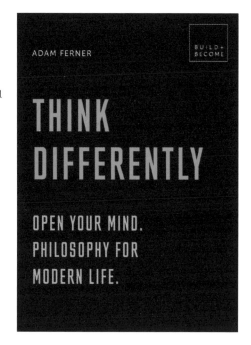

ADAM FERNER

BUILD + BECOME

THINK DIFFERENTLY

OPEN YOUR MIND. PHILOSOPHY FOR MODERN LIFE.

PHILOSOPHY IS ABOUT OUR LIVES AND HOW WE LIVE THEM.

Using a unique, visual approach to explore the science of behaviour, *Read People* shows how understanding why people act in certain ways will make you more adept at communicating, more persuasive and a better judge of the motivations of others.

The increasing speed of communication in the modern world makes it more important than ever to understand the subtle behaviours behind everyday interactions. In 20 dip-in lessons, Rita Carter translates the signs that reveal a person's true feelings and intentions and exposes how these signals drive relationships, crowds and even society's behaviour. Learn the influencing tools used by leaders and recognize the fundamental patterns of behaviour that shape how we act and how we communicate.

Rita Carter is an award-winning medical and science writer, lecturer and broadcaster who specializes in the human brain: what it does, how it does it, and why. She is the author of *Mind Mapping* and has hosted a series of science lectures for public audience. Rita lives in the UK.

Using a unique, visual approach, Nathalie Spencer uncovers the science behind how we think about, use and manage money to guide you to a wiser and more enjoyable relationship with your finances.

From examining how cashless transactions affect our spending and decoding the principles of why a bargain draws you in, through to exposing what it really means to be an effective forecaster, *Good Money* reveals how you can be motivated to be better with money and provides you with essential tools to boost your financial wellbeing.

Nathalie Spencer is a behavioural scientist at Commonwealth Bank of Australia. She explores financial decision making and how insights from behavioural science can be used to boost financial wellbeing. Prior to CBA, Nathalie worked in London at ING where she wrote regularly for *eZonomics*, and at the RSA, where she co-authored *Wired for Imprudence: Behavioural Hurdles to Financial Capability*, among other titles.

NATHALIE SPENCER

BUILD + BECOME

GOOD MONEY

UNDERSTAND YOUR CHOICES. BOOST YOUR FINANCIAL WELLBEING.

WE ALL MAKE CHOICES WITH MONEY – UNDERSTAND YOURS.

Through a series of 20 practical and effective exercises, all using a unique visual approach, Michael Atavar challenges you to open your mind, shift your perspective and ignite your creativity. Whatever your passion, craft or aims, this book will expertly guide you from bright idea, through the tricky stages of development, to making your concepts a reality.

We often treat creativity as if it was something separate from us – in fact it is, as this book demonstrates, incredibly simple: creativity is nothing other than the very core of 'you'.

Michael Atavar is an artist and author. He has written four books on creativity – *How to Be an Artist, 12 Rules of Creativity, Everyone Is Creative* and *How to Have Creative Ideas in 24 Steps – Better Magic.* He also designed (with Miles Hanson) a set of creative cards *'210CARDS'.*

He works 1-2-1, runs workshops and gives talks about the impact of creativity on individuals and organizations. www.creativepractice.com

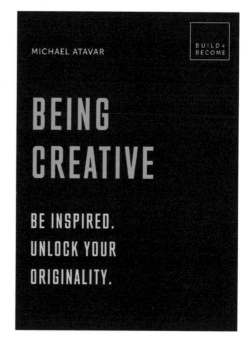

MICHAEL ATAVAR

BUILD + BECOME

BEING CREATIVE

BE INSPIRED.
UNLOCK YOUR
ORIGINALITY.

CREATIVITY BEGINS WITH YOU.

We are living longer than ever and, thanks to technology, we are able to accomplish so much more. So why do we feel time poor? In 20 eye-opening lessons, Catherine Blyth combines cutting-edge science and psychology to show why time runs away from you, then provides the tools to get it back.

Learn why the clock speeds up just when you wish it would go slow, how your tempo can be manipulated and why we all misuse and miscalculate time. But you can beat the time thieves. Reset your body clock, refurbish your routine, harness momentum and slow down. Not only will time be more enjoyable, but you really will get more done.

Catherine Blyth is a writer, editor and broadcaster. Her books, including *The Art of Conversation* and *On Time*, have been published all over the world. She writes for publications including the *Daily Telegraph, Daily Mail* and *Observer* and presented *Why Does Happiness Write White?* for Radio 4. She lives in Oxford.

CATHERINE BLYTH

BUILD + BECOME

ENJOY TIME

STOP RUSHING.
GET MORE DONE.

TIME IS NOT MONEY.
TIME IS YOUR LIFE.

NOTES

ACKNOWLEDGEMENTS

We'd like to thank all the wonderful researchers we have worked with on *Plus* magazine over the years, in particular Chris Budd, Philip Dawid, Sarah Garner, Martin Nowak, Jeffrey Rosenthal and David Spiegelhalter, who were part of some of the mathematical stories in this book.

Rachel Thomas and **Marianne Freiberger** are the editors of *Plus* magazine, which publishes articles from the world's top mathematicians and science writers on topics as diverse as art, medicine, cosmology and sport (plus.maths.org).

Before joining *Plus* in 2001, Rachel was a maths consultant in Australia working for government and industry. She recently edited the *Gazette of the Australian Mathematical Society* and has developed and taught science-writing workshops for graduate students. She also created mathematical walking tours of London and Oxford and a virtual mathematical tour guide of the world for Maths in the City. Rachel obtained her MSc in Semigroup Theory in 1998 from the University of Western Australia.

Marianne joined *Plus* in 2005 after completing a PhD and then a three-year postdoc at Queen Mary, University of London. As a researcher she worked in complex dynamics and held various teaching engagements. In the world of maths communication she has been Editor-in-Chief of the Mathscareers website and sometimes gives presentations about maths to a general audience and about communicating maths to mathematicians.

Rachel and Marianne have co-authored the popular maths books *Numericon* and *Maths Squared*, and were editors on *50: Visions of Mathematics*. Between them they have nearly 30 years of experience writing about mathematics for a general audience.

What readers say abc

Absolutely brilliant! —Bridget, Readaholic book reviews

a thoroughly engrossing time travel story that promises to continue as vividly as it started. —Nan Hawthorne, *An Involuntary King*

a delightfully intricate tale of time travel, life lessons, challenges of faith, and redemption...moving, witty, and captivating...a page-turner...I highly recommend this novel. —Jennifer, Rundpinne.com

Vosika spins a captivating tale.... The pacing flows from a measured cadence...and builds to a climatic crescendo reminiscent of Ravel's Bolero. I become invested in the characters. Both Shawn and Niall are fully fleshed and I could imagine having a conversation with each. Write faster, Laura. I want to read more. —Joan Szechtman, *This Time*

fast-paced, well-written, witty...Captivating! —Stephanie Derhak, *White Pines*

Ms. Vosika wove these aspects...together in a very masterful way that...kept me spellbound. I could hardly put it down. —Thea Nillson, *A Shunned Man*

Original & intelligently written. I couldn't turn the pages fast enough.
—Dorsi Miller, reader

Ms. Vosika spins the web so well you are a part of all the action. If you love history, romance, music and the believable unbelievable...this book is for you. I couldn't put it down until I closed the cover on an ending I never expected.
—Kat Yares, *Journeys Into the Velvet Darkness*

...best time travel book I have ever read. Fantastic descriptive detail and a sweet love story are combined beautifully. —Amazon reviewer

some of the best writing it has been my pleasure to read....
—JR Jackson, *Reilley's Sting, Reilley's War,* and *The Ancient Mariner Tells All.*

a very exciting tale.... —Ross Tarry, *Eye of the Serpent,* and other mysteries

Vosika is a master at creating engaging characters...a riveting plot, well-drawn cast, and the beautiful imagery of Scotland.
—Genny Zak Kieley, *Hot Pants and Green Stamps*

I love books on time travel, but this is so much more. The characters come to life in your heart and mind. —Jeryl Struble, singer/songwriter, *Journey to Joy*

One of the most intriguing stories of Scottish history I have ever read...riveting.
—Pam Borum, Minneapolis, MN

I found myself still thinking about the characters long after finishing the book.
—Goodreads reviewer

Thank you

Elaine White

for your friendship that sprang from these books

The
Battle is O'er
Blue Bells Chronicles Five

By

Laura Vosika

G/H

Gabriel's Horn Publishing

Contact: editors@gabrielshornpress.com

Published in Minneapolis, Minnesota by Gabriel's Horn Publishing

Publisher's Note: This novel is a work of fiction. Names, characters, places, and incidents are either products of the author's imagination or used fictitiously. All characters are fictional, and any similarity to people living or dead is purely coincidental.

Cover photo by Laura Vosika
Author Photo Emmanuel's Light Photography, by Chris Powell

First printing March 2018
Printed in the United States of America

For sales, please visit www.bluebellstrilogy.com

ISBN-13: 978-1-938990-34-2
ISBN-10: 1-938990-34-X

Acknowledgments

No book is complete without acknowledging those who have done so much to make the book happen.

Thank you to the Night Writers—Ross, Lyn, Judy, Genny, Judd, Jack, Janet, Stephanie, Sue, Catherine, and Meredythe—for years of listening to Shawn's story, your great input, and your warm welcome and enduring friendship from the first night I showed up.

Thank you, Dena, for all your help, support, and faith in, the *Blue Bells* story.

Dear Readers:

Blue Bells of Scotland was originally a trilogy:

1. Shawn and Niall trying to get home
2. Shawn in Scotland
3. Shawn's adjustment to his own time

However, as the story and characters grew, they were better served by splitting the second and third parts into two books each. Hence, a five book 'trilogy.'

- *Blue Bells of Scotland,* Book One;
- *The Minstrel Boy* and *The Water is Wide* were originally Book Two.
- *Westering Home* and *The Battle is O'er* were originally Book Three.

Other Books
by Laura Vosika

Fiction
The Blue Bells Chronicles:
Blue Bells of Scotland
The Minstrel Boy
The Water is Wide
Westering Home

Non-fiction
Go Home and Practice: a music record book
Food and Feast: a gastronomic historic poetic musical romp in thyme

The Battle is O'er

I returned to the fields of glory
Where the green grasses and flowers grow
And the wind softly tells the story
Of the brave lads of long ago

March no more my soldier laddie
There is peace where there once was war
Sleep in peace my soldier laddie
Sleep in peace, now the battle's over.

In the great glen they lay a sleeping
Where the cool waters gently flow
And the gray mist is sadly weeping
For those brave lads of long ago.

See the tall grass is there awaiting
As their banners of long ago
With their heads high forward threading
Stepping lightly to meet the foe.

Some return from the fields of glory
To their loved ones who held them dear
But some fell in that hour of glory,
And were left to their resting here.

PRELUDE

Glenmirril Castle, Shores of Loch Ness, December 1317

The explosion of thunder, a rarity in winter, drew Niall's eyes up from prayer just as lightning flashed across the rose window. It lit the statue of Mary, danced across the polished backs of the pews in the small chapel and flashed away, leaving Niall, Hugh, and the Laird once more in only the flickering light of candles.

A chill shot down Niall's spine. 'Twas the chill of the long ride, he told himself. But his eyes locked on the statue. The air wavered before it. Voices cut the silence—a deep voice, angry; a rasping elderly one, that seemed to laugh.

The Laird turned to Niall, a frown flitting across his face. He drew his knife from his belt.

"You hear it, too?" Niall murmured. His heart pounded a sharp staccato. The air shimmered before Mary as a second clap of thunder jolted them all.

Hugh's head shot up. "God in Heaven!" he exclaimed.

Amy, there you are! Thank goodness!

Niall heard the words distinctly, a woman's harried, breathless voice in something akin to Shawn's peculiar English. A baby squealed. Hugh's eyes grew wide, turning to Niall. "What's this, Niall?"

"The crucifix...." Niall began. But he wasn't in the tower! He rose to his feet. Amy stood before him, staring in horror at the statue. Niall turned cautiously. "There," he whispered to the Laird. Before the statue, a bull of a man faced an elderly monk with a few wisps of hair clinging to his bald pate.

Hugh leapt to his feet, dropping into a half crouch. The Laird rose, too, gripping his knife, as the old man, fragile as thistledown, threw himself at the warrior. The air wavered. Amy's eyes widened, frantic; her mouth opened in a silent shout.

A third clap of thunder shook the very stones of Glenmirril. The bullish man wrenched himself from the monk's grasp, stumbled against one wall, backed up, and slammed into another.

"Amy!" a man shouted. The baby's cry split the air.

Lightning streaked across the rose window, and Amy was gone.

In the candlelight, the elderly monk crouched, his hands on the stone floor, wheezing. Niall, the Laird, and Hugh stared at him, too stunned to speak.

The monk lifted his head, looking from one to the other. A smile broke across his face for just a moment before he fell into a fit of coughing. Niall dropped to his knee, touching the old man's back.

The monk drew a deep breath, shaking with the effort. "Niall," he wheezed. "I've waited so long to finally meet ye in the flesh!"

Glenmirril, Present

Shawn explodes into the room, shouting, "Amy!" He's scooping me up off the flagstones, and immediately pushing me away, reaching to his back for a claymore that isn't there. "Stay there!" he orders. He storms to the crucifix, to the wall behind it.

I gather James in my arms, burying my face in his thick, black hair, breathing thanks to Niall, to Eamonn, wherever, whenever, they are; to God. James is safe. Nothing else matters. Not Shawn, not Angus, not Niall or Eamonn. James is safe.

His crying settles to hiccups, that still manage to echo in the empty medieval chapel. He buries his face in my shoulder, and becomes still.

"What just happened here?" Mairi's voice shakes. "Where did they go?"

I take her hand in my free one, squeezing. "I'm not sure," I say.

Shawn emerges from behind the crucifix, his face grim. He is every inch the medieval warrior, in his trews *and* leine, *his red-gold hair tied with a leather thong. "Wait here," he barks.*

In his sharp command, I hear a man who has become a leader, who has ridden with James Douglas and fought at the Bruce's side. He draws knives from his boots. "Mairi, take James," he snaps. "There's a passage. I'm going in. One of the tunnels leads to the water gate. He could circle back."

My heart pounds as I hand James to Mairi. I'm obeying his orders without question. And I realize—I'm doing it because I've grown to trust him. "What if he finds us here while you're gone?"

"I won't be gone long enough for that to happen," he says. But he flips a knife, its blade pointing at himself, and thrusts it into my hand. "Kill if you have to. Up, into the ribs." He guides my hand with the knife, in his, as if up into his own ribs. "Understand?"

I nod. My hand shakes. I'm not sure I can hold the knife, if Simon comes back, let alone kill him.

Shawn glances at my trembling hand. "You'll do whatever it takes to protect him," he says. And settling his own knife more firmly in his hand, strides back behind the crucifix.

When he's gone, Mairi and I back against a wall, as if that will offer some protection. "By the door," I say. "If he comes back, we need to get out."

FIRST MOVEMENT

CHAPTER ONE

Glenmirril, December

Simon stumbled against the wall behind the statue, feeling for the tunnel he'd found on another visit. He turned, scrambling in the dark, deeper into the passageway, a hand on the wall. He stumbled as his feet hit the sharp edge of stairs and slowed, feeling his way, moving with what haste he could. The chill bit more deeply as he descended. The voices faded above as dark pressed in. The dankness grew. His heart pounded in the black void pressing in on his eyes, the cold chilling his bones, and mustiness filling his nose. This was the Glenmirril of Shawn's time.

It was, too, the Glenmirril of Niall's time.

There was no way to know, down here. Anyone might be following him. He pushed through the blinding dark, a hand to the wall. He heard a voice behind him, muffled, agitated, and forced himself to go faster, feeling with leather boots, and with cold hands on rough, damp stone.

The stairs ended, flattening out, and the passage turned sharply. The ceiling lowered, closing in. He fought a sudden panic rising in his chest, the hammering of his heart. He was back in the deep reaches of his father's castle, running from the old healer stirring potions in the dank kitchens, blinded, panicking! He would be lost in this void forever! He would starve here, not even knowing what century he was in!

He stopped, pressing his back to the dank wall, staring up into the black void above. He drew a slow breath and lifted his chin, defying fear and anger.

He was the Lord of Claverock!

He let the breath out and with calm returning, he pushed his mind back through the steps he had taken. He could retrace them. He could find his way out. He touched his jacket, feeling the *notebook* from Shawn's time safe against his chest. One had been left behind in his haste. But he had the one that mattered. Yes, he *would* find his way out. He *would* set their knowledge to his service—but only if he were back in his own time.

He listened. Only silence came to him in the unending dark. Whoever was above—Shawn or the men of medieval Glenmirril—they had not followed him this far. He felt the walls, felt a recess, and backed into it, waiting. He would find his way, somehow, out of this pitch-black maze. He would find what time he was in, and proceed from there.

Somewhere, he felt a cool draft. He smiled. If he but followed it, he would find his way out. He drew his knife, prepared for whatever—whoever—he might find.

Glenmirril, December 1317

"How do you know me?" Niall asked.

"The other one!" Hugh looked wildly around the chapel. "Who was he?"

The old monk, his hand pressed to his breast, wheezed, "Shawn will take care of him."

"You were with Amy." Niall pulled the monk to his feet more forcefully than he'd intended. Shame washed over him thinking of an elderly priest, far away in England. "My apologies, Brother. Forgive me." He ushered him to a pew and helped him sit down.

"Who was he?" the Laird demanded. "Who *is* he?" He turned to Niall. "Is he here or there? Was that Amy?"

"Aye," Niall said tightly. "'Twas Amy."

"He's from Shawn's time, then? But if this monk crossed, did he, too?"

"Did he?" Niall asked the monk. "Who is he?"

The old man doubled over, coughing. The candles flickered. The scent of incense grew. Niall touched his back, torn between compassion and agitation. "The passage," he said to Hugh. "You must check it. 'Tis possible he crossed and went down it."

"How would he know of it?" the Laird demanded.

"We've no idea who he *is!*" Niall nearly shouted. He bit back his agitation, bowing his head. "My apologies, my Lord. We know not who he is. We must at least look."

"Hugh." The Laird took a torch from the wall, handing it to his brother. "Search the passages, to the end, and back to the entrance under the great hall."

On the hard pew, the monk forced himself upright, waving a weak hand. "Shawn will take care of him."

"Is he not in our time, then?" Niall pressed.

"Shawn will take care of him," the old monk repeated.

"He's in Shawn's time, then?" Niall asked.

"Go all the same, Hugh," the Laird said.

Hugh rounded the statue of Mary, and disappeared into the passage at the side of the altar.

"You'll not find him." The monk sat up straighter. He drew a deep breath, and his lungs seemed to clear. "Shawn will see to it."

Glenmirril, Present

Shawn's heart pumped harder as the dark tunnel closed in on him. Brutality, violence, knives—it was all supposed to be in the past!

He moved his leather-booted toe along the edge of the stone stair, feeling carefully before inching down into blackness. His hand tightened on the knife. He shouldn't leave Amy alone up there. But he had to know if Simon was still here. He stopped, straining his ears—for a footstep, a breath. Simon, if he lay waiting in the dark, had the upper hand.

Shawn touched the cold stone wall and lowered himself step by step, silent in his leather boots, listening, his stomach tensed against the sting of a blade.

CHAPTER TWO

Glenmirril, Present

James lets out a short, sharp squeal. I try to shush him as Mairi and I press ourselves to a wall, me with a knife gripped in a hand unequipped to do anything but draw a bow. Yes, I'll Telemann *him to death if he comes back. Certainly that will work. I'll protect my son with the bowing of a* Paganini Caprice *flashed out in a knife, the likes of which I've never even held before. My hand trembles.*

Behind me—because of course my son is behind me, even if it only gives Mairi an extra three beats to run, hopelessly, with him—Mairi whispers to James, shushing him, too. He's trying to reach for me. She's trying to dissuade him. As if she thinks I can actually protect them with this knife. My eyes flicker between the chapel door and the crucifix, watching in terror for Simon's return.

James falls, thankfully, silent. In the quiet, Mairi speaks. "Amy." Her voice quivers. "Who was he? Who were those men?"

My hand shakes. I'm used to making up stories by now. But she's seen. She's no fool. I'm watching the crucifix, dreading Simon re-appearing. But if he does, Shawn will be there, chasing him down. Right? Won't he?

Or will he be dead in the passage, leaving me alone to face Simon?

James lets out a questioning squeal. "Ma..." he says.

"Who were they?" Mairi asks again.

"Niall." I wish my hand would quit shaking. I try to think what I'll do if Simon appears.

"They were wearing medieval clothes."

I nod, watching the crucifix. Who am I fooling? I can't protect my son. I can't protect him! My hand shakes harder. I grip my right wrist with my left hand, trying to steady it. "That was Niall. And the Laird."

"Tonight I am Niall Campbell," Mairi whispers. "Born in Glenmirril on the shore of Loch Ness."

"Yes," I whisper.

"In 1298."

"1296," I correct. Like it matters, I think irritably.

"Shawn was telling the truth," she breathes.

"Yes." I speak more shortly than I intend. "Simon is—was—a knight in King Edward's service. He might come back," I say. "He's from Niall's time. If he comes back, you have to get James out."

"I'm not leaving you," she protests. James whimpers.

"You're getting him out!" My voice is firmer by far than the knife wavering in my hand.

Motion behind the crucifix! My heart kicks into tempo. I jump forward, shouting, "Get him out!"

Even as I recognize him, Shawn is yelling, "It's me. I didn't find him. Move!"

I sag with relief. Shawn will take care of us. Mairi's face is pale. Her arms are tight around James.

"We're not safe!" Shawn herds us toward the door, looking over his shoulder. "If he's in there, he may come out under the great hall or down by the water gate. He may circle back on us. I lead the way. Amy, watch our back. Mairi, get James out, no matter what. Understand?" He slaps car keys in her hand. "No matter what!"

She nods, hugging James tight, her knuckles white on the keys. James whimpers, looking to Shawn with a furrow in his brow.

"What about Brother Eamonn?" I ask.

"You know where he is," Shawn says shortly, even as he glances around the empty chapel. "Niall will take care of him."

Questions burn in Mairi's eyes. She glances at the door.

"Aim to kill," Shawn reminds me, and he leads us into the hall, keeping Mairi close, her arms wrapped around James.

Glenmirril, December 1317

Huddled on a pew in the chapel, the monk clasped his hands around the chalice of hot mulled wine Hugh had brought. "I am Brother Eamonn, a monk of Monadhliath," he said.

"Nay!" The Laird's hand flashed to the knife at his belt. "I know all the monks of Monadhliath!"

Niall held up his hand. "He is from the Monadhliath of Amy's time."

"My *brother* is in those passages!" the Laird roared. "Tell us, Monk, what you know of this man. Is my brother in danger? Man of God or not, I'll no lose my brother, nor my daughter, who is with child, nor a single one of my people, for your refusal to speak clearly!"

"Shawn will stop him." The old man huddled over his wine.

"If Shawn will stop him," Niall said to the Laird, "then he is in Shawn's time." He looked to the elderly man, a fur wrapped around his shoulders. "You *are* from Shawn's time, are you not?"

The man nodded. A shiver racked his frail body as he lifted rheumy eyes to the two men.

Niall lifted the jug to pour more wine. "You'll warm up soon enough," he said. "I imagine our time feels cold to you, being used to warmed houses."

The monk nodded, gratefully sipping the brew from his stone chalice. He lowered it, breathing in the steam for a moment, before saying, "I've long thought what it will feel like. I've long anticipated meeting you."

"How do you know Niall?" the Laird asked.

"We will meet at Monadhliath." His liver-spotted hands tightened around the cup, seeking warmth. He stared into the fire. "*Have* met at Monadhliath, in my own life. I was quite young then. But you'll know me. Brother William will be fine. But do see he has a knife about him."

Niall nodded. "I will. If you would tell me when."

Brother Eamonn stared into the fire.

"When?" the Laird asked.

Niall hated to press the elderly monk. His recent attack on Adam Newton, a man of God, on top of the elderly priest in England, left his conscience discomfited enough. He hardly needed to add this to his list of how he treated clerics. But Hugh had not returned. He had no idea if this stranger was inside Glenmirril, let alone who he was. He knew only that the man had been at odds with Amy, and he did not want such a one around his people.

The monk stared, his watery blue eyes losing focus. His hands convulsed on the wine chalice.

"I believe he's going into shock," Niall murmured. "Call servants to bring furs. I'm going to find Hugh." His nerves trembled, despite the monk's assurances Shawn would deal with it.

Glenmirril, Present

"I want men on this castle."

In the safety of my car, looking down on Glenmirril, I shake my head, listening to Shawn shout into the phone. I hold James close on my shoulder. He sniffles intermittently, his hand wrapped in my long hair, his face nuzzled close in my neck. He's scared. He's upset.

"And you are?" I hear Clive's voice, irritable—and I can't blame him—over the phone.

"You know damn well who I am," Shawn snaps.

"Oh, aye, more's the pity," comes Clive's voice over the speaker. "You're missin' me point, Mate. Some rich American waltzes in telling the Inverness PD what to do and we listen to you why?"

"Give me the phone, Shawn." I squirm in the car, handing James back to Mairi. James protests, shouting, "Ma! Ma!"

"I've got this!" Shawn protests in the same belligerent tones as his

squalling son.

"No, you don't." I take the phone from him. He bursts from the car, slams his palm on the roof of his rental car, parked beside mine, and paces the dark lot under falling snow, watching the castle below. "Clive," I say, "It's Amy."

"Aye." His voice softens. "I know it's you. How're you holding up?"

"Clive," I say, "Please. He's upset. He's not being diplomatic."

"This is not about diplomacy!" Shawn barks through the window. "He's being a butt!"

I wave a hand, telling him to shut up, and say, "But he's right. Someone has to watch this castle."

"Why?" Clive is less irritable than he was—but clearly still sees no reason to buy into our unexplained request.

"The man who attacked Carol." I'm assuming, I'm hoping, they heard about it through police reports.

"What of him?" Clive asks.

His words tell me I'm right. "He took me and Brother Eamonn at knife point to Glenmirril. He may still be here."

Clive's voice becomes sharp. "Abduction? Why did you not...?"

"We just got out. We're calling now."

"But why not....?"

"He may still be there! Please!" In the back seat, Mairi shushes James. I hear her rifle in the diaper bag. How am I going to explain that he may be here now—or he may be here in 1317? How I wish Angus were able to speak to him!

"Angus would have...."

"Yes." But Angus is in a hospital bed, barely clinging to life after falling off a cliff. I watch Shawn pace the parking lot, looking down into Glenmirril's spotlight-flooded courtyard, watching for the evil that walks within. "He knew far more about Simon than he told you. Please, Clive, he's very dangerous."

"I'll have someone there," he says.

"They're coming," I tell Shawn.

"Keep the windows up," he answers. "If he comes, leave."

"I'm not leaving...."

"You're protecting James," he says, and starts a restless pacing of the lot, his eyes locked on the castle walls, hazy behind the veil of falling snow.

I nod. Yes. I will protect James. I watch him pace, fearing Simon appearing. But he's watching the castle, watching the exits, guarding us.

"He fought with the Bruce," Mairi whispers. "He wasna lying!"

Cop cars stream into the parking lot; men pour out.

My heart swells with pride, watching as he meets them, strong and fearless, knowing what needs to be done, ready to go in and face Simon, to risk everything to protect his son. "No," I say. "He was not lying."

Glenmirril, Present

"There's a door down by the grand hall." In the midst of fifteen police officers, with Amy, Mairi, and James safe in her car humming beside the group, Shawn pointed with his knife. "You have to watch that, and...."

"I'll be giving the orders, if ye dinna mind." Under the light of the street lamps, the chief, a burly man with a big mustache, glared into Shawn's eyes.

Shawn's lips tightened. But he gave a respectful bow of the head. "Yes, sir. It's only that I *know* these passages. I know where they come out. If he's in there, he's dangerous."

"*If* he's in there?" the chief demanded. Snowflakes caught in his mustache. "You said he is."

Shawn's jaw clenched. "He was."

"You've been standing here the whole time, aye?" the chief asked. Several of his men nodded agreement. "You can see both exits from up here." He flung his arm wide, indicating the snow-covered crenelations below.

"He has a knife." Shawn avoided the direct question. "Or knives. He's strong. He knows how to fight."

"As do we," snapped one of the officers.

Shawn glanced at him, and his eyes cut back to the chief. "Not like he does. Now will you...." He stopped his curt words, and tried again. "Please. He can get out by the water gate or out through the chapel. If he leaves Glenmirril, he's going to try to kill my son again."

The chief looked for a moment as if he was going to argue. Then he glanced at Amy, in the car beside them with James, and said, "I'm doing this for Angus. I don't know what's going on, but my gut tells me Angus would have believed you. And he's right almost every time." He waved his arm, leading his men down the path.

"Keep the car running." Shawn spoke through the window to Amy. "If you see him, get out."

She nodded, and he jogged ahead, joining the chief, their feet leaving a trail of footprints in the snow.

"Five to the chapel," Shawn said, as they ran side by side. "Spread out from there, check every hall, every floor, everywhere in the castle he might have gotten to. Five start at the water gate. Rest of you into the passages with me. We'll have to split up at times. You have your flashlights?"

"We're ready," the chief said, as they led the group through the gatehouse, and back into the flakes dancing in the spotlights of Glenmirril's courtyard.

Shawn nodded, his lips tight. "You." He pointed to a man. "Up the tower stairs, branch out from there. At *least* two of you together." He looked from one to another. "*At all times*. He's *strong*." To Clive and the chief and the rest, he beckoned as he led the way to the tunnel entrance at the Great Hall.

Soon, the dark engulfed them. The officers switched on their lights, illuminating stone walls, with patches of green. The dankness grew as they moved in. Memories engulfed Shawn—the first time going down these halls with the Laird, later with Christina. *On your left. Same hand as we wear a wedding ring in my time.*

"You know these tunnels?" the chief asked.

"I know them well," Shawn said tersely.

You're in your time. That had been Amy, as he led her, too, seven hundred years later, just months ago in July, through these same passages. A chill shot through him, wondering if Niall might be down here. He had seen just a glimpse of him, as he burst into the chapel, just a glimpse of his best friend, his *brother*, knife raised, looking at Simon, as he faded away. Yes, he, too, would be in these passages, searching for Simon in his own time. He wondered if he would see Niall...feel his presence. If the times would cross. But they couldn't! He didn't have the crucifix. Amy had it, up in the car.

"How do you know them?" Clive asked. His light played over the dirt floor.

"Watch carefully," Shawn returned. "There's a side passage opening up. He could be hiding there."

"How do you know them?" Clive asked again.

"I told you," Shawn said irritably. "I spent a good part of two years here."

"You were only gone a year."

In the close passage, Shawn spun. "Do you not *get* it? There's a medieval madman down here—maybe—who will *kill* you. He'll have his knife under your ribs and through your heart in the blink of an eye, if you're not paying attention! Could we maybe have our coffee klatch later?"

In the shadowed light of the labyrinth, Clive's face remained passive, his eyes dark. He nodded, and spoke to the chief. "Guard the exit. I'll go with Kleiner." Before his boss could object, he pushed ahead in the cramped tunnel, saying over his shoulder, "You're needed at the precinct."

Shawn followed him.

"Clive!" the chief barked.

Shawn and Clive turned as one, to the head of the Inverness precinct glaring at them, his bushy eyebrows drawn in anger.

"Guard the exit well," Shawn said. "Trust me, he may get out this way, if we pass him hiding in a side passage."

After a moment, the man's face relaxed. "Go," he said, gruffly.

♫

Though the dark pressed against his eyes like a black cloth, the faint smell of candle wax came to Simon, on the cool draft. He followed it, the ceiling growing lower. The draft grew stronger, carrying with it the new smell the sea.

And now, behind him, faint voices broke the silence. Shawn's? The men of fourteenth century Glenmirril? He drew his knife, as happy to kill one as another. Still, it was best if no one knew where he was at all—regardless of which century he was in.

♫

Clive moved in silence beside him for a time, flashing his light around the cells that opened on their right, and down another passage.

"That one comes to a dead end," Shawn said. "I'll take a look. Stay here. Be ready for anything."

He followed the passage, quickly, his mind on Niall. His brother would be searching too, neither of them knowing if Simon had crossed or not. He hoped Niall would be okay—not ambushed by Simon. He hoped he wouldn't be ambushed himself, nor Clive, nor the chief. He doubted they could take on a medieval knight with years of brutal warfare under his belt.

He found the end of the tunnel, and it's small branch, hoping the silence from Clive's end was because Simon hadn't appeared—not because Simon had slid a knife in from behind, dropping the man quietly to the ground, hardly aware he'd been killed. He turned and hurried back.

He found Clive waiting, flashing his light one way and another, watching. They proceeded another twenty feet through the dark tunnels before Clive said, "You were telling the truth."

The word *Bingo!* flashed through Shawn's mind, loaded with sarcasm. It flashed away as quickly. Clive didn't deserve it. "Yes, I was telling the truth."

"Angus knew."

"Yes. Angus knew." He scanned the dank passage ahead. Nothing moved. He hoped Clive wouldn't run through the whole list of new realizations that must be dawning on him.

The sword and blood were real.

Yes, the sword and blood were real.

That's why you keep thinking you were gone two years.

Yes, that's why I keep thinking I was gone two years.

The conversation could get really boring really fast.

"Tell me how I best fight a medieval warrior," Clive said stiffly.

"Everything you've got." Shawn's respect for Clive inched up a notch. "Forget this modern nonsense. Forget undue force. Kill him. Through the ribs. The jugular. Because I guarantee he's going to kill *you* if he can."

"Aye," Clive said, and fell silent.

♫

The ceiling lowered, forcing Simon to stoop and, finally, to drop to his knees, crawling with his knife between his teeth. Roots and dirt brushed his hair. The draft grew stronger. There must be an exit.

Behind him, the voices grew louder. Men's voices, at least two. He couldn't be caught like this, unable to fight.

He brushed a low-hanging root from his eyes. Ahead—did he imagine it? The heavy blackness seemed to lift, just a little.

"I've searched everywhere."

He heard the words, muffled, in English. He inched further down the tunnel, wary of alerting the searchers, while straining to hear.

Another voice spoke, unintelligibly.

The first, louder, answered. "The door is locked. The tunnel?"

Simon crawled, feeling with his hands. Branches scraped his fingers and he saw, ahead of him, a black latticework against black night.

"You think a stranger to Glenmirril could find the tunnel?"

"We must search," insisted the first man.

"We don't *know*—we may be hunting for someone who isna here."

Simon prised the branches aside, breathing deeply of the cold, fresh air that touched his face.

"Better that than not find someone who *is* here."

The reply was once again muffled.

A laugh boomed through the passage as Simon wiggled through the tangle of roots, coming out onto a dark, snowy patch of shore. Heavy clouds scooted across the face of a full moon. Snow fell from the sky, dusting the shoulders of Simon's *jacket*. He backed up, knife in hand, glancing from the tunnel to the shore. A large flattened boulder sat in the middle, covered in snow, beside a weathered tree, its bare branches lined with white icing. He turned to the castle walls, and the water surrounding him, searching for escape.

A narrow band of trees climbed up a rise alongside the walls. A path stretched away to a hill. He backed away from the cave entrance, straining to hear if they were coming. The dead branches remained still. He turned, studying the narrow slope. The clouds slid briefly off the moon, lighting the patch of shore. If he ran for the hill—if they came out from the tunnel—he would be easy to spot, in his dark *jacket* against the white snow, under the gleaming moon. The path alongside the castle was too rocky and narrow for any but a single man. It offered hiding, if not escape.

Grasping a dangling tree limb, he pulled himself up through a deep drift, pressing back against the castle wall. His footsteps left tracks in the snow. He ripped a branch from the tree, using it to sweep over them. Snow continued to fall, covering his path. He pulled himself higher, up over a boulder, reaching back to sweep his path once again.

"What say you?" the loud voice broke the night air.

Simon sank down, behind the boulder, wedging himself into a small space between a large tree and the castle wall. He pulled cold snow in over the shoulders of his dark coat. He held still, peering cautiously between the trunk and wall.

"While we search here, he could be heading out the other way," said the second. "If he's here at all. He's nowhere to go if he's made it out here."

"Check the slope," recommended the first.

"'Twould be a trick to climb it in this weather."

The voices moved closer. Simon's fingers tightened on the hilt of his knife. He glanced at the night sky, swirling with white flakes, and drew behind the tree. The clouds crossed again over the moon.

"I see nothing," came the louder voice. "He wore a dark cloak. Surely we'd see it against this snow."

"No tracks?"

"No. Though the snow is coming fast."

"Fast enough to cover tracks?" The voice sounded doubtful. "We don't know if he crossed."

"The monk said Shawn will deal with him."

Simon's brows knit. Was he still in Shawn's time, then? He shivered, as more snow covered his shoulders, pressed behind the tree and boulder.

Their voices sounded at the bottom of the slope, a dozen feet away, a few more muffled words, and they faded away.

♫

They searched the labyrinth, all the way to the Bat Cave. "Careful," Shawn said, pushing through the hanging shards of wood at the entrance to the cave. "There's a recess on the other side."

Clive shone his light around the high walls, up to the ceiling two stories above, and followed Shawn to the recess, where he flashed it across the alcove.

"This was where the Laird prayed," Shawn said. "That crucifix up in the chapel—he made it. It used to hang here."

"Amazing," Clive whispered.

"Shine the light in the side."

The beam flashed over the kneeler. Clive whistled, reaching to touch it.

"That's mine," Shawn snapped. "Come on, he's not in here."

They finished the search in silence, through the remaining passages, up the dark softened stairs to the chapel, where two officers looked back from the hallway. "Nothing," they reported.

Clive and Shawn headed back down, doing another search of the tunnels, the cold air and nervous energy keeping Shawn on high alert. He was relieved to see the officers at the first branch of the labyrinth, unharmed. "There's no way he could be in here," he said.

Leaving a guard at the entrance, he, Clive, and the chief joined the other men in a systematic search of Glenmirril. They met, finally, in the courtyard, with the moon now high overhead and crisp, bright stars twinkling down from the night sky. "Three hours," the chief announced. "There's no way he can have eluded this many of us for this long." He pinned Shawn with a stern gaze. "D' ye care to explain how he was in here, he never left, and yet now, he's not?"

Shawn sighed. "Not really." Either he'd hidden well and would be back out to hunt James, or he was in 1317—with Niall, Allene, and their children. Shawn hated both explanations.

Glenmirril, December 1317

Niall climbed the dark stairs from the underground tunnels up to the chapel, fighting irritation, exhaustion, and frustration. He tried to be grateful: the Bruce had given him leave. "They may be looking for you after the incident with poor Father Newton," the king had said. "Spend a fortnight with your wife, do a few jobs for me, and return by March."

Once again, he'd not so much as seen Allene before being thrown headlong into problems. She was worn with pregnancy and asleep already, the Laird had said when he'd greeted them on the drawbridge, under the early stars. He'd taken them straight to the chapel, only to have their prayers interrupted by the monk, the stranger, and Amy.

And despite their thorough search, Niall wasn't at ease about this stranger who had supposedly stayed in Shawn's time. He worried about Christina, acting the part of surprised foster mother to her own son, supposedly found in the woods. And he desperately wanted to see Allene and his laddies.

The Laird stood at the door of the chapel, as Niall and Hugh emerged, his hands on the hilt of his sword, its point rested on the floor. The old monk huddled under his furs in the front pew, gnarled hands wrapped around the head of a cane. He lifted his head, watching Niall.

Niall's irritation faded. He tilted his head. "You're from Shawn's time." It was not a question this time.

"Aye." The man spoke in Shawn's English—but not quite.

"You can tell me about history...science...the stars."

The monk smiled. It was a sad smile.

Niall's heart stilled. "You can't?" he asked. "You...?"

"You haven't time," the monk said. "Knowing their history, their science, their time, is not your calling."

"Dealing with your own is," said the Laird. "Shawn turned your head."

Niall lifted his eyes sadly to the Laird. There was no point telling him how Shawn's world had opened his own—because the Laird was right. He had duties here: Allene, his lads, the work he must do for the Bruce by March. The wonders of Shawn's time mattered little to the here and now.

"What can you tell us of the man we saw?" the Laird asked.

The old monk cleared his throat, staring up at the crucifix over the altar. "He is called Simon. He believes he is in danger from Amy's son."

"An infant?" Niall scoffed.

"Infants grow to be men," Eamonn wheezed. "And James—Amy's James —will grow to be quite formidable." He turned his gaze on the Laird. "I've a great many matters which require spiritual direction. May I go into the hills tomorrow that I might pray there?"

"You are not of this time," Hugh said. "Do you know your way?"

"I know it well," Eamonn said. "Monadhliath is a thin place. I saw a great deal there. As I saw you there."

"There are wild animals in the forest," Niall said. "Surely 'tis not safe for one of your age."

"You ought go with him," Hugh rumbled in something like a chuckle. "Protect him from boar."

"Perhaps I ought," Niall shot back, "since none of the rest of you do aught but lallygag out there."

Eamonn smiled. "I shall be fine on my own. I'll not stray far. But I wish to be alone in prayer."

"It shall be granted," the Laird said.

"You'll let the guards know?" Eamonn asked.

"I will," the Laird promised. "Now, tell us of this Simon. Is he of our...?"

Eamonn doubled over suddenly, hacking, a kerchief to his mouth. His cane wobbled under his gnarled hand. Niall sprang forward, grabbing his elbow.

The Laird glared, his hands tightening on the hilt of his sword.

"Have you no heart, Malcolm?" Hugh demanded. He took the monk's other arm. "Come, let us get him to his chambers. We can talk tomorrow."

"Not knowing if someone is in my castle?" Malcolm demanded.

"Shawn...." Eamonn managed to wheeze. "You're quite safe."

"We'll talk tomorrow."

"Yes," Eamonn whispered. "But please leave me...to pray...for now. I'll regain...breath...shortly."

"I shall return in an hour," the Laird said. "Niall, stay with him."

Eamonn nodded, his eyes vague, his fingers feeling for his Rosary, as he slid to his knees on a kneeler.

Glenmirril Castle

Cold racked Simon's body. He waited a long time after they'd left. His fingers grew numb on his knife. He pressed one hand at a time inside his *jacket*, warming them. Finally, he dared emerge. Looking up, he saw a guard walking the parapets above. He watched the man pass. Surely a guard meant he was in his own time? He touched his chest, feeling the notebook solid inside his shirt.

He waited for the man to pass before seeing if he could get higher. He quickly discovered that even without the snow, the small slope became sheer rock, and impassable. He waited for the guard's next pass before scrambling down to the shore. Water stretched before him and the path on his left climbed to a hill.

He clenched his teeth against the deepening chill, as understanding dawned: he would freeze to death in the night if he tried to run. The water gates stood high and solid, bulwarks against invaders. The only answer was to go back into the subterranean passages. He could stay in hiding, and slip out the castle gate during the day. Or perhaps return here to cross this hill.

From above came the *chink chink* of chain mail. He pressed himself to the chilly stone wall, waiting for the guard to pass, before dropping to his knees, and wiggling back in through the dead branches.

In the tunnel, darkness enveloped him once more. He stayed for a moment, letting the close walls provide some little warmth—or at least less chill—and listening. He pressed a hand inside his coat, trying to warm his fingers enough to wield the knife. When no sound came to him, he crept forward, stopping every few yards to listen. As the ceiling rose, he climbed to his feet. From ahead came the faint scent of tallow he'd smelled on the way in. There must be a chamber. Maybe he could hide there.

He felt his way, a hand on each dank wall. Soon, he touched a wooden door. The scent came from behind it. He listened, his ear to the wood. When no sound came, he squeezed the latch on the handle. It stuck fast. If he forced his way in, he would give himself away to whoever had put the lock on. It also meant, he realized, that the tunnels were used. There might be nowhere to draw back, if anyone came. He must try to find his way out before that happened.

He moved as quickly as he dared in the dark, feeling with hands and feet. He could get lost forever in here. He knew only one place these passages led for sure—back to the chapel.

It was also the one place they might set a guard.

He stopped in the dark, considering his choices: freeze to death outside or risk wandering, lost forever in the pitch black maze to die of starvation if he tried any other route. He knew nothing of the layout of Glenmirril itself, let alone this underground labyrinth, that might help him find any other exit.

And men had been known to go mad in such blackness.

He would work his way back to the chapel, he decided. He stood a better chance in a fight than wandering in a place of which he knew naught. He took several moments to review the route he had taken to get here. The stairs were two turns from this passage. There should be a faint draft coming down. But he wouldn't feel it unless he found the right tunnel to begin with.

He thought of James, shrieking in the doorway; of that foolish woman, Amy, daring to throw herself at him—*he*, a knight in the service of King Edward, Longshanks himself!—and a frail old monk admonishing him. Anger grew. He *would* find his way out. He *would* prophesy to Edward and take power!

He'd turned right. He felt his way, smiling when his fingers slipped into nothingness. He moved his hand back. Yes, there was the passage, opening on his left. The next one should be on his right—thirty steps. He counted them in his head, naming with each step the men he would place as his advisers and commanders. And there, at twenty-nine, the wall opened up. He turned and moments later, the faintest of drafts, as he'd predicted.

He smiled. He was Simon Beaumont. Lord of Claverock. He would not die in this dark labyrinth.

Another forty steps, and he felt the first stair. Again, he listened, as he took one step, and then another, climbing upwards. All remained quiet above. All remained quiet below.

The draft grew. He twisted up the tight, circular steps, feeling carefully in the dark, prodding each with a toe. The air grew fresher, carrying with it the sweet waft of incense. His hand tightened on his knife, as the faintest flickers of candlelight now wavered on the stone walls. Reaching the top, he stopped, listening. All was silent.

He stepped carefully from the passage, his pulse quickening. He peered around the statue—to see an empty room. It had many pews, no longer just the few. He smiled. He was home. He took another step and saw the monk.

Rode to Inverness, Present

12:00 am, said the glowing green numbers on the dashboard of Amy's red Renault. It seemed days, Shawn thought, since he'd picked up the rental car at the airport. In truth, it had been but hours, driving north, the strange conversation with Eamonn at the hospital, and the race to Glenmirril.

Amy slept in the passenger seat on his left. As the dark waters of Loch Ness unspooled on his right, he replayed that conversation. He hated to think of the old monk in that brutal time. But there was nothing he could do.

Mairi's breath fell deep and even behind him. She had looked at him with something like fear, or awe, when he'd returned to the car after long hours of searching Glenmirril's subterranean labyrinth. Like Clive, she knew. It was a relief—two more people with whom he didn't have to lie or pretend—but he didn't want to talk about it. He had bigger things to deal with than answering questions—questions that served only to satisfy meaningless curiosity; no other purpose.

James breathed softly, every now and again giving a chuckle in his sleep, as if none of the past hours had happened. Of course, Shawn thought, as he guided the car north under starry skies, for James, they *hadn't*. James lived in innocence with no idea of the danger he'd just faced, and no concept of medieval killers or of the destruction Simon might wreak on the world. He had no responsibility to figure it all out, to protect those he loved. Most importantly, he'd had no glimpse, as Shawn had, into his own possible future.

At least the chief had left men to watch the place. Clive had volunteered for the first shift. Shawn felt security in his own thorough search and in knowing Clive and the Inverness Police continued to watch the place. And hadn't Brother Eamonn said Simon would go back?

Beside him, Amy stirred. She opened her eyes, large and shining in the dark, like James's, looking at him. It took him back to the first night she had stayed in his home. In his bed. She had loved him like no one ever had. He'd always felt different with her—as if, finally, what he did mattered. He couldn't stand to lose her. She'd been worried about Angus. It was understandable. It was good of her. But she'd hugged him tightly when he'd returned to the car, and he'd felt her shuddering intake of breath. "I was so scared," she'd whispered.

"I couldn't have protected him," she said now, her voice as soft as the moonlight pouring through the window. She touched his hand where it rested on the gear shift. "Are you sure he's gone?"

Shawn shook his head. He'd thought, through the hours of searching the tunnels, of Eamonn's story, recounted just hours ago in the hospital chapel. And suddenly he second-guessed himself. Eamonn's story didn't tell him *when* Simon made the cross back to the past. There were the scrolls. Surely, if Simon had appeared in Niall's time, they'd have *told* him of their prayers being interrupted.

Maybe they had and he hadn't read that far yet. He'd *have* to read them now. He couldn't keep avoiding it. A chill inched up his spine. Hopefully they would recount a *successful* battle against their surprise intruder, and not tell him one—or more—of them had been killed in the passages.

He glanced at Amy. He should tell her about his conversation with Eamonn —what Eamonn had to say about James. He said nothing.

Her eyes drifted shut. Loch Ness slid away. The headlights illuminated the winding road, and soon the lights of Inverness twinkled ahead. He *had* to tell her! But he still had decisions, he argued with himself. James was *not* going back. And his best protection was to be far away in the States.

The thought crept up on him: there *was* no protection, if Simon succeeded in his plan. If Simon succeeded, all of history would change, as it had when he and Niall charged into battle at Bannockburn.

Still, Shawn thought, easing the car into a roundabout, if Simon must be dealt with, he would do it himself. Not James. Simon was almost certainly back in Niall's time—but James was still better off in the States. He didn't have to tell Amy all Eamonn had said. Why upset her over what had to be the ramblings of a crazy old man? No, he would take care of things, one way or another. But how could he convince Amy to return to the States? To take James back?

First things first, he reminded himself. Was Simon even in this century?

Mairi stirred as he slowed at her street. "We're here," he said, pulling up to her house. He climbed from the car, opening her door, and taking her to her front step, checking over his shoulder. Don't be ridiculous, he chided himself. Simon wasn't here.

But a man *could* still slip through all that surveillance. His agitation grew, uneasiness stirring him, like a cold, curling snake. Simon was either there, *then*, with Christina and Niall, or here, meaning he must get Amy far away.

Niall would certainly tell him in the parchments. And they had been left behind in his own home in the States, in his hasty flight to Scotland.

The problems were supposed to be solved. Instead, they were just starting.

Glenmirril, December 1317

Brother Eamonn struggled to his feet. "Put away the knife, Simon."

"I know naught of the Laird of Glenmirril." Simon kept the weapon pointed at the monk. "But I trust he is no fool. He must certainly know anyone in that labyrinth has only one option for leaving safely—and that is to come back the same way. If I were him, I'd have a dozen guards waiting here." His eyes flickered over the chapel as he spoke.

"Niall has gone for more hot wine. 'Tis my shaking hands—I spilled what I had." A draft from the hall stirred the fine white wisps of hair on Brother Eamonn's head. "We have several minutes."

Simon glanced at the door, but lowered his knife, his eyebrows creasing. "You fancy yourself a seer, do you not? So surely you knew I was here. Why would you protect me?"

"I am in fact the greatest seer of all time." Eamonn smiled. "However, God seems rather determined to do things *His* way, not mine. It seems I must get you out of Glenmirril that you might fulfill your destiny."

Simon glanced at the door to the hallway, before stepping down from the altar. His hand went to his heavy *jacket* from the future, feeling the notebook there. "I've every reason to believe you of all men object to my plans."

"I did not say your plans." Brother Eamonn pushed his hands into his sleeves. "I said your *destiny*. You've work to do. You see, whatever *your* intent, things have a way of coming aright.

"You said you'll get me out." Simon took a sudden step, spinning the frail old man, and putting his knife to his throat. "You have one minute to tell me your plan," he said.

"For tonight," Eamon wheezed, "sleep in the passages below."

Simon pressed the knife tighter. "Do you tell the Lord of Claverock what to do?"

"You are free to find your own way out." The monk remained still in his grip.

Simon's rage leapt. But the old man was right. He had no one else to trust, and no way out. And Eamonn *had* gotten him back to his own time. Slowly, he released the pressure of the blade. "Tell me, monk. How will you take me out?"

"The Laird's men have orders to allow me into the hills to pray. Take a robe from the sacristy, and leave with me on the morrow. The morning guard have not seen you and will presume it quite reasonable that a frail monk such as myself will not go out alone."

"Why would I trust you, Monk?"

Eamonn shrugged. "Who else have you? You are in an enemy castle. Trust me or kill me. I am, as you say, a seer, and I tell you that trusting me will allow you to live significantly longer."

"That would require me to trust you are telling the truth, would it not?" Simon whispered.

Eamonn cackled in childish delight. "Would a monk lie?"

CHAPTER THREE

Inverness, Scotland, Present

"You'd be safer in the States." Shawn is adamant. His anger fills Angus's small row home. The argument has raged, an insistent, growing Bolero, starting with an uneasy, palpable silence, pulsing like the barely audible rhythm of the snare when he got back in the car after dropping off Mairi, growing to the throbbing melody as we walked through Angus's door, till it exploded.

"You said he was nowhere to be found." I head into the kitchen, needing a hot drink, wanting the argument away from James, who sleeps upstairs. *"We both know what that means. It's not* me *who's in danger. It's...."*

"You could *be!"* He charges into the space between words before I can say Niall's name. *"We don't* know. *He might have found a way to hide down there."*

"You know the tunnels well."

As if I studied this score, I see the harmonies and unspoken ostinato pulsing beneath each of our motifs. I want to believe my son is safe. I want to believe I am. Because I want to stay here with Angus, and still believe I'm doing the right thing by my son.

And yet the police swarmed Glenmirril for hours. There are only two ways out, and Simon took neither.

Shawn's whispering counterpoint is a soft plaintive thread hidden by the orchestral swell of his anger: he wants me away from Angus.

There's another, softer whisper. I see it, there at the very bottom of the score, a soft rumble of the basses that almost no one will hear: If I'm in danger, he can do something. He can move me to safety. If Niall and Allene are in danger, if Simon is in fourteenth century Glenmirril with them, he is helpless. I think of Christina. He's helpless to protect her, too.

It is his fear and helplessness that drive the pounding timpani of anger that threatens to drown me. I jam the faucet up, sending a furious spray of water into the kettle. I'm angry because I know he's right. There's that one small chance Simon managed to hide from all of them, wait them out. He might come up in a few hours, in a day or two.

How can I leave Angus when he needs me? When he, too, is helpless?
How can I stay with my son in danger?

And yet, there's a raw edge to his anger that I don't fully understand. A note is missing—I can't identify the chord, except to think it's not right.

I jab the faucet off, dropping the kettle on the burner, at the same moment Shawn's phone rings. Relief and fear mingle in me. I want a breather from his onslaught. But what if it's Clive calling to say Simon has been spotted? I glance down the hall, imagining Simon bursting in, dashing up the stairs before I can reach James.

"Yeah, Clive," *Shawn's professional veneer drops over him. His shoulders relax as he listens. So it's good news.*

I reach for a mug and as suddenly, my heart beats harder. What if Clive is calling to say something has happened to Angus? I watch Shawn's face for a clue.

"Uh-huh, yeah," *Shawn says, and listens another few moments, Clive's brogue coming over his phone. Shawn watches me, nodding at whatever Clive is saying. The kettle begins a slow hiss. I glance again down the hall to the front door. But Simon could as easily burst through the kitchen door. I edge away from it. The helplessness and fear of the chapel, of being left alone to defend James and Mairi against a man even Shawn could barely fight off, return. How can I even think of staying, if there's any chance he's still here?*

"Thanks, Clive. I don't know how you managed it, but hats off to you." *He ends the call, his eyes on mine.*

Not that I like the thought of Simon in Glenmirril with Niall and Allene and the Laird any better. The kettle screeches. I lift it from the stove top, pouring steaming water into the mug. "What is it?" *I ask.*

"Clive stayed with two men, watching Glenmirril. He's convinced them to send men back to continue searching, and to keep it closed to visitors for the next three days." *He bows his head.* "Just agree to me staying here with you."

"Of course!" *The kettle clatters to the stove, and I sink into his arms, grateful, to him, to Clive. He holds me, as the fears of the last twelve hours, of being forced away at the point of Simon's knife, of my fear for James's life, erupt in sobs such as I have not known since I was a child. He strokes my hair, until the sobs die to hiccups, whispering,* "It's going to be okay. I'm going to make sure."

Finally, he pulls away to jam a chair under the kitchen door knob, and drag another one down the hall to block the front door, too. He retrieves four sharp knives from the kitchen, and leads me up the stairs, settling into Angus's bed, next to James's cot, his arms around me. "You'll be fine," *he says.* "James will be fine. Clive and the whole Inverness Police are watching Glenmirril, and I'm not going to let him near you if he's here."

I lie awake a long time in the dark, listening for every creak, every sound in the night. In my heart, I believe Simon has gone back to his own time. Still, sleep will not come. I listen to James's breathing in his cot. My thoughts turn back to Angus, alone in his hospital bed, his future uncertain. If they find Simon, I know I need to take James to safety, no matter how much I hate leaving Angus.

Inverness, Present

Shawn eased his shoulder carefully from under Amy's head. Her eyes flew open. "It's okay," he whispered.

"He's not here?"

Shawn shook his head. "I'm just going to check on things," he said.

She reached for his arm. "Don't leave."

"It's okay," he said. "I'm going to be between the doors and the stairs at all times. I'm taking two knives. I'll block the door. You're safe."

She let go of his arm. With a glance at James, sleeping soundly in the crib, she sank back into the pillow, and almost immediately into the sleep of exhaustion. Shawn touched her forehead, running his finger along the line of dark hair at her temple. *Did you ever watch her sleep,* Angus had asked. He thought of Angus, in a hospital bed, in pain, as they slept in his bed. He pulled his hand away, watching her another moment before snatching up two of the knives and turning for the door.

He pulled it shut behind him. There was nothing to block it with. He went into the bedroom at the end of the hall and walked the nightstand out, a barrier in the narrow hall. It was little protection against a medieval knight, but better than nothing.

He descended the stairs, checked the front and back doors, the chairs still jammed under the knobs, and planted himself, back to the front entrance, watching the kitchen door down the hall. His suitcase sat in the hall where he'd left it after his frantic flight from his mother's house—just a day earlier? Not even? His head spun with exhaustion. There had been that last concert—the sound of his headlights being smashed—Brian's father, his face streaked with tears, Brian's snowy grave in the moonlight—arriving home drunk in the wee hours and waking to find Amy gone, his mother taking him to another cemetery, another grave—the visit to the prison, to Clarence, a text from Angus, the flight to Scotland and the surreal conversation with Brother Eamonn. It all swam in his head.

He couldn't even remember how many hours or days ago it had all started. He drew a deep breath, staring up at the ceiling for a moment. It didn't matter. He had the here and now to deal with.

He pulled out his phone, and, knife in hand, tapped his mother's picture on the screen, even as his eyes flickered over the bisque crucifix hanging on the wall, with the palms sticking out behind it. Niall would pray. *Okay, so help me out,* he thought as the phone rang.

"Shawn?" Carol's voice came abruptly over the line, full of anxiety. "Is everything okay?"

He hesitated, unsure how to tell her all that had happened.

"Shawn?" The pitch of her voice rose. "You flew out of here in...."

"It's fine," he said. "Listen, I'm begging a favor from you. I left a...a package...a long oilskin...in my room. I need you to get it and...." He hesitated. It was asking a *lot*.

"Where in your room?" she asked.

"Top drawer on the right." He watched the kitchen door, the knife half-raised, ready to strike, as if Simon might crash through at any moment.

"What do you need with it?"

"If I had a ticket waiting for you...."

"You want me to fly this over to you?" Her voice rang in disbelief.

"It's important," he replied. "First class, Mom. You can sleep on the plane. Good food, free drinks."

"What is this?" she asked. "It looks old."

"It is." He imagined her turning it over, studying it. "Look, Mom, don't open it, okay. The papers inside are fragile. Your favorite ham and noodle casserole waiting."

"You said everything is fine. I had plans for the next few days," she said. "Maybe next...."

"Please, Mom." Shawn hated the pleading in his voice. "I need this ASAP. It's important."

"A courier...." she started.

"I can't trust it to a courier." He felt the persona that ruled a musical empire rise, putting an edge into his voice. "I *cannot* afford to lose this!"

Silence came back over the line.

"Look, I'm sorry," he spoke into the silence.

"What is it?" she asked, an edge to her own voice, "that is so important you expect me to drop everything?"

He hesitated a moment, wondering if he should tell her what had just happened. "I think," he said carefully, "there's information in there about the man who attacked you and James."

"You said it's old," she countered. "How can it have...?"

"He was back yesterday," Shawn broke in.

He heard her sharp intake of breath. "He was....?"

"Can you *please* just trust me on this?"

"If there's information inside, why don't I just read it to you?"

"It's in a foreign language," he hedged. There was no way she could wade through pages and pages of medieval English. Who knew how deep in their letters a mention of Simon might turn up. Or a story of him appearing and disappearing again. She was already agitated enough at his claim to have been in another century.

"I can take it to someone at the U. I can...."

"I'll make it up to you!" Shawn glanced at the stairs, wary of disturbing Amy, and lowered his voice. "Somehow. Anything you want, I'll do it, but I need that package."

"Anything?" Carol asked.

"Anything," he said.

She didn't answer. A beat. Two beats. His heart pounded. "Please, Mom," he whispered.

"Have the ticket waiting at the airport," she said.

"Take good care of it." Relief coursed through Shawn. "Don't leave it for a second. Take it to the bathroom with you."

"Get the ticket," she repeated. "I haven't lost anything yet."

"Mom?"

She sighed. "Don't take my eyes off it?"

"I was going to say thank you," he said.

"Anything I want," she said. "Just remember that. Text me the flight information. I'm already pulling out my suitcase."

She hung up. Shawn drew a deep breath. *Anything.* There really wasn't anything she could ask for that wasn't easy to give. Albums, sheet music, t-shirts, downloads, concert tickets, the trading cards—they were all selling like arrows before battle. With a last glance at each of the doors, he climbed the stairs, swinging his legs awkwardly over the nightstand blocking the narrow hall, and headed back to the bedroom to try to sleep.

CHAPTER FOUR

Glenmirril, December 1317

As the rose window faded from black to gray in the early morning, Simon found Eamonn in the chapel, as promised, his hands clutching a gnarled wooden staff. He struggled to his feet as Simon entered, and held out a robe of brown wool like his own. Simon wiggled into it distastefully. He was no cleric! What disgrace to wear such a thing! If the wretch betrayed him, it would be difficult to fight in.

And yet—he *could* fight. Had he not ever been the best among the boys in his father's castle? A robe did not hinder monks and bishops from fighting. And he had precious few options. At the moment, it seemed—none.

"Put the hood up," Eamonn instructed.

"Is that how you address your superiors?" Simon demanded.

Eamonn smiled. "I am a man of the cloth. We are brothers, none superior to another. Come, if you would have me save your life." He strode into the hall, tall and thin, and anything but infirm as he swung his staff with each step.

Simon glared, but lifted his hood as he fell into step beside the monk.

"We will collect food for our journey in the kitchens," Eamonn said

"Journey?" Simon asked. "Beyond these walls, I have no need of you."

"You will always have need of me." The monk chuckled, as if speaking to a child.

"I won't," Simon snapped. It was a weak response. A foolish response. He thought, uneasily of his so recent meeting, in Amy's time, with this old man, in a church with lightning flashing outside, of seeing the leer of the ancient healer in his father's kitchens. Uneasiness stirred in him.

"You may wait in the great hall if you prefer," the monk said. "Do you fear kitchens?"

"I fear *nothing*!" Simon insisted.

"I will get the food myself. However, should the Laird come down early, 'twould present a problem for you. And there's someone in the kitchens you would fain meet."

"I've no time for wenches," Simon barked. "And I've no interest in any other I might meet in the kitchens."

The monk proceeded, smiling, at Simon's side, as the gray sky turned the faintest pearl pink in the hall's windows. Simon tamped down the anger, falling behind the old man when they came to the staircase that twisted down, down, down, into the kitchens below the great hall, lit only by the fires roaring in hearths and the gray light winking through the window high above.

Already, they were alive with a bevy of girls, bustling red-faced under the directions of a plump older woman. Eamonn sank back into the form of a decrepit, old man, his hands wrapped together, shaking, around his staff.

The woman turned, wiping sweat from her brow, to see Eamonn, and bobbed a quick curtsy. "Brother! You shouldna ha' come down! Sure an' I told you we'd bring your food to the hall! We're just after packing it up!" She turned, hollering, "Kate!"

"Bessie," the monk whispered.

The woman turned. "Bessie?"

"Aye," rasped Brother Eamonn. "I'd fain meet Bessie."

"How do you know Bessie?" she asked, and almost immediately waved a hand of dismissal. "'Tis neither here nor there! Bessie! Quick, lass! The good brother wishes to meet you!"

At the ovens, a girl turned. Simon watched her—a waif, no doubt an orphan, with mousy hair tucked under a kerchief. She'd be fair in the right clothes. But she was of no import. "What game are you playing?" he whispered, as the girl, her eyes lowered, came hesitantly toward them.

When she drew close, Simon saw the flush on her cheeks, from the oven's heat. Her hair stuck to her forehead in damp curls that escaped the kerchief.

She eyed Simon like a doe caught in a trap, as if seeing through his robe, knowing he was not a monk. She looked to Eamonn, and bobbed. "Good Brother? What do you wish?"

Simon seethed. The wretch had just ignored *him*, Lord of Claverock, in favor of an ignoble fool!

Eamonn took her hand, as if taking the hand of a princess, and bowed over it. "Bessie."

"How do you know me?" she asked softly. "I do not know you."

"Your mistress just sent you over, did she not?" The monk wheezed out something reminiscent of a laugh.

She smiled uncertainly. "But you knew my name."

"Aye." The old man released her hand, to indicate Simon. "Meet my fellow brethren. His son will marry your daughter."

"I have no son!" Simon turned to him in surprise.

Several of the women glanced at him and quickly averted their eyes.

"He is a monk." The girl lowered her eyes. Her hands wrung her dress. "And I...I will have no daughter, good Brother. I am...I am not...I am not wed."

"You think none will have you, after Duncan's cruelty," Brother Eamonn said softly. "You are mistaken. He has a long way to go—but he is on his way."

Bessie's eyes held his a long moment, before heat flushed her cheeks and she stared at the floor.

"What is she to me?" Simon whispered harshly. "Let us be away."

"Indeed, let us. Take the food, Brother Anselm."

The girl lifted her eyes, offering Simon the bundled cloth.

"The guards will not expect a frail old man to be carrying it," Eamonn murmured. His hands clutched the staff, trembling like a leaf in a winter storm and he turned for the stairs.

Simon snatched the bundle from the maid's hands, glowering at her. She cringed and backed away. Simon spun on his heel, going after the old man.

"Best behave like a monk," Eamonn murmured, as they climbed the stairs, "if you wish the guards to believe you are one. Humility. Meekness."

"I am the Lord of Claverock," Simon hissed.

"An' you dinna pretend otherwise, you shall be dead," Eamonn said.

They reached the light of day, and crossed the gray courtyard, Simon furious. But the monk spoke true. He was in an enemy castle. He lowered his head, following with what meekness he could summon, keeping his eyes off the few people in the dawn-streaked courtyard.

A youth with vivid red hair stepped forward. "Brother Eamonn?" he asked.

The old monk stooped. The staff quivered in his hand as he nodded. "I believe your Laird told you I am going into the hills to pray. I am taking my servant with me."

Simon's jaw tightened. He would *kill* the old man.

"Aye, Brother Eamonn!" The boy nodded enthusiastically, as if this were the best news he'd heard in years.

Maybe he'd kill the boy, too, Simon thought. Just to stop the inane good cheer.

"Would you not like a garron?" the boy asked.

"Thank you, Red," Brother Eamonn croaked. "But we shall walk." He laid a frail hand on the boy's shoulder and added, "My Lord Shawn bids you well."

The boy startled at the name. Then eagerness lit his eyes. "You've seen him? They said he'd gone—forever."

"He does not forget you," Eamonn said, and started across the drawbridge.

Simon stalked ahead, unwilling to be treated as a servant a moment longer, now he had passed the guards.

"There are archers on the parapets," Eamonn said, behind him.

At the end of the drawbridge, Simon stopped, waiting impatiently. Eamonn shuffled at a maddeningly slow pace. Simon glanced at the guards on the castle wall, and re-considered his decision to continue in the lead.

"We're alone now," he murmured when the monk finally reached him. "Let me take your arm and help you along. You were ever so concerned about it looking real."

But Eamonn leaned heavily on his staff, wheezing as if he'd run a great distance. "You'll want to know about Bessie," he said.

"A maid servant?" Simon scoffed. "It was but another of your games."

Eamonn lifted his head, his watery blue eyes meeting Simon's. He shook his head slowly. "I play no games. All I do and say is deadly serious."

"I care naught for her," Simon replied.

"You ought," said Eamonn, "for her daughter will marry...."

"I have no son," snapped Simon.

"Meekness, humility," murmured Eamonn. He lifted his head, looking to the parapets.

Simon did, as well. Two of the guards stood together, watching him. He placed a hand on the monk's back in what he hoped looked like a consoling gesture, and leaned close, hissing, "And if I did..."

"You will."

"...he'd not marry a scullery maid's urchin."

"The mighty fall," sighed Eamonn. "And the lowly rise."

"Let us go." Simon took his arm.

"Your son," added Eamonn, his voice suddenly strong, "will be your undoing."

Simon's hand dropped, as if burned. Then he abruptly laughed. "It seems he must join the queue. Young James is to kill me! My own son is to kill me! Did you not tell them *Shawn* would deal with me? Now I know you but make sport. Let us go."

"I think," Eamonn whispered, "that I am in no shape to pray after all."

"Come along," Simon said angrily.

"Do not forget Bessie," was Eamonn's response. "Or your son." He put a hand to his mouth, coughing loudly. "Oh, my! I'll tell the guards I sent you on ahead, but I fear you must go now, ere they come to help this wretched old man!" With that, he went into a fit of hacking, a fist to his mouth, waving Simon onward with his other hand.

Simon glanced at the guards, pointing now, and shouting. He looked at the steep hill rising to the north, and the slope descending gently toward the south.

"South," Eamonn gasped between fits. "Find MacDougall. He will want to help the likes of you."

Simon hesitated, with a glance at the archers. "MacDougall?"

"If you hurry to Claverock," Eamonn said, "you can be on time to give Edward a proper greeting when he arrives." A spasm racked his body.

The boy, Red, started across the drawbridge, calling, "Brother Eamonn! Are you all right, then?"

"Go," Eamonn gasped.

Simon tuned south, trying not to look back, trying not to look like he was hurrying. The moment he reached the cover of trees, he broke into a run.

Glenmirril, December 1317

"You slept well?" The Laird shook Niall's shoulder.

Niall grunted, rolling over and sitting up, from the fur by the hearth in the Laird's room. He squinted in the dim room, the sun not fully up yet, at his good-father. "I'd have slept better in my own bed."

"Allene is worn. I did not want her disturbed." He turned his back, his bushy mane bent over the food on the table. He poured ale into a cup and handed it to Niall. "In truth, there's more. *You* needed a good night's sleep after your hasty journey."

Niall cleared his throat, unwilling to ask just what MacDonald meant by that. He climbed to his feet and stretched in front of the window. The faintest streaks of pink silhouetted the trees on the hill. Gratitude at being home warred with the upheaval of the previous night.

"What brought you home?" the Laird asked.

Niall took the ale, seating himself at the table. He drank deeply, before saying, "A wee stramash involving a cleric. The sight of my face might be a greater than normal annoyance to the English at the moment." He grinned. "So the Bruce has set me to some work—arranging for weapons, bringing a gift to Glendochart—until things mayhap die down a bit."

"How long have you at home?"

"A fortnight."

The Laird grunted and seated himself at the table, glowering at the cheese there.

Warning stirred in Niall. "Is there aught amiss, besides last night's events?"

Before the Laird could answer, a knock sounded on the door. "Enter!" he called.

The door swung open to admit the elderly Brother Eamonn in the doorway, half-supported by Hugh and Brother David.

"Sit!" The Laird rose hastily, offering his seat. "You did not go to pray?"

"My lord," Brother David said as the old monk sagged into the Laird's chair at the table, "Red is in the hall. He wishes a word with you."

"Red summons the Laird?" the Laird muttered. "Shawn's time continues to interfere with our order." But with a glance at Eamonn, he left the room.

Inverness, Present

Shawn insists on driving me to the hospital. Not that he has to insist very hard. I've spent the night and morning imagining Simon breaking through the door or jumping from a bush. I'm grateful, nonetheless, that Shawn stays at one end of the hall—watching, but giving me privacy.

I enter Angus's room in the early afternoon, tired from the late night and broken sleep, worn with the nagging worry that we don't know Simon is gone.

I worry, too, about Eamonn. He's with Niall, I tell myself, and the Laird. Surely they'll take good care of him. But I don't know.

James, perhaps feeling our agitation, has been fussy, slapping his palm in his morning cereal, rolling and wiggling away as I changed his diaper, fighting me every step as I try to slide a clean shirt over his head, over the peculiar wing-shaped birthmark on his shoulder; shaking his head and protesting the car seat with short, angry squeals.

Only as we climb the stairs, does he settle, laying his cheek on my shoulder. The nurse points me to Angus's room, whispering, "I do hope ye can cheer him up, then." I stop in his doorway. He's alone, flat on his back, his face to the window, where soft white flakes drift down. Was it only twenty-four hours ago I stood in this same place? He told me to go home. I don't know what he'll say today. I don't know where Simon is. I'm worried about Eamonn. I feel the shaking begin in my hands. I'm just a musician. I want my normal life back, of playing violin. It was a peaceful life.

"'Twas the meds talking yesterday," the nurse says softly. "Of course, he's still on quite a lot today. Poor man." She sighs, and leaves.

I gather my courage. "Angus?"

He doesn't turn, but I swear he stiffens.

James lifts his head, sees him and squeals in glee. I enter the room, rounding the bed. His eyes are squeezed shut in a determined attempt to look like he's asleep. My agitation melts a little bit, watching him. I feel the love I've always felt, looking at his dark curls and ruddy cheeks. I bite my lip, torn between laughter and tears. "You're a terrible actor," I say softly.

He squeezes his eyes more tightly, not swayed by my vote on his chances for an Academy Award. James squeaks and laughs, and leans forward. I hold him tight, as he lunges for Angus, patting his cheek.

Angus's eyes snap open. Love and hope leap there. As quickly, they turn hard. "Go home," he croaks.

I seat myself in the chair, ignoring his order. James reaches for him again.

A smile flits across Angus's face, before he quickly steels his features. "Simon," he says. "Butcher...."

"We think he's gone." I take his hand in mine, as James leans down, his cheek to Angus's.

"Gone?" he whispers.

"To his own time." He doesn't need to hear the whole story, or know they're still watching Glenmirril. Dark circles under his eyes speak of his exhaustion. He needs all his energy for healing. And if I'm honest with myself— I need to believe it, too. I don't want to remind myself it's not sure. "He disappeared at Glenmirril last night."

He is silent for twenty seconds, thirty. I squeeze his hand, hoping. Even now, Angus has a strength I need.

"Go home," he whispers. "I don't want you."

Glenmirril, 1317

"Your prayer time did not proceed as planned?" Niall addressed the elderly monk as he sat, bowed at the table in the Laird's rooms.

Brother David poured mulled wine into a stone chalice, and stepped back to stand by the window.

"Things always go as planned." Brother Eamonn lifted the chalice, murmuring "Thank you," to Brother David. "'Tis a question of *Whose* plans."

Niall smiled. "Truer words were never spoken."

"Nor will be," said Brother David.

"How does Shawn fare?" Niall poured his own wine and sat down at the table. "Does aught go as *he* planned?"

Eamonn sipped his wine, before saying. "His life unfolds as it ought."

Niall's smile grew, liking this old man. "We both know Shawn. Therefore I take it things have *not* gone as he planned."

Eamonn, too, smiled, their eyes locked in humor.

At that moment, the door burst open, the Laird storming in with his fur-lined cloak billowing behind him and Hugh on his heels. Niall shot from his chair, standing before his lord, while Eamonn struggled to his feet, and Brother David straightened at the window, head bowing.

"Who left my castle?" the Laird bellowed.

The old man did not draw back. "Simon Beaumont, Lord of Claverock."

"That butcher!" The Laird's hand shot out, bunching Eamonn's robe in his fist. "*He* was here? He crossed over? You told us...."

"My Lord...." Niall began.

"I told you Shawn will deal with him," Eamonn said. "And he shall."

"Shawn is not here," Niall objected.

"Malcolm," Hugh admonished. "He is a monk, and elderly."

"He has endangered my people!" Malcolm shot back. "Simon Beaumont, the Butcher of Berwick was in my castle and he lied to me!" But he released the monk, who sagged into Brother David's waiting arms. Malcolm threw himself into his chair, and waved a hand. They all sat, Niall and Brother David helping the old man. The Laird's eyebrows worked furiously.

"May I speak, my Lord?" asked Brother Eamonn.

"Not only *may* you, you had best," the Laird replied. "This time, you'll tell us everything, and you'll not leave until you have, regardless of your age." His eyebrows lowered. "Or your well-timed coughing fits."

Hugh looked from Malcolm to Eamonn with concern.

Brother Eamonn held a hand to his chest a moment, drawing in a long, slow breath, before beginning. "My Lord." He inclined his head to MacDonald. "Niall, Hugh, Brother David." He gave a nod toward each of them. "I am here through no fault of my own. I am here, nonetheless, with *your* best interests at heart." He looked from one to the other. "And the world's."

"What was Simon Beaumont doing in my castle?" Malcolm thundered. "Lord of Claverock, of vast Northumbrian lands?"

Niall glanced between the two men.

"And why," demanded the Laird, "have you set him free? He is *English*! He seeks to harm my people!"

"Oh, aye, he does that!" Eamonn agreed. "And far more than that."

The Laird's hand tightened, knuckles white, on the hilt of his sword.

Niall held up a hand, staying his anger. "Give him but five minutes."

"Five minutes," warned the Laird.

"He is evil," Eamonn said. "However, there are paths that must be followed. Things that must happen, that would not, had he died in Glenmirril."

"Such as?" demanded the Laird.

Eamonn lowered his head, staring at his gnarled hands. "He will seek out Edward, king of England. He will pose as a seer. It will come aright."

They stared at him.

"*What* will come aright?" the Laird exploded.

"All that should and must. You, those you love, Amy, Shawn, the world. It will all come aright."

Inverness, Present

Shawn kicked the kitchen door open. At the table, Amy startled. "My specialty burgers!" He stood in the doorway grinning, a plate of burgers in his hand. "Grilled out back, like a true Minnesotan!" He stamped the snow off his feet and came in, setting the plate on the small counter in Angus's kitchen.

Amy laid her pencil on the score paper that covered the table. She'd been quiet since her visit to the hospital the previous day.

"Hey!" Shawn spreads his arms, showing himself off. "Plus me for company! The Best of Scotland! What more could you want?" He snapped the lock on the kitchen door.

A faint smile crossed her face. "Some things never change."

"Hey." His voice grows serious. "I know it's stressful, waiting. But the front door is blocked, I was guarding the back."

"No word yet on the weather?" Snow at MSP had kept the scrolls, the possible answer to their fears, far across an ocean, leaving them in a state of unknowing.

He shook his head. "But the snow has to let up sooner or later and she'll be on her way."

"You can't keep going the way you are," she said. "You haven't slept in two days."

"I'll get a nap before I head back." When she said nothing, he added, "Hey, no news is good news. They've got the place on lock-down, they're searching with dogs, and they haven't found him."

"I think he's gone."

"You can understand it's important to know for sure?"

"Of course I can!" Her words came out sharply. Angus hovered between them. She shuffled the pages together, muttering, "I'm sorry," and took them to the front room.

By the time she returned, Shawn had the table set. "James is still asleep?"

She nodded, as she took a bottle of soda from the refrigerator, and poured it into the glasses full of ice Shawn had set out.

"Okay, I know it's been rough." Shawn piled a bun with a thick beef patty, tomato, onion, bacon, olives, and lettuce. "But—is there more? Angus is going to get better. Right?"

"We hope." She put a patty on a bun and shook ketchup over it.

His phone beeped at that moment. He glanced at it, and sighed. "Thank you!"

"Clive?" Amy asked in a rush.

"My mom. She's boarded. This time tomorrow, we'll have an answer."

"A lot can happen in twenty-four hours," Amy said.

"It's looking more and more like Simon is gone."

Amy sighed, as she put toppings on her own burger. "If he's not here, he's with Niall and Allene."

The muscles in Shawn's neck tautened. He took a big bite of his burger, his eyes not meeting hers.

"I'm sorry," she said. "I guess it does no good to talk about it."

"Yeah, talking doesn't fix some things," he agreed.

They ate silently for several minutes, before Amy set her food down.

"I was right," Shawn said. "There's something else."

"What was Simon's thing about James?"

Shawn stood abruptly, busying himself with clearing food. "His...thing?"

"He looked at him funny—up at Monadhliath. It was unnerving."

Shawn kept his back to her, pushing tomatoes and onions into plastic containers. He couldn't tell any part of the story without telling it all. He didn't want to upset her!

"He was always asking about him. Wanting to see him. Saying he had gifts for him."

"Obviously he's demented." Shawn searched the cupboard for another container. He felt her hand fall on his shoulder, soft as a hunting hawk lighting. He swayed for a moment between Amy and Christina.

"You know something you're not telling me." It was Amy's voice.

Thoughts chased one another, his hand still lifted to the shelf. He felt her lay her cheek against his back.

"Shawn, I may be gullible still, but I actually believe you never wanted to hurt me. But you're still doing it. You're still hiding things."

"I'll see that you're safe." He turned abruptly. A plastic tub fell to the floor.

He grasped her hands against his chest. "James is safe."

Her eyes clouded. "You're not telling me something."

"He's *safe*," Shawn insisted.

"You know something."

Shawn bowed his head, avoiding her eyes, staring at the plastic tub on the old linoleum. He drew a deep breath and spewed the words out in a rush. "Brother Eamonn said it's his destiny to kill Simon, Simon knows it, Simon is gone, James is safe."

She stared at him, her eyes clearing with understanding. Then she yanked away, spun once, spun back, her eyes blazing.

"See!" Shawn jumped on it. "See, this is why I didn't tell...."

"He's my *son!*" she shouted. "It's supposedly my son's destiny to be tangled up with some madman, and once again you just didn't tell me. Don't upset Amy! Yes, I know, I'm supposed to be grateful. Everyone's protecting me! Well, I'm not. We're talking about my son. When do you think I finally have a right to know the things that affect my life? When do you stop telling yourself you're being kind when you're only treating me like a child?"

"I'm not...."

"And for your own convenience!"

"My own convenience!" Shawn railed. "It's hardly convenient to be left on my own to figure out how to protect him!"

"Then quit doing it all alone!" Amy stepped forward suddenly, grasping his face between her hands, making him meet her eyes. "Share the burden. I'm right here, begging to help!"

He stared at the floor.

"Let me in, Shawn," she said. "Let someone be there for you."

He wrenched away, storming down the hall to Angus's small front room, where he poked at the fire, stirring embers, and stooped to push twigs into the hearth. He felt her, in the doorway, although he didn't hear a sound. He felt her as surely as he'd learned to feel danger in the medieval forests. When flames licked around the twigs, he added a log, and rose.

She stood there, her hair falling to her waist. It pulled him back to the first moment he'd seen her; the first night he'd spent in her apartment—just talking. Every feeling of that night lurched through him. And now she was so much more—and they had a son who might still be in danger.

"Okay," he said. "You're in. What are we going to do?"

She crossed the room, wrapping her arms around his neck, and laying her cheek on his chest. "I don't know," she said. "But at least we do it together, right?"

"Right," he said. But he heard the doubt in his own voice.

"You know Simon will go back—if he hasn't already?"

"Brother Eamonn was quite clear he'd get back and gain power from Edward."

"But if Simon is gone, James is safe." She looked up to him. The fire threw shadows across her face and sheens of light on her black hair. "We've never been a hundred percent sure about Brother Eamonn.

Shawn stared into the fire. The crackle of the flames, the dance of orange and yellow tongues, the smell of burning wood, all brought him back to medieval Glenmirril, to campfires with James Douglas and Hugh, to days of riding hard to protect Glenmirril and the people there.

What good did it do to worry Amy with this? She was right. They had never been sure of Eamonn. Not really. He dropped to his knees, tossing another bunch of kindling into the hearth.

Amy knelt beside him, her hands on his shoulders, her cheek on his back.

"You're right. Brother Eamonn is an unreliable narrator," he said. "Simon *believed* him. That doesn't mean it's *true*."

"Because it couldn't happen," Amy said. "Not if Simon is already gone."

Shawn could feel her body relaxing, her fear draining away. "Right," he agreed. "Brother Eamonn was off his rocker on this one."

Maybe he really was, Shawn assured himself. Did it matter now? Brother Eamonn obviously didn't think a baby or even a child, was going to kill Simon. So he had years to figure it out. And deal with it himself if need be.

Ah, but fairy tales draw their truths from life, do they not? He heard Eamonn's words, sharp in his mind. *If you go, you will not come back.*

She sat down before the dancing flames, patting the floor beside her. "You really think Eamonn is crazy?" she asked.

He sank down beside her, pulling her head onto his shoulder. "Way too much of Brother Jimmy's Special Brew," he assured her. He had years to figure it out. He had years to be with her before it came to that.

But it wouldn't come to that. Brother Eamonn was crazy. He stared into the small flames. Yes, Brother Eamonn was crazy.

Glenmirril, 1317

"Things will come aright!" The Laird shouted the old monk's words, his hands raised to the ceiling. "The Butcher of Berwick interrupts our prayers on his way to seize power from Edward. The English army and clergy have new grievances with Niall, MacDougall and Christina! How can aught come aright?" He glared at the monk.

"MacDougall and Christina?" Niall's voice rose sharply.

The Laird turned to Niall. "I wanted you not to hear it from Allene. You needed rest before I told you."

"Tell us what?" Hugh's eyebrows dipped, reminiscent of the Laird's own.

"MacDougall escaped."

"The letters!" Niall said. "We shouldn't have...."

"He's taken Christina," Brother David interrupted.

"Christina?" Niall looked from the Laird to Brother David. "Why have you not gone after him?"

"I sent...."

"Why didn't you tell me the moment I arrived?" Niall demanded.

Hugh laid a hand on Niall's arm.

The Laird's eyebrows dipped in warning. "You stumbled in half-dead from pushing yourselves like the Devil himself was after you...."

"Yes, we'd *reason* to think the English might ha' been," Niall snapped.

"Listen to your Laird and show respect," Hugh said sternly.

Niall heaved a deep breath, muttering, "My apologies, my Lord." He lifted his eyes. "Yet, my Lord: Christina is out there *alone* with a man we suspect has already done her harm and you take us to the chapel to *pray?*"

"Was either of us in any shape, last night," Hugh asked, "to be out seeking MacDougall, let alone fighting him?"

"Laddie," the Laird said sadly, "your faith has been shaken badly this last year. You fail at the moment to see that prayer is the *first* thing we must do."

"Had you not been in the chapel to pray," the old monk added, "you'd not ha' seen Lord Claverock arrive. He'd have killed several of you in your beds."

Hugh and Malcolm glanced at each other. "It does seem," Hugh said, "that at least *that* has come aright."

Niall spoke more calmly. "Still, I should have known." Shawn's disembodied voice—*it's a trap*—flashed across his mind.

"Had you known," the Laird said, "you'd have pushed to leave immediately. You'd have spent half the night arguing the point if not trying outright to leave without my permission."

Niall stared at the ground, unable to deny it, and unwilling to concede.

"And had all the force of God and angels prevailed on your hard head to see the need for rest, you'd not have slept, regardless, for agitation."

"We've ridden hard these past days," Hugh reminded him. "We will eat and hear what Malcolm has to say."

"Are plans at least begun to set out after Christina?" Niall asked.

"There is no hurry," Eamonn said.

Niall shot him a sour glance.

"Lords Darnley and Morrison set to provisioning the men the moment we heard," the Laird assured him. "Scouts have been out tracking them. The cooks have been up through the night preparing food."

A knock sounded on the door. Brother David went to open it. Bessie entered bearing a tray of food. She set it on the table and bobbed a curtsy. She glanced at Eamonn, and heat flushed her cheeks. "Milord." She turned her eyes swiftly to MacDonald. "Any word of milady?" She looked to Niall. "Surely, Milord, ye'll not leave her to him?"

"We will not," Niall promised. "We are even now discussing it."

MacDonald dropped heavily into his chair, reaching for a piece of meat.

"I've sent scouts out."

"I thank ye, Milord," Bessie bowed her head again. "She was ever so kind —to me. To all at Creagsmalan."

"'Twill stand her well," Hugh said grimly, "if he thinks to harm her in front of his men."

"Aye," Bessie agreed. "But the ways in which milord MacDougall will harm her are those which his men will not speak against."

"'Tis more than his limbs and head will hang on our castle walls," Hugh said darkly, "an' he does. And I think we'll leave him alive to see it."

Niall and the Laird exchanged looks. And Niall understood the Laird's refusal to tell him the previous night, as he said to the scullery maid, "We could not set off ill-prepared. Thank you, Bessie."

She bowed, and left the room.

Hugh, Niall, and the two monks joined the Laird at the table.

"As long as we must eat," Niall said, reaching for a thick slab of cheese, "tell us what you know of Beaumont's plans. More importantly, how we are to stop him."

"'Tis not your place to stop him," Eamonn replied.

"We are to let him head off for England and seize power? 'Tis not forgotten what that madman did at Berwick. We can be sure he will visit the same brutality on all Scotland if he is not stopped."

Eamonn smiled. "You assume he *will* convince Edward of his prophecies."

"'Tis to our benefit that Edward is fascinated with watching men fish," Hugh said. "But 'twill also be to Claverock's. I suspect we'd prefer Edward on the throne."

"Indeed you would," Eamonn agreed.

"Then tell us how to deal with this madman you have set free." MacDonald glared at him.

Eamonn said nothing.

The Laird leaned close, saying softly, "Shawn believed us quite barbaric. Do you not fear to defy me?"

Eamonn smiled, but said nothing.

MacDonald sat back. "Be assured, I've no qualms about throwing a man of the cloth, even an elderly one, in my dungeons, if I am not satisfied with your answer. Or perhaps disemboweling him."

Niall cleared his throat, scowling at MacDonald.

"There is naught to be done at the moment." Eamonn turned his pale eyes to Niall. "The time is coming when you will meet him and his men. You will do what needs to be done. James will take care of the rest."

"James?" The Laird looked to Niall.

Eamonn pressed his hand to his mouth suddenly, a fit of coughing racking his body.

"James Douglas?" Niall asked. "Surely not my James. He is but a child."

Eamonn hacked loudly, his thin body doubling over.

"Stop this!" the Laird ordered.

Brother David rose, in alarm. "My good brother! Some wine!" He pressed a chalice on the old monk, who continued to cough harshly.

Hugh, too, rose, concern on his face. "Malcolm, perhaps...."

"He's prone to well-timed fits," Malcolm replied, watching the drama.

"Milord, forgive me if I speak out of turn." Brother David hovered over Eamonn. "But he has told us all we need know at the moment. Our concern is Christina."

"Lochmaben," Eamonn gasped.

"Lochmaben?" Hugh asked.

"You shall find her at Lochmaben."

"And things will come aright?"

"They will," Eamonn said, "though the road may be long. And the night will grow dark ere they do." He pressed a hand to his thin chest, stilling the coughing, and winked. "I've seen the future. The stars still burn bright in the sky."

Niall looked to the Laird. A small smile touched his lips. "Wise words, my lord."

South of Glenmirril, 1317

Simon stopped at the first village he reached—no more than a collection of rough huts. They seemed small, so filthy, after his long months in Shawn's time. He shoved open the door of the first, stepping into the dim interior. The buildings of Shawn's time had been full of light.

As his eyes adjusted, he saw a woman look up, startled. Gray hair hung in a long, thick braid. A scar crossed her cheek, holding one eye half shut. Simon blinked and swallowed. She looked old, battered—so different from the unscarred, youthful faces he'd known for so long. How had he forgotten this? And why were the people of Shawn's time so...fresh? So undamaged?

His hand went to his stomach, feeling the notebook safe under his clothing.

The woman's eyes flickered over him. "My good brother?" she whispered.

He wore a robe—yet she was scared. He smiled. People knew. People saw beyond clothes. "Get me food," he ordered. "And proper clothing." He ripped the detested robe off over his head, revealing the tight, stiff *jeans*, the boots, and the peculiar *jacket* of Shawn's time.

She backed away, sketching the sign of the cross on the air between them.

Simon laughed. "Surely you don't think that would stop me? Clothes, food."

"I've no men's clothes, Sir," she said.

He smiled. She *knew*. She knew he was no monk. "You live alone?" he asked.

She nodded, even as she scurried to a shelf. "There's food in the pot." She pulled down a bowl, took a large iron spoon hanging by the fireplace, and scooped out pottage. The ladle rattled against the wooden bowl. She set it, with a wooden spoon, on the single table with trembling hands, and backed away, her hands clinging to one another. Her eyes skittered to a door on her left.

He glanced at it, as he lifted the first spoonful of pottage to his mouth. He knew. As surely as she knew he was no monk, he knew her daughter was behind that door. He thought about the old monk's claim. *Your son will be your undoing.* He glanced at the door. Even if he begat a child this morn, the offspring of a peasant girl was no threat to him. He took another mouthful of the stew, reveling in its faint, sweet taste. Flavors burst. He wondered what he'd been eating, all these long months, that peasant fare tasted so good.

"Tell me news of Alexander MacDougall," he said.

"The MacDougalls were here—some time back," she replied softly. "The lot of them, storming through, attacking Glenmirril. Duncan MacDougall was killed, Sir, and Alexander held prisoner these many months. They say he recently escaped."

"There's no love between Glenmirril and the MacDougalls, then?" Simon asked.

"No, Sir. They've ever been at odds."

"Niall Campbell. Tell me of him. An unusual man, I've heard."

"'E's right up again from injuries that should have killed a man, they say." She took up a poker, and stirred at the logs in the hearth, raising the flames.

Because there was a man who looked like him, Simon thought. He took another few bites, scooping the last of the broth from the bowl, before asking, "Where might I find Sir Alexander?"

"Creagsmalan, Sir, his son's home, mayhap?" Her eyes darted to the door and as quickly back to him. "'Tis south, a hard day's ride," she said firmly. "Sure an' you'd do well to make haste." The poker did not leave her hand.

"I'm sure I'd do well to do that." Simon smiled, rising from his seat. His gaze drifted to the door. He had more important things. But a little distraction was always good. "You're sure you've no men's clothes in that room?" he asked.

"None, Sir," she said. "I've but the boy, and he's the plague on him, poor wee thing."

Simon shot back from his chair.

"Sure, an' you'd not like to go near that room, for fear of catching it."

She was lying! But he studied her face, scarred, aged. He wanted to remain like the faces of Shawn's time, smooth, unmarked. His eyes wandered up and down her gaunt figure. She would do. But he found he yearned for the fresh faces he'd known for so many months. And he didn't want to remain in a house that *might* have plague.

"Why did you not tell me, woman?" He considered slitting her throat. But she herself might have the plague. He spun, storming out into the morning sun before she could answer, and down through the village. At the far end, he found an old nag and mounted it, driving it into the Great Glen, heading south toward Creagsmalan.

Glenmirril, 1317

The fire flickered in the hearth even as morning sun grew, lighting the tapestries lining the walls of the Laird's chambers, and the faces of Morrison, Darnley, and Conal as they entered, seating themselves around his table. Hugh poured wine in their chalices as the Laird and Niall watched silently.

Morrison, small and wizened, stared at his cup, not touching it, then raised his eyes to MacDonald. "My Lord, may I speak?"

"Speak," MacDonald replied.

"My brother's daughter, whom I swore to protect, has been abducted. Despite our efforts to keep MacDougall locked up."

Niall stared into his wine, sharing Morrison's sentiments, but unwilling to speak against MacDonald.

"We have forborne his repeated theft of our cattle," Morrison continued, "even his attack on our home, with patience and mercy. We have lost good men to his attacks and deceit."

Niall glanced at Darnley, who stared at his untouched wine. He wondered what Darnley knew of his son's death, or William's dealings with MacDougall.

"Now he has seized Christina!" Morrison's voice rose. "My own charge! 'Tis time for stronger action."

"Conal has been selecting and arming men," the Laird replied evenly. "However, we have learned there are other, graver, matters than MacDougall." Behind him, the fire dimmed. Hugh rose to add another log. The flames danced higher, brightening the room, as he resumed his seat.

"My Lord," Morrison said, "forgive any appearance of impertinence, but for me, having sworn to protect her, there *is* no graver matter."

Niall lifted his eyes, a sad glance at the shrunken old man. He might carry the wisdom of age, but he no longer had any ability, on his own, to fight or defend anyone. He looked to MacDonald, curious how he would explain the problem of Simon. MacDonald and Hugh glanced at each other and to Niall.

Niall resisted the urge to shut his eyes, to escape this nightmare. "My Lord," he addressed MacDonald, "we are but few. Might we not better face the wolf without the rat harrying our heels?"

Conal nodded. "I know naught of the wolf you speak of, but MacDougall is an immediate threat—moreso now than he has been even in years past."

"Think, Malcolm," Hugh said, "'twill take time for Claverock to lay his plans, which are far grander than MacDougall's."

"Claverock?" Darnley leaned forward. "Simon Beaumont, Lord of

Claverock? *He* is the graver matter?"

Niall and Hugh looked again to the Laird, whose bushy eyebrows dipped in annoyance a moment before he nodded, saying, "Aye."

"What has he to do with this?" Morrison queried.

The Laird met Niall's eyes. *Shawn,* he mouthed.

Niall raised his eyebrows, an unspoken question in return: *So I'm the one who now has the skill to make up tales like Shawn?*

"My lord Niall?" Darnley pressed. "You know of this?"

Niall cleared his throat. *What would Shawn say?* "We had word from...." He glanced at MacDonald, who gave a barely perceptible shake of his bushy mane. Eamonn was not to be named. "From one who must remain unnamed for his own protection."

"When did we receive this word?" Conal cocked his head. "We've had no messengers."

"At Berwick," Niall answered. "I came home on the Bruce's orders...."

"We have had word." MacDonald's voice slashed across the table. "We will deal first with MacDougall. Lord Morrison." He addressed the old man. "We leave anon to retrieve Christina. We've cause to believe she may be safe at Lochmaben."

"You've had word too of that?" Darnley lifted an eyebrow.

"My lord." Niall spoke quickly, leaving no space for the awkward question to hang. "Lord Morrison has said it is time for stronger action. This time—we kill MacDougall?"

MacDonald hesitated, his eyebrows furrowing. White flecks of snow pelted the window. A draft curled in, leaving a sudden chill even as it stirred one of the heavy tapestries lining the walls: *David staring up at Goliath.*

"My lord," Niall said softly, "we've granted mercy long enough."

"We collect Christina. Then we hunt MacDougall." MacDonald rose, an imposing stature even in his age. "Niall, how long until Bruce recalls you?"

"I must be back at Berwick by March."

"Then we must work swiftly." MacDonald looked around his men. "We seek Christina at Lochmaben." He addressed Niall, Hugh, and Conal. "Then you go to Creagsmalan to learn what you can of his whereabouts."

"And when we find him?" Niall pressed.

MacDonald's lips tightened. "Aye, this time we kill him."

Inverness, Present

Nerves climbed Shawn's arms, waiting for his mother to come through customs. He should have used a courier. What if she'd left the parchments in the overhead bin? Or at a fast food court back at MSP? Moments later, she appeared, a large black purse tucked tight under her arm.

"You have it?" He wanted to kick himself for not so much as saying hello.

"Of course I have it. You're welcome!" She patted the purse, smiling, and embraced him. He steered her quickly through gathering onlookers who stared and pointed, girls with cameras straining to see him. Another full day searching searching Glenmirril, on top of the previous days and nights, and the resulting exhaustion, didn't help matters. He should have let Amy do the pick up. But every fiber of his being feared the loss of the scrolls.

He steered her through the crowds, into the night and to the parking lot.

"What's happened?" Carol asked. Her smiles for the crowd disappeared as he slid into the drivers seat of Amy's car and sealed them in.

Shawn turned the key and shifted into reverse. "An incident. We've put it to rest, really nothing to worry about." With a glance over his shoulder, he backed out.

She looked stern. "I just spent the last...."

He grinned at her. "Best ham and noodle casserole ever. You get to see James. Vacation in sunny Scotland!"

"Shawn, I went to a great deal of trouble to get here. I had to cancel plans I didn't really want to cancel."

Shawn glanced at her as he pulled out onto the road, his headlights reflecting off the white snow falling in the night. "The old couple on the west coast," he said. "The ones I stayed with for the year I was gone." He pulled the car onto the highway. "Information they needed. Family emergency."

She stared straight ahead. "Right. Of course."

"How was your flight?" he asked.

"Good," she said. "It was good."

"I really appreciate it," he said. "The old couple...they really appreciate it."

"You have the return ticket booked?" she asked.

"Tomorrow morning like you asked. Big date?" The prison warden flashed through his mind, and he instantly regretted the words. He didn't want to know.

"Traveling Broadway production at the Orpheum. No, I didn't want to miss it. And don't forget your promise—anything I want."

He grinned. "Name it. It'll happen."

She leaned back against the seat. Lines appeared around her mouth, watching the falling snow. "I'll let you know."

CHAPTER FIVE

Scottish Borders, December 1317

Christina forced her eyes to stay open as she lay on the cold ground, her back to a long, low pile of stones, listening to the men sigh into sleep around a low fire that did little to dispel the chill. Life had been beautiful for a short time, living in the safety of Glenmirril, seeing the love in Shawn's eyes.

It had all gone so bad since the men left to get Shawn home to his own time—the unexpected appearance of MacDougall in her rooms—loathing rose in her throat at the memory that would not go away—the battle—knowing in the morning that Taran's father was dead, and Shawn was gone to his own time, that MacDougall would stay in the south tower—the growing horror of realizing she was pregnant—the horrible ruse of leaving her wee James Angus under a tree, vulnerable to wild animals to convince the people of Glenmirril he was a foundling, given to her to raise—and finally, MacDougall's reappearance, dressed as Brother Andrew, in her chambers, taking her arm and forcing her at knife point away from her home and son.

She had bided her time, waiting until tonight, when they slept only miles from her kin near the English border. Her eyes went to MacDougall, lying inches from her. His eyes met hers in the dark. He touched her hair. She smiled, pretending as she had for two days, that she'd warmed to him. She let her eyelids droop into feigned sleep. "Sleep well, my lady," he whispered, and his hand fell away.

She prayed silently, trying to ignore the itching of leaves and moss she'd secreted in her boots, and in her bodice, each time she'd claimed the privacy of the forest these past days. Finally, his soft, rumbling snores joined the night symphony of bare branches rustling and the crackling fire. She opened her eyes, watching his face, his beard pressed to his pillowed hands, his mouth slightly open, showing the yellow teeth.

He'd ridden every day at her side, flashing those teeth in frequent smiles to charm her. She'd flashed her own back, disarming him. He'd played the gallant knight reaching for her heart. She'd played the courtly lady, coyly considering, with pleasant conversation and shy flutterings of her lashes. It was enough. He trusted his charms and he trusted his night guard, the doddering old Somhairle.

Most importantly, he trusted she was weak and had nowhere to go.

In the dying firelight, his shoulders relaxed, finally, in trusting sleep.

Christina rolled quietly, looking for Somhairle. The old man had a penchant for disappearing into the woods, if he drank too much. And he drank too much if given a chance. She felt no guilt for leaving her full skin of ale near him, ere she lay down. She watched from the depths of her hood, as he looked over the sleeping men. His eyes fell on her a moment, before reaching for the skin. He lifted it to his lips, drank deeply, and wiped his sleeve across his mouth, letting out a ragged sigh of contentment.

She smiled, feeling no guilt, either, for slipping crushed holly leaves into the ale. Rather, she thanked God she had found them. With MacDougall's snores in her ear, she lay in the dark, watching. Finally, the old man rose from his log and pushed into the undergrowth. The effects of the holly would keep him there a good while. She edged her hood back to watch the men. Not a one moved. She raised herself carefully on one elbow.

MacDougall rolled suddenly, flinging his arm over her. She froze, waiting, listening as his breathing settled back into a rhythmic rise and fall. She could hear Somhairle deep in the forest, grunting. Slowly, so slowly, she removed her father-in-law's hand from her hip. When he settled back into deep, even snores, she rolled carefully, terrified of having to turn her back on him.

As quietly as she could, she dug in her dress, pulling out moss and leaves, dropping them on the stones. A sound from across the camp froze her, hand stretched from her cloak, moss soft in her fingers.

She waited a full minute...another minute...each second stretching out, barely breathing. Sweat prickled her brow. Somhairle wouldn't stay in the forest all night. With trembling fingers, she placed the last of the moss, then inched up the hem of her skirt to reach into her boot for the leaves that made her ankles appear inhumanly swollen. She settled them in place, softening the shape of the stones.

She stopped, listening. Bare branches scraped together overhead. A light dusting of snow sprinkled down. In the forest, Somhairle moaned. She took a dead hare from where she'd secured it to her belt, and laid it at one end of the long pile of stones.

In the faint glow of the embers, she studied her work. It wasn't much. But it was dark. Somhairle's guts would be burning. His eyesight was poor.

Most importantly, he, like the rest, underestimated her.

She strained her ears, listening for him in the forest. Fear held her fast, its grip even colder than the night. But there would be no better chance. Tomorrow, they would cross into England. She listened to MacDougall's steady breathing, and finally, knowing she must move or lose her chance, edged her cloak, inch by inch, off her body, onto the pile of stones and moss, drawing the hood over the rabbit, whose soft fur might pass for hair, at first glimpse, when they pulled the hood back in the morning.

The winter night sank its chill into her bones. She thanked God she had been wearing her dark dress when MacDougall had accosted her in Glenmirril's hall. Not only was it her warmest, but it would serve better than any of her others, to hide her in the night. She rose to a crouch, barely breathing, and backed into the forest, scanning the sleeping men, terrified one of them had been watching her all the while.

On the far side of the fire, one of the boys rolled suddenly, sat up, and spit into the trees on the far side of the clearing. She shrank, crouched, against a tree. Her eyes flew to her handiwork—yes, in the dark, it appeared she was there, sleeping soundly.

He didn't look her way, but laid back down, yanking his cloak over his shoulder. Already shivering, she backed up another, crouching step, and another, and finally stood, listening for Somhairle. To her right, she heard him, grunting in pain. He'd be doubled over. She studied the forest, replaying the way in her mind. The moon shone above the trees in small patches of silver-white peering from behind occasional clouds, just enough light to keep her from stumbling, dark enough to hide her. She thanked God once more. She might have four or five hours start.

She slipped the reins of her pony from the tree, and set it loose. Please God, it would turn for Glenmirril, leaving snowy tracks for them to follow northwest. She turned south, praying.

Inverness, Present

"How's Angus?" Carol asked as they entered his home.

"Stable."

"I take it Amy's there?"

Shawn nodded. "She spends all her time there. Mairi will bring her home." He led the way to the kitchen, though he wanted nothing more than to collapse in bed. "Not as good as I usually make," he apologized, gesturing at the crockpot. "Lots going on here. Didn't have time to run out for my usual spices."

Carol waved her hand. "You're back, that's all I care about."

"Yeah. Well." Shawn slid a couple of plates on the table. "Don't make me blush, okay. I'm twenty-six years old." He grinned at her.

She frowned.

"I'm just kidding," he said, and when she didn't answer immediately, added, "What's wrong?"

"You're twenty-five," she said.

"Yeah." He didn't lose a beat. "That's what I meant. That's what I was kidding about."

She tilted her head. "I'm missing the humor. Usually your humor is witty."

Shawn set forks at each place, and brought the crockpot and serving spoon to the table. "I'm slipping," he said. "What can I say."

Carol served up dishes to herself and Shawn. "I got some news before I left. You might want to sit down."

Shawn hesitated, his hand in a cupboard reaching for a teacup. "That's the kind of news I especially *don't* want to sit down for."

"Do it anyway," she said.

With a grunt of annoyance, Shawn set her teacup in front of her, switched on the burner under the kettle, and sat before his plate steaming with casserole. "Okay, so?"

"How did your talk with Clarence go?"

Shawn took a bite of his meal, savoring the taste that brought his childhood back to him. It had been less than a week since his talk with Clarence, his foster brother, the murderer of his father, through prison glass; since forgiveness had washed over him. He cleared his throat. "I think you'd say it went well. I'm not sure I want to talk about that."

"You're usually very sure what you want and don't want," Carol said.

"I was being nice because you're my mother." Shawn grinned. "I don't want to talk about it. Are you going to eat?"

"I hope it went well enough." Carol's hands remained in her lap. "Clarence is being released."

The fork settled without a sound on the rim of Shawn's plate. His eyes met Carol's for a full thirty seconds, while a hundred thoughts chased one another through his mind. His fingers against Clarence's on the glass. *We're brothers again.* His father's snowy grave.

Clarence belonged behind bars, no matter how much he was forgiven.

He cleared his throat. No words came.

Carol spoke in crisp tones. "I know this has to be hard for you, no matter how well it went. I want you to know he's been entirely remorseful for years."

Shawn nodded, though his head buzzed. Clarence had been remorseful.

"He's done a lot of good with the ministries he set up. He's changed lives."

Shawn nodded again, his head light and fuzzy, unseeing. He had no idea what sort of ministries Clarence had set up.

"Shawn, say something."

"Yeah. Okay." He pushed his chair back, made his way blindly to the front door, let himself out into the street, and started walking under the black night sky. Snow fell, sticking in his hair. His mind spun, wheels in a snow drift, unable to make sense of any of the thoughts that leaped one after another through his head. Except—it felt wrong. The boy belonged in jail.

He walked faster, head bowed against the snow, his hands pushed deep in the pockets of his breeks, his leather boots landing silently in the snow. The boy had killed his father. *He belonged in jail.* A shiver tore through his body, and he realized he'd left without a jacket. He stopped under a streetlight...turn back...keep going? No thoughts would come. A second shiver wracked his body. He wrapped his arms around himself, but his feet remained as immobile as his mind, refusing to work.

Something warm slid over his shoulders.

"Shawn, come back inside," his mother said.

He nodded, followed, pushing arms numbly into sleeves, snow thickening around them. Lights burned from houses along their path; he was surprised how far he'd gone. Carol didn't speak until she ushered him back into Angus's home, to the couch in front of his small fireplace, and put a blanket over his shoulders. She disappeared briefly and returned with a mug of hot tea, pushing it into his hands.

"I know we've been worlds apart on this." She dropped into Angus's chair. "But if your visit went well, I hope you can understand my feelings."

"That he should never be let loose?"

"He was *seventeen*," she pleaded.

"I knew stabbing people was wrong when I was seventeen."

"He was living in hell. He's been as sorry as a human being can be." She shook her head. "No, I'm worried about Clarence's future. How's he ever going to set his past behind him? Will people let him? Will he be able to get a job?"

Shawn raised his eyes to her, staring in disbelief. "You're *worried* about him?"

Carol didn't answer.

"You're worried about him," Shawn said.

"I had hoped," she said, "since you said it went well...."

"Yeah." Shawn gave a humorless chuckle. "It went well while he was locked up where he belongs. While there was *justice* to go with understanding his rough life." He set the tea mug down with a sharp crack on the coffee table, his voice rising. "Wasn't it enough I could have some compassion for him? Now I have to accept that his life just goes on while Dad gets to stay dead?"

Carol's mouth tightened.

"Don't you *care* that he's dead?" Shawn demanded. "How is this okay with you?"

She stared at the fire in the hearth. Her hands tightened around the mug. "I never stop missing him," she whispered. "But we have to deal with what *is,* not what we want."

"What *is*," he shouted, bursting from his seat, "is my father is *dead!*"

"What *is*," she countered rising to her feet, too, "is a man who has lived with remorse for an action that happened in the midst of his own trauma, who has done everything possible to make amends!"

He turned his back on her, hands on hips, staring out the dark window to the snowy night. "I don't want to talk about it."

"Well," she said, "you better either talk about it or tell me what your plan is for dealing with this."

"Plan?" He turned, staring at her blankly.

"Well, your drinking, gambling, womanizing, and disappearing act have cost you your job and the nicest girlfriend you ever had. So let's try something different this time."

"I haven't done any of that stuff in two and a half years," he said.

"You gambled away your trombone barely eighteen months ago." She sank back into the chair, her arms on her knees, leaning forward. "In the year you were gone, I Googled you often enough, looking for anything, and I was lucky enough to come across a YouTube video of you stumbling down the street drunk with Caroline."

"YouTube," Shawn said. "That's on *YouTube?* Where the hell do all these people come from with their cameras?" He shook his head.

"Another proud mother moment," Carol said tartly. "And it was not two years ago."

Shawn dropped back onto the couch, sipping the tea in lieu of answering.

"So you want to tell me where you were for a year?" she asked. "I've given you plenty of space, I figured you'd tell me, but I've waited long enough and I think you owe me some answers."

Shawn stared into the fire. "An old couple in the west of Scotland?"

Carol snorted. "One of many stories you've told. It's the most believable, but if it were true, you would have said so from the start."

He took another long drink, letting the tea thaw out his insides, before he set it down. "Okay, Mom, what has Amy told you?"

Carol shook her head. "Nothing I can make sense of. A ring from Robert the Bruce. You turned up in the tower of Glenmirril. None of it makes sense."

Shawn set the tea mug down, rose, and lifted the linen shirt made by Christina, revealing the scar.

Carol gasped.

"I was cut nearly in half by a sword."

"At the re-enactment? But why didn't the hospital call me? Why didn't Amy...?"

Shawn tugged at the leg of his breeks, pulling it from the leather boot, till he could lift the wide leg high, revealing the scar on his right leg. "From a wolf," he said. "It attacked us in the forest."

Carol's hand flew to her mouth, looking at the long red pucker. "Us, who?" she asked faintly.

Shawn yanked his left pant leg from his boot. "Me, Allene, Brother David."

"Who are they?" she asked.

He pulled the leg high enough to reveal the scar on his left leg, a clean white line compared to the first, thanks to modern medical care. "Compliments of Duncan MacDougall's knife."

Carol's second hand covered the first. "Didn't you report him?" she asked.

"Oh, I mentioned it to the police." Shawn let the legs of the breeks fall down over his boots, and plunked himself back on the couch. "Mainly, they were interested to know why I was carrying a sword covered in blood. I told them it was Duncan's, I told them I killed him, and I told them why."

Carol shook her head. "Shawn, this isn't funny. Why are you saying this?"

"You've heard the story I tell most often."

"You were in medieval Scotland."

"For two years. I can understand this is more than a little hard to accept, but put all the pieces together. It's why I keep forgetting how old I am. It's why Amy went to the monks of Monadhliath. It's why she has a ring from Robert the Bruce. It's why...."

Carol rose from the chair, shaking her head, and left the room.

Shawn sagged back against the couch. He stared at the white scar running across his palm from Allene's knife, wondering if he should have insisted on the old couple story or made up a new one. He missed her, he missed Niall, he missed Christina, he missed the Laird, Hugh, Brother David, Lachlan, Owen, Margaret. He missed Ronan and the widow Muirne and her six children. He missed the boy, Red, and hated being yet another man—another father of sorts —to disappear from his life. And somewhere in 1317, Beaumont was plotting evil against all these people he loved. And MacDougall was still alive, even if he was locked up in a cell.

And somewhere in his own time, the boy who had killed his father was about to leave his cell.

They were two of a kind, two killers. They both ought to be locked up.

His thoughts drifted back to his mother. No sound came from the kitchen. He pushed himself off the couch, and went down the hall. "Mom?" he asked.

She sat at the kitchen table, sipping tea. A bottle of whiskey stood on the counter.

Shawn looked from one to the other. Exhaustion hummed in his brain. "Please tell me you didn't unload too much of that in your tea."

Carol shook her head. "Don't take this as a compliment, but if anyone could drive me to drink, it would be you." She stared at the whiskey bottle, saying, "I considered it, but it would hardly help, would it?"

Shawn dropped into the chair opposite her. "Want me to go back to the old couple story?"

Carol nodded. "For tonight. Yes, for tonight, let's go with the old couple."

Scottish Borders, late December 1317

Beyond earshot of MacDougall's camp, with no sound of pursuit, Christina broke into a run, stumbling down a hill, for the glen below. Running burned away the worst of the fear, and warmed her, chilled as she was without her cloak. She prayed with every step, prayed in gratitude that these were gentle, rolling hills compared to the Highlands, prayed she wouldn't slip in the patches of snow or trip on the stones strewn across the hill, ready to ensnare her. She prayed she would reach forest cover before MacDougall's men thought to hunt south. She prayed they'd go north, searching the way back to Glenmirril. She prayed, as her breath came in short gasps, that if MacDougall caught her, he wouldn't kill her.

She prayed that if he did, he would do it swiftly.

She skidded on a loose pebble, cringing as it clattered downhill, striking others on its way—a raucous cacophony in the still night. She caught herself, slowing enough to guard against a twisted ankle, though every fiber of her being screamed to tear from MacDougall as fast as she could make her legs move.

Skirts gripped high, she reached the bottom of the hill, finding herself in a narrow glen. The moon had passed its crest. Surely if she kept it on her right, she would be traveling south. She tried desperately to remember anything from her visit here, years ago—but she'd been so young. She followed the glen, staying close to trees where she could run freely, but duck in their cover if need be.

A stitch burned her side. Heaviness dogged her legs. Fear and cold pushed her on, while she prayed God would stir some memory of the landscape from that long-ago journey. At the far end of the glen, a river gleamed in the moonlight. She stooped for water, cupping hands to scoop chilly water. A memory floated to her, of stooping by Annan Water, swirling her hand through, while her mother called for her to hurry, that they were near Lochmaben and their kin. She had mounted her palfrey, and they had followed the stream. At a sharp bend, they struck out southwest, and soon the castle had come into view across a loch, banners fluttering on the ramparts.

Christina looked back over her shoulder. No life showed in the forest behind her. Her legs burned with the need to rest. Her breaths came ragged and deep. Sewing altar cloths with Allene and Margaret had been poor preparation for this night. It was worse preparation for being caught by MacDougall or his men. She pushed herself up from the bank and forced herself along the river, glittering on her left. She slowed, her legs crying in pain. A flash of light caught her eye ahead. She stopped, wary, and saw it again. Moonlight glinted off metal. She backed into the trees, trying to quiet her quick, heavy breaths. They couldn't have discovered her missing so soon! Let alone gotten ahead of her! They *couldn't!*

Far down the river, a man appeared, wearing dark clothes, and carrying a sword. She drew back into the woods, trying to stay silent. He peered the length of the water, and edged into the cover of the trees. Christina pressed herself to a dark oak, listening for his every step, searching for a better place to hide.

"'Tis a dark night for a lady to be out alone," whispered a voice in her ear. A hand snaked around her wrist.

Inverness, Present

With his mother in bed and Amy home, Shawn pushed away the issue of Clarence—he'd done it well for many years, hadn't he? He should get back to Glenmirril. Help Clive. But the answer might be here. Exhaustion dogged him. He smoothed the parchments carefully on Angus's bed. His heart hammered. Either answer meant someone he loved was in danger. He picked up where he'd left off, scanning quickly through faded ink. It was only a few pages before he found it, dated December 1317.

To My Brother,

You will want to know. Simon Beaumont is not in your time, for he is in ours. Your Brother Eamonn...

Shawn scanned Niall's account of Eamonn's ruse and his concerns over what Beaumont would do. Amy was safe. James was safe—for the moment.

Niall, Allene, and Christina were not. Relief and fear danced a painful *pas de deux* in his stomach. The words danced before his exhausted vision. He stared out the dark window. There was nothing he could do for them. He closed his eyes—just for a minute—sinking into the muted sounds of Simon's feet pounding through Glenmirril's passages to Allene's and Christina's rooms. The pounding grew, a pounding on the door, a voice calling, "Shawn! Shawn? Clive is here. Shawn?"

Scottish Borders, December 1317

Christina froze, her heart pounding erratically.

The hand covered her mouth tightly. "Don't scream, Milady," whispered the voice. "We know not who else is about, aye? They mayn't all be as friendly as me."

She considered how friendly the owner of the hand might be; considered biting and fleeing. But she was near Lochmaben, she was sure of it. She had nowhere else to run—even if she could escape.

"You'll not scream?" He edged his hand away.

"No," she whispered.

"Aye, then." The hand lowered further, though he still gripped her tight against his body. "Who are you? More important, who do you flee?"

"You'll not help them?" Christina's voice shook, as much from fear as from cold. "I've done no wrong. They abducted me."

"I'm of a mind to believe a lady fleeing in the night has good reason," he whispered.

Another man slid into view around a tree. A brace of hares hung from his belt. A slingshot dangled from one hand. He studied her up and down. "Let her go," he said, and the first man released her. "Who are you?"

She hugged her arms close about her shivering body. "Christina Morrison of Glenmirril. I'm trying to reach my kin at Lochmaben."

The men looked at one another, silent conversation passing between them.

The first spoke. "You keep quiet about us, sure an' we'll help you, too."

Poachers, she thought. It mattered little to her. She nodded furiously.

"Take her there," said the second man. "My good-brother is on guard. He'll let her in the postern gate. Milady, say naught of us, and we'll say naught to any who come looking for you. We understand one another?"

She nodded, almost sagging in relief. He gripped her around the waist, catching her, even as the other man swung his cloak off and wrapped it around her shoulders. She huddled into its rich odor of earth and dead animals, grateful for warmth, as they helped her along the river, to safety.

Creagsmalan, Scotland's West Coast, 1317

Simon ached with the jolting gait of the old nag by the time he reached the next village. There, he entered an inn, where the men and the serving wenches stared doubtfully at his odd outfit. "I've come from afar," he barked at them. "On King's business! Have you no sense but to stand and gawk?" He turned on the innkeeper. "You! Get me clothes befitting a lord! And a better horse. The Bruce will compensate you." His eyes trailed over the two girls, who backed away, behind the man.

His hand on his knife, he was sure, and the implied threat to the girls—the innkeeper's daughters, no doubt—did more to influence the man's decision to jump, than the false promise of Bruce's gold.

Simon changed into more comfortable clothes, tucking the precious notebook, his key to power, securely inside his shirt. Keeping the *jacket* from Shawn's time under his new cloak, he was soon on his way.

The aches did not cease with the new horse, but grew worse, as the sun climbed in the cold December sky. Anger grew in him, as he understood. His months in Shawn's time had left him weak and soft.

He pushed himself on to an inn near Invergarry, taking note of the serving girl who brought him ale. The monk's words flashed through his mind. *Your son will be your undoing.* There would be no son, then. He would see to that. Nonetheless, he fell into a fitful sleep.

In the gray hours before dawn, he helped himself to a better steed from the inn's stable and set out under sprinkles of white flakes drifting down from the dark sky. He pushed the animal as hard as he dared through wooded trails alternately icy, snowy, and muddy, seeking food and news of MacDougall. He was in England, men told him. His wife, still at his son's home in Creagsmalan, might tell him where.

It was a day and a half, and some horses later that he rode down into the windswept glen, past a few rough huts, into the gray-stone town of Dundolam, up its cobbled streets to Creagsmalan.

MacDougall's wife sat in her son's high-backed chair in the great hall, waiting for him, a slender woman with her chin lifted high, steel-gray hair showing under her barbette, and smooth, taut skin, but for the deep lines etched around the corners of her mouth, and sad eyes. "Beaumont, Lord of Claverock," she said. "You disappeared after Bannockburn. Where have you been these three years and more?"

"Three...years?" Simon asked faintly.

"Are you unaware it is the Year of our Lord 1317?" She stared down her thin nose at him.

"Of course," Simon replied. "It is but that time has gone swiftly."

"And why do you come to me?" She raised her chin.

"I am recently escaped from Glenmirril."

"Is that so?" Her eyes narrowed only the smallest bit.

He smiled slowly. He had her. "I believe we both have accounts to settle with them."

"Indeed." She raised a hand, summoning a servant. "Bring food. My Lord of Claverock, join me." She rose, lifting the hem of her skirt, and descended the stair, to sit at a table by the great hearth. "I told Alexander to stop this war with them. I hold him equally to account for my son's death."

"And yet, they *did* kill Duncan," Simon said.

"Were you there during the battle?" she asked. "Surely you've not been held at Glenmirril all this time?"

Simon scoffed. "MacDonald could not hold one such as me."

Her eyes narrowed. "He is perhaps more wily than you—or Alexander— ever credited him. Where then have you been?"

"An amazing journey, my Lady," he replied. "One that has left me with a mission, and the need to meet with your esteemed husband."

"He is none so esteemed in Scotland these days," she said.

"That is to his credit, no doubt," Simon answered. "A loyal servant to King Edward."

"What do you wish with him?" she asked. "Do not be fooled into thinking I didn't notice you have not said where you have been these years past."

"You are wise. The story exceeds that of King Herla and Thomas the Rhymer together." Simon inclined his head in a show of respect. "It is, in fact, this wondrous adventure that has left me with knowledge of Niall Campbell of Glenmirril. I believe your husband and I share similar goals, regarding him."

Her chin rose higher. The pain slipped from her eyes, into anger, as the muscles in her neck tightened. "Alexander's impetuousness aside, Niall Campbell *did* kill my son. If your mission concerns his destruction, I shall gladly direct you to my husband."

Simon smiled. "My Lady, I will indeed destroy Niall Campbell. The more you can tell me about him, the more we might work together to give him his most suitable reward."

She smiled, a tensing of the lips that suggested she was not accustomed to doing so. "Alexander has taken refuge at Linstock Castle. Sit down, and I'll tell you all I know of Niall Campbell."

Inverness, Present

Roused from deep sleep, Shawn's eyes move blearily between me, holding James on my hip, and Clive, as he ushers Clive into Angus's front room, gray in the hours before dawn, and offers Angus's couch. I don't miss the tragic irony, and by the look of embarrassment on Shawn's face, neither does he. Clive's clothes are rumpled. He's been up the better part of three days.

Shawn is equally rumpled, having come in only hours ago, and slept fitfully, jarring awake now and again to sit up, to slap his hand on the nightstand on his knives, check that James sleeps safe in his cot, and sink again into sleep, before rising for his run to the airport.

"Nothing," Clive says. He looks from Shawn to me, and back again. "Chief says...."

"I got the scrolls," Shawn interrupts.

Tension grips my body. I've been walking the small downstairs restlessly with James, waiting, unwilling to wake him. "And?"

"He's there. Then." He sinks onto the couch, arms on knees, staring between clasped hands to the floor.

"That's grand then!" Clive slaps his hands together, grinning.

Shawn lifts his head to meet my eyes.

"Is this not good news?" Clive glances between us, perplexed at Shawn's response. "He's gone where he can do Amy and James no harm."

On my hip, James pats my face, and says, "Ma...ma?" He leans suddenly into me, snuggling into the warmth of my neck. I put my hand on Shawn's shoulder, feeling the melancholy melody flowing beneath the silent bow.

"Oh." Clive stares at his shoes for a moment, then raises his head. "Angus needs her. He needs the will to live."

Shawn is silent for a long time. I squeeze his shoulder. "Shawn, you know I'm staying."

"He needs her," Clive repeats.

"He got you out," I remind Shawn. "He did more than anyone to get you home."

Shawn lifts his eyes to me. He reaches for James's hand, and he kisses the pink skin. "Of course," he says. "He deserves that. But don't ask me to leave, either."

Scottish Highlands, December 1317

Fury scarred MacDougall's features, as the rising sun glanced off the pile of stones. Christina's blue cloak dangled from his hand. He lifted his eyes to Somhairle. "How does a woman build up a pile of stones and slip away under your very nose?"

Somhairle stared at the ground, swallowing convulsively. "I'd my business to do, milord."

MacDougall snatched up the dead rabbit, shaking it under the old man's nose. "A hare?" he shouted. "You mistook a *dead rabbit* for a *woman*?" He flung the animal, its body slamming into a tree and sliding to the snowy ground. A crow lifted, cawing, and flew away in fright.

"I'd no choice, my lord. I tried to keep a close listen."

MacDougall wrenched his sword from its sheath and slapped the man across the back with its flat. The old man fell to his knees, cowering. The others stood, waiting for orders. Even the birds had fallen silent, fleeing MacDougall's wrath. "She can't have gone far," he shouted. "Will we all stand about staring at stones?"

Roger stepped forward. "My lord, she'll have returned to Glenmirril. They'll be well aware she's been taken by now, and mayhap on our heels already. We cannot turn back."

MacDougall glared at Somhairle, whimpering on his knees before him. He recognized the truth of Roger's words.

"It's Niall we want," Roger reminded him.

"She was going to help accomplish that." Macdougall bit out the words in helpless fury.

"We can still do it," Roger said. "But we don't wish to meet MacDonald's army now, do we? 'Tis not part of the plan, my lord."

MacDougall stared into the northern glen, stroking his beard, and wondering how far she'd gotten. He had hoped Christina would want to stay with him. If he was kind. If he gave her all she wanted. In time, she would have come to see his generous and compassionate nature, and she would have loved him. Disappointment consumed him.

But Roger was right. Niall was the real goal.

CHAPTER SIX

Inverness, Present

Bright light woke Shawn. Silence jarring him from sleep, he shot to his feet in the sunny room, clawing at the comforter that tangled his legs and nearly tripped him, as he looked frantically around the empty room, at the sheets neat on Amy's side of the bed, the door wide open. He snatched two knives off the nightstand, shouting, "Amy?"

Silence greeted him. Clive—Clive had been there. He tried to still his heart. Simon was gone. But his mind leapt with images of Beaumont hauling her and James and his mother away. It was time travel. Maybe it was a *future* year in which Simon disappeared from the twenty-first century. He bolted into the hall, his heart pounding.

Simon was gone, he told himself in the scant minutes it took to search the empty bedrooms and the front room downstairs. Niall had described the same event they, too, had experienced. In the kitchen—likewise empty—he found a sheet of paper on the table. He picked it up, his heart slowing.

Shawn, said Amy's handwriting, *Took your mom to the airport, going to hospital with James. Figured you needed the sleep. Your mom says good-bye and don't forget your promise. Eggs in fridge.*

It was then he noticed the smell of coffee. He opened the refrigerator to find eggs on a blue stoneware plate, scrambled with a smattering of cheese and green herbs, just the way he liked them. He smiled as he slid them into the microwave, enjoying the soft sizzle and pop as he poured the fresh coffee she'd made him. *She still loved him!* He was sure of it.

He read her words, again, thinking of the notes he'd once left her, in their lives together so long ago, and of the symbol he'd left carved on a hundred altar stairs for her to find. He picked up the pen she'd left, and wrote, *I'm glad to be back.* His hand hesitated. The words meant more now than they ever had. *I love you,* he added. He sketched the flattened S, a symbolic pun that doubled as a trombone. His eyes stuck on the symbol, thinking of all the times he'd slid notes under her door, or left them in her violin case.

With those warm recollections came the harsh memory, a needle across a

Barry White vinyl, of Dana, of Caroline; telling Amy the electricity had gone out at the liquor store—*yes again! Yes, hard to believe, but wasn't his luck terrible lately!*

He bowed his head, staring at the flattened S. It was supposed to be all good. He was home. Simon was gone. But the flood gates had opened. He thought of Clarence, in a prison across an ocean. *We're brothers again.*

He had felt, in a flash that day, as if he understood Clarence's life. He and Clarence weren't so different, despite the vast gulf in their circumstances, despite the fact Shawn was hailed as Idol and Clarence despised as Murderer.

The microwave beeped.

In truth, Shawn thought, Clarence had been a terrified, traumatized kid, acting in a flash of pain, panic, and fear. He, Shawn—what excuse did *he* have? He'd been an adult, with years to sort himself out, deliberately plotting betrayal against Amy, bullying others to cover for him, lying to her when she'd trusted —and loved—him; even making her doubt her own senses.

He opened the microwave. He stared at the eggs on the plate. Heat prickled his eyes, a truth welling up in him that he couldn't deny. He blinked angrily. He'd fought with James Douglas! He sucked in a great breath, and spun to the table; snatched up the pen, scratching words with staccato, *forte furioso. I'm sorry, Amy. I would do anything for you, anything at all, to make it up to you. Anything.* He scratched three harsh lines under *anything*. They looked like the three lines he'd scratched on the many altar steps, trying to recreate the S in the midst of battle and raiding, flames and women screaming.

Hadn't Amy's soul screamed, too? Cried in the same pain. *There are other ways of killing people,* she'd said. And he saw, with new clarity, how right she was. It was murder of it's own sort—of the heart, the psyche, the very soul.

He shoved himself away from the table, eggs uneaten, coffee cold. He'd made a commitment to be better, and there was nothing he could do about the past. He went up to shower in Angus's bathroom, vowing he would do whatever he needed to, to give her happiness in the future. *Whatever she needed.* He rinsed the conditioner from his hair, swiped body wash over the multiple scars across his waist and down his legs, and climbed from the claw-footed tub, toweling off. He could start by being there for her at the hospital. *Whatever it took.*

Dressed in his *trews* and *leine,* he headed down to put the eggs back in the refrigerator and wash his coffee mug—she wouldn't have to deal with even so small a thing when she came home—and headed out, through the streets of Inverness to find a bus to the hospital.

Lochmaben, Scottish Borders, 1317

Christina dropped to one knee before the lord of Lochmaben. She'd bathed that morning and clothed herself in a fresh gown brought by Joan, the lady of the castle. The poachers had brought her, last night, to the postern gate. There, one of them had lifted cupped hands to his lips, giving the hoot of some bird. A burly guard had appeared, whispering with him before leading Christina in and rousing a servant girl to find her a room.

"Rise, Christina Morrison." The old man's voice rumbled, sounding twenty years younger than his lined face suggested. Beside her, Joan helped her rise. They both looked to Lochmaben, waiting. "I remember your mother as a young girl," he said. "'Twas tragedy, her passing. And your father's. Is it back to Glenmirril you wish to go?"

"Aye, my lord."

"My men will escort you. And MacDougall. There's naught we can do about him. He'll go to England and that will be the end of that."

Christina inclined her head. "Yes, my lord." He had no understanding of MacDougall's animosity. There would be no end to it until he died. But then, it wasn't Lochmaben's problem. She couldn't blame him. "I am grateful for your help and your protection." With a deep curtsy, she allowed Joan to turn her back toward her chambers.

"Parchment?" Joan asked. "We can send a messenger ahead to let them know you're on your way."

"Aye, please," Christina said. Cool air brushed her cheek as they stepped out of the great hall, onto the stairs leading down to the courtyard. Joan led her past a stone fountain with gargoyles and mermaids, and water where fish would swim in warmer months. From a window above came the soft lilt of a lute and delicate melody of a harp. Pain snagged at Christina's heart, thinking of Shawn in the window sill, playing Niall's harp, cocking an eyebrow at her with that grin that caught her breath every time.

She longed to stoop and trail her fingers through the pool, to stay and listen to the music, but Joan led her on. She stopped briefly to speak to a boy, and soon Christina was back in her room, with a sheaf of parchments and bottle of ink waiting on a tall desk in a splash of winter sunlight under the window.

Christina seated herself, quill in hand, undipped, listening. The smell of ink and vellum came to her, and the music drifting across the courtyard carried her back to so many days in Niall's solar, with Shawn in the window ledge, his favorite place, the sun catching glints of gold in his hair, plucking out melodies on a lute, or dancing around with a recorder, playing something he called *jazz*. *Take the A Train*. She smiled, still trying to match the light and happy sound of the song with the smoke-belching, roaring behemoths Niall had told her of, or the dangerous streets Shawn had described in the town called *Harlem*. These months—nearly years now—hadn't lessened the way she missed him, though she'd gotten better at refusing to think of him, except to pray—for him, for Amy, for their son, for herself.

She dipped the quill, smoothing a heavy bead of ink against the side of the clay pot, and gave in to the desire that had not lessened despite her attempts.

My dearest Shawn, I'll not put this one in the kneeler. I wonder if anything we leave there will survive to your day. I pray so. We do all we can to protect them and you know where they'll be. I imagine you finding them one day. Maybe two years ago, when you made the crossing, you went straight down and knew immediately all it has taken us these many months and years of living to witness. Perhaps you know already what becomes of us, if MacDougall will finally kill all of us—things we can only fear at the moment. It is hard to comprehend.

Maybe it took you—will take you?—longer to go down. Maybe it was not possible to get to the cave, or it fell in or was locked or perhaps you were nervous about knowing or simply wanted naught more to do with us, but only to move on with your life.

The quill ran dry. She dipped it, and rested it on the edge of the pot. It was both blessing and curse, having Niall ever before her. Shawn's face could not recede into the past. Or, in his case, she thought wryly, the future. She wrote hastily: *I miss you.*

She stared at the words. They did but little to convey the overwhelming emptiness of his absence. MacDougall, she didn't mind so much. The questions about the child and whispers of her immorality, she could hold her head high against. *God* knew her every act had stemmed from trying to save a man's life —and her own.

But each day brought the same desire to see Shawn again, to watch him play the sackbut in the Bat Cave, laughing in joy at the music; or watch him play chess with Hugh or listen to him argue with Brother David while she and Allene sewed; to watch him cradling James in his arms, warming her heart. The memory of his lips on her fingertips or the look in his eyes that day never left her. She wanted nothing more than to see him hold James Angus, teach him to ride one day, to....

She must stop! It was exactly all of that which she had no desire to convey to him, seven hundred years hence. He had wanted to go back to Amy. To redeem himself. She *wanted* him to have his deepest desire.

She stared at the words. *I miss you.* The letter wasn't for Shawn. It was for herself. She would tear it up, cast the pieces in the fire. He would never see it.

Life feels empty without you. She wrote suddenly, in a rush, the words and flowing. She paused finally, stared out the window several minutes, and added: *I knew it could never be, that you must needs return to your own world, and yet, I miss you, I miss our time together, so deeply, every day, and it never gets better.*

I will say here, in this letter which I'll tear up, which will never come between you and Amy in any way, that I love you in a way I did not previously understand love, and I pray for you and her and your son each day, and that you've found the peace you sought, and happiness you wanted with your family.

I had a son. I called him James Angus. I pray your James gives you the same joy mine gives me.

She hesitated. But the letter was only for herself. She wrote, *I desire peace for you. I want your peace and happiness above all and I know you will have that only in the redemption you sought.*

She considered a moment before telling him. *I have committed myself to marrying Hugh when I return to Glenmirril.* If the Laird would yet allow it, she thought.

She re-read the words. They seemed a line from a romance dropped into a tale of adventure—words from the wrong book.

It feels wrong, and yet I've the greatest faith that somehow, all will be well. With the greatest love, Christina

She sanded the letter, shaking the fine grains back into the bag of sand. She addressed it, sealed it with a shimmering ruby pearl of wax, and slid it into her pocket. Outside, children hollered and laughed, and a woman reprimanded them. The harp and lute stopped briefly, and in the moments she listened, started up again, with a lively melody, perhaps a strathspey. Shawn would have been in his element, playing such a thing. She smiled.

She dipped the quill and set to writing the letter for Glenmirril, telling them she was 'okay,' as Shawn would say, and on her way home.

Inverness, Present

From the hallway, Shawn heard her.

"What were you thinking?" she demanded.

He stopped in the doorway of Angus's room, grateful he hadn't been recognized in the hall, and grateful Amy didn't turn around. Her hair fell down her back in thick, black waves, grown beyond its usual waist length.

She blocked his view of Angus, apart from his legs under the white sheets. He said something Shawn couldn't hear, before Amy spoke again, in agitation. "You risked leaving me, your parents, your brother and sister and nephews, the police force, everyone who loves you and cares about you!"

Angus's words came to Shawn clearly. "You thought I could walk away and leave a boy to die there?"

Amy's back bowed. She spoke words that stayed only with Angus.

A mumble rose from the bed. Shawn watched Amy's shoulders tense. She shook her head, setting her hair swishing across her back in what he knew to be anger. "How can you say that! I *did* love you! I *do* love you!"

A blow to the chest: Shawn's breath caught. But—the eggs? Sleeping in his arms...? Angus spoke more words Shawn couldn't hear.

"I love him, too. How am I supposed to make either of those feelings just go away?"

"I told you to make your decision," Angus said clearly.

"How was I supposed to make that decision!" she demanded. "You walked away. Again. It's what you always do! I'm sorry I was torn, but *you* sent me away. And this...." There was a quick sniff. She leaned over the bed. "This brought it home to me. I want to be here, with *you*."

Shawn backed up a step. Amy tensed. She made a sound of fury, and straightened. Shawn spun and hurried down the hall, listening for her footsteps behind him. They didn't come. He reached the hospital's cafeteria, scanned it for something to pretend he'd been there all along, and bought an apple.

"Mr. Kleiner?"

He spun guiltily, a smile on his face. "Great apples here!" he said. The woman behind him lifted a hand to push the mop of dark curls off her forehead. Shawn's eyes fell on the boy giggling in her arms. The smile slipped from his face. "My son!"

"Aye." She reached awkwardly to offer her hand, still clasping James. "Mairi...Angus's sister. The other night at Glenmirril...."

"Yeah...yeah, of course." Shawn's mind still thrashed with Amy's words. She loved Angus? Still? No—that's not how the story was supposed to end!

"James was restless while she.... And it means so much...." She stopped. "I mean, never mind. Did you want your son, then?"

Shawn reached for James. The apple made it awkward. With a little difficulty, he and Mairi exchanged fruit and boy. James grinned at him, and Shawn's face lit up in return. "How've you been?" he asked. "You didn't give your mom too much trouble?"

He's an angel, he is." Mairi laughed. "Even has the angel wing on his shoulder to prove it." Her laugh immediately died. "Have ye come for Amy?"

"Come for her?" Shawn blinked. "I don't know." He wasn't sure what he'd intended. Only to see her safe.

At that moment, Amy herself appeared in the doorway, looking flustered. She glanced around, saw Mairi, and swung toward her, smiling. Then she saw him. She stopped. "Shawn!"

"Amy." Her words to Angus stung his heart like drops of acid. He couldn't let her know he'd heard. He struggled for kind words. "How is he?"

Shock crossed her face, and he realized she hadn't expected him to care.

Mairi, too, stared at him in surprise.

"Hey," he protested, "Am I green?" Had the man not taken a large chunk of Amy's heart, Shawn had to admit, he might like him. Even admire him. He didn't want him caught here in a hospital bed, facing life in a wheelchair. "Do I have horns sticking out of my head? I'm not a complete ogre, you know."

Mairi shut her mouth, and Amy said, "Nobody thought so, it's just...."

"Yeah," Shawn said. "It's just. I just thought, well, maybe you need—help. Or company on the drive home. I don't know. Or I'll leave now and put dinner in the crockpot for you. What do you want?" Damn this improv. He'd gotten too good at it. But he was committed now.

"Dinner would be good. There are noodles and spaghetti sauce in the cupboard. You can take my car." She fished in her purse, and handed over the keys with a jingle. "You'll come back and pick me up?"

"Just call." He glanced at Mairi. It could only look, to her, as if he'd come to take away someone who might comfortAngus. "Uh, don't rush, okay. Stay as long as you like. When you want to come home, I'll just..." He glanced again at Mairi, conscious of speaking for her benefit. "...go to a hotel or something." He turned to go.

"Your apple," Mairi called.

"Uh, yeah." He turned back, shifting James to one arm to take it, trying to hold the apple and Amy's keys.

"The diaper bag." Amy saw it at the table where Mairi had been sitting. She retrieved it and draped it over a shoulder for him.

"Anything else?" he asked wryly.

Mairi bit back a smile. "Perhaps you could help me with my packages?"

"Do you need a ride home, Mairi?" Amy asked, and Shawn realized he'd just been volunteered.

"Aye, that'd be lovely," she replied. "The boys will be wanting their own dinner in a wee bit."

Soon, he had James settled in the car seat in Amy's car. Red, like his own, he thought. It didn't mean anything. She'd been clear, with Angus, about where she stood.

"Ye look knackered." Mairi slid into the passenger seat on his left.

"Yeah." He gave a small laugh. "Yeah, I'm knackered. Which way?"

But for her directions, an occasional babble from James in the back seat, and Mairi replying to him as she reached over the seats to hold his hand, they rode in silence. As he pulled into her driveway, she let go of James's hand, and said, "He's a good man, ye ken."

Shawn studied her face, her dark eyes intense under the flop of curls. "What can I say?" he finally asked. "She'll go to him in the end."

Mairi shook her head. "Angus is a very humble man, yet his pride will kill him. He's sent her away."

Shawn frowned, trying to fit that information with what he'd seen.

"He thinks he's no longer fit for her, in a wheelchair," Mairi explained.

"Amy's not like that," Shawn said softly. "She'd be the last to see that as changing anything."

"He's stubborn, he is. I'm not sure she'll convince him of that." She reached for the door and climbed out of the car.

Northern England, December 1317

MacDougall surveyed the gathering in his chambers at Linstock. Edward Carnarvon's deputy strode in, glancing from MacDougall to Roger. "His majesty is of course willing to welcome you into his household and perhaps in future grant you lands. He is first, however, interested in your promise that you can deliver another traitor."

MacDougall nodded. "Niall Campbell of Glenmirril. He was on numerous raids in Northumbria. He is even now with Douglas and Bruce besieging Berwick."

"They've tried three times before." The deputy tossed his gloves on the table and stretched his hands out to the fire. "They haven't the equipment or force to take Berwick."

MacDougall gave a slight bow, acknowledging the truth of the man's words. "Nonetheless, he's worked against Edward for years. He is highly favored by James Douglas and the Bruce both."

"There are many," added Roger, "who would be happy to see him dragged through their towns on his way to London."

"Aye, and a good execution would certainly raise spirits amidst our troubles. Still, is he worth the lands you're asking?"

MacDougall held the man's eyes a moment. "I believe Douglas himself will come to rescue Campbell. What is it worth to his Majesty to bring the Black Douglas to London in chains?"

The deputy smiled. "I think you'd find a suitable reward in that case. The agreement starts with Campbell."

MacDougall was silent a moment before saying, "I've heard my good-daughter is at Lochmaben. Has she plans to return to Glenmirril?"

"She is returning to Glenmirril under protection of Lochmaben's men."

"Do you know which route they'll take?"

The deputy shrugged, as he gathered up his gloves. "Send word when you have Campbell."

"My Lord." Roger stood as the man left the room. At MacDougall's raised eyebrows, he spoke. "She will travel with his best soldiers. Are we prepared to provoke Lochmaben's animosity and deepen MacDonald's?"

"They killed my son." MacDougall's lips tightened.

"My lord," Roger said, "we will seize Niall Campbell. I say only, let us think carefully on the best way to do so."

Inverness, Present

At the slam of the door, Shawn sat for a moment. Did Amy's words matter, if Angus had sent her away? He turned to James. "Just you and me now, huh? I'll show you a few things about cooking, how 'bout that? Spaghetti's easy."

He drove the few blocks to Angus's house, and lifted James out into the biting wind. It felt strange to be walking, once more, into another man's house, a house where his girlfriend was staying, making dinner with someone else's pots and utensils, someone who had not invited him. He hoped Angus would be okay with it. It was for Amy, after all.

With James still cheerful, he set the boy down on a blanket in the front room and went to the kitchen. His eyes fell on the note he'd left her. *I'm sorry, Amy. I would do anything for you, anything at all, to make it up to you. Anything.*

It seemed like a cruel joke now, his rush of emotion for someone who wasn't even interested in him. He picked it up, reading the words again. Had he meant it, he asked himself. Or did it just feel good to say? To sway her and move her?

He drew a sharp breath. No, he'd meant it. So he would do what she needed. Right now, that meant dinner. It didn't take much looking to find the noodles, and sauce. He got the meal going, added a few spices he found in the cupboards—it appeared Angus liked to cook, too, judging by the variety—and went back to sit on Angus's couch. There was nothing to do. No piano, no score paper. James, standing against the couch, flashed his little teeth in a big grin, and gave a shout.

Shawn reached out his hands. "You want a story?" he asked. "What do you like to do, now that you're big enough to stand up but not big enough to play ball?"

James grinned, gave a hiccup, and pounded his hands on the couch.

"A story it is," Shawn answered. He looked around the living room, thought there might be a book or two in the diaper bag, and as he squatted down next to it, his eyes fell on the bookshelf. Colorful books and toys filled the bottom shelf. He pulled one out—a child's book. He flipped it open, and saw the handwriting in the front. *To James, love Angus.* One after another, he pulled them out. Although only some had inscriptions, it was clear Angus had bought a number of the books, and, Shawn thought, most likely the toys, for James. He picked them up one after another. A wooden train in primary colors. A toy violin. A stuffed rabbit. "Ba-ba!" James launched himself onto his knees, and scrabbled across the two feet. He grabbed the rabbit and stuffed its ear in his mouth, smiling around it at Shawn. He lowered it and said again, "Ba-ba!"

"Is that your rabbit?" Shawn asked. "Does he have a name?"

"Ba-ba!" James chewed in delight on the ear, watching Shawn.

"Did Angus give you this?" Shawn asked.

"Aig!" James shouted.

Shawn studied the shelf full of books and toys, and, knowing he shouldn't, said, "Stay there a minute." He climbed the stairs. Just a quick look, he told himself. He wouldn't touch drawers or open the closet.

A quick look was all it took to see what he hadn't noticed, in the exhausted days of searching Glenmirril. A framed picture of Amy sat far back on the neat and tidy dresser. A picture of James stuck out from the mirror frame. Despite his promise, he inched a drawer open. A blue baby blanket was folded inside it. A card lay on top of it. Shawn leaned closer, telling himself it was wrong. But it looked very like something Amy would choose. He opened it. The handwriting, and the words *I love you. Amy,* confirmed it.

He turned his eyes, his heart pounding, to the bed. What *had* happened there? What did she mean by *love?* How much did she *love* him?

He slapped the card closed—another blow to the chest, another shortness of breath, *I love you* burning in his mind—and turned blindly, down the stairs.

Snatching a book from the shelf, he scooped James up and sat on the couch with him, reading words that didn't touch his brain. He growled out the voice of the king, as his father would have done, and cackled the voice of the witch, wondering how often Angus had read this book to his son. How had he fooled himself into thinking there hadn't really been anything between them?

James seemed enthralled. He gripped the edges of the pages, his hands small besides Shawn's, and listened intently. And when Shawn reached the end, James looked up with wide eyes, pounded the book, and said, "Aig!" Shawn read it again. And again, until his cell phone jangled in his pocket. James squealed, and looked up at him while he dug it out. It was Amy.

"Hey," Shawn said, answering it. "You ready?"

I love you. Amy burned on his brain.

James climbed to his feet on the couch, holding Shawn's shoulder, and reached for the phone. "Aig!" Shawn tilted it so James could hear Amy's voice.

"Whenever you can get here," she said.

James listened intently.

"Sure, I'll leave right now." She couldn't really be in love with him!

"You've been okay with him?" she asked.

James put his hand on the phone and laid his cheek against Shawn's.

"Of course I have," he said indignantly. "Dinner's on. We found books and toys, and we've been reading something about a king and a witch."

"Yeah, he likes that one a lot. Angus read it to him all the time."

Another blow to the chest. Shawn felt his air coming in short, sharp breaths. *I love you. Amy.* They were just words, just what people wrote in cards. They didn't matter.

"Aig," James said.

Shawn stared at him. "Okay, I'm heading out, Amy. Bye." He pushed *end.* Dates and numbers spooled through his head, adding up the months and days James had spent with Angus.

Of course her words mattered. He'd come back for her and she was in love —really in love—with someone else. James patted the phone. "Aig!"

Lochmaben Castle, Scottish Borders, December 1317

Relief and doubt flooded Christina, soft as the pastels of the rising sun, as she mounted her horse. Hugh was a good man. She should be ever so grateful. But it only felt wrong. In the courtyard, mist swirled around hooves and ankles.

"You are afraid, my lady." Holding the reins of Christina's horse, Lochmaben scanned the fifty men who would escort her to Glenmirril. "MacDougall will not dare to attack you under my protection."

"You have been most generous, my lord," Christina replied.

Two young matrons rode up beside her, guardians of a different sort.

The Lady Joan walked with them, reaching to press her hand in Christina's. "A man has gone ahead to arrange accommodations," she said.

Christina squeezed her hand in return. "I thank you, Joan." She wanted more than anything to be back with James Angus. A shout sounded at the front of the line and with the familiar creak of leather and jingle of tack, Lochmaben's men started moving. Christina let her hand slip from Joan's. Lochmaben took it, touching his lips to her fingertips. "God go with you."

"My deepest thanks, my lord." Christina touched her heels to her mare, and fell into line in the midst of the soldiers.

They passed out under the gate house, into the growing light of day, and she settled in for her long ride, grateful to be going home. She touched the new cloak they had given her—blue like the one she'd left behind, and lined with fur—feeling the letter to Shawn in its inner pocket. The emotions of writing it rushed back. She drew a deep breath, letting it out in a frosty puff that hung in the air, and lifted her eyes to the skeletal arms of trees, reaching to the pearlescent dawn.

She had been strong throughout the marriage to Duncan. She had protected Bessie as well as she dared and passed the message to Niall, in his disguise as Fionn, asking him to get her out. She had been calm, almost serene, through the meeting with Shawn in the confessional and the brazen promise to lure MacDougall from the castle for the trick that would save Niall from hanging. She had done what she needed to do.

No, it was MacDougall who had shaken her serenity.

"My lady," spoke one of the matrons, "you look concerned."

"Lochmaben's are the hardiest of men and fiercest of fighters," spoke the other. "No harm shall come to ye with them at your side."

Christina scanned the dozens of archers and swordsmen surrounding her, sunlight flashing off their helmets. "No, it shall not," she agreed.

"The Bruce has banned MacDougall from Scotland," said the other. "He'll not come near."

"I have been fortunate," Christina replied. They fell into silence as their horses climbed the first rolling hill. She must count her blessings, she thought. She had taken great chances—so many things could have gone wrong with the daring rescue. Yet Bessie had found courage and gotten Niall out. By another miracle, Shawn had forced MacDougall to set her free, too.

Her heart swelled; a smile touched her lips. He had been thinking of her, even then. Her insides warmed at the knowledge. Her hand strayed again to the letter inside her cloak. She must destroy it as soon as she reached Glenmirril.

She had been fortunate in the Laird's acceptance of her child and their willingness to create a ruse that allowed her to claim her child *as* hers. She had been fortunate the ruse had gone well—though her heart had nearly stopped when she'd first looked down from the upper hall and seen Niall ride in through the gatehouse torn and bloodied from the fight with the boar.

Yes, God had been with her, she thought, her gratitude growing, in strewing the path with holly, in leaving old Somhairle as her guard, in granting a dark night to hide her flight, in the poachers He had put in her path. She might, even now, be captive in England with MacDougall. Instead, by the grace of her Lord, she was riding home to her son.

She would resume that serenity that had once felt so natural. She must stop mourning Shawn. She must once again accept life, as she had when married to Duncan, and do serenely what must be done. James Angus was now her life. She must stop this grief in which she had indulged for too long, and do what was right for him.

Around her, tack jingled. Ahead, a man laughed, and somewhere behind, two men spoke quietly. They would be at Glenmirril in a matter of days. She would apologize for flouting the Laird's authority. She would throw herself at the Laird's feet, begging his mercy. She would thank him for his generosity, and accede to his wishes. It was the right thing to do—in respect to the Laird, for herself, for Hugh, for James Angus, and yes, even for herself.

Hugh was not only a *good* man, she reasoned. He was tall and strong of stature and comely. He laughed with joy and would be the kindest and best of husbands. It should feel like opening a cage door, stepping out of an ashen world into fresh, clear air. So why, beyond the threshold, did she see only gray mist?

Claverock Castle, Northumbria, December 1317

Simon stared up with pride at the great stone walls of Claverock, at his banners snapping on its towers. His steward had kept the place up. It was good to see after the long ride, the men of MacDougall at his side. MacDougall's wife had given him all he'd asked. They'd driven hard, down into Northumbria, through sleet and damp snow, their horses splashing through icy streams, and sleeping short hours each night, as best they could, huddled close in squalid inns.

It was all worth it, Simon thought, as he looked on his home. He drew breath. This was his moment! "Open the gates!" he bellowed up at his towers. "Your Lord of Claverock is home!"

Men looked down from the walls above the gatehouse. One pointed and shouted. Two ran, and soon the portcullis creaked, lifting. His steward raced through, falling to one knee, as he cried, "My Lord! My Lord, is it really you? We thought you dead, my Lord!"

He led his master's horse across the drawbridge, and Simon swung his leg over, dropping to the ground. He landed awkwardly, sore from many days' ride. His steward reached to steady him. Simon glared, flinging his hand away.

"My apologies." The man bowed his head. "I've sent men to prepare your chambers. A feast is even now being laid. My Lord..." He lifted his eyes. "You disappeared at Bannockburn. More than three years hence. Where have you been?"

Simon strode toward his great hall, speaking as his steward fell into step beside him. "I was badly injured. Someone pulled me from the field, near dead, and spent many months nursing me back to health." He'd had plenty of time to weave a story, culled from the books of Amy's time. He told it now with relish.

Why Simon, an imagination! The mocking words of the horrid *professor* with snake hair came to him. Of *course* he had an imagination! He was Lord Claverock. He had anything he pleased.

"It was an old crone," he told his steward. "She asked one favor of me—to bring her a golden lamp from the lands of the Moors." As he climbed the stairs, he substituted a real country for the unnamed land in the tales he'd read. "I felt honor bound to do so, after she had saved my life. I encountered many difficulties on the way, but persevered."

The steward hurried to throw open the huge wooden doors, revealing Claverock's great hall. Fires blazed in all four hearths. His chair stood at one end, polished and well-cared for. The thick wall hangings that kept out winter chill were cleaned and in good repair. Sweet scents rose from the rush mats on the floor. The trestles were in place, and pairs of men carried long wooden boards to lay across them. The head table was already set up.

Simon nodded approvingly. "Well done." He watched the servants bustle, two women flinging out a long white cloth over the head table. They could not have done all this so quickly on word of his arrival. He could smell the venison and boar and sauces cooking. They were well into preparations for some great personage. This was his first test. He smiled. "King Edward dines with us tonight."

"My Lord!" The steward's jaw dropped. "How did you know? We received word ourselves but this morning. We understood his journey was to be made quietly."

Simon's smile grew. The old monk had told the truth—this time. "The woman to whom I brought the gold lamp. She granted me the gift of prophesy."

The man stared at him, his face going pale. "Prophesy?"

Simon rather liked this imagination thing. "That and perhaps...." He let his features turn cold. "Reading men's souls. I hope you've run the place well and honestly in my absence."

"Of course, my lord!" The steward sounded aghast.

"Send the chamberlains to draw a bath," Simon said. "I shall be here to greet our king on his arrival."

Inverness, Present

"Your mom must be well on her way." They were the first words Amy said on the drive from the hospital back to Angus's house. She'd been unresponsive to all his attempts at conversation.

"Mid-Atlantic, I'd guess," Shawn answered. "Maybe we could have lunch tomorrow? Or work on that album?"

She shook her head. "I'll be at the hospital." She fell silent again.

With the week's events, it didn't seem like the time to tell her about his visit to Clarence. He wasn't sure he wanted to, anyway. He wasn't sure where her words to Angus left him—but he was pretty sure he didn't like being there.

Seems a little unfair, he thought irritably at the god Niall trusted so implicitly. *I'm finally trying to do the right thing, and now you decide to slap me down over and over?* And yet, he thought, as he pulled up to Angus's home, a small piece of him rejoiced that Angus had told her to leave. That wasn't nice. Not exactly turning over a new leaf, was it?

He wanted to feel bad for her—he really did, he thought, as he climbed from the car and reached in the back for James. But they'd be a family now. She'd visit Angus a few times, see it was over, and come back to the States with him, and they'd have more happily ever afters than any princess in any medieval castle.

Not that princesses actually had happy ever afters, he thought irritably. Look at Bruce's daughter, spending most of her life locked in isolation, only to fall off a horse and die just when life was finally getting good.

At Angus's kitchen table, Amy twirled her spaghetti aimlessly on her fork, not meeting his eyes, while he juggled his own dinner, and James on his knee. "So what's the verdict?" he finally asked.

"Verdict?" She glanced at him and back to the orange-stained noodles twined in her fork. "Oh." She set her fork down. "Better than when he got there."

"Wheelchair?"

Amy nodded, still giving a great deal more attention to her plate than it deserved. James shrieked and pounded the table. She glanced at him with a weak smile.

Shawn looked down at James, on his knee, to see him grinning, his small teeth showing through his spaghetti-orange goatee. Shawn smiled, too, but he looked back to Amy. "How's he taking that?" he prodded.

Amy shrugged.

Shawn wondered why she didn't just tell him Angus had told her to leave. "A bit of a blow, I take it," he hinted.

"Mmhm. He'll be starting physical therapy. He might walk again."

"That's great." When she kept her eyes on her plate, and didn't answer, Shawn busied himself with spooning bits of spaghetti into James's mouth.

"Better than mashed carrots," he said. "But messier."

"Get! Get!" James announced, grabbing a handful of noodles.

"Yes, spaghetti," Shawn said. *Aig. Angus.* James hadn't said *daddy* yet.

He glanced at Amy, pushing a meatball aimlessly around her plate. He wanted to ask how often Angus had sat at this very table, holding James while they ate. But he knew: Enough for his son to be saying the man's name. He thought of the other James, Niall's son. For all the time Shawn had spent with him, he wouldn't be saying *Shawn.* It stung.

But he'd done some good for them, he consoled himself. Wasn't that what mattered? Not which child said whose name? He'd taken a step forward with Clarence, which would please Christina and Amy. Maybe, since she had nothing to say, it was a good time to tell her. She'd see he'd really changed—especially if, maybe, he left out that last conversation, with his mother, about Clarence—it wasn't like he ever had to *see* Clarence again—and things would get steadily better between them, now that she saw there was no future with Angus.

He cleared his throat. "So you don't want to talk about Angus. At least, you don't want to talk to *me* about Angus. I have something...."

Amy lifted her eyes. She looked miserable. "You haven't exactly been his biggest fan."

Despite his intentions, defensiveness kicked him in the gut. "I think it's easy to see why. But you know, I'm trying to look past that. I wanted to tell...."

"Look past *what*? That someone loved me better than you ever did?"

Shawn snorted, his thoughts about Clarence sliding away. "Past the fact that my girlfriend is in love with another guy. You know, it's a guy thing, I guess. Guys have a problem with that sort of thing."

"I'm not your girlfriend," she reminded him. "Not really."

"Not my girlfriend? Not really?" He stared at her in shock for just a moment before anger kicked in. "Then what have these past weeks been about? Iona. Spending all my free time here with you. Christmas shopping like a happy family?"

She shrugged. "It was about trying. Because I loved you. Part of me still does. We have a son together."

"So doesn't that make you my girlfriend?" he asked in disbelief.

She shrugged. "I don't know what it makes us."

He stared at her. "Either you are or you aren't."

"Okay, then I'm not."

The words stunned him. Angus had told her no. She'd take *no one* over him? His words came out more sharply than he intended. "I guess that was obvious by the way you took off from my mother's house without even saying good-bye." Wrath curled inside him, as it had the night he'd climbed to the tower, begging God to take him home. He'd done everything *right!* So why was everything going so *wrong*?

"You were just a little bit drunk," she snapped. She lowered her head, poked at the meatball, and added. "I guess my need to get here made it obvious to me, too."

"These last few nights...."

"We thought Simon could still be here."

He fell silent, wrestling with his own words. *I would do anything. Anything at all.* He closed his eyes, calming himself, before asking, "So when were you going to tell me—we're not actually anything?"

She shrugged. "You're not always easy to talk to, you know. When it's things you don't want to hear or believe."

"So it's a go with the two of you now?" He almost held his breath, wondering if she'd tell him the whole truth.

"Aig!" James said. "Get get!"

She didn't answer, but instead, asked a question. "Can you accept that he's a good man? Very good?"

He answered cautiously. "Yeah, I can see that. Did you get back together with him?"

"No."

James slapped his hand in the spaghetti, sending sprays of sauce spattering onto Shawn's *leine*. He jumped, catching himself on the edge of a curse, and softened the words to, "Look at this mess!" He brushed at his shirt, smearing the sauce in worse.

"It needs cold water." Amy rose, hovering, to help.

Still holding James, Shawn turned on the water, and, one-handed, damped a cloth and blotted at his clothes. "No?" he asked Amy. "You didn't get back together with him, but you're dumping me. You want to explain that? Why aren't you getting back together with him?"

"Because he's being an idiot," she snapped. "Is this any of your business?" She turned her back to him, her shoulders tense.

"Considering we—a we that includes me—have a son who would be spending time with him, yeah, sort of."

"I'm sorry," she mumbled. "It's been a long week. I'm tired. You didn't deserve all that."

On his hip, James kicked and let out another ear-piercing squeal. "You done eating?" Shawn asked him, and turned back to Amy.

He didn't want to consider anything from her position. He wanted to tell her about Clarence, he wanted everything to be right between them, he wanted all he'd come back for. But he wouldn't be feeling great, either, after an abduction, a murder attempt, and days at her bedside, worrying about her. He'd come back to do it *right* this time.

"Let me put James down." He touched her shoulder as he edged past her in the small kitchen into the front hall, but on second thought, called from the front room, "Amy, come on in here." He set James down and handed him the rabbit. "Is there a bottle of wine or anything?"

"Don't even start that." She appeared in the doorway, arms across her chest.

"Start what?" Then he understood. He shook his head vehemently. "No, Amy, no, this wasn't a come-on. You made it pretty clear where I stand. You've had a rough week. I'll light the fire, you put your feet up and have wine. I'll leave. I'll get a hotel room. I'll take James if you want, I'll leave him with you. Whatever you want."

"How are you going to get to a hotel?"

Shawn pushed a hand through his hair. "I don't know, Amy. You just informed me I've been dumped, and only because I bothered to ask, but I'm still trying to help. I can rent a car."

"At this time of night?"

"Then drive me to a hotel. Or I'll drive myself and come back in the morning to take you to the hospital."

Amy waved a hand. "Just stay here. I'm too tired to deal with this. There's a bed in the other room." She stared miserably at the empty fireplace.

Shawn crossed the room in two steps and knelt to start a fire. *Don't look for flint*, he reminded himself. "Matches?" he asked. He found himself wanting more and more to talk to her about what had happened in the prison, of the miracle of the pain and hatred slipping away. And of the new pain, of Clarence's release, of his confusion between the two.

"On the mantel."

When he'd gotten the fire going, found a bottle of wine, and brought her a glass, Shawn dropped into the recliner in front of the lace-curtained window. He closed his eyes, listening to the crackle of the fire, and James babbling half-words while he played, calling his rabbit *ba-ba*.

Eventually, James's sounds died away. Shawn opened his eyes to see him sound asleep on his stomach. Amy sat on the floor beside him, rubbing his back through a blanket. Her wine glass was half empty.

"Want to talk about it?" Shawn had to force the words out. "How's he being an idiot?"

Amy sighed, and looked up at him. He recognized the look. He'd seen it on her face every time she found a phone number or thought his story was off. It was a look that asked, *Can I trust you?*

His heart melted. She looked young and vulnerable and scared.

And very beautiful. He pushed the thought away. "Amy," he said softly, "consider me a friend, nothing more." He *wanted* her as a friend, someone he could talk to. "I can still listen, can't I? For the sake of what we had?"

"You've never been able to be just a friend." She twirled the wine glass between two fingers, watching the blood-red liquid swirl up the sides.

"The night I brought the tickets," he reminded her. "On the Ferris wheel. When we went skating. All the times we worked on the arrangements. Yes, we were friends, too." When she didn't answer, he asked again, "How's he being an idiot?"

She sighed. "He thinks he's no use now." Her hand stilled on James's back. "I get it. *I* wouldn't mind losing the use of my legs so much. I could still play violin and do everything that really matters to me. But to him, his legs are everything. He can't even play with the pipe band, if he can't walk again."

Shawn snorted. "Someone can push him down the street in a wheelchair. So what, he's telling you there's nothing left between the two of you?"

She didn't meet his eyes. A delicate pink blush stained her cheeks. He wanted to reach out and twine her hair around his hand. He remembered again how beautiful he'd always found her. He hadn't said so often enough when he had the chance. And now was not the time.

"He says there's no future for us." She sipped her wine, and a calm seemed to come over her. She spoke with certainty. "It's not what he wants."

Shawn watched her, his emotions doing battle as harsh as James Douglas against the Gascons. It was early days, the Gascon captain whispered in his ear. It was easy for her to say, now, that she was devoting herself to someone who wouldn't have her. But in a week...in a month or two....

He could almost see the stern shake of James Douglas's leonine head and hear the reprimand in James Douglas's thick brogue: *Ye said ye'd do anything. Anything a' tall.* He watched her rub James's back, her hair falling to her waist, thick and shining black. He didn't want to do anything, anything at all. He just wanted Amy.

Scottish Highlands, December 1317

They were two days into the ride, Niall fretting at the delay in seeking MacDougall, Red elated to be riding with the men and once more beyond Glenmirril's walls, Hugh silent, the Laird and Morrison grim, when a scout came racing back to them, spurring his garron. "Lochmaben ahead!" he shouted joyfully. "Milady Christina rides with them!"

Cheers rose from the men of Glenmirril. Relief washed through Niall—not only at the news of Christina's imminent arrival, but to hear the men cheer her safety. Their ruse with James Angus had not entirely silenced the whispers, though Margaret had been stalwart in looking down her nose at those who did so, reprimanding them and silencing them.

His thoughts flickered to a dinner with the orchestra in Shawn's time. *There's an interesting local saint,* one of the musicians had said. *Margaret Morrison.* His first memory of Margaret was the day she had dropped a dead fish down the back of his shirt. He wondered what her future held that led her to sainthood after such an inauspicious start, even as he appreciated what she was doing for Christina. He was grateful the rumors had not diminished the love of the people of Glenmirril for Christina.

"I see them!"

Hugh's roar smashed through Niall's thoughts, yanking him back to—the present. The past.

There, rising up over the hill, he saw Lochmaben's banner fluttering.

"Unfurl our colors!" the Laird shouted. The words were barely out of his mouth before Glenmirril's pennant snapped in the cold breeze.

A call sounded from the troops on the hill, and suddenly, a rider broke from their ranks, her blue cloak billowing behind her as her mare cantered down the hill. Hugh and Niall slid from their mounts as Christina's horse pranced to a stop before them. She threw herself from the animal, dropping to her knees before the Laird, still astride his garron.

She glanced at Hugh, and back to MacDonald, before bowing her head. "My lord," she said in a rush, "I have been ungrateful."

"Christina." Hugh laid his hand on her shoulder.

She looked up to him, and back to MacDonald. "My lord, I'd no right, ever, to tell you no." She turned her eyes back to Hugh and again to the Laird. "I am honored and blessed you have seen fit to give me such a good man as husband. Please, my lord, if I have not displeased you too much, I beg your forgiveness and wish to marry Hugh."

Hugh pulled her to her feet. Their eyes met. "Will this make you happy, Christina?" he asked softly. "I wish only your happiness."

Niall's horse pranced a step backwards, stayed only by his hand on its reins. *No.* He heard the voice almost as clearly as he'd heard the words *It's a trap* so many times. It wasn't only Shawn he was thinking of, he knew. Like Christina herself, if he watched her marry Hugh, he would have to accept that Shawn was gone forever.

Christina bowed her head, but she spoke clearly. "You are a good man, Hugh. I am humbled, I am *honored* that you would have me."

The Laird spoke. "Christina, my men will take you home. Hugh and Niall have work to do. When they return, you and Hugh shall wed."

Niall wanted to speak; to shout, *No, it's wrong.* He held his tongue.

Hugh wrapped his arms around Christina, around her blue cloak and flowing black hair. She laid her head on his chest, whispering, "Thank you."

Niall's heart sank. It was right. He knew it was right. It would silence the whispers. James Angus would have a father. They would be good for one another. But his insides screamed that it was all wrong. She didn't belong with Hugh. It wasn't the way the story was supposed to end.

"Lachlan, Owen, Hugh, Niall, Conal," MacDonald said. "You've much to do."

Hugh pulled away from Christina, touched her hair, her cheek. "I'll see you anon." His voice was husky with emotion.

Her hand lay on his chest another moment, and she said, "God go with you," before she took the reins of her mare.

Hugh mounted his horse and they turned for Creagsmalan. Niall glanced back as they rode past Lochmaben's men. Christina, beside her mare, watched them go. He realized with a shock that even now, in some part of his heart, he had expected Shawn to return.

Linstock, Northern England, December 1317

"Lord Claverock!" Edward, King of England, stopped in surprise on seeing Simon, arrayed in his finest clothes, at the doors of the great hall. "You were believed dead at Bannockburn!"

Simon smiled, enjoying the king's shock, a repetition of the shock of each of his servants and guests as they'd flooded in. It was good to be back in his own clothing, befitting his station, his cooks and chamberlains serving him; to have a proper meal, at his own head table. "I've a fascinating tale, my liege." He bowed deeply, and extended his hand, ushering Edward in, to the sound of his minstrels in the gallery. "I have, by the grace of God, returned on time to offer you my hospitality personally. Eat your fill of my best food, the deer of my forests. Drink of my finest wine. And we will talk."

Edward swept down the long room, to the deep bows and curtsies of his subjects at the tables filling the great hall, with Simon at his side. Simon waited while the king seated himself, before taking his own richly carved chair. Finally, Edward's entourage and his own people sat, and women and boys began streaming from the kitchens bearing platters piled high with food.

Simon told his story as they ate, as wine flowed.

"Prophesy?" Edward laughed. "You claim to see the future?"

"Some of it," Simon said. "Possibilities. No seer sees everything."

"He knew you were coming," interjected the courtier on Edward's left. "They've been speaking of it."

Edward's smile slipped. "One might suppose, on seeing such preparations, that the king is coming."

"One might." Simon dipped his bread into a sauce, relishing and fearing this—his next test. He lifted his eyes. "One might have more difficulty, however, guessing that your daughter will be born on the 18th of June."

"My daughter?" Edward stared. "My wife is not with child."

"You will find she is," Simon assured him. "And it is a girl. Your kin, Elizabeth de Clare and Roger Damory, will also have a daughter, in May."

Edward's eyebrows furrowed. "We will find the truth of this soon enough."

"We will." Simon put the bread in his mouth, savoring the fine tastes after such long subsistence on paltry meals.

Edward set his chalice down, studying Simon. "What can you tell me of my enemies?"

"Within the month," Simon said, "a man named Deydras will appear at Beaumont Palace in Oxford, claiming to be the true king of England. You will know I speak the truth when you hear he is missing an ear."

"How could anyone claim such a thing?" Edward asked. "Especially a man missing an ear? It's preposterous! All here have known me since my birth."

"He will say that as a small child, he was attacked by a sow in the palace courtyard. The servant, fearing punishment, switched him with a carter's child —by which, he will mean you—who grew up to be king."

Edward's mouth twitched, as if about to laugh. At Simon's solemn gaze, all hint of humor fled. "Surely none would believe such a story!"

"Many will." Simon lowered his eyes. "Forgive me, your majesty." Aware of the courtiers leaning in to listen, he lowered his voice. "They will believe it is a carter's son, and not a king's, who likes to dig ditches and thatch roofs."

Edward's face darkened. He opened his mouth to speak. Heat flared up his face, and he shut his mouth, grabbing for the chalice. Around them, the courtiers turned their eyes quickly to their own meals.

"Forgive me, my king," Simon said again. "I tell you only what the future holds, not what I myself think."

"If it does not come to pass," Edward said tightly, "you will answer for such insults."

"And when it does," Simon said smoothly, "when you find your wife is with child, and Deydras appears at Beaumont Palace, you will believe and take me into your counsel that I might stop his evil plots against you?"

"*If* these things come to pass," Edward said, "yes. If not, you shall die. You will wait in the tower in London while we see."

Simon smiled, hiding his discomfort. He hoped nothing could happen to change the history he'd read, before it happened. "I shall be happy to, your majesty. Might I ask one favor?"

"You might ask," Edward answered. "Whether I grant it is another thing altogether, is it not?"

"Indeed, your highness." Simon bowed his head a moment. "I only request that I might be given leave to meet with Alexander MacDougall on my way to London. Please—send as many of your men as you wish to guarantee my arrival there. I will take none of my own."

Edward hesitated.

"What harm may come from such an agreement?" Simon asked with a smile. "Your majesty, I have ever been, and always will be, your faithful servant. And I believe MacDougall is, and will be, your faithful servant as well."

CHAPTER SEVEN

Creagsmalan, Southwest Scotland, January 1318

Niall waited in the hills outside Creagsmalan as Conal, Lachlan, and Owen rode in. He and Hugh sat on a pair of boulders, watching sunrise spill light over the water beyond the town. The very sight of the hills, of the castle rising on the edge of the sea—the place where he had waited in Duncan MacDougall's dungeon, listening to his own gallows being built—made his insides turn over. "You think this is a good idea?" he asked Hugh.

"'Tis a *very* good idea for you not to be seen there," Hugh replied.

"Not me." Niall rubbed his neck. "Them. What if *they're* recognized."

"Who will recognize them?" Hugh asked. "MacDougall and his men are gone. We've a good story, have we not?"

"Worthy of Shawn himself." Niall pulled his cloak close as the wind picked up.

"Are ye *sure* she'll believe it?" Hugh's eyebrows suddenly furrowed. "Surely she knows the Bruce will not return his lands until he swears fealty."

"Hope," Niall said. "She wishes to believe it. Moreover, Bruce *is* known for mercy. She will count on that, for she does not wish to leave Scotland."

They fell silent, watching the town below.

"How long do we wait?" Hugh finally asked.

Niall shrugged. "There are three..."

A shriek erupted from the trees around them. Niall and Hugh leapt to their feet, spinning to confront the intruder. Joan stood under the nearest tree, her hands to her mouth, staring at Niall. A basket lay at her feet, with firewood, mushrooms, and bunches of herbs splayed around it.

Niall stared in shock, as Joan emitted a second high-pitched squeal and pelted across the frosty meadow, crying, "Fionn, you're back, you're *back*! I knew you'd never leave me, Fionn!" and barreled into him, gripping him in a fierce embrace of fleshy arms and attempted kisses. He floundered in her enthusiasm, trying to extricate himself from the drowning sea of her affection, above which sounded a roar like the surf.

He stumbled backward, tripping on the boulder behind him. Joan landed atop of him, her face aglow, as Hugh bellowed in laughter.

Niall glared at him. "As much help as you were with the boar," he snapped, scrambling out from under Joan, and to his feet. He swatted at the dirt and frost covering his tunic, even as he reached a hand to help Joan up. Words tumbled from her mouth, rushing faster than the *trains* of Shawn's time. He put his finger to his lips, with a desperate glance down the hill. "Aye, I'm back," he reassured her. "Stop, just stop a moment! *Please,* ye dinna want MacDougall to find me, aye?"

Joan, steadying herself with a firm grip on Niall's arm, shook her head fiercely. "Oh, no, no, Fionn, no, he's gone away! 'Tis but milady left, and...."

"*Where* has he gone?" Niall broke in.

"...there's none but me up in the hills, I was gathering mushrooms and...."

"Where has MacDougall *gone?*" Niall repeated.

"...winter burnet...." Joan stopped momentarily, as if suddenly aware he'd spoken. She glanced at Hugh, and gave a start, saying, "Och! I dinna see ye!"

"'Tis my wee stature," Hugh said with a grin. "Sure an' you're no the first to overlook me!"

Niall emitted a short, hacking cough, forced any semblance of a smile off his face, and said again, "Do you know where MacDougall *is?*"

Joan shook her head. "No, Fionn. Nobody knows where he's gone. He up and disappeared one day, with half the men in the garrison, not a word to a soul. His wife has said naught...."

"Joan!" This time, Niall pressed his finger to her lips. "Joan, who *might* know where he's gone?"

"Now, does it matter?" A smile broke across Joan's face, and she threw herself on him again. "We're *together,* Fionn. He's no here to do ye harm and we can be *happy*...."

"No, Joan." Niall scrambled for a story to tell her. "There were things I couldn't tell you at the time." He wondered what Shawn might have said to her. He had no idea if he might contradict something she'd already heard from the other 'Fionn.'

"What, Fionn?" she asked breathlessly, but she was off on another cascade of words before he could even try to answer.

He led her, as she talked, to her basket, and began replacing the firewood and mushrooms. She kept talking the whole time. Niall glanced back at Hugh, who grinned at him, and shook his big head in amusement.

"Joan," Niall interjected. "You must stop a moment. You cannot tell anyone you've seen us, you ken?" He turned her, still talking, to the hill.

"You're coming to the inn?" she asked, as he led her down the slope.

"No," he said, though no decision had been made. The innkeeper liked 'Fionn.' It might be a place to seek information. "Perchance." Then again, having been in MacDougall's dungeon they might all assume....

"Ye're not planning on killing him, Fionn?" Joan stopped abruptly, her eyes wide, looking up to Niall. "Surely...."

"Joan!" Niall gripped both her arms. "Dinna say such things! Do you wish me dead?"

"Surely not! I *love* ye!"

"Then you must return to the village," Hugh said sternly, "and say naught of seeing Fionn."

"But, Fionn, sure you came back for me?"

"I've business," he said, "before I can speak of such things." He lifted the basket. "Mayhap you know someone who can tell us where MacDougall is?"

"Mayhap," she said.

He considered saying, *I could hurry back...if you could find out for me.* His conscience wouldn't let him lie so blatantly. "Would you try?" he asked.

"What do you want with him?"

The truth, Niall thought. Or close to it. In this instance, it worked well. He was starting to think like Shawn, he chastised himself. Hadn't he always told the truth? But these were unusual times. "He abducted your lady, Christina, from Glenmirril," he told Joan.

Her face paled. She shook her head as if in denial. "Oh, no, now 'tis bad, that is. He's a lech, that one, like his son, though he thinks himself better. And all could see what he wanted...."

"You'll help us, then?" Niall asked.

She nodded vigorously. "Aye, if it should help you rescue mi'lady! It pains me—it slays me, it does, to think of her with...."

"Meet us here tomorrow at sunset," Niall said. "Tell us what you've learned."

She nodded, her words already flowing again. "I will, Fionn. If it should help her, I'll *do* it. I know some of the men—those left behind when he fled."

"Don't mention my name," Niall reminded her.

"Aye, sure an' I'd not do that!" She looked hurt. But, with one last hug and stretching on tiptoe to kiss his cheek, she took her basket from his arm and turned down the hill, looking back once.

"She dinna stop talking, does she?" Hugh said.

Niall shook his head. "I dinna think so."

"How long until she tells all of Dundolam that Fionn is back?"

"We can't leave her alone," Niall said. They pelted down the hill after her, Niall pulling up the hood of his cloak and Hugh trying to make himself smaller inside his own.

Inverness, Present

Angus stared belligerently at Shawn, standing in the doorway. Mairi had helped him up against the pillows and left the room only moments before.

"I brought you licorice." Inviting himself in, Shawn tossed a bag on the bed. "Amy said you like it." *What would Niall do,* he reminded himself. *What would Christina do? James Douglas?* At the moment, he silently cursed them all.

"Get out of here, you eejit," Angus said. "I've had enough of you. Take Amy and James and go back to America."

You have no idea how badly I want to do just that! Shawn smashed the thought down with the steel-armored fist he'd once wielded in more than mere thoughts, and plastered a nonchalant grin on his face. He *would* put Amy first. He'd promised himself. "You don't mean that."

Angus shoved irritably at the bag of candy. "Don't tell me what I mean."

"It's early days. In a month, you'll feel different."

"Don't tell me what I'll feel."

"She wants to be with you."

Angus stared out the window a long moment, before saying, "I sent you an e-mail. Why did you let her come back?"

"I only saw it after she'd flown out." Shawn paused, hating having to say the words again. Saying them once had been hard enough. But Christina would want him to. Niall would say this was what love meant—putting someone else first. "She wants to be with *you.*"

Angus rolled his eyes. "I dinna need her playing Florence Nightingale."

Shawn's irritation flared at his callous dismissal of all Amy felt. "I suppose that's not what Claire is doing?"

"That's different."

Shawn pulled up a chair, ignoring Angus's glare. "Why?"

Angus reached for the call button.

Shawn smiled.

"What are you smiling at?" Angus asked bitterly.

Shawn cocked an eyebrow at him. "Seems I found a noble purpose for my gift with women."

"Meaning what?"

"Jenny's got beautiful eyes and a date with me tonight. Now that I'm single."

Angus made a noise of disgust. "So she's going to ignore my call?"

"So she didn't notice me lean over and disconnect it."

"You can't do that!" Angus strained forward in agitation.

"I'm Shawn Kleiner. I can do anything I want." Shawn grinned. "No worries. I'll hook it back up when I admire her eyes again on the way out."

Angus pushed himself off the pillows, bellowing, "Nurse!"

Shawn scrambled off his chair and bolted for the door. Jenny with the beautiful eyes—and they really were quite nice, he thought—stuck her head out from the room next door. "Hey, Jenny," he crooned. "It's okay. He just needs water. I got it. I don't want you worn out before I even pick you up tonight."

"Aw, thanks, you're a dear, now, aren't you?" she said in her charming brogue, glowing all over—her brogue made him glow all over, too—and yanked her head back in the other room.

Shawn returned to his seat, pushing Angus's door shut this time. He didn't try to wipe the gloat off his face. "We talk, I leave."

Angus thumped his arms across his chest and turned his head away.

Shawn shrugged and reached for the bag of licorice. "Have it your way." He ripped the bag open and selected a black piece from the jumble of colors.

"Give me those," Angus snapped, stretching out a hand.

Shawn held it back out of the way. "Why do you want Amy to go away?" He popped the licorice in his mouth, making sure its good taste was evident on his face, and grabbed for another one.

"I don't need her here out of guilt or pity."

Shawn lowered the licorice. The look on her face, in their conversation, was painfully etched in his mind. "I don't think she's here out of either of those."

"She has a better life—there. Take her back to your fifty acres and all she's used to."

"It's only twenty."

The door flew open. Shawn bolted from his chair, barely catching the licorice, and prepared to charm whichever nurse barged in.

Amy, with James on her hip, glared at him. "You didn't!" she erupted.

"No, I'm sure I didn't," Shawn said quickly.

"Aig!" James squealed, and leaned, beaming, his arms stretched for Angus. Angus turned away.

"The two of you!" Amy marched to the bed, dumped James on Angus's lap, and lit into Angus. "Doing this to me is one thing, but you are *not* going to do it to him. If you let him fall off that bed, you are the lowest of the low and you deserve everything you got on that mountain. And you!" She rounded on Shawn, circling the bed, and snatched the bag from his hand. "What is this?" She shook it in his face.

"Licorice." At her angry glare, he wiped at his mouth, wondering if he'd managed to smear some. "I offered to share!"

"Uh-huh. And why did you suddenly get such a big heart as to want to bring Angus licorice? What's really going on here?"

"You have such a suspicious mind," he said.

"Especially when I find out you have a date with a nurse you can't have known for more than an hour."

Shawn threw up his hands in a gesture of innocence. "You dumped me. I'm single. I'm free. You're jealous now?"

Amy studied him. Her anger didn't abate. "I'm not jealous. I'm *angry*. Maybe you missed that? I'm angry partly for her sake. You're blatantly using her just like you've always used everyone."

"Who, me?" Shawn opened his eyes wide. *"Me?"* He looked around the room, searching for anyone else about whom she might be speaking. "How can buying her a nice dinner be *using* her?"

"It seems someone—and I can guarantee it was none of the nurses—disconnected Angus's call button up at the nurse's station."

"No, I didn't do that."

"Did you give even the first thought to his well-being? What if he'd actually needed a nurse?"

"I wouldn't have a clue how to do a thing like that," Shawn protested.

"I have a vague memory of you dating a nurse somewhere in your sordid past," she snapped. "Maybe even while you were supposedly faithful to me. I'm sure you do know. And it wouldn't be so hard to do if she's entranced in your compliments about her beautiful eyes."

"What beautiful eyes? I couldn't even tell you what color..."

A cough behind him caught his attention. He spun. Jenny stood in the doorway, her hands on her hips.

"...that nurse's eyes were. That was years ago." He turned back to Amy, who narrowed her own eyes at him in fury. "We were talking about *that* nurse, right?" He put his hand on Amy's shoulder, and gestured at the bag of licorice. "Maybe you can convince him. Maybe...."

He stopped at the sight of Angus holding James on his lap, clapping their hands together and singing in Gaelic, oblivious to the noise around him. He looked happy. James made sing-song noises along with him. It was clear they'd sung this song before. When it ended, James grabbed Angus's hands and made an awkward attempt to get him to clap, saying, "'Gen!"

Angus obliged, starting the song again.

"What were you doing here, anyway?" Amy demanded. "Rubbing it in his face? He already tossed me out on my ear, so it's not like you need to be here convincing him I'm better off with you. As usual, you come out on top."

Shawn backed away, his eyes on Angus and his son. "I have to go," he said. "Stay here with them." He turned to Jenny, leading her out of the room. "I hear Nico's is the best. You're off at six?" The sight of James with Angus had shaken him to his toes.

Dundolam Town, outside Creagsmalan, 1318

"Now, then," Joan announced, once they were inside her small home, "we dinna want you looking like a Norseman again, do we?" She studied him.

Sitting on a three-legged stool in her front room beside the cooking hearth, he resisted the urge to fidget—to get up and run. Her mother, she said, had passed on since he'd last been there. She seemed unconcerned about such a momentous event, but quite happy to have the two small rooms to herself.

"Aye, black will do nicely." She dug in a drawer and pulled out a packet of herbs, which she mixed into a small bowl of water. Soon, she had Niall leaned back over a bowl, her hands, like they had months ago, pulling some mixture through his hair, as she leaned in far too close for comfort. "Now let it sit a wee bit," she said. "I'll pull some turnips from the garden and have a nice stew for you!" Cleaning her hands, she turned to the cauldron hanging over the fire.

"I'll see what I can learn in town in the morning," Niall murmured to Hugh, when Joan headed out to the garden for vegetables and firewood. "While Conal and the others go into Creagsmalan."

"With what story?" Hugh rumbled almost as quietly. "Will you tell them about Christina?"

"I think he's not so well liked here," Niall replied. "Christina was. But if we name her, they may guess we come from Glenmirril. And he *will* still have supporters here, as well."

"Aye," Hugh agreed. "Then stay with the story we've news of his lands."

Joan burst in the door, her heavy face glowing with joy, her heavy arms full of vegetables, already talking, and the discussion ended.

Linstock Castle, Northumbria, 1318

It was six days later, six days of hard riding in bitter winds and driving sleet and snow, Edward's men pressed irritatingly around him, as if he were a common prisoner—as if they'd no memory of the days he'd ridden at the side of the great Longshanks himself—that a steward ushered Simon into an upper chamber at Linstock.

A middle-aged man, black hair hanging to his shoulders, sat at a small table laden with food. Wind flung itself against thick leaded windows, rattling them and spattering them with stinging rain. Tapestries lined the walls, scenes of a hunt, a man playing a harp, something Biblical. A fire roared in the hearth behind him, casting sharp shadows over his face.

Simon bowed. "My condolences on the loss of your son."

"Simon Beaumont, Lord of Claverock." MacDougall's steel gray eyes raked over Simon, not acknowledging his words. "You disappeared after Bannockburn. Never found, dead or alive. Imagine my surprise when Roger told me you had arrived, seeking my audience."

"I've quite alive, as you can see." Snow and rain dripped from Simon's cloak, pooling around his boots. "Your wife was good enough to tell me where to find you."

"Was she? She'd best have good reason." MacDougall's eyes pierced. "Where have you been?"

Simon glanced at the empty chair; his eyes swept over the food he had not been invited to share. "On a long and wondrous journey. One that has left me with knowledge which I—and your wife—believe will be of interest to you."

"Will it?" MacDougall twisted his wine goblet between his fingers, a skeptical eyebrow arched.

Simon removed his cloak, revealing the *jacket* from Shawn's time.

MacDougall stared at it, then lifted his eyes to Simon's. "This—*garment*—is from his land? I'm well-traveled and have never seen its like."

"Nor will you, for many years to come." Simon smiled, seating himself at the table. "The garments are of little import. The *weapons....*" He lifted his eyebrows. "I believe you will find *them* far more interesting." He reached for the jug of wine.

"Indeed?" MacDougall watched silently as Simon tilted the jug, a ruby cascade of his best wine flowing into Beaumont's goblet. "Are you offering me these weapons? Or merely telling me?"

Simon took a deep draft of his wine, savoring it after the long, cold days on horseback. He set the chalice down and leaned back in his chair.

MacDougall scowled.

Simon smiled. "In my peculiar journey, I was given the gift of prophecy. Like Thomas of Erceldoune, whom I believe you knew."

"I did know him," MacDougall confirmed.

"I offer you the chance to work with me," Simon said. "Give me your support, recommend me to Edward. In exchange, I will bring you Campbell."

"I can get Niall Campbell myself." His lips pursed, watching Simon help himself to bread.

"Then why have you not done so?"

MacDougall glared. "I am devising a plan."

"We've a mutual interest regarding him. Why not work together? I myself *have* a plan." Simon leaned forward, the bread still in his hand. "Had you guessed there are two men?"

MacDougall set the chalice down sharply. "What do you know of them?"

"I have met the other," Simon replied. "It was he who killed your son, not Niall."

"Is he called Shawn?" MacDougall asked.

"You've made his acquaintance?" Simon asked.

MacDougall shook his head. "I believe I have. I'll have him as well as Niall. They have played me for the fool and will pay. Where is he?"

The fire cast shadows across Simon's face as he said softly, "Far away, in a land beyond your reach."

MacDougall leaned forward, brows dipping over his beaked nose. "When Edward rewards me with lands, I shall have the resources to track him to the ends of the earth."

"It is not so simple as that. Think beyond the ends of the earth, Alexander." He indicated the *jacket*. "Think to the land that produced a garment such as this. It is where Shawn lives."

MacDougall's eyes flickered over the odd *jacket*.

"Give me your aid," Simon said, "And I will bring you *Shawn*, who you cannot reach without my help." He smiled broadly. "I will even bring you his son. Surely you would like to kill his son as he killed yours."

"I'll disembowel his son before his very eyes," MacDougall rasped. The fire crackled behind him. The storm beat at the glass. "However, you said he is beyond my reach. How, then, do *you* propose to bring him here?"

Simon smiled. "What are the two most powerful forces?"

"England and France," MacDougall replied.

Simon let out a loud laugh, his head thrown back. "England and France! Surely you jest!"

MacDougall stabbed at a piece of meat on his trencher and jabbed it in his mouth. "I don't believe I invited you to share my wine."

"Ah, but you should have." Simon lifted his cup in salute. "For not only will I give you Niall and Shawn, in exchange for speaking on my behalf to our king, I offer you the chance to wield power greater than Edward himself." He smiled broadly. "But first, you must know what those two great forces are."

Irritation flickered across MacDougall's face.

"Love and hate!" Simon leaned across the table. "For love and for hate will a man do anything." His voice dropped to a whisper. "*Anything at all.*"

MacDougall stared. The flames flickered, settling lower in the hearth.

Simon sat back in his chair. "For which do you act?" He swirled the wine in his goblet, and took a slow sip. "For which, do you think, will Shawn and Niall do our bidding?"

Inverness, Present

Shawn let himself into Angus's house. Angus had insisted, over Amy's objections, that he stay there as long as he was in Inverness. "There's no point him going to a hotel," Angus had argued. "He can be with James, and you two can work out your plans." Amy had harrumphed and rolled her eyes.

Now, as Shawn slipped into the front hallway, Angus's home was dark but for the faint flicker of shadow and light showing through the crack of the living room door. The smell of wood smoke reached him. He went to the kitchen to pour a glass of brandy, and came back. He inched the door open. Sitting in the armchair, James asleep in her arms, Amy looked up. "You're home early," she said.

"Yeah." Shawn set the brandy on the coffee table. He crossed the small room to stir up the fire and add another log. He'd surprised himself.

"I thought you might be out a lot later."

Shawn dropped back onto the couch. He lifted the glass to his lips, tasting the warm glow of liquor. "You thought I'd go home with her." Her assumption irritated him.

"It wouldn't be unheard of. For you."

He'd considered it. She'd seemed inviting. But Christina, Amy, and his promises to them both, had hovered over his shoulder, watching, waiting. Talk about a mood killer. "I'd think you of all people could see that a couple years in medieval Scotland brought a few changes." Bitterness tinged his words.

She stared into the fire. He thought he saw color climb on her cheeks.

"Is that why you walked away from me?" he asked. "You still believe I'm the same jackass I was?"

She shook her head. "No. I can see you've changed. I'm sorry, Shawn, I shouldn't have said that." The fire snapped. "I walked away because Angus needed me."

He swished the deep amber liquid in the tumbler.

"It's just," she added, shifting under James's weight. "The timing was all wrong. If Angus hadn't come into my life in that year, it would be different. But he did."

"Yeah, well, sucks to be me."

"Yeah, good-looking, talented, rich, and famous," Amy scoffed. "Women falling all over you. Life is rough."

Shawn took another sip of the brandy and gave her a sour look. "If only that summed up all that matters in life." He rose, thinking he could have gone home with Jenny. But all he really wanted was to kiss his son. He held out his arms. Amy lifted James up, handing him over. Shawn returned to the couch, sinking down. James's dark eyelashes fluttered against his cheeks. He heaved a high-pitched sigh and snuggled closer to Shawn's chest. "I should have stayed home," Shawn said softly.

"Then why didn't you?"

"I kind of worked myself into a corner," he admitted.

"Getting at his call button."

Shawn watched James. In his sleep, he sighed more softly, and smiled. "She really does have beautiful eyes." He looked up. Amy leaned forward in her chair, watching him. "Not as nice as yours, though." Her hair fell over her shoulder. The shadows of the fire flickered across the smooth planes of her face. "I always meant it when I told you that. I never lied about that."

Amy sank back into the armchair. "It's nice to know. Are you going out with her again?"

"Are you jealous?" He grinned at her.

"I thought I might be, but no. I hope you like her." She laughed, and took a sip of her wine. "Ironic, isn't it, using your ex-girlfriend's car to take your new girlfriend on a date."

Her answer sent his spirit plummeting. He took a quick gulp of the brandy. "She's not my new girlfriend."

"You and Niall must have been quite a pair. Poor Allene." She regarded him for a moment, before asking, "Have you read any more of their letters? Are they okay?"

Shawn shrugged. He wondered how she could have gone so quickly from flying home to spend Christmas with him to not caring if he dated another woman. But he answered her question. "There's nothing I can do about it if they aren't, you know?"

"Survivor's guilt when maybe nothing has happened."

"It's medieval Scotland," he said. "I can guarantee something has happened. Or will."

"Why don't you Google him?"

"Why don't you?" he shot back.

Amy shrank back into the chair, pulling her arms around herself. "Touchè," she said.

"It took all of half a second to realize that even knowing him for a week, you don't want to know he died some awful death, do you?"

Amy shook her head. "I'm sorry, Shawn. It must be terrible for you, wondering."

"The funny thing is," he said, his eyes locked on James, "I *know* they're all dead. So what am I worrying about? That I'm going to find out they died?" He forced a laugh at himself.

"No, Shawn, it's not the same!" She leaned forward. The firelight played over her hair, throwing a high sheen down its length. "You know what you're worried about. Finding out it was some awful way to go."

"I know they all survived that night." He knew he didn't have to elaborate on which night. "But I keep putting off reading further in the letters, because every one I read, I feel like they dodged another bullet, and their luck can't possibly hold out. I'm always afraid the next one is going to tell me about the blade falling. So to speak."

Amy took a quick breath. "Do you want me to read them for you?"

Shawn looked up from studying James. He felt his eyes narrowed, and felt the old thoughts: she wanted to get at his secrets. He shook the feeling off. She didn't look as she had all those times, trying to trick him into revealing a lie. She looked as if she genuinely didn't want to go near them. "No," he said. "You'll end up feeling the same way."

"But you're killing yourself wondering. Bruce died in his own bed. Maybe you're worrying for nothing."

"Maybe I should turn them over to a medieval professor who has no personal interest in the outcome."

Amy relaxed back into the chair, letting her breath out, and he knew he'd been right. She didn't want to find bad news any more than he did. "You can't do that," she said. "You'd never get them back. Besides, how would you explain seven hundred year old documents addressed to you? You'd be charged with trying to pull an elaborate hoax. Speaking of which...." She frowned. "Who drew the picture of me?"

"What picture of you?"

"There's a drawing of me known as the Glenmirril Lady. I'm sitting with my violin, wearing the Bruce's ring you threw to me. You can't draw. Did Niall do it?"

"I doubt it. I never saw him draw."

"Well, he must have," she said. "Because there's no one else there who ever saw me."

"I don't know when he'd even have had time. And Allene would have twisted his ear off if she'd caught him drawing pictures of another woman."

Amy smiled. "I wish I could have met her."

Shawn, too, smiled, resting for a moment in a cloud of memories. "Do you want to go to bed?" he asked. "I can stay here with him."

Amy rose from the chair. "He'll be asleep for the night. I'll put him in his crib." She leaned over to take their son from his arms. The smell of her shampoo teased Shawn's nose. Her hair fell, brushing his arm. His stomach tied in knots, wanting to take her in his arms and beg her to let Angus go, to come and be a family with him and James. He swallowed. "Sleep tight, then."

He watched her slip out of the room, taking his son, *their* son. His arms felt empty. His heart felt empty. He tried to think of Jenny, laughing with hope and admiration in those beautiful blue eyes, across a candle-lit table. He couldn't interest himself. His mind stayed on Amy, James, Niall, Allene, Christina, even on Angus in the hospital. He poured another half glass of brandy, and sipped it, watching the flames dance, a smaller version of the great braziers at Glenmirril.

Amy was right. Maybe he was worrying about nothing. He sipped the brandy, slowly, only delaying the decision he knew he'd already made. When he'd drained the last of it, he went to the room Amy had assigned him upstairs, and took the thick oilskin of parchments out of his suitcase.

His phone beeped a notification at that moment. He sighed, sliding a finger across its face. An e-mail from Ben. Following up on the last concert. *Let's get to work planning the next one. The Broadway album—have you started on it?* Shawn skimmed the lengthy e-mail and turned off the phone.

He sat on the narrow bed, staring at the tight roll in his hands for some time, thinking of them all. He was destroying himself with worry. In the end, there was nothing he could do for them. He had a life to rebuild, his musicians, an audience counting on him. Amy. James.

He slid the fragile roll back into the oilskin and zipped it into his suitcase.

Dundolam, 1318

In the morning, Joan produced her dead father's tunic and trews, rough home-spun pieces. With dirt on his face, one eye half closed, and an unnatural slouch to his gait, none would think of Sir Niall Campbell of Glenmirril.

"You'd best clean up afore Allene sees ye like that," Hugh snickered, when Joan left the room.

"Aye, and the sooner the better.," Niall said. "Stay here and make sure she dinna tell the whole town I'm back!"

The disguise served well, though the winter wind was chilly without his warm cloak. Niall limped through narrow, cobbled streets with none looking twice at him. In the town square, protected from the stiff breeze blowing off the sea, he asked questions of the fishmonger and a man trying to sell a horse, with no luck. Through the morning, he tried the baker, the weaver, and the tanner.

Finally, half frozen, he headed to the inn for a drink. His disguise had served so far, but they knew Fionn at the tavern. He would stay in a dark corner, and throw inflections from Amy's twenty-first century English into his speech.

He no sooner ducked into the warmth of the place than he recognized the cadences of Joan's solo repertoire, on another of her long ballads. With his eyes still adjusting to the sudden dimness, he slipped into a dark booth, searching for her. She stood by the bar, her back to him, talking to one of the tavern maids.

"Hugh," Niall muttered to himself, "you were supposed to keep her there."

As he watched, Joan suddenly glanced around, then leaned in close. With the girls' heads close together, he could hear her sibilant whispers.

"No, Joan," he muttered. He didn't need to hear her words to know what caused her so much excitement. He hesitated—stop her? It was already too late. He would only draw attention to himself.

The bar maid glanced at him and Joan turned. A broad smile broke over her face. He shook his head, but she snatched two large mugs off the counter and hurried to his booth, beaming.

"Joan," he hissed, "you just told her."

"Och, Maura's me dearest friend!" Joan pushed the mug at him. "'Tis a hot posset. Sure you'll be wantin' one after a mornin' in this cold!"

"Lower your voice!" It was hard to be stern while keeping his own down.

Joan laughed. "We're among friends, Fionn."

Niall stifled a groan.

The smile fell from her face. "Really, Fionn, you can trust her. She means ye no harm."

Suddenly a boy in curly-toed shoes skidded up to the table. "Fionn! You're back! Maura said ye were!"

"He's a friend, too," Joan said.

Niall looked up at the bar maid to see her smiling and glancing at him as she talked to her sister. The skinny innkeeper came from behind the bar, wiping his hands on a towel, nearly as happy as Joan to see him, while the panic began to rise in Niall's chest.

"Fionn!" the man cried. "Have ye my son's recorder? Will ye play for us?"

As more patrons turned to see what the commotion was, Niall glanced at the door, gauging his way out. A lank, dark-haired woman stood in the doorway, her gaunt face lean and angry. She stared at him, their eyes meeting. She held his gaze a long moment, before a hard smile glanced across her lips. She turned and left.

Niall leapt to his feet. "Who's that?" he demanded.

"'Tis Osla." Joan said. "Martainn's wife."

"Martainn—the jailer?"

Joan nodded, looking perplexed.

Martainn had paid dearly for Niall's escape. Niall guessed by the smirk on Osla's face she was not the forgiving sort. "Joan." He gripped her arm. "Get Hugh. Get him out of this town as fast as you can."

For a moment, Joan had no words. She nodded silently. Then asked, "But where are *you* going, Fionn?"

"Get him back to where we met you," he replied.

She stared at him a moment, her jaw set, before suddenly turning to the innkeeper. "Frederic! Milord MacDougall abducted our Lady Christina," she said in a rush. She glanced at Niall with trepidation and raced out the door.

Niall rolled his eyes to the ceiling, cursing her. But there was nothing he could do. He turned to the innkeeper, who stood stunned by his booth.

"Milady is *alive*?" Frederic asked. "We feared for her when...."

"I'm seeking a friend," Niall interrupted. But relief flooded him. These people loved Christina. "He came here to try to help her." He glanced at the door, deciding whether to chase Osla or seek Conal. "He's a wee bit shorter than me, dark hair, a beard, a brown cloak." He described Conal. "He calls himself Hamett."

"He stayed here last night." The innkeeper scratched his beard. "He broke his fast and left." He looked to his daughters.

One of them blushed. Niall would have smiled but for the urgency.

"He sought an audience with Lady MacDougall," the girl said.

"How? Who did he ask for an introduction?"

"He asked me to take him to the stable master," the girl said. "He's horses for Creagsmalan. You mind how proud milord was of his stables."

Niall's mind spun. Conal had left the inn hours ago. He was certainly inside the castle walls by now. But maybe Lady MacDougall had not yet granted him an audience. "Where are the stables?" he asked.

"Are you here to help the Lady Christina?" Frederic's voice was sharp.

Irritation reared in Niall, to hear an innkeeper speak to him so—and when he had to find Conal! *The great Sir Niall has been bested by a kid.* The words popped into his head, slowing him. Shawn had accused him, all the nobles, of looking down on others. *Even I am not too proud to carry my own dishes.*

The innkeeper stared up at him, waiting, and Niall realized he had abandoned his slouch, standing now like the knight he was.

For Christina, this man defied a knight.

Humility settled on Niall's shoulders. "Aye," he said. "We are here to help Christina. Martainn's wife blames me for his punishment. Once she tells Lady MacDougall...."

"Alis," the innkeeper said, "see if you can slow Osla."

"I need to find Hamett," Niall said.

"Jep," the man said, "haste, lad. Ask if Hamett has left the stables. If not, tell him to stay."

The boy bolted out the door.

Niall started to follow, but the innkeeper gripped his arm. "She'll call the first guard she sees. She'll tell them what you look like. Come, now!" He dragged Niall behind the bar, calling another of his daughters. "Get my old clothes." He pulled Niall up a dim, narrow flight of stairs, into a small room with a bed and a water basin on a stand. "The hair must go," he said, already pushing Niall into a chair and reaching for shaving soap.

"Not again," Niall muttered.

But he saw the sense in it, even as he chafed at the delay. At least the man's hands were steadier than Bessie's as he drew a blade over Niall's head, over jaw, lip, and throat, in sure, quick strokes. The chill of winter hit Niall's bare scalp and crept down his spine. The girl entered, leaving trews and a new shirt on the bed, and ducked out.

The innkeeper made a final swipe and drew a towel across Niall's head and face, wiping away the last of the soap. He shook out the new shirt as Niall yanked off what Joan had given him. Soon, he was dressed as a tradesman in a bell-sleeved shirt and leather vest.

As they headed down the stairs, Jep appeared at the bottom. "He's gone into Creagsmalan," he gasped. "He's an audience with Lady MacDougall after the noon meal."

Alis burst through the tavern door, her skirts gripped up in one hand, breathing heavily. "Da," she gasped, "She'd not speak to me. She says she'll warn the Lady we're harboring enemies, she said...."

"Is she in the castle?" Niall asked.

"Aye," the girl almost sobbed. "As is Hamett. Fionn, is he in danger?"

"Aye," Niall said grimly. "I believe he is."

Inverness, Present

Angus stared at fat white flakes sliding in white streaks down the window of his hospital room. Pain gnawed at his spine, as it did each time the drugs ran low. Only slowly did he become aware of the faint reflection of a woman in the window, watching him. He hoped it was Amy. Maybe Shawn was right and she could still love him as he was.

He hoped it wasn't. He'd only tell her to leave again.

Maybe it was Claire. She was pleasant company. He hoped it wasn't her.

He pulled his eyes from the window, turning to see Mairi. Relief and disappointment mingled.

She beamed—but he saw the tension behind the smile.

"I need your help," she said—too brightly.

"Now I'm not much use to anyone at the moment." Calling the nurse with his call button was about his limit even for helping himself.

"Oh, ye're more use than ye think!" She laughed, and laid a small black case on his lap, along with a slim book. "See, Hamish wants to play, and ye know I've no ear for music."

"Flute? But I don't play the flute."

"But ye play the pipes, so sure ye can work on this a wee bit and teach him something!"

"I can't even sit up on my own, Mairi," he reminded her irritably.

"Aye, well, but ye *can* sit up, and ye *can* read music, and sure he'd love it if ye'd work with him."

Angus sighed, relaxing against his pillows. He stared at the wet streaks on the windows. His whole life had been built on water and mountains. He couldn't face life in a wheelchair. He heard the scrape of wood on linoleum and turned to see her pulling up her chair close against his bed.

She laid her hand on his. "I was thinking—you were so interested in that family tree you found—where was that? There's something you could do now. I could bring you your laptop and...."

"That particular history isn't online," he said. "Honora's quite the character. 'Tis all hand-written on a large scroll." He smiled despite himself. "With a great many cats."

The worry disappeared from Mairi's face as she broke into a broad grin. "There ye go! We'll get ye to Honora's. Or we'll bring her here. Sure ye'd like to see it again and learn more? Did ye not say ye found our family in it?"

"I did. However, I can't get there."

"I'll see what I can do," Mairi said, rising to her feet.

"Mairi, stop trying to push things on me," he snapped. "I'm looking at life in a wheelchair, and ye want me to be excited about who gave birth to who seven hundred years ago!"

Her normally cheerful face creased into the closest thing to anger he'd ever seen. "Aye! I would! I'd much rather that than watch ye stew in self-pity," she said. "Ye're hardly the first to have a head-on collision with life!" She spun on her heel and left the room.

He turned his head, glaring at the snow-streaked window. His back hurt. What did she know? He drew a deep breath. She knew plenty. She'd ever supported him, encouraged him. He'd no right to be angry with her now. He dropped his gaze to the case on his lap, and snapped it open. A silver flute gleamed—three pieces nestled in blue velvet. He lifted the head joint out, turning it, and opened the book. The chart at the back showed him the fingering was not so different from the bagpipes. It was just a matter of blowing. Maybe —maybe it would be something to pass time.

CHAPTER EIGHT

Creagsmalan, Scotland, 1318

Walking through the gatehouse was easy enough, Niall thought, as Frederic raised a hand in greeting to the guard on duty.

"Ralf!" he called, "yer man's come from England. News for our Lady."

"One just came through." Ralf's hand rested on his sword.

Niall felt naked without his own.

"What business have you?" Ralf asked.

Niall stepped forward. "A message regarding milord MacDougall's lands."

Ralf grunted and waved them in.

"I'll take ye to the antechamber," Frederic said, as they passed through, "but I daren't go in." He glanced back at the guard in the gatehouse, edged Niall into the tower whose stairs led to the great hall, and slid his own knife into Niall's hand. "We'd do anything," he said, "anything a' tall—for milady."

"Thank you." Niall pushed the knife into his waistband, and gave the man's shoulder a firm pat, before turning and hurrying up the narrow, twisting stairs. The night in Glenmirril's tower came to him, and Shawn's disappearance. He pushed it away, as he burst into the great hall at the top of the first flight.

Two guards stood at the door to the hall where the Lady met with complainants and petitioners. Two disheveled men, three merchants, a young girl hanging her head beside a man who must be her father, and a boy with tears on his cheeks stood waiting. Niall's heart hammered a moment, before he spotted Conal at the head of the line. His pulse settling, he strode forward.

"Halt!" commanded a guard, stepping up.

Niall gave a quick bow. "I've word for the horse merchant," he said.

Conal turned, confusion crossing his face at the sight of Niall.

"There's been an accident with one of the horses," Niall told him. "We must go anon."

As Conal turned to follow Niall, the doors to the throne room burst open. Lady MacDougall stood framed between them, her head lifted high.

At her side, Osla's sharp eyes scanned the room.

Run or bluff? Niall's mind spun. Neither of them had ever seen 'Hamett.'

But then—they would *know* which man among those in the ante chamber they had never seen.

As if reading his mind, Lady MacDougall's eyes settled on Conal. *What would Shawn do?* As Conal reached him, Niall edged in front of him. Osla barely glanced at him.

He bowed to Lady MacDougall, even as he took a step back, hoping she wouldn't notice—calculating the distance to the door. "My Lady." What next, his mind screamed. "One of the horses has been wounded. His Majesty will be most displeased if aught were to prevent us tending them at once." He inched back another step.

One of the guards took a step forward. The men along the wall tensed.

Lady MacDougall hesitated at mention of Edward. She leaned toward Osla, murmuring. The woman's eyes flickered over Niall and she shook her head, whispering back.

"Come to the throne room," Lady MacDougall commanded.

"My Lady," Conal protested, "the horses...."

"Now," she said.

Niall made his decision. The throne room was death. He gave another bow. "My apologies, my Lady, I cannot displease my king." He turned, striding for the door, Conal at his side, the door looming far beyond them.

"Guards."

He heard the single word. He bolted, slamming through the doors, down the short hall to the circular stairs, Conal at his heels.

"Guards!" Behind them, Lady MacDougall raised her voice. Footsteps pounded behind them. As they reached the courtyard, three men raced from the gatehouse, swords swinging.

"Now what?" Conal muttered.

"Can you think up an explanation?" Niall backed up to Conal, his knife raised. It wasn't much against five swords.

"None," Conal answered, as guards surrounded them. Around the courtyard, men gathered—a beefy man wearing a blacksmith's apron, one who must be a baker. Small and skinny among them, Frederic pressed against a wall, fear on his face.

Suddenly, a shout arose, and a youth in MacDougall's colors, carrying a sword in each hand, dashed from one of the towers. "Stop!" He looked around the growing crowd. "Who of us has forgotten Duncan's treatment of Lady Christina?" he asked.

"Orders, Ninian," snapped the oldest of the men.

"Orders be damned," the youth hissed. "If Fre...if 'tis true, these men rescued Lady Christina. And Bessie." He looked to Niall, questioning.

Niall gave the smallest nod of his head, afraid to admit, afraid to deny.

Another man stepped closer to Niall, looking him up and down. "If ye're Fionn, as Osla says, 'tis your doing, the stripes on Martainn's back."

Niall opened his mouth to speak, but Ninian cut in, even as he backed up to Niall and Conal. "'Tis his doing Lady Christina and Bessie are free and Duncan dead."

Niall felt a sword press between his body and Conal's. Conal's fingers close around its hilt.

Ninian looked from one of his comrades to another. "Martainn himself escaped to warn her rescuers. Martainn himself would stand with these men."

"Go quietly," spoke the oldest guard, ignoring his words, "and it may go the better for you."

"We did naught," the young guard insisted, "for a lady's honor, and I for one am shamed. *Shamed*!" He looked around the courtyard, challenging them to disagree. "So I will help these men now. I shall kill any who threaten them."

"Kill them!" Lady MacDougall's voice rang from the great hall.

The guards closed in. Niall tensed, with only his knife. Ninian shouted to the crowd, "They rescued our Lady Christina! Who will let them die?"

One guard hesitated. Four raised their swords.

"Come along quiet," the oldest said.

"Go!" Ninian shouted. He and Conal flew at the nearest guards, while Niall dove with his knife under an uplifted sword.

"Come along!" roared a voice from the crowd, and Niall caught a glimpse of the blacksmith charging, poker in hand.

In the moment's distraction, he was able to skirt his attacker's sword, and wrestle him from behind, pressing his knife to his throat. "Drop your weapons!" he shouted.

"Kill them!" Lady MacDougall screamed from the stairs. "Dinna mind Donnel!"

The man wrenched in Niall's grip, slamming his elbow into Niall's abdomen. Blood trailed his throat, but he threw himself to the ground, rolling, as two others closed in on Niall. He couldn't hope to get past their swords. Adrenaline pulsed. Men fought all around, swords and pokers and scythes swinging. A woman screamed. Guards circled him. Sweat prickled his brow.

"Niall!" Hugh's voice rang over the melee, sunlight glanced off a long blade as it sailed through the air.

Niall laughed out loud as it thudded into his outstretched hand, flashing instantly at his attackers. He spun, swinging at both, dodged, lifted the sword with both hands and lashed it through the air, connecting with one of the MacDougalls.

"Get out, get out!" The youth grabbed his arm, already running.

"Stop them!" Lady MacDougall screamed. Niall turned for just a moment to see her storming down the stairs, her skirts lifted high, sailing like a ship amidst men tussling, sparring, swinging swords, and rolling on the ground all around her.

He stumbled, caught himself, as someone shoved him through a passageway and out into an alley.

"Get them out!" the young guard shouted.

"Hurry!" Joan was pushing him at her laundry cart, grabbing Conal's arm. "Ninian, you'll not...?"

"The smith will hide me," the boy said, as Niall and Conal stumbled up, clambering awkwardly over the sides of huge wicker baskets. Linens dropped over his head, Joan's hand pressed him down, so that he crouched, breathing heavily, and a lid closed over him, dulling sound and light. He could hear Conal's heavy breath in a basket beside him, and Joan saying, "Between the baskets, now, haste!"

The cart swayed and creaked dangerously. "Easy, Hugh!" he whispered. His own basket rocked as Hugh wedged himself between the rows of big bins. Cloth whispered and scraped against the wicker.

"Be still," Joan hissed, "and the linens ought hide ye!"

The cart lurched. Niall could feel the poor beasts straining to pull their weight. It held fast a moment, before shooting forward. Iron wheels rattled on cobblestones, the horses grunted, sounds of fighting came to Niall as he huddled, trying to slow his breath.

"My sword?" he hissed.

"With me," Hugh rumbled back.

"Hush!" Joan muttered. Fear shook her voice even in the single word. The cart clattered on, rocking them, and suddenly Joan screamed, short and sharp.

Niall tensed in the stuffy space, ready to spring, trying to think how he'd fight with sheets tangling and blinding him, but her words came quickly. "Mungo! Let me out! I was only tryin' to get the linens for washing!"

"No one leaves!" It was the voice of a boy, still cracking.

"Aw, now Mungo," Joan said, "You *know* me! Let me out before the fighting gets worse!"

He grunted something Niall couldn't hear in his confines.

"Tonight." Joan's voice took on a soft hint of promise. "Come by tonight. We'll sit in the gardens, aye?"

The boy laughed and whispered. Joan, too, laughed, while Niall's heart pounded. *Go, Joan, haste!*

The cart jolted, the wheels clattered. They bumped onto cobblestones, the cart now angling downhill. His heart slowed only somewhat. They might still come after Joan, demand to check her baskets.

"Be still now," she whispered nearby. "'Tis but moments."

The ride dragged on interminably. He feared his legs, cramped in the tight crouch, would be useless should he need to fight. He strained his ears, in the dark, for sound of pursuit.

None came. And suddenly, the cart jolted again. Niall's basket toppled, falling off the cart. He hit the ground with a solid *thud*, knocking his head and dazing him momentarily, before he was able to crawl out of the tangle of sheets, scrambling to his feet in Joan's dusty courtyard. Hugh was grabbing his hand, grinning, and Conal was snatching his sword from the laundry cart. A chicken flapped its wings wildly at their intrusion, scolding with sharp squawks.

"Haste!" Joan pointed to the slopes rising behind her home, south of Dundolam. "You'll find a road through the hills, if it's milord MacDougall you're wanting. 'Tis three days' hard ride to England. They'll not be able to mount men to follow you until the morrow." She launched herself suddenly at Niall, her big arms wrapped around his neck, her lips pressed to his.

"Joan!" he gasped, managing to turn his head.

She stepped away, leaving a blade's width of space between them. "Oh, Fionn!" Tears sparkled in the corner of her eyes. "Ye're not Fionn, air ye?"

He hesitated just a moment before shaking his head. "I'm sorry, Joan. I'd no intent to hurt you."

"If ye did it for my lady, I can forgive ye."

"We must get out!" Conal grasped Joan's hand, tearing her from Niall. He kissed her fingertips and looked up. "My lady...."

"I'm no lady." Color flushed her cheeks, as she lifted her eyes to Niall.

"Courage and honor make a lady," Conal said. "You have shown both. I thank you."

The color deepened in her cheeks as she locked eyes with Niall.

"Let us be away!" Near Joan's small hut, Hugh led their ponies to the gate.

"I'm sorry, Joan." Niall squeezed her hand. "And I thank you. From the bottom of my heart." He jumped on his small garron, touched his heels to its belly, and the three of them bolted for the southern hills.

Glenmirril, 1318

Christina shed her cloak and tore off her gloves as she entered Niall and Allene's solar, flying to scoop James Angus from Margaret. He shouted in joy, lunging from Margaret's arms into hers. A sigh escaped her lips as she pressed his cheek to hers.

"Mamama!" he babbled, wrapping his arms around her neck.

She laughed in joy, holding him back a bit to see his face. "I've missed you so," she whispered, and settled him on her hip, saying to Margaret, "He's looking bonnie, aye! You've taken such good care of him."

"Milady!" A shout came from Allene's room. Bessie burst out, running to drop to one knee, taking Christina's hand. "How I feared for you! Has he harmed ye, mum?"

"Get up, Bessie." Christina pulled her to her feet, squeezing her hand and giving her cheek a kiss.

"Bessie has helped a great deal with James Angus." Allene appeared in her doorway, her hands tucked under her swollen abdomen.

"What joy to see you again." Tears shone in Margaret's eyes. She gave

Christina a tight hug, and stood back. "How I've prayed from the moment we knew!"

"We all have, milady," Bessie added earnestly. "We were that fearful we should never see you again."

Four men came in, carrying a large tub to Christina's chamber—the chamber that had been Shawn's. A portly matron with steel-gray hair showing under her barbette and three young women followed, carrying steaming jugs.

"The water's heating in your room," Margaret said. "We started the moment we heard you were on your way."

"I've no wish to be away from James Angus even a moment," Christina said.

"Come, milady, you've had a long journey. Have a wee soak and James Angus will be here waiting." Bessie scooped Christina's cloak and gloves off the floor, whisking them away to her room.

Christina went to Allene, her son still on her hip, and held her close.

"He dinna hurt you?" Allene asked softly.

Christina shook her head. Her words came out with a touch of sad irony. "He believes himself an honorable man. And so, for the days we rode—with his men always around—he was, hoping he might sway me to willing dishonor."

"No good ever came from a MacDougall," Allene whispered angrily.

"Mamama," James Angus said.

Christina's arms tightened around him. Perhaps this once it had. "He's gone to England," she said. "Niall and Hugh have gone after him."

"Are we done with mercy?" Allene asked.

Christina nodded. "We are. Set yourself to prayer and be on your guard. For MacDougall does not give up and I myself shall not leave Glenmirril until I hear they have succeeded."

A ripple fluttered across Allene's stomach. She gasped, her eyes widening, and pressed a hand more tightly to it. "Surely another boy," she said with a laugh, when it had passed.

Christina smiled. "You are blessed."

Allene laid a hand on her arm. "You shall be, too."

"Aye," Christina agreed. "I shall marry Hugh when he returns. Please God, we will be blessed with braw laddies and bonnie lasses."

Doubt flickered across Allene's face.

"Are you not happy?" Christina asked. "Was it not what ye all wished?"

"I am," Allene said. "We did, but...."

"Shawn is not coming back," Christina said softly. "I have been foolish."

Before Allene could say more, the men filed from her chamber, the eldest saying, "Your tub is ready, milady."

"I thank ye." Christina turned, handing James Angus back to Margaret. "You are right. I must take that hot bath." To Bessie, she said, "I shall prepare for dinner in an hour."

With a kiss on her son's cheek, she glided to Shawn's chamber—hers. The women were pouring jugs of steaming water into the tub. One of them sprinkled in dried rose petals. Christina smiled, breathing in the sweet scent. "Thank you," she murmured. The women bobbed and filed past her and guilt flashed through Christina's heart. She had been thanking God for returning her to her son. She had barely noticed them. She turned, saying more loudly, "I thank ye."

"Aye, milady," said the matron. "Sure we're so grateful to have ye home safe!" She gave Christina's hand a squeeze. "'Tis our joy to welcome ye home!"

Christina's lips trembled, touched. She turned back into the room, remembering as she did, the letter still in the lining of her cloak. She opened the wardrobe, where Bessie had hung it. Her eyes flickered over Shawn's shirts and trews, many sewn by her own hand, hanging in neat rows. More guilt curled through her stomach. She must send them away. There were men who could use them. More—she had committed to Hugh and no matter where her traitorous heart insisted on turning, she *would* be a faithful and loyal wife to him—even in her thoughts. He was a good man. He deserved that.

She slid the rolled parchment from the deep pocket inside the cloak, and stared at the single word, *Shawn*, on the outside. She touched its edge to her lips, thinking of the words inside. She must burn it. She should have done so before leaving Lochmaben. She stared at the tub, steaming in front of the flickering fire. A quick toss and the letter would be in the flames, consumed and turned to ashes.

"Milady." Bessie edged the door open. "Will I help ye with yer dress?"

"Aye, thank you." Christina crossed the room and pushed the parchment into a pigeon hole in her desk under the window. The letter did no harm at the moment. She would burn it when Hugh returned, just before their marriage, and remove Shawn's trews and shirts then, too.

She gave herself up to Bessie's help, and soon reveled in the hot rose-scented waters, grateful to be home.

Inverness, Present

With Amy spending days at the hospital in hopes of seeing Angus, Shawn worked through e-mails from his publisher, his lawyer, Ben wanting answers about his appearance with the Arkansas Symphony. *Mike loves your duet of that Duck in the Corner piece,* Ben wrote.

Maybe nothing has happened. Amy's words rang in his head.

Duck? Shawn shot back. *It's mandrake! How do you get duck out of mandrake? Yeah, tell him let's do it. Set a date, I'll be there.*

Maybe nothing has happened. Yeah, but...maybe it *had.*

While James played happily on a small keyboard, Shawn worked on a brass quintet. When James went down for a nap, Shawn pushed his hand through his hair. *Maybe nothing has happened.*

Simon hovered constantly on the edge of his consciousness. But they were still here—so it was safe, right? Simon couldn't have succeeded. His mind strayed to Christina.

He sat for a long time in Angus's chair, staring out the lace window. They'd left him the letters, he finally decided. He headed to the kitchen to pour himself a glass of bourbon, climbed the stairs to Angus's spare room, and pulled out the parchments. He slid the last one he'd read to the back. Once again, he stared out the window, to the gray Scottish skies, for some time, before lowering his eyes to the new letter.

1 January 1318—by your accounting, read the date at the top of the manuscript. "By my accounting?" Shawn didn't recognize the script. *Those around me will call it 1 January 1317.*

Recognition jolted Shawn. "Brother Eamonn." Of course. In Niall's time, the year would not change until Lady's Day, March 25. Questions flashed through his mind—and relief. The old man was well enough to write. Shawn smiled. Amy would be glad.

Let me tell you a story, Brother Eamonn wrote. *I have spent my life reading other men's stories. This once, I will tell my own.* Shawn could almost hear his voice, cracked and thin, as it had been in the hospital chapel, just weeks ago. He took a sip of the bourbon and began.

I heard the music first, a thin, piping sound, thinking 'twas my dreams, as I lay half awake in my cell, pleasantly chilly on the brink of dawn. I thought it odd. Nobody played a flute here in the monastery. I was young, new to the place. But I'd been there a month, and nobody played flute.

I rose, pulling on my brown robe. It was still a novelty and a joy. It still left me feeling as if I was slipping into a character, the monk I'd long wanted to be, like the saints I'd read of growing up. I pulled on their skins when I pulled on my robe, and while I still sometimes felt I was playacting—waiting for someone to come and tell me I was just the boy from the west, the boy who fought with his brother and once threw a rock through old Mr. Hamilton's window— perhaps that playacting, if such it was, brought me ever closer to the truth: this was who I wanted to be, and by virtue of wanting it, in some small sense, I already was.

I stood at the window in my woolen robe, looking out over hills all covered with mist, and purple in the gray dawn, the sun still drowsing behind the hills, and the stars still bright in the dusky charcoal sky. And I listened to the reedy sound. 'Twas no flute, I realized, but a recorder, playing an ancient melody.

I felt a frown crease my forehead. Nobody here played recorder, either. And if they did, they'd not be doing so in these, our few hours of sleep. I turned, dropping to my knees under my crucifix, and saying, as I always did, "Use me, God. Let me be your servant."

I felt for a wee moment that I heard a chuckle. Not audibly, ye ken, yet as clear as anything I've ever experienced. Given the monks have believed for fifty years I'd a penchant for drink, 'twill mean little to most. Nonetheless, it is so. It was there, sure as this quill and parchment. I chuckle as I write that. 'Tis my private joke, for sure no sane person believes I write, in the 1300s, of my life in the twentieth and twenty-first century. Yet it is true.

Shawn smiled. Though he'd railed at the old monk, he liked his humor. He suspected Niall would like him, too. He sipped the bourbon and continued.

I looked up to the crucifix, and words came strong to my mind. "Do you really wish to be my servant?"

Whether 'twas my own thoughts or God's voice, I answered the same. Of course I do.

Be careful what you ask, the Thought warned me, for I may surely grant it. I've a rather large job.

Let me be your servant, I repeated. I thought of the Little Flower and other great saints, who did naught of importance by the world's standards. Do small things in a great way. I would be the best, most godly porter to the monastery, in the way of Saint Conrad. I would tend the gardens, if such were my calling, pruning my soul of vice and tending prayers for the world, as I pruned and tended the blooms.

No, no, the Voice said. Not little things in a great way, but aught that is truly large, of great import.

Let me be your servant, I repeated. But I was becoming disturbed by the Voice, for I could no longer pass this off as my own thoughts. 'Twas not how I spoke, for one thing. But perhaps—perhaps God summoned me to travel and preach! Disturbance gave way to excitement.

You will suffer indignities, It warned.

I was not so quick to answer this time, and I thought I felt the Voice laugh.

May it please You, I replied hastily, I am your servant. I thought of the martyrs, and my heart leapt a bit. Was I truly to be counted among the great?

Oh, you will suffer a martyrdom, the Voice said. But there are many kinds of martyrdom. Will you do it?

I am your servant, I replied. I was a wee bit relieved, for I didn't fancy a painful death. I would be a living martyr like Mother Teresa, working among the poor, hailed by the world for my devotion and the kindly wrinkles that would show on my face one day. Yes, I smiled, liking this idea. The best of both worlds. Living martyrdom, drawing people to Christ, and hailed by all!

Shawn smiled, recognizing the self-mocking humor.

The Voice fell silent. I felt my fiat had been heard. The music returned, and without thought, I rose, donned my sandals, and followed it out into the ancient stone walls. I walked down the hall, always feeling 'twas near, yet I never got closer. I slipped into the cloisters, where it sounded equally near, equally faint and far away. I wandered through the kirk yard, and still it seemed everywhere and nowhere. And then, a strange thing: This ancient melody became a very modern song. Take the A Train.

Another jolt! Shawn set down the glass. Niall. It had to be Niall. He shook his head. No, it could have been anyone. He scanned the next words.

So it was one of our monks. Old Brother Edward, perhaps, who had spent some years in America and loved their big bands? I heard a laugh, and the music stopped. As you can guess, this was my first experience of times overlapping in Monadhliath.

So it was Niall.

As you have seen, my martyrdom was not as I imagined. 'Twas to see what few see, to question my own sanity, to struggle alone, for there were no books, no prayers, no precedents, no leaders, and none who would believe these things were happening. 'Twas to be called drunkard in my youth, doddering fool in my age, and still to have this mission, regardless of how it caused me to be seen.

I have written elsewhere of how I came to understand what I dealt with and what I must do. I learned that one event never changed. A young knight would come over that hill and kill Simon Beaumont, English knight, in the fourteenth century.

Somewhere in my soul, I understood this was how it must be. I understood 'twas not my calling to stop Simon, but that of a young knight called James. James Kleiner.

Shawn shook his head. No. He closed his eyes, wishing he had not opened these parchments. No, it would not be James Kleiner. With a hand to his mouth, he opened his eyes. If there was something to be known about his son—hadn't he better know it? He lowered his eyes to Eamonn's ancient words. But he was already telling himself...*Brother Jimmy's Brew.*

This meant little to me for fifty years. Then Amy appeared at Monadhliath, a wee bairn in her arms. I learned the wean's name was James.

And there sat Simon Beaumont with them. I saw how he looked at the child, and I knew—that he knew. Somehow, he knew, as I did, that this child's fate was linked to his own, that their lives were tied together, that one of them would kill the other.

I had read of a long-ago knigh, a faceless young man, no background, no mother nor siblings nor father to worry about him. Just a name. Now they stood before me: A mother and son.

I knew who would win. But now I also knew, seeing a lass before me, cradling her child, the cost of that victory. A mother who loved her son, whose purpose in life, as for all mothers, was to protect her child. A mother who would learn one day that she could no longer protect him, that she must give him up to the world.

In my humanity, I could not face it. I had so much knowledge. What was it worth if I couldn't do this one wee thing? I knew what would be attempted, what would succeed, what would fail, I knew each player's moves. Did it matter how Simon was stopped?

He had underlined *how.* "No." The word slipped out on a breath. *No, Shawn agreed. It does not. And my son is not going there.* He had sworn, too, to protect Amy. *She is not going to suffer, either.* He leaned in, though he told himself, more strongly: *Brother Jimmy's Brew. Just because he believes it....*

Did I not have more to guide me, with all my knowledge, than a boy born in the twenty-first century? How would such a one fight a medieval knight?

So I took it into my own hands, to fight Simon myself. I was tired of a history so big, so vast it no longer had a human face. I wanted, finally, to do this one thing, for this one lass before me, holding her child. I finally wanted to see the face of one person and know I had made a difference, rather than being ever the fool.

The ink grew faint. Shawn adjusted the lamp, shining it on the spidery script.

But is that not vanity, like any other, to demand to see results and take credit? It should not have mattered. But I had spent years being thought the fool, never seeing how my life mattered to anyone, in any way. Oh, I supposed standing here in between times was doing something. I knew God had a purpose. But I couldn't see it. Just this once, I wanted, I felt I needed to see it.

We overestimate our own needs, do we not? I wanted it. So I grasped. And here, I address my words to Amy. Please see she gets them:

I lured Simon to the Glenmirril chapel, sure I could send him back into a time and place where he would be swiftly dealt with and I could finally look at one person and know I had made a difference.

I was vain and overconfident. Perhaps I'm not so different from Shawn and Niall in that respect.

Shawn smiled.

I believed I, alone, could change the history that never changed, that I could change another's fate, and take on the task reserved for another. In the end, perhaps the reasons don't matter.

As I knelt, dazed, in the chapel, not knowing what time I was in, the Voice from years ago sounded in my heart, full of grief, telling me Simon must play out his fate. And his fate was not to die in Glenmirril.

All things must work together, the voice gently chastised me, and so I must get him out, that he and Niall and MacDougall—and yes, James—might play the parts as they must be played.

I tried to save you from this, but in the end, I am naught but a foolish, vain, old man. I could not. This is how it must be. 'Twas always James's story to stop Simon, not mine. 'Tis yours to prepare him for that moment. His life rides on it, but you know that. More importantly, history rides on it. He will leave you, and he will face Simon. You know the story ends with James's victory over Simon, but knowing the ending doesn't always make it easy to live through the story.

'Tis sorry I am I could not save you from that.

Brother Eamonn believed it.

One more thing. There's one letter that never changes. But there are many that do. One letter tells of the young knight dying of infection. A victory, a vital victory, but a costly one. Perhaps that letter might change.

Shawn stared at the words for long moments. Suddenly, he shook his head sharply. No, James was not going to die from a medieval injury! It was ridiculous. Eamonn was crazy. Wouldn't a man go a little crazy living such a life? Brother Eamonn believed it. That didn't mean it was true. He had been right in what he said to Amy. James was safe.

Hands shaking, Shawn rolled the parchments and shoved them back in the oilskin.

Scottish Borders, 1318

"We still don't know where MacDougall *is.*" Irritation hung on Niall's words as he scratched his flint beneath the twigs in a hastily-constructed fire pit. Night was falling swiftly. They'd eaten nothing and slept only an hour since leaving Dundolam the previous afternoon, picking their way cautiously even through the night in a hard drive for the border.

"We ask again at the next town." Conal came into the small clearing, dangling two hares by their ears. He tossed one to Hugh, who laid it out and set to skinning it.

"They're most certainly coming after us," Niall replied. He disliked having to stop, knowing they were being tracked, but their ponies needed rest. "We're slowing ourselves by asking at every town, *and* we're leaving a trail. They'll be asking if we've come through. We're marking ourselves." The flint sparked, a tiny glow in the darkening forest.

Hugh's hands paused in his work. "We've been eejits."

"Eejits?" Conal looked up from the hare he worked on. "How've we been eejits?"

"They'll follow us." Niall saw immediately what Hugh meant.

"So ought we not keep going?" Conal asked. He resumed flaying the hare, slicing fine strips of meat.

"No," said Hugh. "We ought wait. Let them pass and follow *them*."

"But we dinna ken which road they'll come down any more than they know which road we've taken," Conal said.

Hugh stuck a thin slice of rabbit meat onto a whittled stick. "Will they not go to the first city in England?"

"MacDougall was at Carlisle when I was there with...." Niall stopped. He couldn't name Shawn in front of Conal, Owen, and Lachlan. "'Tis where MacDougall will likely have gone to start. So mightn't his men go there first?"

"The roads converge at the Firth of Solway," Conal said. "They must needs pass through there."

Niall took the stick Hugh handed him, three strips of meat stuck on its short branches, and held it low over the fire. "So we get there first and hide. We follow them when they pass. They'll lead us to MacDougall."

They fell silent, roasting their meat over the low flames that sparked and crackled and did little to keep the cold at bay.

"Could he not have kidnapped her in July?" Hugh grumbled.

Niall pushed a larger stick into the fire. The flames climbed, spreading a little more warmth.

"If they go into Carlisle," Conal asked, "how do we watch all three gates for their departure? They might stay for days."

"'Tis the best we have at the moment." Hugh lifted his stick off the fire, studying the meat, and held it away to let it cool.

"'Tis *all* we have," Owen said.

"*If* they go in," Niall said, "and *if* they leave again in search of MacDougall, they'd take the south or east gate. We can watch both from the southern hills."

"They might stay for days," Conal said doubtfully.

"We'll watch in shifts," Niall said.

They busied themselves with picking at the cooling meat, tearing it off the skewers with teeth and fingers. At last, with the stars twinkling far above the trees, and the flames dying to embers, Conal glanced over his shoulder into the dark woods, and asked softly, "When we find him, what is the plan?"

"We are not to be seen or associated with the deed," Niall said. "Wherever we find him, we seek a way in, a disguise, mayhap we hire someone in a nearby village to assassinate him. Think on it." He gathered his cloak about himself. "I'll take the first watch."

The other two nodded and rolled into their cloaks, huddled close to the low orange flames, while Niall started a patrol of the dark forest.

CHAPTER NINE

Inverness, Present

"Angus?"

Angus turned at the sound of a voice vaguely familiar. The meds left him drowsy, stumbling up from dreams of Amy, James, and Shawn...shame at Amy seeing him like this swirling in a helix with a wonder that maybe she could still love him—that maybe, *maybe*, somehow, they had a future. He should relent— let her in. Take the risk. Yes, he would. She was here now....

"Angus?" The voice spoke more sharply, pulling him back to the hospital room. His eyes focused drowsily on the dreadlocks, the swirling gypsy skirt in shades of pink, blush, and rose, fading out to white points at the hem. It wasn't Amy. He knew her. But he couldn't remember....

"Helen," she said. "Helen O' Malley, professor of medieval history. I met you and Amy in the graveyard below Stirling."

Something clacked, a small metallic click and clack, as she spoke. He remembered it—but not what it was. He tried for a smile. "Of course." He couldn't place why she'd be here. His back hurt.

"Sure, you're tired. I can come back later."

"No, I'm grand," he lied. He pressed the button that raised his bed.

"I heard about it on the telly," she said. "I wasn't sure—you know, should I come down. How you were feeling. And then your sister rang me." She held up a large, thick yellow envelope.

"Mairi?" He tried to think what the envelope had to do with Mairi.

"The genealogy," Helen explained, her eyes brightening. She came into the room, confident now of his enthusiasm and began pulling sheets from the envelope, covered in names and dates, spreading them across the white blanket that covered his legs. "She said you hoped to look it over and see...."

"'Twas not *exactly* what I said."

"...where the MacLean family comes into it all. You remember the MacLeans link up with the Stewarts and the MacDougalls?" She lifted pink glasses from a rhinestone-studded chain and perched them on her nose.

"The thieving MacDougalls," Angus murmured. His back hurt. The drugs made his head swim. He closed his eyes for a moment, remembering Honora and cookies and Amy, her hand pressed to her swollen, pregnant stomach. He missed James. He opened his eyes. "There was only the one copy."

Helen beamed. "I visited her again. She was happy to let me take it all down for posterity. I've the entire MacLean branch for you!" She indicated the sea of papers covering his legs.

He stared at them, wavering between irritation at Mairi and amusement. He pulled one closer.

"'Tis hard to follow," she apologized. "But see here—and here." She moved the sheets swiftly, pointing. "I've highlighted the path so you can see your ancestors—back to James Angus." She sat down, her face suddenly becoming serious. "'Tis maybe not the best time, but it's been on my mind."

"James Angus?" Angus asked. He shuffled the pages into a neat pile. His back hurt. His head swam with the medication that kept worse pain at bay.

"Not him," Helen said. "Simon Beaumont."

Angus's hands became still, the pages clutched between them on his lap. "Simon Beaumont?"

"An odd man."

"At the least," Angus replied.

"I almost feel I shouldna bother you with it. You've enough worries."

"Something happened?" His voice rose, along with the fear in his chest. Shawn believed Simon was gone.

"Oh, not recently," she said hastily. "'Twas when he visited me—ages ago. It's been worrying me, and lately—even more. And I keep thinking—I don't know why—but I keep thinking I must tell you. Or maybe the musician—Shawn Kleiner." She removed her glasses, staring at him intently. "But I *know* you, so when Mairi called, it seemed the Universe was telling me to tell you."

Angus twisted, trying to sit up straighter, but pain shot through his back. And his legs were useless. He settled for, "What happened?"

"'Twas not so much that anything *happened*." She scooted her chair closer. "'Twas a letter written in the thirteen...teens. The date was obscure. In truth, it may be the twenties. 'Tis too faded to say."

"What was in it?" Angus prodded.

Her nervousness dropped and she spoke, suddenly, as she might before her classes at university. "'Twas written by an unknown source, addressed to a Lord Douglas. We presume the Good Sir James, though we don't rightly *know*. It told of a battle. I have it here." She dug in a voluminous bag sparkling with fuchsia spangles, and pulled out a folded sheet of crisp white printer paper. "Are you good to read, Love?" she asked. "Or will I read it to you?"

"I'll read it," he murmured.

She unfolded the paper and handed it to him. He glanced at the winter light pouring in the window, and, with a sense of dread, turned to the New Times Roman re-printing of ancient words.

Linstock, Northern England, 1318

"The Bishops' palace?" Hugh scratched his chin through his beard.

"Brilliant, both of you," Niall murmured. MacDougall's men had ridden hard through the crossroads, just as Conal had predicted they must, mere hours after they themselves reached it and hidden. MacDougall's men had looked neither left nor right in their haste, leading the group from Glenmirril another day's ride, past Carlisle and directly to Linstock.

Niall and his group now watched from forest cover, across a wide field as MacDougall's men rode for the castle, two dozen of them clattering across the drawbridge. It reminded Niall uncomfortably of watching those same men, so long ago, crossing Glenmirril's drawbridge.

"Why would he go to the *bishop*?" Lachlan asked.

"'Tis often let to gentry," Owen told him.

Niall dismounted, watching the last man cross. "He won't stay long."

"No," Conal agreed. "Sure Edward will grant him lands in England."

"So we strike fast, else risk having to find him all over again," Hugh said.

"Which we've not time for," Conal said.

"Aye, we've much to do before March." Niall's breath hung in the air. He wished he could be riding into a castle, himself, to a warm bed. It would be nice to believe he could kill MacDougall swiftly and race home to Allene, his sons, and warmth—something he had begun to think he would never feel again. He gave his head a sharp shake, dislodging the fantasy. There was work to be done.

"Think, though," Owen said. "His men are even now begging his audience, to warn him we were seeking him at Creagsmalan."

"Will he know 'twas us?" Conal asked.

"They'll say 'twas Fionn, so aye." Hugh slid off his horse and squatted beside it, staring at the tower across the field. "So unless we can get in and kill him in the next few moments, he'll be alert for us."

Niall frowned. "He'd not think we followed his men. MacDougall has always believed himself smarter than others. That leaves surprise on our side."

"What surprise," asked Conal, "have we in mind? Do we find a way in to kill him in his bed?" He nodded at the drawbridge, wide open, and the farmer even now crossing it with a cart laden with hay. "Getting in will not be hard."

"No," Niall agreed. "But his guards may object to finding their master dead. Coming out may not be so easy. We'd have little enough time to find his chambers before nightfall and his men will recognize any of us wandering about the castle asking for it."

"Poison," suggested Hugh. "We get into the kitchens and poison his food."

"The food may not go to him," Conal pointed out.

"One of us takes a job as a servant and serves him." Hugh studied Conal. "We can change your hair and looks. He's less familiar with you."

"Is poison worthy of a knight?" Niall asked. "Where is the honor in poisoning a man?"

Hugh snorted suddenly, loudly.

Niall cocked his head. "You find humor in the question of honor?"

Hugh glanced at Conal before addressing Niall. "We know *one* who would argue there is no honor in killing a man a 'tall."

"That *one*," snapped Niall, "lives in comfort and safety, does he not, with no fear of Amy being abducted or his son murdered. He may speak of honor or killing when these matters once again impact him."

"Once again?" Hugh asked.

It discomfited Niall to realize he'd spoken as if Shawn were returning.

"Of whom do we speak?" Conal asked.

"Of whom *do* we speak?" Niall looked to Hugh, one eyebrow arched.

"Brother Andrew," Hugh replied.

Confusion crossed Lachlan's face. "Brother Andrew has a son?"

"He spoke metaphorically. We'd best find food and a place to stay." Niall turned to survey the area beyond their cover of trees. More than two miles outside Carlisle, it held little but the castle itself, surrounded by fields and a cluster of rude homes,

"A wee hamlet like this?" Hugh shook his head. "Five strangers will be quickly noted."

"We stay in the forest, then," Niall decided.

"Do we go in or wait for him to come out?" Owen asked.

Lachlan cast an eye to the gloomy sky. "It could be a long wait."

"Perhaps we can hasten his leaving." Niall grinned suddenly. "Let us find a horse and a boy."

Inverness, Present

The Year of Our Lord 13— Angus squinted. It was, as Helen said, hard to be sure of the date. *To My Lord Douglas, Greetings and Good Health. We met Sir Simon Beaumont of Claverock on the road, coming with his men as you predicted. We engaged him in battle, though he outnumbered us.*

He looked up. "What was Simon's reaction when he read this?"

"I'd gone for coffee," she said, "so I didn't see his immediate reaction, though he seemed angry when I returned. He said no, only perplexed."

Angus dropped his eyes back to the letter.

The battle went poorly, My Lord. I fear young Wat will not be coming home. Jamie MacPherson will not be fighting again, such is his injury. We lost several others, too, early on.

I fear, My Lord, the outcome would have been different, but that a

young man suddenly burst over the ridge on a great charger, letting out one short, sharp, piercing scream, and threw himself into battle, hacking and slashing. He rode straight at Beaumont. My men, whose hearts had begun to falter, took courage, and threw themselves back into the fight. Soon our fortune turned.

Grim relief filled Angus.

When finally I saw no more of the enemy, I turned and saw this young knight fighting with all his strength against Beaumont. They were afoot now, their helmets off. He had long hair, as black as your own. But for its color, I would have thought 'twas Sir Niall Campbell of Glenmirril, for the face was his.

Niall's son? Angus's stomach knotted, as if he read of his own nephew.

The young knight fell to his knee, but as Beaumont moved in to kill, the man thrust his sword upward, piercing Beaumont's chain, and Beaumont fell dead.

Relief returned with a force. Angus felt weight lift from his chest. He realized his muscles had tensed, as if he'd been gripping a sword, himself. He looked to Helen. "Did he comment on the man in question having the same name?"

"*I* did," she said. "I joked, saying, *Bad news, it seems you're dead!*"

"Did he seem angry at that?" Angus asked.

Helen nodded, the dreadlocks moving vigorously, and then immediately shook her head instead. "He laughed. Said he'd not only survived, but lived another seven hundred years."

"But you first nodded," Angus pointed out.

"Aye, for he *did* seem angry. He seemed tense, coiled, ready to spring. He laughed. But his laugh was...."

"Not genuine," Angus said.

"Like a poor actor," Helen agreed.

Angus lifted the letter, reading the last of it.

The young man, despite his victory, fared poorly. We sat him on a rock, noting his injuries. We asked his name, that we might send word to his people. He said he was called James, and would say no more.

A still small warning stirred in Angus. Niall had a son called James.

We removed his mail and saw a distinctive brown mark on his left shoulder blade, flaring out in the shape of flames or a wing.

"No." Angus shook his head, staring at the words. His insides felt cold.

"No?" Helen leaned forward. "In ten years working with medieval records and artifacts, I've never seen anyone so personally distressed." There was no trace of her usual humor in her voice. "Sure this Simon Beaumont who died was no friend of yours?"

"Of course not," Angus said. "He was the most vile of men! His death was no doubt celebrated in the streets."

Helen took the letter from him, scanning it. "James? The young man?"

Angus stared out the window. The snow had stopped, leaving heavy gray skies. His back hurt.

Helen rose. "Shawn Kleiner says he was in medieval Scotland. Him— I'd expect a publicity stunt. You—there are only so many reasons, are there not, to be personally agitated about an event?"

Angus turned back to her. "We'd like not to be thought mad, now, aye?"

She tapped her rhinestone studded glasses against pursed lips a moment, frowning, before saying, "Aye." And then, "It may be pure coincidence, but some of our research around this letter suggests Simon Beaumont was working with Alexander MacDougall when this happened. You'll remember his son, Duncan, of Creagsmalan, is thought to be the father of James Angus—your ancestor."

"Duncan and the raging Christina who somehow ended up at Glenmirril." He tried to remember if Shawn had said anything about Christina. He couldn't remember.

"Another interesting question." Helen cocked her head. "Perhaps Shawn might tell us how that came to be?"

Angus smiled through the pain in his back. "You'd have to cite sources other than the word of a madman, would you not?"

Helen grinned. "Ah, but perhaps a madman can tell me where to begin looking for less reliable but more *trusted* sources. Will you see him again?"

"More's the pity, I'm sure I will," Angus said.

Helen laughed, setting her dreadlocks to shaking. "Still, put the meeting to good use and ask him for me, will you, Love?"

"I'll do what I can." James needed Shawn, if he was to fight Simon.

Helen rose, smoothing her wild pink skirt. "Was it good I brought you the letter then?"

"Very good." He would have to make sure James had what he needed.

"Let me know what Shawn thinks happened." She grasped his hand, giving it a squeeze. "Get well, Love."

Linstock Castle, 1318

"Fine animals." Simon moved through the stables, examining MacDougall's horses.

"The best of Scotland." MacDougall couldn't restrain a smile, though Simon rankled him. Showing off his animals relieved the most noxious of the man's company. And his beautiful horses could almost make him forget Christina for a time. And Duncan. "Angelo here—you'll appreciate him on our hunt on the morrow. Swiftest horse I've ever ridden."

At that moment, a clatter arose outside. MacDougall turned from his animals to stride through the straw into the chilly courtyard. A dozen of his men thronged the cobblestones, icy breaths floating in the winter air. His lieutenant slid from his horse, already bowing before him.

"My Lord, I bring news from Creagsmalan."

MacDougall's face grew dark as he listened.

"Fionn, Niall, Hamett, Sydney Carton," Simon said, as bits and pieces of the story came out. "Shawn. Just how many men are there?"

MacDougall scowled at him. "I've no doubt there are but the two we discussed. Hammet may be another of the young men of Glenmirril." His eyebrows drew down fiercely over his sharp nose. "I want them both dead."

Simon smiled, the corners of his mouth tight. "It seems Sir Niall feels the same of you. However, it is Niall who will draw Shawn. Be cautious in meeting him ere the trap is laid."

MacDougall's eyes bored into his uninvited guest for another moment before he turned to his man. "How long ago were they at Creagsmalan? Did anyone tell them anything? Where have they gone?"

"They escaped, my Lord," the man replied. "We know not where, but sure you'll find Sir Niall either at Glenmirril or at Berwick."

"So he'll not be here looking for me?" MacDougall clarified. "None told them where I am?"

The man shook his head. "My Lord, only Lady MacDougall knew where you were. She was warned of the ruse and said naught."

MacDougall slapped his hands together, beaming. "Then we have once again outwitted the young fool." He turned to Claverock. "Is it Angelo you'd like for the hunt?"

"My Lord!" Before Simon could speak, a young boy ran into the courtyard, calling, "A wild boar in the forest! The likes of which ye've never seen! The governor's promised five pounds to any who bring it in!"

"It seems God smiles on us," Simon said. "Come, let us leave behind these problems for the day and bring in this monster! I have missed the hunt." He smiled. "And I believe you and I could each use some extra gold at the moment."

SECOND MOVEMENT

CHAPTER TEN

Inverness, Present

Alone in Angus's front room, with James and Amy asleep upstairs, Shawn stared at the damped fire in the hearth as he sipped his bourbon. *Survivor's guilt.* Shoving the parchments away, into the oilskin, deep into a closet, hadn't stopped Amy's words rolling through his mind. *Maybe nothing has happened.*

But maybe it had. His world would be complete, he thought, if it could just be this—soft embers and Amy and James happy and safe in his home, his music. No worries over what happened to people he'd loved. *Did* love. Rational or not, it didn't matter that they were dead.

He sighed. They *were* dead. There was nothing he could do. He stared at the glowing embers. It wasn't just them. No matter how he hid from it, there was Eamonn's letter. And he couldn't quite win the wrestle within his own mind and believe Eamonn was crazy. If past events impacted his son, he had to know.

No! He set the bourbon on the end table with a sharp crack of glass on glass and yanked out his phone. Eamonn was demented from a life of seeing what no one else saw! He had to just get back to work! He opened his e-mail.

His accountant had sent statements—royalties, ticket sales, payments to venues and musicians—questions about tax shelters, and a request for a meeting. *Thanks,* he spoke softly into the phone in the dark. *Let's look at some donations to....* He stopped, considering. Aaron, he thought. *...to that program in Los Angeles. Give the money in Celine's name.* He hit send before he could consider the ramifications of that. It was one small thing he could do to apologize to her.

His publisher wanted another book. His arrangements were in demand. *We need to have a meeting,* the e-mail concluded.

I need a couple months to finish the next set, Shawn said into the phone. *I'll look at my schedule.* He hit send, and stared at the soft red glow in the hearth a moment, wondering what the next series should be.

Laughing Brass, he decided—and the other pieces that had disappeared when he and Niall changed history. *The Lost Pieces,* he would call it. He took another drink of his bourbon. He hadn't settled the moral question. But the first thing was to put that music back. And no one but him *could* do it.

A notice had come about the kids he sponsored—the one decent thing he'd done before Niall, he thought ruefully. He opened the digital photograph of Rogers, now seven, squinting into the camera, his head cocked to one side. Shawn read through the short letter from the boy's mother telling of his progress in school, hesitated a moment, and dictated a short letter into the phone, encouraging him.

Ben's voice came next, in strident pixels. *Shawn, I'm not sure what's going on there. But we have business here. BSO and CSO want to book you. I need an answer ASAP. You've got people waiting on meetings. You have to come home and talk to them if you want to keep this thing rolling.*

Shawn got up from the chair, tightening the tie on his trews, and headed into the kitchen to pour another glass of bourbon. He drank a quarter of it, wrestling desires and commitments. But he'd always made decisions quickly. He snatched up the phone, bought the tickets and dictated an e-mail to Ben. *Just bought a ticket. Set up the meetings.*

He stirred the embers, tossed in some kindling and another log, and sank back into Angus's chair. There were no more e-mails. No more distractions. He stared at the flames as he sipped the bourbon.

Who was he fooling? He could no longer bury himself in business as he once had. He couldn't even bring himself to enjoy a night with Jenny—to run away from it all in her bed—as he could once have done so easily—to pretend none of it mattered, and life was about having fun.

Survivor's guilt.

He nursed the bourbon.

Maybe nothing has happened.

The flames crackled. Shadows danced across Angus's front room.

If something impacted his son, he had to face it.

Shawn finished his drink, set the glass down, and went upstairs for the parchments.

Inglewood Forest, across the river from Linstock Castle, 1318

Niall scanned the trees rising thick around them. The forest started just south of the river, cut by a single dirt road—the only path for MacDougall to take, should he buy the boy's story.

"We could be waiting for days," he murmured to Lachlan. "He may not care to hunt boar."

Lachlan shook his head. "He's lost his lands in Scotland and a good deal of his income. He's cooped up with naught to do at Linstock. He'll come."

"Hopefully today," Owen muttered.

Hugh looked to the gray sky overhead. Flakes of snow were beginning to drift down.

"If not in the next hour," Hugh said, "we rest for the night and rise early."

Niall nodded in acknowledgment. His spirit was anxious to act or move on. But he could only wait under the snow-frosted trees, reviewing their plan: Lure MacDougall into Inglewood with the story of the boar, shoot to kill, ride hard for the eastern hills, and from there, up into Scotland with none the wiser.

"Look." Hugh's deep voice broke into Niall's thoughts. He pointed at the ground, where their ponies' hoof prints showed plainly in the gathering drifts.

"Make tracks going south." Niall looked around the small group, at Conal, Owen, and Hugh. "Separate. Leave three trails, and come back by different paths." He turned to Lachlan. "You're the best shot. Wait for them; we'll be back in minutes."

The rest of them spurred their ponies into the swirling flakes, spreading out and quickly fading to dim shapes in the growing snowfall. It muffled the thud of their hooves. "Haste," Niall hissed into the whirl of white. "We dinna want him meeting them alone."

"If they even come." Conal appeared out of the snowy mist, and a moment later, Hugh's form emerged on his other side.

"Mayhap they left immediately on hearing the story," Hugh said, "before it grew so thick."

"Owen," Conal whispered.

"Here." A horse's head appeared, a dark shape, and then the man atop the animal. They swerved in and out among the pines for several minutes before turning their animals north and weaving back through snow-veiled trees.

A shiver racked Niall. They wouldn't come out in this. But he kept the thought to himself—in case they *were* here in the forest to overhear. "Have we lost our way?" he whispered.

"He's still ahead, I think," came Conal's voice from his right.

Then, somewhere ahead, a voice cut through the white veil. "A fine day for a boar hunt! We'll be fortunate to get home, never mind find this beast!"

"'Twas not snowing so hard when we left."

Niall's adrenaline spiked. It was MacDougall! He slowed, veering to his right, deeper into the woods, and sending up a silent *Ave...*two *Aves...*that they might find Lachlan.

"I hear something!"

It was Roger, Niall guessed. He hoped it wasn't himself and his friends who had been heard. He looked back, a finger to his lips, and slowed his garron. Snow muffled its footsteps. But Hugh's tack jangled behind him.

He uttered another *Ave,* desperate for a sight of Lachlan, even as he slid an arrow from his quiver and nocked his bow, gripping the pony with his knees. He heard the whisper of three more bows from the men riding behind him.

"We came this far," MacDougall shouted. "We'll flush it out."

"What if the snow gets heavier?" asked a man with an English accent.

Unintelligible words reached Niall's ear, as he urged his pony slowly forward, and then, "...path...easy to see...."

A dark shape wavered against the white cloud of snow. Niall's heart lurched in the split second before he saw the outline of Lachlan's raised bow, three arrows pointing toward the voices.

"How far can...." The words were lost as more voices rose, arguing.

Niall raised his own bow. The others fell into place beside him. He jerked his head, and they all nudged their ponies back a step. "Three flights," he whispered, "and retreat."

A burst of swearing erupted from the path, and now they heard the dull tromp of hooves. No more than a dozen, Niall guessed, and likely fewer. They had the advantage of surprise. But how much was that worth when they could barely see?

"It's not out and about in this!" someone shouted, and in between the last two words, they materialized, just beyond two trees, leaning forward to peer down the shrouded path.

Niall gave a sharp nod of his head and released his arrow. It hissed between his fingers, his arm already cocking back for the next arrow, letting it fly. A scream erupted. He grabbed a third. It shot into the white cloud to an ostinato of curses filling the air, a shout as Hugh grabbed his arm, yanking him 'round to the sound of another scream, and the five of them pounded through the white forest, in and out around ghostly trees, leaning hard over their horses' necks.

From the path came the sound of yelling. Niall kept his eyes on the trees, his mind on all he knew of the land, praying he was still driving east. He could feel Conal's horse just behind him, Hugh at his side. He disappeared around a tree and veered in again.

Cries rose behind them, the thud of footsteps, and a man bellowing, "I can't see a thing!"

A muffled voice answered.

Niall dodged around a great oak, pushing his mount to greater speed. "Lachlan!"

He heard Owen's shout and slowed enough to look over his shoulder.

"I'm here!" gasped his friend. "Go!"

Niall kicked his heels into his garron, spurring it forward. They pressed through the blinding curtain of white that stung at faces and bit at hands clutching reins, skirting trees. At last, they slowed, their animals resisting, the snow blinding.

"Do we even know where we are?" Conal asked softly.

"Look ahead." Hugh, like a dark wraith in the forest, pointed. The trees appeared to thin, and light to grow in the distance.

"The edge of the forest," Lachlan said.

"Let's hope 'tis the edge we were aiming for," Niall muttered. He should be grateful, he thought, that they hadn't simply run in circles and stumbled back into MacDougall's men.

"What if we've come out at the north," Conal asked, "and they have, too."

"Go quietly," Hugh advised.

"Bows ready," Niall ordered.

The weapons rose. And suddenly, they passed the last dark pine, the five of them side by side looking out over a plain toward the Pennines rising in the distance. Niall crossed himself.

"Thank our good Lord," Lachlan murmured.

The snow was thinning.

"Seventy, eighty furlongs," Hugh said. "We can make it by nightfall." He glanced back over his shoulder. "I hear them. We've no time to spare."

They leaned in, driving the animals toward the safety of the mountains.

Inverness, Present

Shawn unrolled the parchments across the coffee table. Their old smell, musty and a little moldy, soothed him, drew Allene and Christina and Niall to life. Eamonn's letter lay on the top. He shuffled it to the back, hating even the sight of it. The old man was crazy!

His stomach unknotted as he ran his finger down the next one, recognizing Allene's light feminine script. He squinted now and again where it had faded.

21 January 1317. In his own time, 1318, Shawn noted. *The others may also tell you in their letters. MacDougall escaped in December, wearing the gray robe and calling himself Brother Andrew.*

Shawn stared at the stark words.

So much for *Maybe nothing has happened.*

He took Christina with him.

He shot to his feet, drawing sharp breath; raking a hand through his hair. MacDougall would claim what he thought she owed him. He took a step in the small room. Turned, stepped. Let his breath out. He shouldn't have read! It just got worse every time he did!

He drew a slower breath. *Okay, it got worse. But maybe it got better again.* He looked from the parchments on the table to the empty glass. He went to the kitchen and re-filled it, taking a long draft as he seated himself once again at the coffee table.

Your Brother Eamonn assured us she was safe at Lochmaben, having escaped MacDougall, and so she was, and is home safely.

She'd gotten away? Shawn re-read the words, emotions washing through curiosity, pride, fear, and anger. She'd gotten away! How? By her wits, no doubt! What else did she have against him?

He grinned. *Take that, MacDougall!*

But MacDougall *had* taken her from the safety of Glenmirril. Plenty could have happened before she got to Lochmaben. The fear sickened him. He forced his mind from the ugly possibility, turning to the easier—safer—question of how she had gotten to Lochmaben. Anything to distract himself from that fear of what MacDougall had done to her.

He opened his laptop, his mind churning, fingers tapping till a map filled the screen. He found Lochmaben several days' ride from Glennmirril, near the English border. Shawn's mind spit out possibilities as quick as thirty-second notes. So MacDougall must be making for England. And with his hatred of Niall, and taking Christina, he was without a doubt plotting against Niall.

Shawn switched the map to a satellite view. Clouds, like flocks of Highland sheep, obscured his view of the countryside. But he scanned the route they must have followed, across Rannoch and the tip of Loch Tay. He zoomed in, studying the hills she would have traveled with MacDougall.

He'd traveled those hills himself. It would have been cold in late December, and difficult. Yet she'd made it to Lochmaben. How much the worse for wear? And had she had her son with her, or had MacDougall taken her alone?

He studied the parchment. The answers might be there. Or perhaps there would be worse news. He turned reluctantly from the image of the mountains Christina had traveled so long ago, back to Allene's elegant handwriting.

Father storms about his chambers, muttering, No good could ever came of a MacDougall, till I beg him to stop.

Modern notions of judgmentalism aside, Shawn thought, he rather agreed with the Laird on this one point. His blood burned hot in his veins, thinking of the man abducting Christina. She wouldn't have gone willingly.

Then he disappears to the Bat Cave. We've two new benches, a recorder, the shell of a currach, and the frame of a harp, since MacDougall left. Father says working with wood helps him think and he must think on what will come of this. Christina frets that he will return, though he has lost his lands here.

And too, we worry about Simon, whom Brother Eamonn says is a threat at the same time he insists all will come aright. If all will come aright, then he is no threat, aye? I do not understand his riddle and father rages at him.

I assume Father also prays, but I rarely go to the Bat Cave. I am too busy with the bairns and the needs of Glenmirril. Moreso, 'tis empty without the music and laughter that filled it when you were here.

Niall and Hugh have gone to seek news of MacDougall, and to put a stop to his misdeeds. They've work to do for our dear king and will then return to besieging Berwick with King Robert and Sir James.

Shawn typed *siege Berwick* into the search engine. The first site confirmed what he thought he knew. Bruce had tried to take it in March of 1312, and again in January of 1316. He remembered that one well. They'd beaten a hasty midnight retreat when Douglas had been injured, scaling a ladder up the city walls. That they would try—*had tried*, he corrected himself—again in August of 1317 was news. He'd left before then. In April of 1318—Shawn scanned the article only long enough to find out—Bruce finally succeeded,. He turned back to Allene's words:

It is too much to bear some days, all the evil that falls upon our land. Even as good news arrives, we fear what has happened since the messenger left. Niall will return to the siege—if all goes well in his search for MacDougall—and we have failed repeatedly in our attempts to reclaim Berwick. I know not what will befall him there. The waiting is hard.

Shawn wished he could write back to her, reach across time to hold her and reassure her the Scots would reclaim Berwick. He touched her next words, wanting to be with them all.

Ronan spends ever more time with the Widow Muirne. We guess there will be a marriage. Adam's widow is wed again and happy. Christina's son is nearly a year, and walks as we hold his hands. He fretted so with his mother gone.

So MacDougall had taken Christina alone. It had no doubt been better for the child, and easier for Christina to escape, if that's how it had played out.

He takes after her, with thick, black hair, and a peaceful nature.

He wanted desperately to see the boy, Christina's child, wanted to watch Christina by the fire, cradling the child in her arms, smiling up at him in the firelight. The suddenness and the depth of the desire took him by surprise. He glanced guiltily at the door, as if Amy might be standing there, seeing the thought flash through his mind. He closed his eyes, fighting the guilt, confused at the strength of his feelings for both Amy and Christina.

Cautiously, he opened his eyes and read again.

The castle accept him as her adopted son. I see Margaret's eyes on him at times, and suspect she knows. She does, after all, sew with us daily and is observant. Still, she says naught, and I bless her for it. As if to confirm my suspicion, Lachlan, when he was here, seemed to pointedly re-tell the story of finding the wean under the fir tree. Christina's peace returned after the ruse in the forest, finally able to be seen with him, to be acknowledged as his mother. She takes great joy in him, though she often sits in the window where you sat. She was given your chamber, and that, too, seems to comfort her.

We think of you often. Give my love to Amy and James. If Father were here, instead of grumbling and pounding his tools in the Bat Cave, sure he'd send his, too. Niall persuaded him not to rush to judgment on you.

All my love. Allene

As he drifted from the world her words had woven around him, Shawn's senses heightened. He turned—and jumped at the sight of Amy, silhouetted in the doorway, clutching the peach robe at the neckline. Their eyes met and held, the minutes stretching taut as a drum head. Christina hovered in his mind.

"Is Niall okay?" she asked.

He let out his breath, yanked from relief back to fear. "I think so. He's—on business for the Bruce. Looking for MacDougall. Allene doesn't know what will happen. The waiting is hard on her." He didn't mention Christina.

Amy came to the chair where he sat, wrapping her arms around him. He rested his head against her stomach like a child.

He didn't want to leave them. He didn't want to go to the States. "I abandoned them," he said. "I miss them." It was all he could do to hold back a sob. James Douglas's men didn't cry.

Inglewood Forest, 1318

"Wake up!"

A slap stung Simon's face. He floated in darkness, to the sound of moaning. Cold seeped into his legs; then damp. He realized he was lying in snow. His arm burned with sharp pain.

"Wake up!" He recognized MacDougall's voice. "We must go!"

He remembered then: The flash of pain, the thudding impact, falling from his horse. He fought to open his eyes. MacDougall's black beard and yellow teeth and sharp nose hovered close, his hand raised to strike.

Simon glared.

MacDougall lowered his hand, and shouted, "He's awake. Help him up!"

Hands grappled at him; Edward's colors swam before his eyes; jarred his arm. The shock of the pain drove the breath from him before he heard a scream and realized it was himself. He was on his knees in the deepening snow, unsure how he'd risen, MacDougall's men closing in.

He flailed at them with his left hand. "Get away!"

"My Lord Claverock," MacDougall said, "we must *go*. We know not how close they are."

Simon's fury rose. "Are we to run? Why have you not given chase?"

"Look about you," MacDougall snapped.

Simon looked. One of Edward's men was draped across a horse's back, hands and wrists tied. Blood streamed from his neck. Another man sat on a horse, his face ashen, a hand pressed to his side. Five more horses bore men dead and wounded.

"Few of us are in any condition to give chase. Roger and Donnel went and Donnel has not returned." Alexander restrained his irritation with Claverock. He gave no answer where he'd been these years since Bannockburn, missing and presumed dead, yet he demanded coddling of his ego and rage.

"There were many tracks," Roger said.

"We are greatly outnumbered," added one of Edward's men.

MacDougall's irritation mounted. "Besides which, we can barely see in this snow. Even now, they may be surrounding us. We'd hardly know."

"Come, my Lord Claverock." Roger offered a hand. "You are injured."

"A flesh wound! No more!" Simon brushed away Roger's attempt to help. He grasped his horse's stirrup, his injured arm falling awkwardly,

"We wrenched the arrow out while you were unconscious," MacDougall said. "'Twill heal but it is perhaps a wee bit more than a flesh wound."

"Let us pursue them!"

"Look at the snow!" MacDougall flung a hand out, indicating the swirling curtains around them. "We shall count ourselves fortunate to find our way back to Linstock at all and I for one will not risk my life or that of my men blundering about in this mess knowing they might surround us at any moment or we might simply be lost in it and freeze to death!"

They glared at one another for tense moments. "You *are* still under our charge," spoke Edward's guard. "We return to shelter."

"Bind his arm up that we might the better ride," Alexander ordered.

Simon stood silently, seething, as Roger took a rope, intended for tying the boar, and carefully strapped his arm to his chest. The pain eased a bit.

"My Lord," Roger said, "might I help you mount?"

Simon cast an angry scowl at MacDougall, but he let Roger and one of Edward's soldiers boost him awkwardly onto his horse. Indignation burned in his gut; pain burned in his shoulder, shooting the length of his arm. He gripped his reins in his left hand and shrugged off further attempts at help.

His animal fell into step behind Roger, alongside MacDougall. The horses bearing the dead and injured followed behind.

"If we but stay the path," Alexander said, "we'll reach either Carleton or Carlisle." To the man behind him, he said, "Have your sword ready." His hand tightened around his own weapon.

"Do we know who did it?" Simon asked. "Their heads shall rot on my castle walls."

MacDougall shook his head. "Arrows appeared through the snow. We saw only dark clothing. No colors."

"Brigands," Roger called back over his shoulder. "Seeking gold or food."

"Likely the very men the lad overheard," MacDougall said, "hunting the boar themselves."

"I want them dead," Simon gritted. His horse stumbled, jarring his shoulder. He clenched his jaw, holding in a cry of pain. Edward's men did not cry in pain.

"We will deal with them." MacDougall peered through the swirling flakes, not particularly caring what he promised Simon, as long as the man would shut up and worry about the immediate problem. The cold grew around him.

"Ahead!" Roger pointed. A hulking gray shape hovered beyond the white tunnel in which they traveled. They pressed on, Simon falling silent, though a moan escaped his lips. The dark shape grew, filling their vision, until a voice called, "Who goes there?"

"MacDougall!" Alexander called. "I am with Lord Claverock and King Edward's men. We were caught out in the storm. Let us in!"

A guard stepped out into their vision, materializing from the white world to grasp the reins of Simon's horse, noting the bound arm. "He's injured."

"Aye, we need lodging and a physician anon!" As Roger and a guard led away the sagging Beaumont and their injured and dead companions, MacDougall asked another guard, "Have others come in from the storm?"

"No, my Lord," the man answered

"Alert me immediately if any arrive." MacDougall turned, as they passed through the gatehouse arch, to stare into the blinding storm. Whoever they were, they were as likely as not to die out there in it. He turned his horse into the snowy cobbled streets, hoping he was in Carlisle. He had no more love of Andrew Harclay than of Simon Beaumont, but at least he might get good wine and a hot meal tonight, which was more than he'd hoped for but an hour past.

Glenmirril, January 1318

Having gone into her confinement, Allene waited uncomfortably in a high-backed chair, dining alone in the solar while the women attended the evening meal in the great hall. She stared out the window, her food untasted, considering what it must have been for Christina, who had been quiet since her return. How might she herself feel in the wake of such an event? Christina was not one to weep or even to speak of pain. So what might be helpful?

An idea came to her. Shawn's time had one thing of which Christina spoke wistfully: Bibles—Bibles available to all, thanks to the *printers* that would one day be built. It took the monks months to copy just one. Christina would never be able to have one of her own.

Allene was already pushing herself from her chair, pleased with her idea as she crossed to her writing table and took up a quill and a smooth, clean sheet of vellum. She had but little to do, waiting here in these few rooms for her time to come. She could certainly use it to write.

Calling to mind a verse of hope, she dipped the quill, carefully wiped the nib on the rim of the ink bottle, and began writing.

Over the Atlantic, Present

Shawn leaned back against the headrest, staring up at the sterile white of the overhead bins as the plane sped past giant fluffs of white clouds that strained, like Niall's garron, to catch them. He hated leaving Scotland. It felt like abandoning them again.

He had buried himself, since buying the plane ticket, in re-writing *The Lost Pieces*, as he thought of them, in calls with Ben—in trying not to think about Clarence's upcoming release—what a lousy time to be back in the States —and trying not to think how much he missed having Amy's love all to himself. She'd cradled his head, stroked his hair—and the next day took James to the hospital, still hoping Angus would see her.

"A drink, Mr. Kleiner?" He lowered his eyes to the stewardess, beaming at him. The skirt—Bessie wouldn't be caught dead in such a skirt! He blinked. Bessie *was* dead. She was dead even to history. She was dead to everyone except him. What if MacDougall had somehow taken vengeance on her, too, for her part in helping Niall escape? Had he ever even known?

He had to stop this! He pushed Christina forcefully from his mind.

"Mr. Kleiner?" The woman's smile slipped. "Are you all right."

"Uh—yeah." He cleared his throat, trying to cover his confusion at the switch of centuries. "A drink."

"The drink menu is on your tray."

His eyes lit on her name tag. A jolt flashed through him. "You're kidding."

"Excuse me?" Her slender eyebrows knit.

"Your name is Christina?"

She nodded, the smile slipping out again. "Can I get you wine?"

"Uh—yeah. Merlot."

She and her bare legs disappeared. He should open his laptop. Work on the set list for BSO; the arrangement of Cillcurran. Her name—it was just a dumb coincidence, that was all.

Cillcurran, Cillcurran, there by the sea.

He smiled, remembering Niall's anger at Monadhliath.

It would be a miracle if there weren't mist in Scotland! And Brother William—and Simon would be there. Had been there. Or so Eamonn said.

But who could trust him? Eamonn was crazy.

He pulled his laptop from the case at his feet and opened it, pulling up the nearly-finished score of the *Cillcurran* overture. He stared at the notes. The

Celtic melody drew him. It was an ancient air.

The parchments drew him. Maybe there would be more about Christina or her child or good news about Niall. Still, he thought, the good news about Christina had done nothing to allay his fears. MacDougall *was* still free, as was Simon. It did no good to stay away from the scrolls, fearing the worst.

No! He yanked his thoughts back. He had work to do! He had a life *here,* now.

"There you go, Mr. Kleiner."

A glass of red slid onto the tray in front of him.

"Let me know if I can get you anything else."

"Yeah, sure. Thanks." He averted his eyes from her legs, Christina's legs, not sure why this, of all things medieval, was so hard to shake. He should have ordered bourbon. He had no idea why he'd asked for Merlot. He sipped it, thinking of Amy, and returned to the score. The urge to read more would pass.

He finished the wine and the overture, ordered a second Merlot and pulled out a new piece. He rubbed his eyes, undecided if he wanted a vocalist on this one, figured there was no reason not to, and typed in *vocalist* by the top staff. He added a bass clef. How long an intro? He answered his questions as fast as he asked them, tapping keys, and brought a flock of notes singing to life on the screen. He smiled, hummed, hearing it all, added a nice counter-melody in the trumpets for a few bars—he realized, on re-humming that it was a piece Niall played on the harp—and finished the score.

The parchments called to him. He stared up at the bin that held them.

A laugh yanked him back to the plane. A woman stared at her small movie screen, giggling. No. He had a life to live here.

"Are you ready to order your dinner, Mr. Kleiner?" Christina stopped by his seat. The plane hit a pocket of turbulence. She swayed toward him and touched his headrest, steadying herself.

"Chicken," he said.

"We don't have chicken," Christina said. "We have..."

"Yeah, bring that."

"The beef risotto?"

"Yeah." He was already tapping at the keys again, trying to push Christina, and the parchments, out of his mind. *God,* he thought, *You're a real funny guy. Do you want me to read these or something?*

He resisted halfway through the piano line of *The Land that Was* before slamming the laptop shut and shoving it into the bag at his feet.

"Cool shirt." The girl across the aisle grinned at him.

"I bet your name is Christina," he said.

She looked confused. "No, it's Kristin. But I mean, how did...?"

"Of course it is." He shook his head as he pulled the oilskin from the overhead compartment and dropped back into his seat.

Inglewood Forest, Northern England, 1318

"It's getting thicker!" Niall lifted his voice over the wind that gusted down the vale between mountains and forest.

"Scotby ought be close," Hugh shouted. He had dismounted, shielding his pony from the wind, and pulling it along.

Conal leaned close over his garron's neck, trying to duck under the whistling wind. "We can't stop as near as Scotby. They'll be looking for us."

"We'll die an' we try to reach the Pennines," Hugh said.

"Wetheral Priory is nearby," Lachlan suggested. "'Tis a sanctuary. MacDougall canna touch us there if he follows us."

"Think you," Conal said derisively, "that MacDougall will honor the right of sanctuary? Even if he does, we'd be trapped there, for the moment we left, he'd be upon us."

"And," said Owen, "he cannot know at the moment who shot at him. If he finds us at Wetheral, he will."

"*Will* they follow us?" Hugh asked. "Sure we hit a few by the sounds and they must needs see to their wounded, not to chasing us. Regardless, they'll have all they can do to find shelter. Wind and snow have long since cleared our tracks."

"'Tis no matter," Niall said. "We've no choice but to take the first shelter we find." Ice blew across his cheek, stinging his skin. He tugged his cloak up, sheltering his face. The pony shuddered under him, bowing its head, too, its shaggy mane whipping in the storm, ice clinging to it.

Owen gave a sudden shout, as his pony stumbled. He slid off, and like Hugh, tried to give it shelter and help it along.

"Shawn said they build *igloos* in his country," Niall shouted at Hugh.

"Who?" Conal hollered.

"Huts of snow!" Niall deliberately misunderstood the question. He glanced at Owen's garron, struggling along. "If we canna go on..."

"A hut of snow!" Hugh scoffed. He grabbed the reins of Owen's garron, pulling it, with his own, in his wake. It bowed its head, whinnying pitifully.

"Look," Lachlan said.

"We'd freeze inside the snow," Owen said. "Are ye mad?"

"Look," Lachlan said more loudly.

"Certainly MacDougall would...."

"The priory!" Joy filled Lachlan's voice. He slid suddenly from his exhausted animal and ran, stumbling, through the driving snow and they all saw the walls, materializing slowly through the sea of white.

Hugh whooped, Conal spurred his mount, and they pulled themselves as fast as they could through the rising drifts.

Glenmirril, 1318

At a soft knock on the door, Allene looked up. "Enter"

The door swung open to admit Brother David. "My lady, your father bids me ask is there aught more I might bring ye." He glanced at the uneaten meal.

"What good fortune," she said. "Come and look over what I've done and tell me if you've more to add. They're for Christina."

He pulled up a chair and studied her work.

"I want to give her verses of comfort and hope," Allene explained.

"Psalms," he suggested. He closed his eyes a moment, before quoting, "*All who are weary, come to me, and I shall give you rest. Take my yoke and learn from me, for I am gentle and humble and you will find rest for your souls.*" He opened his eyes. "I believe that is another she would find comforting." He took a parchment and wrote several, himself. "A fine gift," he announced, as he finished.

"She will be pleased," Allene agreed. With the parchments sanded and rolled together, she rose carefully, the child once again kicking. With a hand to her heavy stomach, she took the scroll to Christina's room. "Where shall we leave it?" She glanced around.

"The desk," Brother David suggested.

Allene crossed the room. Her eye fell on the scroll sticking out from a pigeon hole. *Shawn,* read the single word on the outside. She slid it out, replacing it with the new scroll, and handed it to the monk. "Will you put that with our letters to Shawn in the kneeler?"

"Surely, my Lady." Taking the letter to Shawn, he left Allene's chambers.

Over the Atlantic, Present

It was hardly ideal conditions. Still, Shawn slid out the thick sheaf and unrolled it carefully, shuffling Allene's letter to the back. The handwriting on the next one jolted him. He couldn't know, he warned himself.

He glanced at the girl across the aisle guiltily, and back to the vine-like curls of words. The script—somehow—was her, serene and strong.

Guilt licked through him. But Amy had broken up with him. Besides, how could he feel guilt over a letter written seven hundred years ago? Hell, maybe Christina was breaking up with him, too—dumping him for abandoning her. He deserved it.

He ordered another Merlot and stared at the overhead bins, until it arrived, the feel of her parchment electric under his fingertips, savoring the anticipation. He took a quit gulp of the wine and read. *My dearest Shawn....*

His heart thumped. This wasn't Allene. It was Christina. From Christina, this was *not* just a casual *Hey how ya doin'?* He missed her. He missed her low, melodious voice, and her sweet serenity. *I'll not put this one in the kneeler.*

But it *was* here. He smoothed a finger over the ink laid by her hand, perplexed—and glad—that it was here. *I wonder if aught we leave for you will survive to your day. I pray it will.*

He smiled, answering her silently: *It did.* Perhaps her god had a miracle or two up his sleeve after all.

We do all we can to protect our writings and you know where they'll be. I imagine you finding them one day. Maybe two years ago, when you made the crossing, you went straight down and knew immediately all it has taken us these many months and years of living to witness. Perhaps you know already what becomes of us, if MacDougall will finally kill all of us—

"God, no," he whispered. He couldn't answer this one for her. He didn't know if he could bear to make himself finish reading. But they'd had to live it. Surely he could walk beside them in spirit, share their burden if only by reading.

—things we can only fear at the moment. It is difficult to comprehend.
Maybe it took you—will take you?—longer to go down. Maybe 'twas no possible to reach the cave, or it fell in or was locked or perhaps you feared to know or simply wanted naught more to do with us, but only to resume your life.
I miss you.

His heart thumped hard. Something squeezed tight, hearing his own thoughts whispered back to him in faded, ancient ink.
Life feels empty without you.
No! He steeled his heart. He'd promised to do right by Amy.
But she had broken up with him!
But Christina was seven hundred years gone anyway!
His thoughts scrambled like a carnival ride, shooting him up and down. Nothing made sense! He took a gulp of the wine, drew a deep breath, staring at clouds drifting lazily beside the plane. He should roll these up. Skip to the next one. Not read them at all. Or finish reading them some day when he was old, when he had waited patiently for Amy.
But he missed her. He lowered his eyes to the page, wanting her voice, if it was all that was left of her.

When we go out with the falcons, I think of you ducking and shielding your head as it plummeted down from the sky, looking for your wrist, and I still laugh at the look on your face, and the poor bird's attempt to land.

Shawn smiled, remembering the day; the two of them on the shore, hating lying to her, hating the grief in her eyes when his stories didn't add up.

Niall's harp seems silent without your songs, without you playing recorder along with him. I never go to the Bat Cave more. There is no reason, and it is painful, as it draws out so many memories. I know we felt for one another, though we should not have.

Shawn stared at the words. *We felt for one another.* He read them again, feeling the power of the statement and the harsh contrast to all those other women. He'd kissed Christina's fingertips. No more. Yet—*they felt for one another.* Yes, he thought, they did. Indeed, they had! And he missed her with an ache that ripped his insides in two more surely than the sword that had left the raw, red scar puckered and ugly across his midsection. His eyes felt hot. He blinked. Blinked again. He brushed at his eyes and took another, hasty, sip of wine. *They felt for one another. How he felt for her!*

But he felt for Amy, too. He'd come back to apologize, to ask her forgiveness, to make things right. Guilt welled up in him—here he was, instead, reading love letters from another woman.

But why was it in the kneeler if she'd determined not to put it there? He gulped a quarter of the glass of wine and continued reading.

I knew it could never be, that you needed to return to your own world, and yet, I miss you, I miss our time together, so deeply, every day, and it never gets better.

It never gets better for me, either, he whispered to her in his thoughts.

I will say here, in this letter which I'll tear up...

But you didn't, he thought. *Why not? What changed your mind?*

...which will never come between you and Amy in any way, that I love you in a way I did not previously understand love, and I pray for you and her and your son each day, and I pray you've found the peace you sought, and the happiness you wanted with your family.

Shawn studied the words. Re-read them. If it weren't for Angus—this little plot twist where *Amy loved Angus* when she was supposed to love *him*—yeah, he thought, there would be some peace. Except, of course, for that *other* plot twist, the one where he kept missing a woman who had been dead for seven centuries. *Lousy plot twist, if you ask me,* he thought. But for once, nobody had.

I had a son. I called him James Angus.

He wanted to see James Angus. He wanted to see what likeness he bore to

Christina, with her thick glossy black hair, if he would grow up to have a low, melodious voice like she did, if he would carry himself with her aspect of peace. It surprised him how much he wanted to know James Angus.

If he was right, after all—wasn't James Angus also the offspring of one of the thieving MacDougalls, father or son? Might he look like *them* one day, with black, greasy hair and yellow teeth? Might he carry their hateful, arrogant spirit?

I pray your James gives you the same joy mine gives me, she wrote.

No, Shawn decided: Because Christina was raising him. He would be a polished mirror, reflecting all that was good in her.

"More wine, Mr. Kleiner?"

The voice yanked him from the past. He stared blankly at the woman, Christina, in her short skirt. She set a glass before him, and disappeared down the aisle.

Shawn stared at the back of the seat in front of him, feeling claustrophobic after having been, just now, in Glenmirril's halls, its great rooms, its gardens spreading across the southern bailey. He drank half the glass and pressed a hand to his mouth. How had he let faded ink suck him away into a different time? Was he going to let faded ink turn his heart unfaithful to Amy and all his promises to her?

But she'd broken up with him! He was only reading a letter! And he *did* love her. In a flash, he understood her dilemma with Angus. But still, he wanted to give her the complete fidelity he had never given her. He gulped the wine. He couldn't make sense of it, except to know he missed Christina with a passion, a pain, that burned. And her letter lay before him, unfinished.

He set down the empty glass and read Christina's words, written centuries ago, on the parchment spread out across his tray in first class.

I desire peace for you. I want your peace and happiness above all and I know you will have that only in the redemption you sought.

His fingers touched her ink. She wanted his peace. Who had ever wanted peace for him? Who sought—who even *thought*—of peace, these days?

He wanted peace.

For himself. For her. His eyes drifted to the top of the letter, to the date. She must have written it while at Lochmaben, after escaping MacDougall. He closed his eyes and felt as if his spirit was whisked away, to stand beside her as she wrote. She had just been abducted. Made a daring and fearful escape. Found help at Lochmaben. She would have been missing James Angus, fearful of MacDougall.

Yet she thought of him. Wished for his peace and happiness.

A shuddering breath racked his body and he swiped at his cheek, at the hot sting in the corner of his eyes. The girl across the aisle looked at him curiously.

He turned to the window, to the clouds galloping by, and down to the sea far below. In the midst of her crisis, she wished for his peace.

Clarence's release was fast approaching. He had no peace about it. Maybe Christina could put in a word for him, up there. He dropped his eyes back to her words.

I have committed myself to marrying Hugh when I return to Glenmirril.

The words hit Shawn like a claymore.

"No!"

He heard the word come out, an angry grunt. He didn't care if the girl across the aisle stared at him. She didn't belong with Hugh! The plane shook suddenly, with a wave of heavy turbulence. Shawn steadied his glass on the tray.

The flight attendant swayed down the aisle, her hand touching each seat. "All will be well," she said. Her voice rolled melodically, peacefully. "All will be well." She touched his shoulder and moved on. The plane settled. Shawn's eyes fell to Christina's last words.

It feels wrong, and yet I've the greatest faith that somehow, all will be well.

The words jolted him. *All will be well.* He glanced behind him, at the stewardess coming back down the aisle. *You're a funny guy, God,* he thought. *Are you trying to tell me something?* He glanced at the parchment.

With greatest prayers for your peace and happiness, and the greatest love, Christina.

He stared at the words, wanting her peace and happiness, too. Hugh was a good man. But he wasn't a good man *for her.* He rolled the scrolls up and shoved them in the oilskin. They hurt. He had work to do in the States.

Glenmirril, January 1318

James Angus sighed in his sleep, and turned himself more deeply into Christina's arms. She smiled, stroking his black hair back from his forehead.

"He fretted so while you were away." On the other divan, Allene looked up from the baby shirt she sewed. "He is usually such a calm and cheerful bairn."

"I worried so about him," Christina said softly. Pain welled up in her chest. She'd feared, more than anything, never seeing him again. She'd feared who he would be left with, should MacDougall hold her forever in England, or kill her.

"Your bed's turned down, milady, and your fire stoked." Bessie appeared from Allene's room. She stepped to Allene's side, offering a hand to help her up. As Allene headed to her room, Bessie addressed Christina. "Would you like me to take him to bed?"

Christina shook her head. "I'll just hold him a wee bit longer."

Bessie bobbed a curtsy, and headed to Christina's chamber.

Left alone but for her son in her arms, Christina's eyes traveled around the solar, to the table where Shawn had played chess with Hugh, or argued religion with Brother David. She smiled, remembering. And she wondered again that such affection could arise between two from such different worlds; such different ways of seeing the world.

Her gaze moved on to her drawing easel by the window, dark now with the late evening. In the morning, perhaps, she would resume work on the portrait she'd been drawing before MacDougall had taken her. It was of Allene. She strove for the realism of the *photographs* Shawn described of his time. It still looked too much like the drawings of women in her own time, whose features all seemed similar, pallid, expressionless.

She tilted her head, studying it, trying to think what she could change to make it look more real, to bring out the expressions that were Allene, and no one else. What was it about a face that betrayed a particular emotion? A slight lift to an eyebrow, a tilt to the corner of the mouth, the eyes narrowed just so? She would work on it tomorrow.

In her arms, James Angus sighed. She smoothed his hair. Amy had thick, black hair, Shawn had said. She wondered if his son shared that trait. She wondered if he was happy. She hoped he was.

"Milady?"

Her eyes flew open at the feel of James Angus being lifted from her arms.

"'Tis but me," Bessie said softly. "You drifted off. Let me put him in his cradle. 'Tis in your room by your bed. He'll be safe by your side."

Christina blinked, coming awake, and nodded. Bessie lifted James Angus from her arms, and she followed the girl to her chambers, warm with the fire crackling in the hearth.

"I'll sleep here as long as ye like."

As Bessie laid James Angus in his cot, pushed up to one side of the bed, Christina noticed the furs and blankets on the other side of his cradle. "Thank you," she murmured. It *would* be good to have someone in the room with her. A desire burned in her heart for it to be Shawn. What would she and Bessie do against an intruder, after all? But little, she answered herself.

Still, 'twas good to have *someone* there. "I am grateful," she said. She glanced at the desk, where the scroll, the letter to Shawn, protruded from the pigeon hole. Her heart jumped, thinking of the words she had written. She couldn't risk leaving it until Hugh returned. So much could happen between now and then. And really, there was no reason to keep it.

She drew it from the desk. With a glance at Bessie, occupied with settling James Angus, whispering soft words as she patted his back, Christina tossed the scroll into the fire. She bit her lip, watching the flames touch its edges and reach up, melting its sides.

She must let her own passions likewise burn away to ash. Shawn had gone back to Amy and she wished the best for him. She would marry Hugh when he returned and they would live in peace and mutual respect, and give James Angus the best possible life. Perhaps there would be more children.

She lifted the poker and pushed the scroll into the heart of the flames. The black edges curled in, creeping toward the center. Shawn had a life. He had love, and the mother of his child. She would do nothing to interfere in that.

Inverness, Present

James, and the letter that told of his battle with Simon, of his disappearance into the past and a serious injury, were still heavy on Angus's mind days later when Hamish and Gavin burst into his room, greeting him with kisses and hugs that shot pain through his back even as they made him smile.

"Now, have I not warned ye about that," Mairi demanded, sailing in behind them. "Hamish, get your uncle's flute and let's see has he been learning it."

"Aye, mam!" Hamish snatched the flute up off the bedside table. He grinned at Angus, saying, "Ye'd best ha' done what Mam says!"

Angus tried to cast Mairi a sour glance as his nephew opened the case on his lap. It only came out resigned. "Can a man not escape your determination even by throwing himself off a cliff?" he asked.

Mairi pushed the flop of hair out of her eye and laughed. "Now that's the Angus I know, and I *have* been determined to see him back! What can you tell Hamish about how to play his flute?"

"We've not got *my* flute yet, mam," Hamish piped up.

"Whisht!" Mairi pulled up a chair. "'Tis on its way. See how your uncle fits it together?"

"No flute for Hamish but one for me?" Angus couldn't resist poking fun at her. "And how is that going to help our Hamish play?"

"Show him how to blow," Mairi replied.

"Aye, mam." Angus winked, and showed Hamish.

"Now show him a few notes."

Angus and Hamish passed the instrument back and forth for a few minutes, before the boys bounced to the window to look down on the snowy city.

Angus sighed, leaning back against his pillows. For just a few minutes, he realized, he'd not noticed the pain in his back.

Mairi glanced at the envelope on the table, as he cleaned the flute and fit it back in its case. "Helen came by then, did she?"

Angus nodded, unsure what, if anything, to say about the genealogy.

She dropped into the chair by his bed. "I was there the night they disappeared," she said softly.

"Simon Beaumont?" Angus asked.

"Aye, himself and the auld monk."

He frowned. "You dinna tell me."

"You'd bigger worries," she said.

"Not too big to override giving flute lessons."

She smiled just a bit. "Shawn's telling the truth."

"Aye," Angus said. He was silent for a moment, staring at his hands on the white bed cover. James, too, had slipped briefly from his mind as he played. He didn't want to talk about James. It was too disturbing. He looked up, meeting her eyes. "Do you know we're descended from the MacDougalls?

"The clan that owned your mate Brian's castle?"

"Aye. Duncan, it seems, may be our ancestor." Angus glanced at the boys, assured they were occupied with the city below. "Shawn killed him at the top of Glenmirril's tower—the night he re-appeared in Glenmirril."

"Why?" Mairi asked softly.

"What I know of him from Creagsmalan suggests he's not an ancestor to be proud of. And his father, Helen tells me, was possibly working with Simon Beaumont."

Mairi blanched. Her eyes dropped quickly to her toes. "I barely saw him," she whispered. "It all happened so fast, as I came into the chapel with James. But he left a presence in his wake. Of evil."

"You're quite safe," Angus assured her. "He's gone back to his time."

"Am I to be reassured that evil is now visiting someone else's home?" Mairi asked.

Angus reached for her hand. "We'd not be human if we wished it still in our own, would we now?" He glanced again at the boys, peering through the window, and said, "It troubles me, knowing he's near Niall. Yet I'm grateful he's away from Amy and James."

She squeezed his hand.

"What does it mean," Angus asked her, after several moments, "to be descended from men like Duncan, from his father, who would work with a man like Simon?"

"Does it change who *you* are?" she asked. "Seven centuries on, does it impact who you are?"

Angus smiled a little, watching his nephews kneel on the wide window sill, pointing to cars and people below. "I like to think not," he said.

"Sure not," she answered. "Do you and I and Ken not carry Mam and Da in us? Are their values and beliefs not imprinted on our souls?"

Angus cleared his throat. "James Angus—our direct ancestor—was a bit of a mystery. His father is believed to have been Duncan MacDougall—who Shawn killed. A cheater and abuser."

"And his mother?"

"The mysterious Christina, later at Glenmirril, for reasons unknown."

"Certainly Shawn can tell you why she was there."

"Mm." Angus grunted.

"Aw, go on now and set the pride aside and ask him," Mairi reprimanded. "You'll have no better chance." She grinned suddenly. "And I think he's not quite the ogre you'd like him to be."

"Has he addled you, too, with his charm?" Angus shook his head. "Me own sister and a married woman, no less."

Mairi laughed. "He may know, too, what Christina was like."

"She's none too highly regarded at Creagsmalan."

"Ask the one who knew her," she said. "Regardless, haven't we all a mixed lot in our family tree? Some MacDougalls, some saints. Is it not up to each of us who we follow?"

CHAPTER ELEVEN

Wetheral Priory, Northern England, 1318

Niall woke to a dark cell. His mind spun, trying to remember. Was he at Monadhliath? Melrose? Cold seeped through his thin blanket and the whirl of last night's blinding snow came back to him.

The monks had swung the gate open, and ushered them in, whisking their ponies away to be cared for and the men themselves to hot food. They did not practice the silence of Monadhliath, but laughed and sang at their meals, and the prior had shown them to rooms.

Niall ran a hand over the stubble where he'd so recently had hair. He grimaced.

"There's none to be Niall with hair this time." Hugh rose from his narrow bed, squeezed against the other wall just a foot from Niall's own. He stretched his arms, and looked around. There was little to see but the two cots, a kneeler on either side of the door, and matching crucifixes on the wall above each.

"Go tell the others to meet in the stable." Niall swung his feet from the bed, reaching for his trews.

"Surely we're not leaving before breaking our fast," Hugh said.

Niall shook his head. "No, but we've much to discuss, best not overheard."

Hugh pulled on his trews and boots, and ducked out under the door.

Niall yanked on his own clothing, his mind thrashing in the problem of MacDougall as he knotted the drawstring of his trews. It went against his nature to seek to kill a man. But it was certain MacDougall was not through trying to kill *him,* and now he'd begun to abduct the women of Glenmirril.

He strapped on his belt. Sure an' MacDougall would kill his sons—James and William with the fiery red hair—if he ever found the opportunity.

They didn't know if they'd succeeded last night. MacDougall might be seeking them, closing in on the Priory even now. He dropped to the bed, lacing his boots swiftly, his eyes lifting to the crucifixes. Anger wrestled in his gut. Where had God been when Taran had been hit on the head, when MacDougall had taken Christina at knife point past Gil, pretending to be Brother Andrew?

As his fingers stilled on the lace, the memory came to him, of Roysia, holding her son up to the Bruce, in the midst of an army, tears streaming down her face. He could not grasp why some prayers were answered and some not; why some children lived despite all odds and some died despite all prayers.

But he could not deny the miracle on the battlefield.

He rose from the bed and dropped to his knees on the kneeler, staring up at Christ on the cross. Memories gathered around him: His father leading him into the chapel so long ago, pointing up to the crucifix and telling him the story. His mother holding his hand, tears sliding silently down her pale cheeks beneath the same crucifix years later, as they watched over Alexander's body. The Laird's life-size crucifix hanging in the Bat Cave. He smiled, thinking of Shawn, of the crucifix he himself had bought in a small cluttered shop in far-hence Inverness and hung on the wall of a castle that would not be built for centuries yet.

It had been a miracle, what had happened. It had saved Scotland. His hand went to his throat, where his own crucifix, given to his father by the monks of Monadhliath, had once hung—the crucifix he had left with Amy.

How could he experience these miracles, large and small, yet doubt God? He bowed his head over folded hands, his eyes closed. He might die this day at MacDougall's hands. James or William might die years hence, MacDougall's knife in their guts, or MacDougall's noose around their necks. He wanted promises from God. And he knew they weren't coming.

But he had seen miracles. He had seen a future where England and Scotland lived again as neighbors; friends, even. He must trust the end was good, regardless of the part he himself was called to play in that outcome. It wasn't about *if* he lived. It was about *how* he lived.

He closed his eyes, murmuring, "God, make me a better man today than I was yesterday. Whatever comes." In his mind, however, he saw James as a young man, his hands gripped around the hilt of MacDougall's knife as it sank into his gut—as Duncan's hands had wrapped around Shawn's.

Whatever came—except that. He couldn't—he would not—allow that.

He rose, and headed out to the stables.

Carlisle, England, 1318

MacDougall looked up as Roger entered the dining hall. "How has our friend pulled through the night?" His hopes wavered between wanting to be done with Claverock and the dangled promise of finding the mysterious Shawn.

"Raving." Roger dropped to the bench beside his lord. "In anger," he clarified. "Not in delirium."

"He's well then?" MacDougall asked in surprise.

"Far better than one would expect after such a wound."

"Or hoped," MacDougall snapped. "I expected a week's reprieve before he's well enough for his usual ill humor." MacDougall spooned porridge into his mouth. "What does he rave about this fine day?"

"He wants the heads of those who shot him, of course."

"He is in no condition to search for them and will not be for days. Weeks, if fever sets in, which, if there's any God in Heaven, it will."

"On the contrary, my lord." Roger shook his head. "He has made a remarkable recovery. He spoke of a book of healing, brought back from his journeys." He smiled. "I found it while he slept." He held up a small volume with a hard cover—yet not of wood—and no gilding.

"Who would make such a plain book?" MacDougall took it, turning it over, studying front and back. "What do you make of it?"

Roger shrugged. "He says it contains healing ointments."

"Surely Harclay's physicians have ointments."

"Open it," Roger advised.

MacDougall quaffed his ale, set the cup down.

At that moment, Simon strode into the hall, his right arm strapped to his side with cloth bindings. "Roger," he said, loud enough to turn heads, "surely you've good cause for having my book in your possession." He snatched it from MacDougall's hand.

Roger bowed. "Forgive me, Lord Claverock. I thought only to make more of the healing ointment for you."

Simon stared down his nose. "It would go poorly for you were I ever to have cause to think you might steal my belongings."

"It is as he says." MacDougall rose. "He was, but now, asking after Harclay's physicians to make more. It seems they work. You're looking hale."

Simon scowled. "To the misfortune of the miscreants who shot me. Have you learned their identities?."

MacDougall stared in disbelief. "How should I have learned that? I went to bed straightaway and have but risen, myself. How would I know who was about in the forest yesterday?"

Simon's face darkened. "I expect Harclay to find out."

"I expect you've let him know that." MacDougall spoke impassively.

"I have," Simon assured him. He looked up to see one of Edward's guards.

"You're looking hale," the man said. "Are you ready to ride for London?"

Simon sketched a bow. "To London and the no doubt well-appointed quarters that await me." He lifted an eyebrow. "All I have said will come to pass." He smiled. "I will remember how I am treated on this journey."

The man swallowed. "Yes, Lord Claverock."

To MacDougall, Simon said, "Find the brigands. We will speak soon."

Wetheral Priory, 1318

As Niall's eyes adjusted to the dark interior of the priory's stables, he saw Conal in a far corner, combing his garron and talking softly to it. Niall smiled, thinking of Shawn's first reactions to horses. He picked up a curry brush and ran it over his own pony's coat. The stables were warm after the wind blustering across the monastic courtyard, and the smell of hay and horses brought the comfort of a lifetime of memories. He wished he could spend every day in stables and playing music, instead of fighting. Hugh, Lachlan, and Owen appeared minutes later, searching the dim interior. Niall hailed them.

Hugh strode down the aisle. The four of them crowded into an empty stall, the farthest from the door. "If anyone comes," Niall told Lachlan, "encourage them to leave."

With a nod, Lachlan led a horse into the aisle between the stalls, grooming it there to keep watch, and keep listeners at bay.

"They'll be looking for us," Conal said. "Which way do we flee?"

"Flee?" Niall repeated in surprise. "We came here to kill him."

"Which we may have done," Lachlan said. "We heard enough screams to guess we hit several."

Hugh shook his head. "They were at least half a score. It may have been any of them. It must be *MacDougall.*"

"Will we go back to Linstock?" Lachlan looked from one to another, in the tight space. "Or Carlisle?"

"We've limited time," Conal reminded him. "The Bruce requires us back by the first of March. And we've a great deal to do before then."

"We expected to have a fortnight at Glenmirril." Niall looked around the small group. "Instead, we left immediately for Christina."

"It might take those two weeks just to find out if he's dead," Hugh warned.

At that moment, a pair of happy voices reached them. They slid down by unspoken agreement, sitting in the hay with their backs to the wall. In the aisle, Lachlan backed the horse up, blocking the stall door, and set to work cleaning its hooves as the monks greeted him.

"A brisk day, is it not?" The deep voice sounded aged.

"'Tis," Lachlan replied. "Surely a day ye'd rather be inside than out here."

"And so we would," replied a second, younger voice, with good cheer. "But we've work to do." Above the half wall, Niall saw the tines of a pitchfork lift off the wall and sink out of sight.

"I'm caring for our ponies," Lachlan said. "Sure I can pitch your hay, too."

"Oh, the abbot would not like to see us return so soon," spoke the first. There came a scraping sound and the soft *pfffft* of hay being tossed. Its sweet scent filled the air.

Niall, Hugh, Conal, and Owen looked across their own hay-filled stall at each other. The horse shifted its body in front of the door.

"'Tis the least I can do in return for your hospitality in our troubles last night," Lachlan insisted.

The men laughed. "Taking in travelers is what we do," said one of them.

Niall leaned against the wall, staring up at the rough beams overhead.

"Another came in after you."

Hugh straightened.

"Ask him who," Niall murmured.

"Caught in the storm, too?" Lachlan said. "Imagine another out in it."

"'Twas quite the blizzard." This was the deeper voice. "Such winters we've had these past years! Sure God has turned his back on this world!"

"The other traveler," Lachlan asked. "He fared well?"

"Ah, now not as well as your group. He was in poor shape, he was."

A horse whinnied, and Niall heard the renewed soft scratching of the hoof knife. "He'd come quite a way?" Lachlan asked.

"Only from Carlisle," said the young monk. "It seems you were fortunate."

"How so?" Lachlan dropped the front foot and moved to the back, lifting a rear hoof to his knee.

"There were brigands in the forest last night. More'na dozen, he says, by their tracks."

"Mary, Joseph, and Jesus!" Lachlan dropped the hoof to the floor. Peering under the horse's body, Niall could see him straighten. "Dozens of brigands!"

Niall's lips tightened against laughter. Either Lachlan missed that *they* were the brigands, or he was quite the gifted actor.

"At least a dozen, he guessed."

"Dozens!" Lachlan breathed. "Our small group would ha' been no match!"

Niall saw what he was doing. Repeating the word until it stuck. The monks would repeat it. The tale would grow and none would associate the forest horde with the five of them. His estimation of Lachlan rose.

"He ran into them in the late afternoon," the older monk said.

"Was he injured?" Lachlan asked.

"Not but for being half frozen and shaken with being lost in the snow so long. We've given him plenty of whiskey, you can be sure."

"He's a citizen of Carlisle, then?"

Niall drew in slow breath, waiting for the answer. He watched Lachlan's legs under the horse.

"Where did yer man say he's from, Brother Oswald?" asked the older voice.

"Creagsmalan," came the answer. "In Scotland."

"He's with the MacDougall who's just come down from there—on Bruce's bad side, they say."

"MacDougall is here?"

"He is. At Linstock while Edward decides what lands to grant him."

"If any," added the other monk.

In the stall, Niall and Hugh glanced at one another.

"Was anyone injured in the attack?" Lachlan asked.

Niall heard the tension in his friend's voice. The horse skittered backwards. The men of Glenmirril pressed more tightly to the wall.

"Aye, Lord Claverock," said Brother Oswald. "Strange doings. D' ye mind the Lord of Claverock—disappeared at Bannockburn?"

"I've not heard of him," Lachlan said.

"Aye, well, 'twas all the talk in these parts, Claverock being not far from here. He's turned up after more than three years."

Here in Linstock? Niall hoped the force of his thoughts would press them straight into Lachlan's own.

"In Linstock?" Lachlan rounded the horse and lifted its other rear hoof onto his bent knee, once again scraping. "Is Claverock not on the coast?"

Niall smiled. Not only had the man read his thoughts, he was a thespian. He sounded for all the world as if it mattered not at all to him.

"It is." The young monk glanced around and suddenly leaned close, dropping his voice. "Our guest last night was rambling. Said Claverock is being escorted to the Tower of London!"

"Why?" Lachlan looked up, pausing in his scraping.

Niall strained to hear.

"Seems he prophesied to Edward. Predicted our queen is with child. And that a one-eared man would appear in Oxford, claiming to be king."

Niall and Hugh glanced at each other, frowning.

"Hush, Brother Oswald!" came the deeper voice. "These are dark doings."

There was a brief silence, the sound of hay being tossed, and Lachlan scraped three more times before asking, "What has Claverock to do with these scores of brigands in the night then?"

Niall's smile grew. Lachlan had again inflated the size of their small band.

"Claverock was struck by an arrow. But our guest was separated from the group in the storm. He does not know Lord Claverock's fate."

Niall closed his eyes, Hugh and Conal pressed on either side of him, their backs against the wall. *Please, my Lord and Savior, make him think to ask....*

"Were Claverock and MacDougall about in the storm *together*?" Lachlan sounded perplexed. The pony's hoof thumped to the dirt floor, and Niall watched Lachlan's legs round the animal, and disappear from sight.

"The story was garbled," said Brother Oswald. "He was in a bad way, was he not, Brother Aethelfred?"

"In desperate need of whiskey!" agreed Aethelfred. "Half frozen."

"And in shock, no doubt, after escaping so many," Lachlan said. "Three dozen? Two score?"

"Now I'm thinking no more than three dozen," said Aethelfred.

"Closer to four," countered Oswald.

"Let us hope they do not come to Wetheral!" Lachlan sounded fearful.

"Oh, let us pray not!" Aethelfred repeated, with a gusto that suggested he would relish such adventure.

"Edward's had a job," Lachlan said, "with these incoming Scottish lords. Has MacDougall lands here or will Edward grant him some?"

Niall and Hugh glanced at each other, smiling in approval.

"MacDougall?" Aethelfred asked. "Do you know, Brother Oswald?"

"Around Keswick, methinks," said the young monk.

"No, now he had lands farther east," argued Aethelfred. "Closer to Ripon."

"Ripon!" Oswald spoke in disbelief. "Surely not! They'd no land there."

"Surely MacDougall himself is well," Lachlan interrupted.

A brief silence was followed by Aethelfred's deep boom. "Oh, it seems he was only irritated about not getting some massive boar they say is running in Inglewood. But no, he's well. Not hit at all. Now Keswick, I'd not think Edward would give lands there...."

"My good brothers," Lachlan interrupted. "You're blue with cold. Let me finish your work. Sure you've been gone long enough the abbot will not fret?"

"Ah, now, we couldn't...." Oswald began.

"Really, I wish to thank you for your help for me and my small party," Lachlan insisted. "I'll have it done in no time. And you must warn the abbot to set a watch. Such a horde might not hesitate to attack even a house of God."

Midwest United States, Present

Lousy timing for being home, Shawn thought, as he lifted a glass of bourbon to his lips. It would be easier to cope with Clarence's release from across an ocean. Here, it was too close and all the flurry of meetings, the dinner with Ben—bursting with ideas for his next tour—practicing, arranging—none of it pushed away the hovering dark cloud of Clarence's impending release.

All will be well. God's humor again—a rather sadistic humor, Shawn thought—that the flight attendant should say those words just before he read them in Christina's letter?

He lounged in his favorite leather chair, the one where he'd sat, his feet propped on the coffee table, the night Aaron brought him home. He stared, unseeing, at the flames flickering in the fireplace. It didn't *feel* like anything would ever be well again.

Dawn was still at least an hour away. *We leave before dawn.* But there was nowhere to leave *to.* Nowhere to run from all that plagued him, that brought the insomnia roaring back full force, as the contents of the scrolls, of MacDougall's release and Christina's love and agitation over Clarence all swirled relentlessly in his head.

He'd spent the night drinking coffee from the singing coffee maker Amy had given him, writing arrangements, practicing, and sending an e-mail to Amy. Now he stared, exhausted, restless, and wide awake, past the flames.

All will be well.

His mother was right, he thought. In everything. Clarence had been a boy. He'd turned his life over. Shawn had even touched his hand, through the glass, saying *We're brothers again.* He'd *meant* it. So why did it cause such agitation that he was leaving prison? Why *couldn't* it all be well?

Because James Kleiner is not leaving his grave! Shawn jolted to his feet. *Because no matter how sorry Clarence is, his life is going to just go on, no problem, and Dad's never will. Game over. No release! No reprieve. No forgiveness from Death.*

He pushed a hand through his hair. *Forgiveness.* Wasn't it a constant theme in his life, repeated over and over in the sweet sound of the violin, echoed in the low laugh of the clarinet, sparkling softly from the harp, pounded out on the tympani, and calling from the horns? Didn't he need his own forgiveness?

But he'd never killed anyone!

Well, no one who didn't deserve a good killing, he amended. Duncan had needed to die. Shawn leaned to scratch at the scar on his leg, then paced to the hearth, poking up the flames and tossing in another log, and moved to the kitchen, throwing sausage into a frying pan. He was tired of arranging. He'd been going steadily for days. He was tired of playing trombone. Even tired of the sackbut he'd been so excited to acquire.

He cracked an egg into the pan beside the sausage, and went to his office, where he brought up *The Stillwater Gazette* online. Nothing on the front page. Of course not. Why did he think Clarence's release would be front page news? It was of no consequence to anyone but him. The rest of the town, the state, the country, slept just fine, with no knowledge of Clarence leaving prison or James Kleiner not leaving his grave. The rest of the world rolled along as if nothing momentous, horrible, painful, was happening today.

Still, he scrolled and clicked through each section, veering between relief at not seeing Clarence's face and anger that it meant nothing to anyone. No one cared. A school was hosting a fundraiser. Mrs. Brown won a quilting award. No one *cared* that his father was dead. A church wanted....

The smell of burning came to him. He swore, jumping up, and raced to the kitchen, where smoke thickened over his charred sausage. A hard yellow yolk and curling black egg white stuck to the pan. He wrenched the skillet off the burner even as he snapped the knob, turning off the heat, and spun to pour cold water into the pan. The hot surface sizzled angrily, puffing up a thick plume of black smoke.

Shawn swore again, dropping it into the sink. The anger left him as quickly as the sizzling died down. This was doing no good. He had to get a grip on himself! He had to do something productive instead of stewing!

He returned to his office. First thing was to shut down the damn site. There was nothing he could do about it. As he reached for the mouse, his eyes fell on the screen and he became still. *Medieval.* The word jumped out. He leaned in, scanning. The church wanted a program of authentic medieval music for its Easter services. They didn't have the resources to hire researchers or arrangers.

He stared at the article. Niall would say God was working. *He* would say he happened to know a little about medieval music—especially medieval music in churches. He would call it fortuitous—and exactly what he needed to take his mind off Clarence.

He crossed the foyer to his music room, slapped score paper down on the table and, remembering the music he'd played for the monks of Monadhliath, began writing.

As the music flowed, the world began to feel right again. Hope returned. Plans for the next tour were shaping up. The new t-shirts Ben had designed were selling. The trading cards of the orchestra had once again picked up.

He'd be back in Scotland soon, with James and Amy. He'd find a way to make them all a family. Christina *had* escaped. Things *would* be well with them. Yes, all would be well.

Wetheral Priory, Northern England, 1318

Niall scrambled out of the horse stall on Hugh's heels, already asking, "Carlisle? Keswick? Ripon? We need to *know*...."

"Think, Niall!" Conal glanced around, assuring himself the stable was empty. "We've orders from the Bruce."

"We've a fortnight at most for MacDougall," added Hugh.

"'Twill no take two weeks to kill him," Niall insisted. James filled his mind. James must be protected. They were so close.

"If we knew where to find him," Hugh said. "But we do not."

"He's in Carlisle *right now!*" Niall almost shouted. "But miles away!"

"How are we to just walk into Carlisle?" Hugh demanded.

Lachlan's pony snorted and shook its head.

"Sh!" Conal hissed. "D' ye want them coming back?"

"A disguise." Niall lowered his voice. "Shaw..." He stopped, glancing at Conal and Owen. "I've done it before."

"And I believe," Hugh said pointedly, "MacDougall figured it out, did he not? They are on watch for Scots. What's more, he was staying at *Linstock*. He may not be at Carlisle by the time we get there but Harclay may be watching for *you*."

"He abducted Christina," Niall reminded him. "You *know* I and my bairns will be next."

"He'll not dare show his face in Scotland again, aye?" Conal said.

"No, Niall." Hugh shook his big head. "Ye mustn't let fear drive you to rash decisions."

Niall's hands went to his hips. He glared at the timbers holding up the thatched roof.

"You must *live* to protect them," Conal said.

"We've *two weeks,*" Niall replied.

Lachlan led his pony back into its stall, saying over his shoulder, "We've two weeks *if* all goes according to plan. 'Tis winter. Roads are bad. Things will assuredly *not* go according to plan."

"He's the right of it, Lad," Hugh said.

With a glance at Lachlan, now grooming a second pony, Niall took a brush from the wall and began on his own. "We hit Claverock yesterday."

"Claverock, again. What has he to do with this?" Conal, too, took a curry brush and began working on his animal.

Niall and Hugh glanced at each other, before Hugh took a bucket from the wall and headed out to the courtyard.

"If they were in the forest together, it seems he has aligned himself with MacDougall," Niall replied. "And perhaps MacDougall is occupied with his injuries. Which makes it a good time to strike."

"The question remains how." Owen tossed a forkful of clean hay into a stall. "Mind, MacDougall will be in a rage. He'll be searching for these brigands." He glanced at Lachlan with a grin. "Hordes of them."

Lachlan laughed. "Aye, quick thinking on my part, was it not?"

"Watch it," Conal said with a wink. "Ye'll be getting as vain as our Niall!"

Outside, day began to break, spilling its first faint light in the stable door.

Niall smiled. For a moment, it felt like having Shawn back. Perhaps he'd underappreciated the friends he had still—had *always* had. He thought of Conal slapping Shawn's shoulder down in the courtyard so long ago. To Conal, that day was part of his friendship with 'Niall.' And in truth, Niall thought, it was. It revealed how Conal saw Niall—a friend for whom he held concern, not a coward to be pitied. It mattered little that it had been Shawn that day.

"This good-looking and not vainer yet," Niall quipped, as he'd have said to Shawn. "I call that great humility indeed!"

Conal laughed aloud, his head thrown back. "There's our Niall!" His pony nuzzled his cheek. He patted its jaw. "And we're waiting on your decision."

Niall sank back into dark reality. "Back to Linstock. He must return, aye?"

Hugh entered, blocking the wee bit of light in the doorway. A bucket of water sloshed in each hand. "We're taking great chances," he warned.

Tension coursed through Niall's jaw. "Aye, as we did at Bannockburn. MacDougall will be looking for us—for those rampaging hordes, at any rate—and his man is here. He himself is likely still at Carlisle. So we get into Linstock while he's out and lay a trap."

Owen's brows furrowed. "At the moment, MacDougall does not know 'tis *us*, what hit them last night. If we fail at this plan, sure he'll know."

"And be angered to do what?" Niall challenged. "Try to kill us? Attack Glenmirril? Abduct Christina?"

Hugh hoisted the water bucket to pour it into a trough, looking over his shoulder to say, "Aye. Niall's the right of it. He's done all that and as long as Duncan is dead, will continue to."

"We must break our fast and leave anon," Niall said. "We'll work out a plan on the way to Linstock."

CHAPTER TWELVE

Inverness, Present

"Welcome home, more or less." With James on her hip, Amy unlocked Angus's front door.

Swinging his trombone, Shawn followed her in. The aroma hit him. "Burgers!" he said.

She turned, a corner of her mouth crooked up. "With your favorite spices."

"Spi!" James added helpfully, grinning at him.

"Hopefully in the right amount," she said.

"It smells great." Something fluttered in Shawn's chest. She'd picked him up from the airport. She'd made him dinner. His mind flashed to Angus, wondering how he was doing—and if he was welcoming Amy in or still pushing her away.

"I'll get them on the table." She set James on the floor in the front room and headed down the hall.

James crawled to the coffee table, pulled himself up, and reached for Shawn, shouting, "Da!"

Shawn left his carry-on, with the parchments zipped inside, in the hall, set the trombone case in front of the fire, and squatted before James, taking his hand. His heart felt warm and tight. "Dad? Did you say dad?" This was his son! He couldn't stop the smile bursting across his face. "Did you miss me?"

"Da!" James shouted. He grabbed the handle of the trombone case, and dragged it closer to himself.

Shawn stared in amazement. "Strong, aren't you?" he asked.

"Dinner!" Amy called.

Rising, Shawn leaned to let James grasp his fingers, and followed his son as he toddled down the hall, babbling in delight at his accomplishment. He glanced up at Amy as he took his seat and lifted James onto his knee. Brother Eamonn's prediction flickered through his mind. Such a destiny would call for strength. He pushed the thought away—Eamonn wasn't reliable—and fixed his eyes on the plate of burgers Amy lifted to the small table. "Mmm, good!" he declared. "Look at that, James!"

"Boo-boo! Boo-goo!" James lunged for the plate.

Amy snatched it back just as Shawn caught his son's hand. "He's strong," she said. "He would have yanked it right out of my hand." And to James, "You wait patiently, right?"

He laughed and shook his head. But he sat quietly until Amy cut up a patty and placed him on a booster seat on his own chair.

"How's Angus?" Shawn ventured, when she had piled her hamburger high with vegetables.

"Sitting up on his own." She lowered her eyes and glanced up again. "Are you seeing Jenny again?"

He almost gave a quick grin as the words *Are you jealous* flashed through his mind. But he knew what she was getting at. He set down his food, avoiding her eyes.

"You shouldn't have played with her heart," Amy said softly.

"No, I shouldn't have," he agreed.

James strained for the plate of sliced tomatoes, babbling.

Amy cut a tomato into small pieces on his plate, and after another moment of silence, asked, "How did everything go over there?"

"Great." Shawn bit into his burger, and as the flavor hit him, said, "Wow! You've been watching what I do!"

She smiled. "Ben was happy to see you home?"

Shawn laughed. "Ben can't decide if he loves or hates seeing me."

They fell into easy conversation, as they ate. He sent her to the front room with James to play her violin and cleaned up, thinking of Christina's wish for his peace and thinking equally of Amy's dark eyes and the flush on her cheeks as she laughed.

He'd survived Clarence's release. He had the music nearly finished for the church—he would bring it to them next time he was home. Christina had escaped and was well. Hugh would be good to her. Amy had made him dinner, cast him furtive looks that told him he still affected her. Yes, all was well.

♫

I lie awake in Angus's bed, James breathing softly beside me. Angus refuses to have me in his room. 'James needs Shawn' is the only answer I get. I thought he would relent, but he hasn't. Rather, I've seen Claire's car in the lot and the flash of her tomato red and baby blue sweaters in the hall. He's clearly not refusing to see everyone. I miss him.

And I missed Shawn while he was gone. We were happy together, over Christmas—even as I missed Angus. How can I feel so much for both of them? Valkyrie and Pachelbel.

He stood in the narrow hall, between our two rooms, and touched James's head and kissed him and awkwardly kissed the top of my head and said, "Sleep well." I wanted to invite him in. But how could that be right when I'm telling

Angus I want to be with him? It would have made a lie of all I've said to Angus. But I wrestled with it. And I lie here awake, thinking of both of them, and wondering who finally gives in. Me? Angus?

In the next room, Shawn's bed creaks. I become still, listening. A pause, and it creaks again. A moment of silence and his footsteps cross the floor and softly, down, down, down the stairs. I think about following. But we have gone to bed so recently. I know him so well. I suspect he's been waiting for me to become quiet.

There is no sound from the kitchen. I sit up in the dark, listening. James sighs, snuggling deeper under his blanket in the moonlight spilling in. And downstairs, I hear the sigh of a single low viola string, swift and short, and a soft rustle. His carry-on sat in the hall. Have I gotten this good at guessing, fitting together the pieces of what he won't say? Am I still doing it, even now, knowing he's changed?

The door to the front door shuts softly—one gentle tap of a drumstick on a snare.

The parchments were in the carry-on. They had to have been. Moonlight silhouettes the tree outside, its branches like an old crone's fingers splayed against the night sky, grasping for moon and stars.

I rise from my bed—as careful as Shawn not to make a sound—and stroke James's back, thinking of Simon's peculiar behavior, of Brother Eamonn, of Niall, whose fate Shawn may even now be reading.

Is he protecting me from bad news? I smile, remembering Niall running his hand down my hair. French shampoo, he said. He was kind and it tears my heart, fearing for him.

Or maybe there are letters from Christina. Yes, it's foolish to be jealous of someone so long gone. But her words are not long ago. And his heart is not. And aren't we a pair, in love with each other still, and in love with other people.

I want life to be simple again.

Maybe it's simpler than I think. Maybe two years in medieval Scotland was so intense, so incomprehensible to anyone in our world that he can't bring himself to open that chapter even to me. There was brutality. He would have witnessed things I can't—don't want to—imagine.

James sighs, turns his head and smiles in his sleep. He was born with Shawn's joy, his energy, his zest for life. I stroke his thick, black hair, and his back where his shoulder bears the birthmark—an angel wing, Mairi calls it. A flame. It's what caught Simon's eye, when I changed his shirt at Monadhliath.

I wonder what Shawn is reading, in the room right under my feet.

But James is safe. One thing I have felt with Shawn, since his return, is safety in his presence. He will protect us at all costs. And in that knowledge, I feel his love for me.

I sink softly, quietly, back into bed, staring at the grasping fingers outside Angus's window, and wishing I had a life to live with each Angus and Shawn.

♫

Shawn pulled to the far corner of Angus's front room, turning on a lamp there, hoping it wouldn't send light onto the lawn. But Amy hadn't stirred. All was silent above him. He stared for a moment out the front window, at the tree, its trunk rising strong and tall in the soft glow of the street light outside.

Guilt trickled around the edges of his heart. But there had been only one letter from Christina. He wasn't going to read it again. The knowledge of her love was a soft, warm blanket around his guilty heart. But he loved Amy, too. And he was going to take care of her and do his best for her, for James.

He was only going to continue the story, that was all, and be reassured, finally, that things were looking up. He might read about Niall's children. The war *did* end. And maybe he would walk beside Niall and Allene and their children as they grew. And maybe he would read of Christina's marriage to Hugh and see their children and he and they and Amy would all grow old together, happy for one another, raising their children and laughing across time at the jokes and stories they would tell.

It would all be well.

He thought about going for a glass of wine. Or a bourbon. To relax into talking with an old friend. *Seven hundred years old,* he thought with a smile. But he didn't want to disturb Amy. He slid the parchments from their oilskin, moved Christina's letter to the back, refusing to look at it, and smoothed them across Angus's coffee table.

He adjusted the lamp and lost himself in the first letter, penned by the Laird. The ink was strong. He spoke of the quiet in Glenmirril, with Hugh, Conal, Owen, Lachlan, and Niall all gone. He outlined the places Niall must go for the Bruce, before returning to Berwick by March. For just a moment, Shawn wished he were riding with them, traveling, working—laughing together. Irritating Niall. Niall irritating him. Hugh's big guffaws. Even Conal's friendly pat on the shoulder, that called him *friend.*

Taran had been sent to Berwick, the Laird wrote. Muirne's oldest lad had asked leave to go to Monadhliath for a time of prayer. Red took grand care of the garrons and was a favorite of the young boys, as he taught them to ride.

Shawn smiled, missing Red. He wondered what it was for the boy—to lose his parents, to lose the farrier who had raised him, for Shawn to disappear on him, and Niall to leave. But Red had ever been affable, cheerful, with a seemingly inextinguishable enthusiasm. And, Shawn thought, medieval times probably didn't share their concept of abandonment trauma.

Still, he missed him. He hoped to read of him growing to a man, marrying, having children of his own. He looked up to the ceiling. All was quiet upstairs. He rose—softly, so softly—and turned the knob oh so carefully, looking toward the stairs. All was silent.

He tread to the kitchen in his stocking feet, let the bourbon slide soundlessly against the edge of a glass, filling it, and returned to the front room, shutting himself in again with a click of the door. He wondered why he couldn't tell Amy about any of it. It wasn't her world. He had no more answer than that. Except, he thought—it was too intense, too personal, the way he missed a world he should have been glad to escape. He smoothed his hand across the new parchment and began reading. All would be well.

Linstock Castle, Northern England, 1318

Even in the heavy drifts, they reached Linstock before noon, a plan in mind. Four of them waited under the heavy fringe of trees across the field from the bishop's palace, while Owen hastened to the nearest village, seeking an apothecary and the old woman skilled with herbs that every village seemed to have.

"Carlisle is closer than Wetheral," Lachlan commented, as they watched the bishop's manor once again from under the trees. "What if he's already returned?"

"Several of his men, and Beaumont, are injured," Niall said. "He'll not be back so soon. You go in and ask for work. In a few days, he'll return. He'll take no notice of a new kitchen lad or stable hand." Niall wondered, as he said the words, how much he himself had ever seen the servants at Glenmirril. Some of them he knew. But if a stranger appeared, would he notice?

"Learn what you can of his future plans while you're inside," Conal said. "If we fail in this, at least we may learn what will help us."

"We'll be on constant watch," Hugh said, "ready to fight and ride, should you need to escape fast."

"And you *do* leave, the moment he's dead." Niall pulled his second knife from his boot and handed it to Lachlan. "You've plenty of weapons?"

Lachlan patted the knife hidden in a sheath under his arm. "And one in each boot," he said.

"What are the plans for getting out?" Conal asked.

"The river is the best way." Hugh glanced at Niall.

Niall stared straight ahead. "Will we be able to find a boat in the hamlet?"

"We'll find aught on a river, sure," Conal said, and turned back to Lachlan. "I come down the river each night to the water gate." He indicated the copse up the bank a ways. "We gather there."

With a grim grip, hand to wrist, and "God be wi' ye," Lachlan left his pony and headed through the small wood to the Bishop's Manor to seek employment.

They watched and waited, the day growing colder, and the gray clouds heavier, before Owen finally returned, grinning.

"Poison?"

"Aye." Owen patted his cloak. "'They were quick to tell me where to find the woman in the woods."

"Whatever it takes," Niall said, grimly. "There will be an end to his plots and thieving. Haste. Lachlan's been in for some time."

Hugh, Conal, and Niall watched him go. "Are we condemning them to death?" Niall asked. "What if they're found out? What of Margaret and Roysia waiting for them? I should have gone, not them."

"What of Allene and your bairns?" Hugh asked.

"They've seen you and me and they couldn't help but recognize Hugh." Conal leaned against a tree, his cloak tight around him, watching Owen pass under the snowy trees at the edge of the pasture. "They were the only choice." He turned, meeting Niall's eyes. "They are smart and strong and you yourself taught them to fight. Trust them, Niall."

"'Tis not trusting *them*. So much can go wrong." Shawn once more crossed Niall's mind. He sighed. *Shawn's* friends would be playing music, not going in to possible death. Shawn fully expected his friends to be alive at the end of the day, at the end of the month, and even years hence—didn't even think about it. Niall closed his eyes, praying for Lachlan and Owen's safety.

Inverness, Present

The date jumped out at Shawn for what it lacked: *October 18, 1318.*

In Angus's front room, the warning of danger prickled the hairs on his arms. She did not write *The Year of Our Lord*, as had been her wont.

I feel I understand you better, Shawn, she wrote. *Never has my faith been tried so. I cannot say 'tis the Year of Our Lord. 'Tis the year of evil.*

He had to squint and shine the light to make out the rest of the faded words. *He is nowhere to be seen in these wicked days. I have tried several times to tell you, and I can not. Father refuses to try. Hugh....*

The words trailed away. A newer, stronger hand seemed to grasp the pen. The words stood out boldly, even so many years later. *Shawn, this is Brother David. Evil has befallen us. You will have seen in prior missives that MacDougall escaped....*

Sickness turned Shawn's stomach as he read. It rose in his throat, mixing with blind hatred. MacDougall was not content merely to kill. He must wage psychological warfare. He must terrorize and torture.

Shawn took a swift gulp of the bourbon, rose to his feet, paced. His hand went through his hair, as he stared at the sturdy trunk of the tree outside, trying to calm his mind, trying to think. Feeling sick. There was nothing left to fear. His stomach heaved as it had after Skaithmuir, staring numbly at skewed bodies spilling out organs and bones and brains. He swallowed hard.

No, he wouldn't give in to cowardice this time. He pulled up a search engine, fingers tapping impatiently: *Niall Campbell.* In quotes, without quotes, various spellings, MacDougall, Alexander MacDougall, Simon Beaumont—any and all the words that fit that day.

It took half an hour, but he found it on the twelfth page of the tenth search.

A tiny blog that hadn't been updated in three years, with a referring link back to a site that no longer existed, confirmed Brother David's story—in vivid detail. Shawn bolted from the chair, one hand on his mouth, the other pressed to his stomach, almost feeling it himself. It couldn't happen. *It couldn't!*

He forced himself to calm. *Think! Information!* The first thing was information. Hating going near the ugly story, he forced himself to sit down again and re-read it carefully.

He pulled up a new tab, and soon had the date. Niall would be at Berwick from the beginning of March until MacDougall launched his attack. Shawn stood again, told himself he was not thinking clearly. He couldn't just drive to Berwick in the middle of the night. He sat down again.

Why not, he asked himself. *It's not illegal.* He stood again, pacing. But *why* would he drive to Berwick? It would do nothing to stop this happening. No, he decided, he needed to take a deep breath. Find out what he could and go from there. He pulled up a third tab and spent ten minutes reading all he could of Peter Spalding and the siege of Berwick.

Douglas, Keith, and Randolph had gathered at Duns Park. Wherever the hell that was. He opened a fourth tab to a map, a fifth tab showed the causeway. He opened more, flipping among a dozen tabs, seeking every detail of MacDougall's plot.

He pushed himself up abruptly from the computer and slammed the lid. He planted one hand on his hip and raked the other through his hair. "What am I doing?" he muttered, and louder, careless of Amy sleeping upstairs, "What the hell am I doing?" He pressed his fingers to his forehead, took a deep breath, and announced to the room, "There's nothing I can do. What am I *doing?*"

But the agitation wouldn't leave him. He opened the laptop again, and delved into all he could learn of Holy Island, the priory, the causeway, the towns between Lindisfarne and London, feeling sicker with each detail he managed to scrape up.

Angus! The answer hit him like the screech of a piccolo. Angus had walked up and down Niall's chambers in Glenmirril, calling out the news of James's birth. Shawn would do the same. Shout until Niall must surely see him across the years, and warn him. And it would be fine.

In a flash, heedless of their age, Shawn pulled back the Laird's letter, scanning the places Niall had gone. He would go there, too. And all would be well.

CHAPTER THIRTEEN

Linstock Castle, 1318

The porter waved Lachlan through with barely a glance. "We're needing more help," he grumbled. "What with the two of 'em here."

He didn't sound happy, Lachlan noted. A wise man would not speak more plainly against visiting lords. It left him only slightly less nervous about being alone inside the enemy compound. Owen would wait an hour or so—long enough to not be associated with him, but soon enough, they hoped, that MacDougall could not possibly have returned with any warning. They hoped he would stay away a day or so, dealing with his injured men. They hoped he would leave those men at Carlisle and thus have fewer here.

Lachlan crossed the snowy courtyard, to the kitchens indicated, where he found a crowd of cooks at work, chopping vegetables and stirring pots. The smells of meat and fresh bread filled the air.

A man wearing a white apron and floppy white hat looked up even as he slid a loaf of bread into a great brick oven. "What d' ye want?" he called.

"Work," said Lachlan. "The porter said you need help. I can fetch, carry...."

"Do you bake?" the man asked.

"A...bit." Lachlan evaded the habitual *wee.* "I learn fast." It was true enough. He'd never been inside a bakery, but he was already watching—flour, eggs, hands plunged in up to the elbows, kneading.

"Hmph." The man yanked the breadboard from the oven and slammed the small door.

"Sir?" Lachlan glanced at a girl carefully adding oil to a big bowl, unsure what the grunt meant.

"Two farthings a week, a blanket and a pallet of straw, you sleep in the kitchens with the other men, no dallying with the women. We need wood for the ovens. You'll find the ax out back."

"Thank you, sir!" Lachlan grinned, as pleased as if he'd just been knighted. He'd succeeded in the first step of the plan!

"Hob." The man offered a meaty hand. "You?"

"Jackin," Lachlan shook heartily.

"Hurry it up!" Hob yanked his hand back.

Lachlan dashed off. Chopping would keep him warm and pass the time waiting to see if Owen or MacDougall showed up first.

Glenmirril, Present

Glenmirril had fewer tourists in the cold months. Shawn waited thirty minutes after closing before vaulting the padlocked gate and dashing down the snowy walk, over the arched stone bridge, and into Glenmirril's courtyard. "Niall!" He cupped his hands and shouted into the empty bailey at the top of his lungs. "Hugh! Christina! Can you hear me! Tell Niall it's a trap! Stay away from MacDougall!"

He looked around the forlorn courtyard, to the spot that had cradled the small kirk, where he'd stood in Niall's wedding finery with Allene at his side; to the place where MacDougall's men had sat on their black steeds behind the crowd that day, demanding Niall's life, over the heads of the silent castle inhabitants. Frost coated each blade of grass. Heavy mist hovered around his ankles. He fancied, for a minute, that it might take him back, as it had before. And suddenly, he wanted it, with a ferocity that rivaled James Douglas in battle. Just for a moment, he wanted it.

He drew a breath, stunned by the force of the desire. He had to take hold of himself! He had a *life* here, a son he loved! The blank windows of the two wings looked down on him. He searched them, hoping for the impossible, for a glimpse of MacDonald, Red, Ronan, Muirne; a flash of Allene's red hair or Christina's sapphire-blue gown. Even their ghosts would do!

But the windows stared down, blank and empty.

He turned, searching the ramparts over the gatehouse, where Taran's father had died the night he crossed back, hoping to see Taran looking down, or Conal, Lachlan, Owen. Any of them!

"MacDonald!" He bellowed at the top of his lungs. Silence answered. He would be in his chambers. Shawn climbed the stairs, feeling his way in the gloom, down the hall to the rooms he'd shared with Allene and Niall.

"Niall! Allene?" He walked the empty space, shouting over and over. "Stay away from MacDougall. It's a trap! Niall, Christina, Allene, anyone! You have to hear me!" Fifteen minutes of shouting left his throat sore, and his ego bruised and feeling foolish. There was nothing to indicate anyone had heard him.

He went, next, to the Laird's chambers, shouting the message again and again, his voice becoming steadily raspier, till he could barely speak, till he shivered in the cold night. He circled back down the worn, winding stairs, along a ground floor hall, till he found the entrance to the Bat Cave in the dark, but he didn't have a flashlight. It wouldn't help anyone to get lost in the dungeons of Glenmirril on a freezing winter night. He cupped his hands, and shouted down the dark stairs. "MacDonald! Hugh! Tell Niall it's a trap!" He waited for any reaction, for them to materialize, ghostly wraiths affirming they'd heard.

Nothing happened. The steep stairs remained dark and dank and silent but for a far-off drip of condensation.

"MacDonald, do you hear me!" He tried one more time. "Don't let Allene leave Glenmirril! Don't let Allene leave! Tell Niall it's a trap!"

Reluctantly, he turned back into the courtyard, now swathed in moon glow illuminating each frozen blade of grass. A young couple stood arm in arm under the silver moon, staring at him in shock.

He laughed. "Just a game," he said. "Just jesting. I mean, kidding." He cleared his throat. "Joking."

The man slid his arm around the woman's shoulders, and backed her away, not smiling.

Outside Linstock Castle, 1318

The watch from the woods was as long and dull as any siege—and colder than most. "Could he not have abducted her in July," Hugh muttered for the second time. He brushed in vain at the frost in his beard, and rubbed a hand under his red nose.

"At least the snow has stopped." Conal hadn't moved from his place by the tree, his eyes locked on the castle. "Owen hasn't gone in. Should he not have gone in by now?"

"He's giving it time," Niall replied. He grinned at Conal. "Did you not tell me to trust them?"

Conal glanced at him. "'Tis the rest of them I don't trust. I find you are right on that matter." He suddenly straightened. "He's going! He's approaching the gate now."

Niall peered across the field to see Owen leaving the wood, moving west toward the castle walls.

"So are they," Hugh murmured.

Niall turned. Hugh pointed to the western horizon, over which thundered a small group of horsemen, riding hard for Linstock. The jutting castle wall blocked Owen's view of their approach. "*Ave Maria, gratia plena.*" Niall whispered. Helplessness settled heavily on his shoulders.

Inverness, Present

As the elevator opens, I see Jenny and Mairi chatting at the nurse's station. Jenny looks up as I come down the hall, James on my hip. Her face is somber. "He's still refusing to see you," she reports.

I sigh. Of course. I had no reason to expect any change. "He won't say why?" I ask of Mairi.

"He'll say nothing except, no," Mairi says. "Or, 'James needs Shawn.'"

"He can't possibly still believe I'm here out of pity."

"There's never been a man less to be pitied." Jenny looks down the hall to his room. "He's quite to be admired, he is. Did ye see on the telly, he's been given a medal for it?"

"I saw it online," I say. "He won't even see James?" I know the answer.

"Aig," James says. "Aig-oo. See Aig." He twists his hand in my hair and lays his cheek on my shoulder.

"He'll not," Mairi confirms.

At that moment, his door opens and Claire comes out in the cornflower blue sweater that matches her eyes, looking over her shoulder and laughing. There's a flush on her cheeks. She turns, sees me. Her smile falters. She has no choice but to pass me on the way out. She lowers her eyes.

Mairi reaches for her hand, and she stops, glancing up at me.

"It's okay, Claire," I say. "I just want him to be happy."

"What must you think of me," she asks. "Calling you to rush over and now he..."

"It's okay," I say again. "We both want him to be well."

"Nobody blames you," Mairi says.

"I don't understand why he'll not see you," she says in a rush. "I know...I know how...." Her face flushes, she mumbles, "I'm sorry," and hurries to the stairs, not waiting for the elevator.

"She's a lovely girl." Mairi stares at her hands a moment, before pushing the curl back off her forehead and meeting my eyes. "She's all wrong for him, but she is lovely and very sweet and we all like her quite a great deal."

"She is lovely," Jenny agrees.

Mairi speaks in a flurry of words. "My heart breaks for her, for him, for you. He believes no one else could love him now. I don't understand how this could have happened, how his gear could give way as it did. He took such good care of his equipment!"

I squeeze her hand. My heart breaks for all of us, too.

"Aig-oo?" James lifts his head from my shoulder. "Ang?"

"Not today," I tell him, and to Mairi, "Call me if he changes his mind. I'll be back tomorrow." And the next day and the next.

I say good-bye to Jenny and take the elevator back down. James twists his hand in my hair, asking again, "Aig?" When the elevator opens, I set him on his feet, moving slowly as he clings to my hand, toddling carefully, foot in front of foot, beside me. We pass through the glass doors and move slowly to my car.

The house seems oddly empty since Shawn's abrupt announcement, days ago that he's going to his house in Bannockburn.

Shawn and Angus are equally inexplicable. Shawn has behaved oddly since his return from the States. Well, I amend, since the morning after. He's edgy, sitting down to work on his arrangements, and suddenly bursting up again, for coffee, for a phone call, for a burst of agitated walk around the *block, to throw steaks in a marinade.*

He laughed off my comment—usually he's so focused—filled in a few measures, glanced at his watch and shot abruptly to his feet, saying he was going out for bourbon. He was gone two hours, returning after dark, looking guilty.

Creeping down the stairs again when he believes I've fallen asleep—he shuts himself in the front room. I hear him pacing below me, as I lay in the dark, knowing he doesn't want me in that room with him. I hear the front door whisper on its hinges and the soft hum of my car leaving each night.

He's lying to me again. He came back a new man. Yet he's lying to me.

It can only be the parchments. Whatever it is happened seven hundred years ago. Yet it scares me.

Linstock Castle, Northern England, 1318

The steady chopping kept Lachlan's tension at bay. He could do nothing, he reminded himself, but wait for Owen to arrive—and pray. Had the Laird not taught him so since he'd been a wean? Had he not *seen* the miracle of wee Robert's birth on a battlefield under the very shadow of the enemy covering the hills around them like locusts—and turning to leave? Did his bonnie Margaret not say she saw miracles in answer to her prayers? Her eyes had flickered over the child found under the trees when she said it. And he'd retold the story of finding James Angus. Yes, God watched.

Sweat gathered under his arms, though his breath hung in frosty puffs in the air. He swung the ax again, to the beat of the *Pater Naster,* and, *Father, protect me you. Know I do this for the. Safety of those he. Seeks to harm.*

The ax smashed down a final time, cleaving a log in two. He tossed the logs in the scuttle and stooped to heave the smaller pieces in, when, from around the building, shouts arose.

Lachlan's heart tripped and thudded hard. It could only mean one thing. MacDougall was back. But MacDougall had no cause to seek out a kitchen lad. Lachlan steadied his breath. One who had come here for work would think naught of the din. He must act the part—but neither did he care to stay in sight.

With an arm strong from drawing bows, he hoisted the scuttle of fresh-chopped firewood, hoping Margaret was praying for him even now; hoping Owen was inside, too. He did not fancy being alone in the enemy compound.

As he hauled the wood across the back bailey, a boy dashed around, ducking into the kitchen entrance ahead of Lachlan. "Brigands in Inglewood!" he shouted as he flung himself down the stairs. "Three score! Milord MacDougall wants food on the table anon!"

Commotion rose from the kitchens, and when Lachlan was halfway down, the boy scrambled back up, squeezing past him without a glance as he shouted for the stable boys. Lachlan entered the kitchen to see the cook, Hob, grumbling. He looked up to Lachlan and barked, "Get those in the oven!" then turned to shout orders at a girl, and the other men.

Lachlan shoved the logs in, wondering how to get to the stables and find out if Owen had made it in.

"Food and ale!" someone yelled. It was Owen!

Lachlan smiled as he heaved another log in, and shut the door. "I'll take it," he called.

Hob turned from an oven, wiping the back of his wrist over his ruddy brow. "Do that, aye. Fresh bread, cheese, dried meat in the butter cellar. Put it in the sacks you'll find there. Ale in the buttery. You!" he barked at Owen.

"Me?" Owen jumped, his eyes wide.

"You, ye eejit," Hob answered. "Help him now, don't be standin' like a ninny!"

"Yes, sir." Owen hurried after Lachlan into the cool chamber at the far end of the kitchens.

Lachlan handed him a canvas bag. "I'm called Jackin," he said. "You?"

"Gawain," his friend snapped. "They want it fast."

Side by side, they shoved rounds of cheese, tack, and dried beef into the bags, and carrying two in each hand, squeezed back through the ruckus in the kitchen. "They're mounting a hunt for the brigands on the morrow," Owen murmured in the confines of the stairwell.

"We're safer in here than they are out there," Lachlan whispered back.

Owen shook his head. "I said a farmer heard travelers say the brigands went west. Spread it."

"But we can't let MacDougall leave," Lachlan hissed.

"Take this." At the top of the stair, Owen stopped, dropping one of his canvas bags, and reached into his pocket for a small packet. He pushed it into Lachlan's hand, whispering, "Get it into the trough of the big black stallion." He grabbed his bag and marched out into the gray day of the courtyard, shouting over his shoulder at Lachlan, "Can you not move faster, you dullard!"

In the courtyard, men moved with purpose, unsaddling horses, removing helmets. Owen and Lachlan moved among them, handing out food, watching for MacDougall. He was nowhere to be seen.

With a last glance around, Lachlan slipped into the dark stables. The sounds of whinnying reached his ears. He peered down the length of the center aisle. As his eyes adjusted to the dim light, dozens of stalls took form, lining the walls on either side. Horses looked over several. From the far end came the sound of a horse kicking its walls. He started down the aisle, looking for the black stallion.

"You!"

Lachlan stopped, his heart pounding, and turned. At the far end of the stables, a soldier scratched the ears of a horse leaning over a half door.

"Sir?" Lachlan thought about the knives in his boots.

"You one of the stable lads?"

"I'm bringing ale to the soldiers." Lachlan avoided an answer that would make it easy to find him. He came back down the aisle between the horses—no big black stallion among them—and offered the aleskin.

The soldier took it, tipping it and drinking deeply.

"Beautiful animal." Lachlan nodded at a big bay.

The man lowered the skin and wiped his mustache with his sleeve, grinning. "The grandest, aye?"

"The bishop's?" Lachlan's heart beat faster.

The man shook his head. "Nay, MacDougall's. Milord knows horses. He's already working on re-building his stables. "

Lachlan feigned ignorance. "Re-building?"

"Aye, he'd the grandest beasts!" The man scratched the horse's ears. "His pride and joy—over there." He nodded at the stall across the way. "Pelops."

Lachlan turned. A great black stallion gazed over its door at him, before tossing its head and kicking its back wall. Lachlan crossed the aisle to admire the animal. He scratched its nose. "Well deserving of the title." He palmed the packet from Owen. Draping an arm over the half door, he tipped its contents into the trough.

As the soldier boasted of the stallion's breeding and training, Lachlan's heart hammered, not knowing what the poison would do. "What of his other horses?" He wanted nothing more than to be away from whatever was to come.

"The mare, down here," the soldier said. "She'll be bred with Pelops."

His heart slowing, Lachlan walked the length of the stable with the soldier, stopping to admire a number of horses. The man scratched ears and noses, commenting on personalities, speed, love of apples, who was prone to nipping or kicking. They made their way back down the other side, once again passing the man's own mount.

"Seems you've a good day's work on the morrow," Lachlan said. "Some brigands to hunt."

"So they're saying," the soldier replied.

"I heard in the kitchens," Lachlan said, "that a farmer met gypsies who overheard them. Heading west to Michael Scott's tower, they said. Three score. Maybe four."

"Aye, well, we've men from Carlisle joining us," the man said. He clapped Lachlan on the back. "Ye're a good man. I can tell, for the horses like ye."

Pelops kicked its stall. Guilt fluttered in Lachlan's stomach.

"What are you called?" the soldier asked.

"Jocelin." Lachlan hoped it was sufficiently different from Jackin that the soldier would be unable to track him down, yet sufficiently similar he could claim the man had misheard if he *did* find him—if he connected him at all with whatever was to come.

"I'm Hutch. Come talk with me about horses anytime." He thumped

Lachlan on the shoulder. "Well met!"

"Well met," Lachlan replied, and, gathering up his ale skins, headed out, grateful to be away.

Berwick, Present

Shawn wandered above Berwick, where the Scots would have spent the months of the siege. He walked the ridge in the evening and at night, waiting until nobody was around, before hissing, "Niall! Niall, tell me you're there! Stay away from MacDougall! Tell them not to let Allene leave Glenmirril!"

He hadn't been able to pin down where the Scottish troops had actually bivouacked, much less where Niall himself might have been. Telling Amy he needed work and practice time at his house in Bannockburn, he'd roamed Lochmaben for three nights, shouting for Christina, until he'd come face to face with a late-night tourist, punching numbers on his cellphone in alarm. He'd turned and disappeared before the local cops could show up.

Now, he called in the quiet outskirts of Berwick, choosing a different place each night, always checking the coast was clear of residents, tourists, cops— anyone who might find his behavior hard to understand. "Niall?" he called into the empty park. "Stay away from MacDougall. Niall, it's a trap."

His phone burbled *Unchained Melody*. He glanced around once, hoping by some miracle to see Niall, before sliding it open. "Amy?" he asked.

"What have you been doing, Shawn?" Her voice held a note of panic.

"Doing?" He asked first in confusion, and then, intuition kicking in, gave a hearty laugh. "*Doing? Who, me?* I'm home practicing. That Czardas! Still a killer after all these years!"

"You can play that in your sleep," she said. "So who do you expect me to believe it is that looks like you walking around Lochmaben, Glenmirril, and now Berwick, shouting, *It's a trap?*"

"Uh, Niall?"

"Niall." Exasperation burst over the phone. "What is going on, Shawn?"

He felt his eyes narrowing. "How do you know about this anyway?"

"You're on the news right now."

"The *news*?" A groan erupted from him. He looked around the dark streets, wondering if someone was live-streaming him even now. He didn't need Ben or his musicians or publishers hearing about this. There was no way to explain.

"Just a small online thing, really, but yeah, the American musician who disappeared for a year and showed up again last June has been seen at two castles shouting, *It's a trap!* A nice picture of you right there on the screen. A ten second sound bite from a psychologist speculating on your mental state. You'll be happy to know they're pretty sure you're harmless, if a little deranged. It took me ten minutes to get over the shock and call you. What is going on?"

"What is going on?" he repeated.

"You know, if you'd just tell the truth, you wouldn't have to stall like that."

"Um, well." He looked around the empty park, with the housing developments shining their lights into the dark. He didn't want to upset her. He wracked his brains for an explanation. "See, I read this thing on the internet."

"It has to have something to do with Niall. Why Lochmaben? I've just googled and I can't find anything saying he was ever there."

"Um. No. I don't know."

"Just tell me the *truth*, Shawn!" she exploded.

"No." The word slipped out.

"No?" She sounded dumbfounded.

"I mean, it's not that I don't want to. Oh, hell, Amy," he said in frustration, "I *don't* want to!"

"Is it about Niall?"

He nodded, still walking the length of the park, peering this way and that into the growing mist, as if he might finally see the ghosts of the men who had camped here centuries ago.

"Shawn?"

"Uh, yeah. Yes, it's about Niall. You don't want to know, Amy."

"Well, now that I know there was a trap, you need to tell me the rest."

Shawn shook his head. The grass crunched under his feet. "Forget it, Amy. It was centuries ago."

"Not in my life, not in yours," she objected. "What *happened* to him, Shawn?"

He sighed, continuing his search of the park. Reaching its end, he strayed into the neighborhood. It hadn't been here seven hundred years ago, he reasoned, at the same time chiding himself about his expectations. But he couldn't just walk away and leave Niall to his fate. "MacDougall," he said. "But I've warned him. It's fine." It would be, he told himself. "They're all dead, anyway, Amy. What difference does it make now?" He hoped she wouldn't ask for details.

"Obviously enough that you're risking your reputation shouting for him in ruined castles."

"I've taken care of it. It's fine." He slid her picture away, ending the call. He peered down the dark street, with the rows of shoulder to shoulder houses rising on the ridge from which the Scots had once besieged the town.

"Niall," he hissed. "Come on, I'm making an ass of myself. Can't you appear or something? Get someone, *right* now, to go leave me a message at Glenmirril telling me you heard! Tell me you're not going anywhere near MacDougall! Tell me you heard me and didn't let Allene go to the widow Muirne's house!"

He turned in a slow circle, scanning the misty road, the bushes, the row houses lining the street.

A man stood in his doorway, smoking a cigarette, watching him. "You all right?" he asked. "Can I get someone to help you?"

Shawn turned away, shielding his face. He wondered if the news had managed to make its way to the States yet, and how long he could get away with not answering Ben's phone calls.

CHAPTER FOURTEEN

Inverness, Present

A girl brushed past Shawn, coming out of Angus's room as he reached it. She glanced at him with cornflower blue eyes dancing with joy and hurried by. She was familiar. He had seen her....

"Sit down. We need to talk." Angus's voice snapped Shawn's attention back to the hospital room. Angus sat in a wheelchair by the hospital window, wearing jeans and a heavy fisherman's sweater, a book in his hand. He laid it down on his lap.

"You're up!" Shawn stopped in the doorway, feigning energy. In truth, he'd had multiple late nights, on top of a heavy load of arranging for the album Ben wanted out yesterday. He wanted nothing more than to be in bed, asleep. But when he collapsed in bed, he turned and rolled restlessly through the dark hours with nightmares—if he slept at all.

"Shut the door."

Shawn did so, and tossed himself into the hard excuse for an armchair in the corner. "You can imagine I was a little surprised to hear from you. What's up? Did you finally come to your senses about Amy?" He hoped not. A few more weeks, they'd go home together. He'd solve this problem about Niall, and they'd go home.

"My senses are grand, thanks." Angus gave him a sour look. "'Tis yours that concern me. They say some madman's running about Glenmirril at night shouting, *It's a trap.*"

Shawn coughed into his hand, covering his surprise. "Um, no. There's no madman at Glenmirril. Who's telling you these stories?"

"Clive. He was on duty when a young couple called."

"Yeah, well, he got it *all* wrong. I wasn't *running*." If he'd been prone to blushing, he'd be doing so with a vengeance. Fortunately, he'd never seen the value in blushing. He smiled, cool as Scottish mist.

"Are you an eejit?" Angus slapped the book onto the window sill. "Running or no, he thought it best if I talked to you before he lets it get into the data base."

"Is it against the law to walk through castles yelling *It's a trap?*" Shawn cocked an eyebrow.

"No, they've not thought to make a law so specific, seeing it's never before been an issue. But there've now been two reports, in addition to a wee news program online." Angus cocked an eyebrow right back at Shawn. "They'll not be able to ignore it forever."

Shawn heaved a sigh. He supposed Angus was right, and no smart comeback would change that.

Angus wheeled his chair around to stare out the window. "I suppose it has to do with the parchments they left you?"

"I suppose there's no other explanation," Shawn shot back.

Angus's next words came out with less the authority of the Inverness police, and more the concern of a friend. "What's happening?"

Shawn relaxed back into his chair, his eyes closed.

"I met him," Angus said. "In a room just down the hall from here, when we thought he was you. Believe it or not, I care what happens to him. And those he loves." He cleared his throat. "Loved, aye? 'Tis past now. But still, I feel I know them."

The weight of it settled on Shawn, the ugly images that haunted his dreams, jarring him awake drenched in sweat, of Niall's bound hands, the jeering crowds. Holding back the words kept the worst of it at bay.

Angus's voice broke through his dark thoughts. "Whatever it is, have you told Amy?"

"No."

"It often helps to talk about it."

"What's there to do?" Shawn pressed the heels of his palms against his eyes. The sick feeling swept over him.

"What happened to Niall?"

Shawn pursed his lips, staring at the monitors beeping near Angus's bed.

"I wish I could do something for you." Angus frowned. "I wish I had a story to tell them why you're going about Glenmirril night after night shouting for Niall Campbell." He paused and added, "You know they think you're mad." He tapped his head. "In need of help."

"Yeah, I need help!" Shawn burst from his chair, pacing, a hand driving through his ginger hair and snagging in the leather thong that held the pony tail. He yanked it free in agitation, stopping himself from saying, *Can they help me send a message back seven hundred years?* "I think you can make some guesses. You *know* I don't need..." He made air quotes, spitting out the word. "*Help.*"

Angus lifted his hands in resignation. "Ye canna keep doing what you're doing. They'll take you in on the next complaint. You're scaring people."

Shawn sighed. He squeezed his eyes shut, seeking some brilliant idea in the darkness behind his lids. The sick feeling climbed higher.

Angus rolled his chair past the foot of the bed. "If there's anything I can do, Shawn, I will. If ye could only think of a story I can tell them."

Shawn shook his head. "Even *I* can't think of a rational story for shouting *It's a trap* every night in a castle ruin." He sagged back in his chair.

Angus picked up his phone, hit a few keys, and laid it down. "Not that I've enjoyed the company of the great Shawn Kleiner, but I've another visitor."

"Yeah, sure." Shawn barely registered the insult. His mind was on Glendochart. He'd just have to be more careful there, calling for Niall.

Linstock Castle, 1318

Lachlan lay awake much of the night, curled on the kitchen floor with the other cooks, waiting for someone to call him out as an impostor, worrying over Owen's safety, and what he had done to the stallion. He'd barely drifted into sleep, when someone shook him awake. He rolled, scrambling in the dark for his knife.

"It's me, Hob. Come now, who'd ye think it was?"

Lachlan's heart slowed. Around him, others were rising, stretching, yawning, in the weak light of several lanterns. An older man was already punching dough in a bowl, while a boy pulled a fresh loaf from the oven. "Gather cheese and hard tack from the pantler," Hob said. "Milord MacDougall's men are already gathering to ride after the bandits. Haste now!"

"They know where they've gone?" Lachlan asked.

Half a dozen voices filled him in with the rumor he himself had spread.

Minutes later, Lachlan climbed the stairs carrying bags of hard tack, cheese, and dried meat. Gray skies glowered down on the courtyard. Men poured from the great hall, some of them wiping their lips. Boys led horses from the stables.

Across the way, Owen came from the great hall, carrying a large jug toward the kitchens. Lachlan threaded through soldiers and horses, delivering food and letting Owen reach the kitchens and disappear down the stairs, before heading back.

"Men! Mount up!"

In the shadow of the doorway, Lachlan's heart pounded, thinking of Niall, Hugh, and Conal outside the manor's walls. But they were on the east side. MacDougall's army would ride west, never seeing them. He glanced back into the courtyard to see MacDougall striding from the stables, leading the black stallion. Its head drooped. Lachlan's heart pounded—with relief or fear, he couldn't say. He hated to harm the animal, but MacDougall couldn't leave!

"Brigands attacked us in Inglewood," MacDougall shouted. "Three score or more! Dunkeld was killed and more lie injured at Carlisle. Has any more word come in of the brigands who did this foul deed?"

Lachlan backed further into the passageway, down the first step, as he watched. An old man stepped forward, giving a bow. "Milord, the farmer who brings hay—he heard from travelers that the band has headed west toward Michael Scot's tower."

"How did the travelers know this?" MacDougall demanded. At his side, Pelops swayed.

"Sir, they came upon them in the night, and stayed back, wary of such a large group, and overheard them. They were laughing amongst themselves, the men said, about having shot at someone in the forest during the blizzard."

Anger flushed MacDougall's face.

Lachlan took another step down.

"Let us ride!" MacDougall shouted. "Let us give them aught to laugh about!" He pulled Pelops around and hoisted himself up into the stirrup.

"Jackin, come away." Owen's voice was soft. "We mustn't seem too interested."

"He's going to leave," Lachlan whispered back.

"Wait a moment."

Lachlan took another step and turned back to see MacDougall's horse buckle under him.

Inverness, Present

My heart jumps, a hundred dancing pizzicato notes, shimmering in excitement, as I hurry up the hospital stairs, glancing at Angus's text. He has relented!

I warned myself his words were short, as I picked out the royal blue shirt I know he loves. But it's a change, I argued, as I brushed my hair to a gleam. My insides flutter, thinking of the first time we kissed behind Glenmirril, wanting to see him again, smiling at the archaeology and police jokes he used to send at the bottom of each e-mail. What do you call a midget clairvoyant escaped from prison? *He's remembering these things, too.* A small medium at large! *He's finally believing this is not pity!*

I burst through the door at the top of the stairs, waving to Jenny. "Aye, he's said send you in!" *She beams!*

He sits in a wheelchair, staring out the window.

"Angus?" His name, his beloved name, comes out breathlessly.

He doesn't move. I cross the room, dropping to one knee in front of him.

He takes my hand in his, but he doesn't meet my eyes.

The excitement dies. "Angus?"

"I'm being released in a few days," he says. "I've made some decisions."

He tells me, the words roaring over my head, dull and echoing in my ears. My heart pounds as I climb to my feet, backing away. "You're wrong," *I say.* "You can't do that."

He stares steadfastly at the wall, not meeting my eyes. "I've already done *it."*

    I shake my head, denying.

    "You need to see," he says, "you'll not be happy with this life for long. James needs Shawn. He needs *him." He turns his head, looking out the window. "You must leave now, Amy."*

    I back out of the room. I spin in the hall, moving, unthinking, and find myself in the chapel, in the silence of the incense and the dim light and the candles flickering. How did I not see this coming? I drop to my knees before the crucifix. Why didn't I believe what he's been telling me for weeks now? My hand goes to the crucifix, Niall's crucifix, always around my neck. I've been a fool. Just like with Shawn—seeing what I wanted to see. I bow my head, trying to find peace, breathing deep, trying to think what comes next.

    I can't. I can't see any next. I can only see I've been a fool and Angus has meant what he said all along and I don't know where to go from here, from all I loved in him, from all I wanted and hoped for with him. The tears slide down my cheeks.

Linstock Castle, 1318

    Owen touched his sleeve as a clamor rose in the courtyard. The two of them hurried back to the kitchens. Owen disappeared out a far door, leaving Lachlan to busy himself with chopping vegetables, his heart hammering. Soon, murmurs arose among the kitchen staff. "Back to work!" Hob barked. The kitchen fell silent but for the *snip snip* and *chop chop* of knives through greens and leeks, the shuffle of feet, the snapping of flames in the ovens.

    The day passed slowly in the dim kitchen, with little information, Lachlan wishing to find Owen, fearing every moment that MacDougall would come storming in, and recognize him. But MacDougall had never lain eyes on him, except—maybe—in the midst of the battle in Glenmirril's courtyard or maybe the following morning. That had been long ago. Sweat prickled his brow nonetheless.

    The moment Hob headed up the stairs, murmurs swelled in a swift, eager *crescendo.* "The leg is swollen," a tall young man said.

    "Will it live?" Lachlan asked.

    "Oh, sure 'twill live." The girl spoke in hushed tones. "But milord MacDougall is in a fury."

    Lachlan held his tongue, wary of appearing too interested. The chatter grew, speculating, but no mention of what MacDougall had *done.* If he'd gone with his men, he might be gone for days. Their opportunity would be lost.

    "The horse but took ill," rasped an old man. "He sees enemies all about."

    "Hush," said another. "Would ye like your words to get back to him?"

    A momentary quiet fell over the room. A boy came down the stairs, clutching the feet of three dead, plucked chickens in each hand.

"Has Milord MacDougall stayed or gone?" the tall youth asked, as he flung them up on the counter.

"Why would he not have gone?" asked the girl. "There are other horses."

"It's his favorite." Hob strode back into the kitchen, a canvas sack bulging with flour slung over his shoulder, followed by the miller carrying two more. He flung it down on a table. White puffs fluffed up into the smoky air. "The only one he managed to bring from Creagsmalan when he fled. He'll likely sleep in the stall with it until it's well. Did I not tell ye to stick to your work!"

Lachlan sliced his knife through a head of cabbage.

"Thinner," the girl whispered. "Milord likes it sliced quite thin." She took his knife, showing him, and handed it back, saying, "He gets irritable if things aren't just so."

"Difficult to live with?" Lachlan asked softly.

"Oy, very. But the bishop wants his guests happy."

"I've heard grumbles," Lachlan said.

"The stable lads are none too fond of him." She laid out a row of leeks on the board beside his and started peeling them, long white curls rolling off under the swift motion of her knife. "He wants all their ways changed, now he's here. And the pages who serve dinner—he's forever finding fault."

"Eda." The tall young man spoke softly as he set out the chickens on the opposite side of the big counter. "Watch your tongue." He glanced at Lachlan as he took a grip on the first chicken's body. "I trust you'll not be repeating all you hear. I'd not like it if his anger were to come down on our Eda." His cleaver crashed down on the neck. Blood spewed out.

Lachlan met his eyes. "I've no reason to spread tales. I've no wish but to earn my way."

"See, well, we don't *know* ye," the man said.

"Jackin," Lachlan replied. "I fled Berwick when the fighting started. I want but to live in peace."

"Don't we all." The man's glower faded abruptly into a smile. "Renfrey. I'm called Ren."

"Well met." Lachlan returned the smile.

They chopped and diced for a time in silence, two chickens going into pots which a boy carried to the hearth. Eda finished peeling and chopping her leeks, gathered them in a bowl and took them away.

"'Twas her brother milord MacDougall had whipped," Ren confided, the moment she left. "Nearly killed the poor lad. He's but ten."

"What did he do?" Lachlan asked.

"Spilled milord's chalice." Ren sliced into a third chicken. "It was the night milord arrived. In a fine fettle, he was. Saw the lad and demanded he serve." Ren pulled the chicken apart, digging inside and pulling out the organs. "We tried to tell him the boy's a bit simple, see, but he'd have none of it. And the poor lad was terrified of milord's raging, and shaking with fear, and sure the wine sloshed."

"It sounds," said Lachlan, "as if the lad is well-loved." That did not bode well for MacDougall.

"Oh, he is." Ren dropped the chicken into a pot. "His mother, now she was a saint, she was, good to all of us, and when she died last summer, sure, we'd do anything for Eda and young Robin."

Lachlan took a bowl from under the cupboard and lifted the sliced cabbage into it. "No doubt he's angered many with such an action." He didn't dare ask if one of them might have done something to injure MacDougall's horse.

"You'll find no friend of his here, sure." Ren wiped his hands on a towel and met Lachlan's eyes. "You didn't hear that from me."

"How long will he stay?" Lachlan asked.

"A fortnight? More, less? It all depends on Edward now, does it not?"

"Where will he go?"

Ren shrugged. "There's no reason they'd tell us such a thing. Judging by his mood, I'd say he doesn't know, himself. Mayhap he'll become a member of Edward's household." He lifted the pot, biceps flexing with its weight, and carried it to the hearth, calling, "Eda, come tend the fire now!"

St. Fillan's Chapel, Scotland, Present

"Niall, can you hear me!" Shawn shouted into the quiet forest glade. Anger raged in him. It was cold! It had taken almost three hours to get here, a story to Amy to explain his disappearance—he needed to work at his keyboard, down at his house in Bannockburn, he'd be gone a few days—and another half hour waiting for some tourists to leave.

And he didn't even know if he was here at the same time as Niall.

He got the usual silence. "Niall?" He headed down the forest trail, between skeletal winter trees. Snow crunched under his foot. The morning had passed quickly in searching the internet, until he'd found a blog with three hundred and forty-five posts about the times of the Bruce. One sentence in one article said it had been one Niel Cambul who brought news to the abbot of Glendochart of Bruce's endowment for the church of St. Fillan. Late January.

Close enough. To save Niall's life, it was close enough. He'd been in the car, shouting the story back at Amy, tamping down his guilt at lying to her again, as soon as he'd double checked the way to Crianlarich and Tyndrum. It was to save Niall, it was to spare her worry, that he'd lied. Wasn't that different from those previous lies?

"Niall?" His voice fell flat in the winter wood. A crow lifted off a branch with a raucous *caaaaw, caw,* and flapped away against heavy gray clouds. With a glance after it, he turned back to the stark white trees and the ripple of the water against the sharp winter air. He looked around to be sure he was alone before cupping his hands and shouting, "It's a trap, Niall! Stay away from

MacDougall! Tell Allene to stay inside Glenmirril!"

He stood, waiting. For what, he asked himself. Angus hadn't seen *him*, when Angus had called at Glenmirril. Shawn moved down the path, stopping at intervals to shout his warning. He came quickly to the priory ruins, a light dusting of snow over its mossy stones.

He stood still, feeling for a moment as if he hovered between two times. Niall must have stood on this spot, studying it with the abbot—whoever the abbot had been at the time—discussing what would be built there. Or had there been a church already, something the Bruce's gift would refurbish? It had been impossible to tell from the sites.

Had Niall seen a church stand here, fresh and new, some time after delivering the Bruce's good tidings? Had he stood here, maybe on this very spot, in 1318, and envisioned what the church might look like? The possibly empty spot of 1318 wavered with the possible broken-down church that needed restoration in 1318, and it all wavered with the ruins that stood before him.

"*Niall!*" He shouted it, suddenly, surrounded by all times, all possibilities. "Niall, you can't go there! Get a message to Allene! Tell her not to go!" He cupped his hands to the mossy snow-flecked walls, shouting, "It's a trap! Niall, you have to hear me! Leave a letter in the kneeler, tell me you heard me!"

He stood in the middle of the ruins, rising on two sides, trees and fields around him, shouting, "Come *on,* Niall! Answer me, dammit, *answer me!*" He stormed to one wall, to another, shouting, "It's a trap! Come on, Niall!" He listened a moment, hearing only silence, and exploded, "*Dammit, Niall,* let me know you hear me!"

He spun, trying to catch any glimpse of those who had walked there.

A teenage girl with running pants, a warm jacket, and a long red braid stood on the path, a tawny Irish Wolfhound beside her, as big as any of the Laird's hunting hounds. Her lower lip tightened. She backed away two steps—three—turned and sprinted, dragging the dog behind her.

CHAPTER FIFTEEN

Inverness, Present

"He's coming home tomorrow." Amy, wearing her peach bathrobe, looked up from the armchair in Angus's living room, as Shawn came in.

"That's good news, right?" Shawn frowned, dropping his briefcase full of music on the end table. He'd spent another two days in Bannockburn, where he still had his keyboard, writing arrangements, practicing, making business calls, Skyping Ben and some young new talk show hostess about his upcoming appearance on her TV show, and letting his mother know he'd be home.

More importantly, his stay covered the trip to Glendochart—not that it seemed to have done any good. He'd spent the morning drive considering his next step, if those parchments didn't change.

"It's great news," she said.

"You don't sound happy." He wondered uneasily if Angus had mentioned their conversation to Amy.

"He told me I need to go to Bannockburn."

"Well, yeah, you have students wanting lessons there," Shawn said. "So get them lined up one or two days a week and come back again."

"No." She stared into the fire. "He means go and don't come back."

Shawn dropped onto the couch. He saw now that red rimmed her eyes. "He'll change his mind. He'll feel better when he's home from the hospital."

Amy stared straight ahead. Her voice came out flat, deadpan. "He asked Claire to marry him."

"Claire?" Shawn's voice rang with disbelief. "Cretin though I am, even I can see he's not in love with her. He thinks she'll go for that?"

"She will. She *has*."

"Okay, then." He rubbed his hands together, wanting to help. He frowned, wondering why he didn't feel great about this. It was all working out beautifully for him. Angus would marry Claire. Amy's feelings would fade into the past, and she would give him her whole heart again. He already had a full half of it—he knew that. They'd be a family.

She stared into the fire. Of all the times he'd hurt her, he thought, he'd never seen her so despondent.

"You'll feel better if you play your violin." Even to himself, it sounded woefully lame.

Amy didn't react.

"Amy, you're always happier when you play."

"I don't feel like it."

He leaned forward, the first niggle of worry chewing at him. "How long have you been sitting here?"

She shrugged.

"Is James sleeping?"

"He's with Mairi. I asked her to take him for a bit."

"How long is a bit?"

She shrugged.

Shawn got up, feeling uneasy. He looked in the kitchen first. The same plate and cup that had sat in the sink when he left two days ago still sat there. He opened the refrigerator tucked under the counter. The same plastic tubs of leftover potatoes and turkey he'd packed up were still full. He opened the breadbox. The bag of bread had not been touched.

He went upstairs. The bed hadn't been made. Amy always made her bed. The towels hung where they'd been when he left. He touched one of them. She'd been wearing her peach robe when he left. That didn't mean she'd been wearing it for two days.

He went back downstairs, glanced in the front room where she sat as she had been, and turned back to the small kitchen. Warmed-up potatoes, livened up with a few spices, bacon he'd bought before he left, a salad, and a mug of hot tea all made their way onto a tray, while his mind jumped from Angus's decision, to Amy sitting blankly on the couch, to the parchments. He carried the tray into the front room.

He used his foot to push the coffee table closer to the chair, and set the tray down. She stared blankly at him. "Food. Eat." He spoke with the voice that had ruled an orchestra, that had ordered men into battle and down into English towns on midnight raids.

At the crack of his voice, her head shot up. Shock flashed across her face, but she unfolded her legs and reached for the fork. He watched for a moment before saying, "I have to go back to the States soon. I need to know you're going to take care of yourself and James while I'm gone." He turned on his heel, storming out of the house.

Inverness, Present

"He's asleep." Jenny looked up from the desk as Shawn stepped out of the elevator. She set down her clipboard, and rounded the desk, coming out to stand on tiptoe and give him a quick kiss on the cheek.

"He needs to wake up." Shawn walked past.

"Now, Shawn, you shouldna...."

He stopped, turned. "No, really. He's making a big mistake. He and I need to talk. Claire isn't here, is she?"

"She left half an hour ago. She's coming tomorrow to help him."

"No, she's not." He marched into Angus's room, dim with the blinds pulled, and flipped on the light.

Jenny hurried behind him, hissing, "Shawn, you canna do this! What if security comes?"

"They won't come if you don't call them," he pointed out. "Trust me, okay?"

"You're not going to hurt him?" she asked doubtfully.

"What the hell do you take me for?" he asked in disbelief.

"Well, I mean your girlfriend, ex-girlfriend, whatever she is, is in love with him. I thought...."

"Nothing so dramatic," Shawn said. "I'm shooting myself in the foot trying to help him. He's being an ass."

Jenny gasped, at the same time Angus stirred and rolled in the bed, squinting in the light. Jenny flipped the light off, and reached for a softer bedside lamp. "If he wants you out, I've no choice but to call security."

"Of course I want him out," Angus grumbled. "What now?"

"You're not marrying Claire! Are you insane?" Shawn demanded. "It's the drugs talking."

"Take her back to America." Angus yanked the blanket up over his head.

"Do you want me to call security?" Jenny asked.

Angus grunted from under the blanket, in lieu of an actual answer.

"If he wanted you to," Shawn said, yanking a chair up on the far side of Angus's bed, "he'd say so."

Angus rolled over, turning his back to Shawn.

"He's still not yelling for security," Shawn pointed out to Jenny. "Just give me a minute, and I'll leave."

"Just leave the big light off. And keep your voice down." She backed out of the room, still looking doubtful.

"What in the hell do you think you're doing?" Shawn hissed.

"Go away," Angus said.

"Call security if you want me to go away. You think Claire is going to be happy for more than six months on these terms? You think you will be?"

"It's what she wants," Angus snarled from under the blanket. "'Tis no business of yours."

Shawn stood and yanked the blanket off him. "*Amy* is my business! And you're devastating her."

Angus threw the covers down and pushed himself upright. He glared at Shawn. "She and I only spent a few months together. 'Tis Amy who will realize in a few months she'd not be happy with me. Take her back to America. She'll realize by this time next year how foolish this was, thinking she wanted to be with an invalid."

"You're not an invalid."

"She'll be happy with you."

Shawn dropped suddenly into a chair. "Look, we're on the same team, right?"

"Only if the other team is Simon Beaumont," Angus grumbled.

"Team Amy." Shawn pumped a fist. "Rah rah! We both want what's best for her, don't we? Let's go for a walk."

Linstock Castle, 1318

Lachlan climbed the tower stairs, lit only by oil lamps mounted on the wall, clutching the wine jug. He'd never come face to face with MacDougall, he reminded himself. There was no reason MacDougall should recognize him. In addition, MacDougall believed himself safe inside his moat and walls. Men generally saw what they expected to see. Reaching the door at the top, Lachlan knocked.

"Enter." The voice was weary.

With nerves tingling, Lachlan squeezed the latch and let himself in. Candles and oil lamps sent soft shadows wavering over the tapestries. MacDougall looked up from a parchment that lay on the table, a quill poised in his hand. A goblet of wine sat before him. "Who are you?"

"I am called Ralf," Lachlan said.

"What do you want?" MacDougall's eyes flashed over the jug.

"I've come to replenish your wine," Lachlan said.

MacDougall glanced at the jug on the table.

Lachlan moved in, refilling it without waiting for permission. "Milord," he said as he poured, "there is a woman in the village who may help your horse. Shall I bring her to you?"

MacDougall straightened, hope coming into his eyes.

Guilt washed through Lachlan. They shared a love of the fine animals. Maybe the man wasn't so bad. Maybe...

But MacDougall's words were strident. "What does she know of it?" He jabbed the quill into its holder.

Lachlan gripped the half empty jug, ready to use it as a weapon. "She knows herbs and it is said she has a way with animals in particular."

MacDougall rose, his brows furrowed, and turned abruptly, marching to the window to stare out into the dark night.

Lachlan's eyes fell on the parchment. Niall's name was clear on it. The twinge of guilt fled. He poured the last of the new wine directly into the goblet on the table.

MacDougall turned from the window. "Aye, bring her to me."

"Yes, my lord." Lachlan lifted the cup, offering it. "Your stallion will be well by the morrow."

MacDougall strode to the table, taking the cup from Lachlan's hand. "Tell the guard in the gatehouse to summon a soldier to go with you."

Lachlan stared, waiting for him to drink.

"Are you daft?" MacDougall snapped. "The woman will not bring herself, will she?"

"No, my Lord." Lachlan bowed and hurried out.

Inverness, Present

"What if I *weren't* here?" Shawn demanded for the second time. The idea that had begun in the car had grown after seeing Amy. Snow crunched under his feet as he and Angus moved down the path of the hospital's garden, Angus pushing the wheels of his chair with gloved hands.

Don't be crazy, Shawn told himself. The half-formed thought was not reason enough to be putting even the possibility out into the open. But hadn't he always started with possibilities, he thought, at least finding out what the possibilities *were?*

"But you *are* here," Angus said for the second time. "I was unfortunately right in thinking I'd not so easily be rid of you."

"Yet you came out with me."

"I had my reasons."

"So what if..."

"You *are*," Angus snapped.

"Are you totally incapable of hypothetical thinking?" Shawn raised his voice in frustration. "Don't you hear the words *what if* at the beginning of that sentence?" Anger remained from his attempt at Glendochart. The parchments had not changed. The websites that told the story had not changed. And for all he knew, the girl had dragged her poor dog straight to the nearest police station and reported him. He supposed it was a good sign Angus hadn't said anything.

"What's the point?" Angus stopped, wheeling the chair around to face Shawn. "Where are you going to go? America? It doesn't change anything. You're still there for her and James and that's where she belongs. She needs you. *He* needs you."

"Would you take her back if I weren't available anymore?"

"You're not going to marry Jenny after two dates, so this is a foolish conversation." Angus turned the chair and pushed it forward again. "If you're going to inflict yourself on me, at least tell me about Niall and Allene."

"Inflict myself!" Shawn shoved his gloved hands deep in his pockets and hunched into his scarf. "Do you know how many millions out there would love to be walking alone with me?"

"Aye, well, do I look like a screaming fan girl?"

"Anyway," Shawn said, "you're the one who walked into my life."

"Funny thing to say to the man in a wheelchair," Angus said acerbically. "But if we must be so thoughtless of my lack of use of my lower limbs, I walked into Amy's," he corrected. "And I shouldn't have. Tell me about Glenmirril or get out of here."

Shawn didn't want to talk about it. It hurt. Like talking about a deceased child, someone beyond reach, someone to whom your soul had become tied. "It feels like losing a limb."

Angus stopped the chair again, under the bare branches of an oak, and Shawn realized he'd spoken out loud.

"Look, I didn't mean—that was tactless."

"I didn't lose a limb," Angus pointed out.

"You lost the use of two. Same difference."

A black bird lifted from the tree above, shaking down a clump of snow. It landed with a *plop* next to Angus's chair. "The great Shawn Kleiner apologizing."

Shawn laughed. It was exactly the sort of thing Niall would have said. "Yeah, well, don't worry, I won't do it again."

"Maybe it's time you told someone how you really feel," Angus said. "You were close to them."

"Like nothing I ever had in this life," Shawn admitted. "Everything I did mattered."

"Tell me."

Shawn did—about, meeting MacDougall in the forest outside Dundolam, the nights outside Berwick playing the harp around the fire for Douglas's men, and discussing yet another strategy for breaching the walls. "It ended when Douglas got shot—I mean, with an arrow—climbing a ladder. Amazing man. He and an engineer came up with the idea of a rope ladder. It had never been done before."

Angus stopped before a statue of an angel. "What was Christina like?"

"What?" Shawn leaned closer. "Sorry, wind took your words away."

Angus cleared his throat. "What was Christina like?"

"Christina?" Shawn's senses shot sky-high. "Why?"

Angus glanced up at him. A frown fluttered across his face for just a moment before he said, "Amy and I did some research—trying to get you back. It seems I'm descended from her."

Shawn jolted. "Her and...?" He studied Angu's face, trying to see MacDougall—or Hugh—in his features. He was big, like Hugh. He lacked MacDougall's yellow teeth or greasy hair.

"Her son. James Angus."

Shawn was silent. All the irritation he'd ever felt of Angus sprang back. *No good ever came of a MacDougall.* The Laird's oft-muttered words jangled in his mind.

"Which makes me a thieving MacDougall." A corner of Angus's mouth quirked up. "I think the MacDonald's cattle are quite safe from me at the moment, aye? Perhaps there could be peace in this century?"

Shawn's humor bubbled to the surface. "I wasn't going to say it," he said.

"The poker table is the only place you've a poker face," Angus replied. "But sure if I'm to be counted a thieving MacDougall—I am also all that Christina was."

"Was?" Shawn blinked then shook his head. "Yeah. Was."

"Tell me about her," Angus repeated.

Shawn turned away, staring at the statue. "She's good," he said softly. "Courageous. Strong. She always puts other people first and thinks of their good." He stared at the statue, relaying the story of how he'd met her in the confessional to save Niall, lost in the story, in time, remembering the sound of her voice soft through the lattice grill work in the dim confessional.

"You miss her," Angus said.

Shawn didn't answer. She and Hugh must have been happy together. They were both *good*.

"Are you sorry you came back?"

"I spent two years trying to get back." He turned, finally, to look at Angus.

Angus glanced up at the snow lining the slender oak branches above his head, and rolled his chair down the path again, not bothering to see if Shawn followed. "Which doesn't answer the question."

"It's a question better not answered." Shawn walked silently at Angus's side, the cold air burning hot and bright in his lungs, bringing back dozens of memories of skiing and ice fishing with his father, of riding through mountains and forests with Niall. "There are those," he said, "who won't believe I'm different. There are those who can't forgive me."

"Amy isn't one of them."

"Maybe not. But I've hurt a lot of people. That will follow me forever here. There, I was someone different. Someone respected. Someone who did a lot of good for other people."

"You can be that here."

"There are those who won't let me be. Ever."

Clarence sprang to mind—Clarence who was trying to be better. Shawn pushed him away. He'd done well enough keeping him out of mind since his release. "And where do I fit in this world anymore?" he asked, forgetting he had no desire to be friends with Angus. "I killed men in cold blood."

"Killing in war is not killing in cold blood," Angus said. "You're no different than any soldier or copper who's killed."

"Pulling a trigger, throwing a bomb," Shawn said. "It's not the same as driving the knife in, watching a man's eyes as he dies. It's up close and personal. And...I *wanted* to see Duncan MacDougall dead. He *needed* killing and I'm not sorry I'm the one who did it. Is that the same?"

"Maybe not," Angus admitted.

"So where do I fit into this civilized world?"

Angus rolled to the edge of a small pond. A pair of ducks shook their wings in the cold, and waddled to his feet, squawking. He laughed, pulling a hard roll from his pocket. Slipping one glove off, he broke off bits of bread and tossed them into the powdery snow. "Have you ever really fit into this world? One way or another, you've always been one of a kind. Why do you find it harder now?"

Shawn brushed away the snow coating the wrought iron bench at the water's edge, and dropped onto it. He instantly regretted it as damp and cold bit through his breeks. He shrugged. "It's different. Being the master of my own ship, you know, being feted and people loving me. It's different from having this secret I can't talk about, this whole world inside that people would condemn, if they even believed it at all. Knowing I'm someone I can't ever show to the outside world, knowing I'm a fraud because they wouldn't be fawning all over me if they knew the things I've done."

"You were living as Niall Campbell. Wasn't that the same, being someone you could never show?"

"But there *were* people who knew the real me. Niall, Allene, Hugh, MacDonald, Christina. Even Owen and Lachlan, even though they called me Niall, somehow I felt like myself with them. It still felt like real friendship."

"There are people here who know the real you. Amy, me, yer man Aaron in your orchestra." He tossed a handful of crumbs at the ducks. "Clive. Mairi."

"It's a short list," Shawn said.

"So after a few months here, would you come back again? If you had a do-over?"

"Yeah." Shawn reached out his hand, and Angus handed him a roll. He shredded bits off, tossing crumbs to the ducks. "I had to come back. I had things to set right." He stared at the gray ripples of the pond. "Like maybe thanking you for coming to Glenmirril to tell me James was born."

Angus sat up straighter. "You heard me?"

"Yeah." Shawn chuckled. "You caused quite a stir. Allene, Niall, Christina, we all saw you. You were like a ghost, saying over and over that I had a son, and we kept answering, but you couldn't hear us. It meant a lot to me, knowing. I never thanked you for that." They sat in comfortable silence, until Shawn added, "But now I'm lost. I'm not sure what's left for me here."

"Amy. James. Your work. They need you. *James...*" Angus stopped a moment, before saying forcefully, "He *needs* you."

"Amy dropped me like a hot potato and got on a plane for you when it came right down to it. James has spent more time with you than with me." He tossed a handful of crumbs. The ducks squawked and abandoned Angus as quickly as they'd flocked to him. "My work? It doesn't really seem important.

It seems meaningless and self-indulgent, when I think about Niall's life." *And death,* he thought.

With his bread depleted, Angus leaned over, resting his forearms on his legs, hands once more gloved and clasped together. "Something in particular bringing this on?"

Shawn shook his head. He tossed the last chunk of bread at the ducks, lifted his eyes to Angus, and said, "Oh, hell, I've told you this much. I don't know what the point is in not telling the rest. Yeah. They built me a kneeler."

"Like in church?" Angus laughed. "You said they were your friends."

Shawn laughed, too. "Yeah, I know, right? But they built it with a secret compartment and left me letters. Things aren't going so well. I feel like the kind of scumbag who would leave his own kid in a burning house to rescue himself."

"'Tis not quite the same." Angus rubbed his hands together and yanked his collar higher. "What's happening? I mean, what was happening? This tense thing is a pain in the arse. Does Amy know you're reading these?"

"More or less." Shawn stared at the pond and the ducks waddling away. "She knows about them. I didn't want her reading them. I don't even know why."

"So let her read them. Talk to her about them. What's happening to Niall?"

"You know I killed Duncan that night."

Angus nodded.

"They locked MacDougall up, but he's escaped and he's...." He stopped abruptly. He didn't want to tell the whole, ugly truth.

"You're worried about Niall."

Shawn forced a shrug. "I mean, there's nothing I can do about it, right?"

"No, there's not." A duck squawked at Angus. "Go away," he told it. "You ate my last crumb, greedy guts." He turned back to Shawn. "They cared enough to get you back. You need to quit feeling guilt for things you canna help." He gave a low chuckle. "Who would ever think those words would ever be spoken to the great Shawn Kleiner?"

"Go to hell," Shawn said. But he smiled. "I'm not great anymore. You ready to go inside?"

"Aye." Angus backed his chair up the path and turned it. "It's not our typical balmy February. And I'm due for physical therapy."

"Walking?" Shawn stood back.

"The pool." Angus wheeled himself around, and back up to the path, stopping for a young woman escorting an elderly man leaning on a cane, and exchanging greetings. Shawn's breath frosted the air. He thought about pushing the chair for Angus, who seemed to be breathing harder than before, but decided it would be an insult. Amy would have known better than to even think of helping him, he suspected.

"Tell me something," Angus said, as they approached the glass doors back into the hospital. "Why are you suddenly so determined to have me take Amy back?"

The doors swished open. Shawn started. Medieval Scotland swept around him, where such a thing was shocking to the senses.

"You all right, Mate?" Angus looked up, his eyebrows furrowing.

"Yeah, fine." Shawn frowned, disturbed at his reaction. "Amy, um, yeah, it's just, she doesn't want to be with me. You know—I just want her to be happy."

Angus wheeled through the second set of doors, past a pair of orderlies, and into the empty elevator, waiting for the doors to slide shut before saying, "She wasn't completely happy with me, either. She was torn."

"I have to go back to the States next week," Shawn said. "I really think you need to reconsider this engagement thing."

"'Tis no business of yours." Angus stared straight ahead, and Shawn felt the brief friendliness between them snap closed. The elevator rose and opened its doors to reveal Angus's floor.

Jenny looked up from the nurse's desk, smiling at Shawn as he followed Angus out.

"Hey!" He raised a hand in greeting. Her eyes twinkled, a distinctive bright blue. They really were attractive, he thought, no matter how much he'd abused the line. He was lonely. "You busy tonight?"

"Aye, now, I might find a wee bit of time." She grinned at him as she rounded the desk. "There you are now, Inspector MacLean. We've some forms to go over with you!" She reached for his wheelchair.

"I can manage," Angus snapped, surprising Shawn with his irritation. He invariably spoke with respect to the staff.

"I'm sorry, Inspector." Jenny took her hands off his chair, unflustered. "Can I get anything else for you? Dinner's soon if you'd like to call in your order."

"Thank you, yes." Angus looked at the floor as he wheeled the chair to his room. "I'm sorry," he said.

"It's all right, Inspector." Jenny turned to Shawn, sliding her hand into the crook of his arm as he watched Angus go. "Dinner would be lovely," she said, looking up at him. "I'm off at six."

"I'll be here." Shawn watched Angus wheel into his room, where Claire waited.

CHAPTER SIXTEEN

Linstock, 1318

MacDougall glanced at the wine. It had been a terrible two days—a terrible month. The last year, in fact, had been terrible. The anger at Duncan's death, the pain glazing Duncan's eyes, never left him. And Christina—she had once more lied to him, and run off, making a fool of him.

Worse—he was disappointed. He'd believed her. He'd really begun to *believe*, as they rode together those few days, that she had relented, had seen him for who he was. Had he not treated her with the utmost respect and kindness? Had he not ever been tender and gentle with her, from the first time she came to Creagsmalan, a scared young girl? His anger grew, pondering her ingratitude.

He lifted the goblet to his lips. His head swam. He lowered it. He'd had too much tonight already. Pelops was even now in his stall, in pain. Did he really need to stumble down the stairs and across the courtyard to meet this woman?

He hesitated. One more sip would change but little. He lifted it again. And swayed. No, he had Pelops to see to! He set it down firmly, and, grabbing his cloak, followed the scullion out the door.

Inverness, Present

It felt wrong. Every part of Claire being there, hovering while his brother and father eased him from the car to his wheelchair, felt wrong. Standing between his mother and Mairi, she glowed, she fretted, she beamed, and when she thought he didn't see, she brushed tears from the corners of her eyes. "I thought...." Taking the handles of his wheelchair, and aiming it up the newly installed ramp, she choked on the words, and tried again. "I thought I'd never see this day. I thought...that first night we got the call...."

"Not our Angus," said Ken, behind her.

Angus moved his hands with the wheels of the chair, chafing at the unwanted assistance—as if he were helpless. He was grateful the street was unnaturally deserted in the evening. At least he didn't have to suffer the neighbors seeing him reduced to this chair.

"Far too ornery for the devil to want him sooner than necessary," Mairi teased. She inserted the key into his front door.

Angus's heart sank. His house would be empty, when it should be Amy waiting for him, Amy at his side, Amy filling his home. Of course, it was his own fault. He'd sent her away. Made sure she'd stay away. But really, what choice did he have, now he knew about James?

Claire was chatting away behind him, a rush of words trying to hide her emotion, that only gave it away. His front door swung open, Mairi stepping in to pull it wide, revealing an explosion of colorful helium balloons filling the entry, and what sounded like a hundred voices shouting, "Welcome home!"

Mrs. MacGonagle pushed through the bobbing balloons, her sour face as close to a real smile as it had ever been, as she all but pulled his chair through the door.

Angus stared in shock, at the color, the noise, the voices, after weeks in a quiet ward. He put his hands on the arms of his chair—*get up, walk away*—and remembered just on time that he would fall if he tried—fall here in front of all these people.

The neighbor girl—the one who saw the ghost in the house next door— danced up and down, making balloons bob and hit his face and arms. He wanted to brush them away, but Claire pushed him into the narrow hall, skillfully twisting the chair into his front room, where half a dozen more people —the couple with the new baby from the end of the block, the ten-year-old blonde twins from three houses down, Joe in his kilt who spent his mornings weeding his immaculate garden—applauded, among a hundred balloons of every color of the rainbow, every one saying *Welcome Home!* in big black letters. Black. How appropriate. What was he coming home *to*?

He felt Claire's breath by his ear, and she kissed him on the top of the head. He resisted the urge to pull away. It was like having Mairi kiss his head. It felt wrong.

"They've all missed you so," she said.

"Aye, 'tis not the same without your cheery good morning!" Joe declared.

Nods went round the room, and more applause and shouts of "Welcome home! Welcome back!"

"Isn't it *grand*!" Claire asked.

"I'm...thank you," Angus frowned at them, wanting to be in his bed. The day, the forms, the drive home, had all been exhausting. A pot of tea sat on a tray on his coffee table, surrounded by bickies.

"There now, he's knackered." His father looked around the crowd, beaming as if he'd just paid them a great compliment.

"What a grand welcome," Ken added, and Mairi and his mother were wishing everyone well as they ushered them one by one out the door, and Claire was on her knees in front of Angus, pouring him tea.

"Brandy." He stared at his fireplace, at the spot where he and Amy had lain on the floor, the night before their trek to Monadhliath, rolling onto the printed sheets of Shawn's family tree, as he kissed her.

"Oh, now, you're just out of hospital," Claire protested. "Is that a good idea?"

"'Tis a *very* good idea," he said. His mind lingered on that night, on Amy, on the family tree, on Honora in the moldering castle above the sea, and the mysterious James Angus who might or might not be a MacDougall. He smiled, thinking of those months of searching with her, driving across Scotland, into England, of her spinning in St. Oran's on Iona, singing, her voice magnified and beautiful in the small chapel. He loved her singing.

"You're feeling better, then?" Claire worked her hand into his. The tea sent up tendrils of steam from the cup in her other hand. She beamed. "'Tis good to see you smile again."

Angus's eyes drifted from his memories, from Amy on the floor, her hair spread beneath her, the fire playing over her fine features, and James sleeping nearby—James whom he loved like his own—back to the bare hearth, to Claire, with her freckles and cornflower blue eyes and bobbing pony tail, looking hopefully up at him, with a cup of tea in her hand, so earnest, wanting so badly to love him and make him happy.

"Brandy...please," he said. "'Tis in the cupboard over the sink."

Linstock, 1318

Lachlan's heart pounded. He had just given poison to a lord, a friend to the king of England, no less. He crossed himself swiftly, reminding God, *He's kidnapped Christina, he's killed our people, stolen our cattle, attacked our home. He'll kill Niall if he can.*

Clouds shrouded the moon, as he slipped out the servants' door into the gardens behind the great hall and kitchens. He stole through the vegetable patches, listening for any cry of alarm from the tower, wishing he could go faster. But the clouds that hid him also hindered him and, he reminded himself, MacDougall might not be discovered until morning. He and Owen would be well away by then.

As he rounded the back corner of the kitchens, the clouds slid away, letting the moon spill silver light over the stone paths between the herb gardens, covered now in frost. He glanced through the arched columns on his right. They held up a covered walkway between the stables and kitchen, giving a view of the central courtyard. All was silent.

As he watched, a man left the tower, striding toward the stables.

He eased himself against the compound wall, moving cautiously to the water gate. Owen was there, looking anxiously through the walkway. A guard lay at his feet, his cheek on the frosty grass.

Inverness, Present

In Angus's moonlit front room, a frown flitted across Claire's face. But she rose and headed down the hall. Angus closed his eyes, filled with self-loathing. He was making Claire happy, he told himself. He was doing the right thing for Amy, for James. Without his legs, he had nothing to give them. He couldn't provide for them, support them. He couldn't climb into the mountains with them, let alone protect her from someone like Simon.

Most importantly, he couldn't *teach* James all he needed to know. Shawn could do all of that and more, and she deserved better than to spend her days caring for an invalid. James deserved better.

So why did he hate himself? He opened his eyes, watching the flames dance in the hearth, a miniature of those at Glenmirril and he knew. Niall would be more honorable. Niall would never be thinking longingly of Amy while engaged to Claire. It did nothing for his spirits that it had been Shawn Kleiner, of all men, to tell him he was doing wrong by Claire.

Claire appeared in the doorway, the amber bottle shining like a jewel in her hand. She knelt at the table, trickling smooth golden brandy into his tea cup, and handed it to him. She was lovely—in a girl next door way. A little sister sort of way. He touched her cheek, wanting to make her happy. He truly did. She put her hand over his, smiling.

"It's been a long day," he said.

She nodded. "Aye, Inspect...Angus." She laughed. "I suppose I shouldna be calling my soon-to-be husband Inspector?"

He smiled, despite himself. She was kind. She was good. "Call me whatever makes you happy."

He heard his mother's voice at the front door, bidding a guest farewell. He sipped the tea, feeling the relief as the heat of the brandy flowed through his body and relaxed him. He thought of Amy at Monadhliath, in shock; how he'd urged the brandy on her, unaware of her lack of experience with alcohol. "It's been a long day," he said again. "I'm knackered."

"Oh, of course," Claire said hastily. "What can I do for you?"

He heard his father's voice at the door. Guilt washed over him. He took her hand. "I'm sorry, Claire."

Pain flashed across her eyes, quickly replaced by a smile. "For what?"

She knew. And he knew she knew. "I'll see you...tomorrow," he said. He wanted only to sleep, tonight, tomorrow, and every day, and dream of time where he hadn't fallen off the mountain, where his legs worked, where he was the man he'd been, where he still had something to offer Amy.

She squeezed his hand, as his father's and brother's voices floated down the hall, to the kitchen. She kissed his cheek. It felt sterile, forced. He smiled, vowing he'd find a way to do right by her, to love her as she deserved. Love could grow, could it not?

Linstock, 1318

MacDougall pulled the cloak close as he descended the stairs. In the courtyard, a sprinkling of snow glittered in the air. He saw the guard in the gatehouse, huddled in his own cloak and looking miserable.

"Has yer man left so quickly?" he called.

The guard came to abrupt attention as MacDougall approached. "What man, my lord?"

"The kitchen lad," MacDougall snapped. "He was going out to get a woman. To help Pelops. You were supposed to...." He stopped, noting the guard's blank face. "There has been no man?"

"No, my lord."

"If he should come," MacDougall said slowly, "do not let him out. Are you familiar with a lad called...." He hesitated, trying to remember. "Ralf? Rolf?"

"No, my lord."

"He brought me wine from the buttery."

"I know of no such man."

MacDougall turned suddenly, charging for the kitchen, shouting, "Hob!"

Inverness, Present

Ken, Mairi, and his parents joined him in the front room after Claire left.

"Things will get better," Mairi assured him.

"See how far you've come already," his mother added.

"You've never been one to mope," Ken pointed out.

"But then, I was never one to be stuck in a chair." Angus tried for a wry grin. "Things have changed a wee bit now, have they not?"

"You've a brilliant mind," his father said sternly. "You've *two* women who want to be with you, legs or no."

Angus stared straight ahead, feeling his jaw tense.

"Claire's lovely now," Mairi added, "but you...."

"I'm knackered," Angus interrupted.

"Now, Angus," his mother began.

"Dad, Ken." Angus maneuvered his chair into the hall. "I'll have to ask for your help." Amy had accused him of walking away from problems. Perhaps it was his just desserts he could no longer walk at all.

Mairi stared at the floor, her mother murmuring to her, as Ken and Mr. MacLean began the arduous task of getting Angus up the stairs.

Ken helped him in the loo. Couldn't even use the loo, Angus chided himself. He certainly couldn't teach James to use a sword. And Amy was better off without this. She'd understand one day.

Ken and his father wheeled him down the hall, helped him into pajamas, and into bed. Ken would sleep on a cot squeezed into the room, in case Angus needed anything in the night.

"You good now?" his father asked.

"I'm grand," Angus assured him. He wasn't grand. He couldn't even use the loo without help.

"We'll be back in the morning." His father snapped off the light, but stood another moment, before saying, "You've a brilliant mind and a great deal to give to the world, Angus."

"Aye," Angus agreed, only for the sake of respect.

Linstock, 1318

"Is he dead?" Lachlan whispered, looking at the man on the frosty ground.

Owen shook his head, working at the bolts on the huge doors. "A potion is all. He'll appear to have fallen asleep at his post." At the same moment, they heard a shout from the courtyard. Owen eased the door open a crack. They slipped through, onto the dock that ran alongside the castle wall and pulled the door shut behind them.

Lachlan scanned the river, whispering, "Conal!" His breath hung in the air.

"Sh." Owen pointed up. "Guards. Every quarter hour."

Lachlan looked frantically down the winding river. Clouds once again slid over the moon. And in the quiet night, they heard the soft splash of oars. Lachlan crossed himself, praying.

They would see, if they came looking, that the water gate had been unbolted from the inside. They would know. Another splash sounded, far down the wall and a moment later, from above, two men's voices. Lachlan's heart pounded. He and Owen pressed themselves to the wall, hoping Conal had also heard and would wait. Above, one of the men laughed. The voices trailed past, fading away toward the far end of the stables. Lachlan let out a slow breath.

"Lachlan?" The whisper came through the dark.

"We're here," Lachlan called back.

Through the door, they heard a commotion rise from the courtyard.

"Hurry," he whispered.

The splashing of the oars quickened, and a moment later, a small fishing currach slid into view. Conal was reaching up, grabbing his hand, and Owen was climbing down beside him, trying not to sway it. Owen reached for a pair of oars, while Lachlan gripped a knife in each hand, wishing he had his bow.

Water dripped from the oars, sounding like the thunder of war horses in his ears. But they couldn't possibly have found MacDougall so soon. And if they had, they'd have no way of knowing who to look for—or where. Still, the shouts grew from the castle compound, unnerving him.

"Can you go no faster?" he whispered.

"Not an' we wish to go quietly," Conal returned.

Lachlan knelt, tense, between the two of them, each rowing slowly.

And suddenly, they slid past the stone walls, past where the river eddied into the moat, and hands were reaching out, grabbing the rope Conal tossed and the boat was crunching through the thin sheet of ice edging the bank.

"He's dead?" Niall whispered.

"I gave him the wine."

"Let the boat go!" Hugh gave a soft rumble and pushed the currach into the river to float away, back toward the castle.

Conal was rushing them into a copse; the horses were surrounding them.

"It worked? He's dead?"

"Get up, go!"

Owen was throwing the reins of a pony into Lachlan's chilly fingers, and Conal was on his own mount, pressing Lachlan's bow back at him. The pony was under him, the land flying beneath him and somehow, his fingers closed around the bow and he was reaching for the quiver as they pounded down the dark forest path, through trampled snow, leaving Linstock behind.

Inverness, Present

As he listened to the two men descend the stairs, Angus's soul burned with shame. He couldn't remember a time his father had sounded disappointed with him. He pillowed his head in his hands, staring at the ceiling. His father was right. His mind was strong and intact. He'd let a lesser nature take over, these weeks in hospital. Shawn's constant presence, didn't help, he excused himself.

Or Amy's, for that matter. He sighed, as he wiggled his toes—about the most he could do with his legs. She kept coming back. But James needed *Shawn,* if Helen's ancient letter was true.

He thought about that climb to Monadhliath, about Simon. He'd been right, in those days and hours just before his fall—right in thinking Simon was from Niall's time. Something else niggled in the corner of his mind. It wasn't only Simon he'd been thinking about, just before the accident. He stared at the gray shadowed ceiling, trying to remember. There'd been Mrs. MacGonagle, the ghost in the house next door, the broken chair, the chipped step, the light fixture, phone calls.... He stared harder at the black shadow. Who had he been calling? He couldn't have been calling Simon.

He pushed the covers irritably off himself.

Brian!

He'd have sat straight up—if doing so had not been so difficult.

He'd been trying to call his mate Brian, at Creagsmalan, *about* Simon. Brian hadn't answered calls or e-mails. Angus rolled his head to stare out the window. It was unlike Brian to ignore his calls—especially multiple calls and contacts. Something was wrong.

He stared out the window into the night. His father was right. There was plenty he could do. He would find out why Brian wasn't returning his calls.

Linstock Castle, 1318

MacDougall found five girls in the kitchen, cleaning up. One of them he vaguely recognized. Her brother, he remembered, was the simpleton who had spilled his wine. She'd begged and pleaded. "Where's Hob?" he barked.

The head cook emerged from one of the pantries, followed by the butler. "My lord?"

"Ralf," MacDougall snapped. "Where's the man Ralf?"

"My lord?" Hob looked perplexed.

"There is no Ralf here, my lord," said the butler.

"Ralf, Rolf!" MacDougall's words lashed out. "He just brought me wine! You're in charge of the buttery, are you not? Surely you know who goes in and out with the bishop's wine."

"The lads brought wine to dinner hours ago. There was neither Ralf nor Rolf among them."

Pressure grew in Alexander's head. "My lord Beaumont and half a dozen of my men are injured and left behind in Carlisle," he said. "Most of my army is gone looking for a horde of brigands who shot at us, and now you tell me I did not just meet a man named Rolf who replenished my wine?" His voice rose. "Get all the scullions, all the kitchen lads out here, or we may entertain ourselves with another whipping in the courtyard on the morrow!"

The two men spun on their heels, shouting for the kitchen staff.

The girl—the one who had begged for her brother—pulled back in a corner, shaking. A plate slipped from her hands, shattering on the floor. A tall young man came running in from the cellars, touching her shoulder, stooping to gather the shards with her. Other boys scrambled down the stairs from the courtyard, or came in from various storerooms, hastily arranging themselves in something like a line before the giant hearth.

MacDougall marched up and down the line. Some looked past his shoulder, while others looked at the floor. "Is this everyone?" he asked Hob.

Hob glanced at the kitchen help, more than two dozen men and boys. "Aye, they're all here."

"Jackin," one of the boys said.

The tall youth, still on one knee, looked up. "Was he not called Jacob?"

"Jacob?" repeated MacDougall.

Hob looked a moment at the lad on his knee before looking up and down his crew, and saying, "Aye, Jacob is not here."

"He'll be here in a moment," said the youth, half a plate in his hand.

"Why did you not say so?" MacDougall asked.

Hob scratched his head. "Well, he was new. He arrived but a day ago...three days mayhap...looking for work. A hard worker, he was, a good lad."

"And what does Jacob look like?" The pressure grew in MacDougall's head, as Hob's description matched that of Ralf. "And did you send him to replenish my wine?"

"I did not, my lord." Worry crossed Hob's fat face. "I'm sure the lad meant well."

"I want him found." MacDougall spoke through clenched jaw and tight lips. "Search the entire complex. I will be in the great hall."

He had not crossed half the courtyard before a shout reached his ears. "Milord MacDougall! It's Hutch! At the water gate. And it's unlatched."

At the same time, the tall lad from the kitchen ran up to him. "Milord! I know where Jacob has gone! You must go after him anon!"

Inverness, 1318

"Need anything before I crash, Angus?" Ken appeared in the doorway in sleeping pants and a t-shirt. "Water?"

"My laptop," Angus said. "And my phone."

"It's past nine, Mate," Ken said. "Who are you going to call?"

"Please," Angus said. "Laptop's in the office, phone should be downstairs with the bag from the hospital."

When Ken had brought them, Angus logged into his e-mail. It took only minutes to find the messages he'd sent out just before the accident. Thomas Ritchie had replied. It was a busy season—maybe in the new year he could meet?

Angus resisted a wry smile. Aye, well, it had been a busy season for him too—surviving. He typed quickly, as Ken punched pillows into shape on the cot beside him, and hit *send*. He closed the laptop, feeling better than he had since —well, since the moment before he'd tumbled off that cliff.

And with feeling better came another thought: there was something else he needed to do. Something equally, or perhaps more, important than talking to Thomas Ritchie. He owed her that.

CHAPTER SEVENTEEN

Road to Dumbarton, Scotland, 1318

They slowed only enough for caution when the moon disappeared behind clouds, and sped the more quickly on the forest path when it once again sent its silver light down through the branches.

"Our tracks will be easy enough to follow, an' it dinna snow again," Hugh pointed out.

"There's naught to be done about that," Niall commented tersely.

As day broke, they stopped long enough to make oatcakes over a small fire, feed their tired garrons, and be on their way again. "He drank it?" Niall asked, as the sun stretched rosy fingers over the eastern hills.

"I poured it all into his jug," Lachlan said. "I filled his goblet directly and handed it to him."

"You *saw* him drink it?" Niall persisted.

"Well, now...I *handed* it to him," Lachlan repeated. "He *took* it."

The alarm grew slowly in Niall. "You dinna see him drink it?"

"Why would he *not* drink it?" Owen asked. "He was sitting in his chambers drinking."

Niall stopped his garron. "We have to go back."

Conal wheeled his mount on the path ahead, the hills rising behind him. "We *canna* go back. We haven't time."

"And think, Niall," Hugh added. "If he drank it, there's no reason to go back. If he *didn't*, then he is surely aware summat was wrong with the wine."

"All was quite *right* with the wine." Lachlan touched his heels to his garron, joining Conal. "'Twill do exactly what it ought."

Hugh snorted. "I daresay MacDougall would have a differing view on what was right or wrong about that wine."

"We must be sure," Niall insisted.

"We must be *alive*," Hugh countered.

"And that is the problem." Niall's voice rose. "I will *not* be alive as long as I'd prefer, if we have failed. Nor, more importantly, will my bairns, and perhaps Allene. You can be sure he will avenge Duncan's death on my own sons."

"He is like to have drunk the wine." Conal spoke sharply. His pony danced under him, betraying his rider's impatience.

"Come away now," Hugh said, "for if, on the Devil's own advice, he dinna drink, he may even now be coming through that forest behind us. And we are *not* three score. We are but five."

"Why did you not wait to see that he drank?" Niall persisted, rounding on Lachlan.

"'Twould have raised his suspicions," Lachlan said. "Moreover, a kitchen lad could hardly disobey a direct order from a lord to leave."

"We must *go!*" Hugh grasped the reins of Niall's pony and led it, Niall protesting, after Conal, urging it ever faster, until once again, they raced through the hills.

Inverness, Present

Angus wheeled his chair to his kitchen table. White china and new silverware looked out of place in the small room, even with the new tablecloth in royal blue. He was pleased he'd managed to get dinner in the crockpot, with the help of a low work table Ken had set up for him. But the whole endeavor left him exhausted. He was in no mood for dinner with Claire. He wanted to be with Amy.

His father's chastisement echoed in his ears. He had not proposed to Claire for the right reasons. He'd managed to convince himself otherwise, back in the hospital. But his father was right. He was still capable of so much. Proposing to Claire had been an act of giving up.

The doorbell rang. He squeezed his chair down the hall, and let her in. She wore a baby blue cardigan. Her hair bounced in its usual short ponytail. "How's things, Angus?" She beamed. She was lovely and sweet. The ring flashed on her finger as she lifted a ruby bottle. "I brought wine!" She leaned down to kiss him.

His stomach knotted in a tight coil. Shawn was right. No matter how he tried to twist it, he'd been lying to himself as badly as Shawn had lied to Amy. "I'm good," he said. "Dinner's on." He turned the chair, wheeling back to the kitchen, where the crock pot waited on the table.

"Chicken?" She opened his drawer for a corkscrew. "It smells grand."

Angus smiled. He'd used two spices Shawn had suggested, wondering what in the world he was doing. He'd hardly intended to become the man's friend. Shawn managed to be likable, though, despite his arrogance and pushiness.

Claire seated herself, and served up chicken, roasted red potatoes, and peppers to both plates, while Angus took a salad from the refrigerator and pulled the cork from the wine. Claire lifted her glass. "To us."

Angus clicked the rim silently.

She lowered her glass. Her smile slipped. "Angus, you're *not* all right."

He lowered his head, wrestling his conscience. She'd be happy. This was all she wanted. He forced a smile, raising his eyes. "No, I'm grand. How was work?"

She set her fork down. "Something's wrong."

"No, it's all grand," he insisted.

She slipped the ring off her finger, studying it in the candlelight. Small sparks flashed off the diamond. She laid it down on the table between them. "Why did you ask me to marry you?"

Angus stared at the ring. "I knew it would make you happy."

"Does it make *you* happy?" she asked. When he only stared at the ring, she said, "It doesn't."

"Nothing makes me happy," he said. "That's naught to do with you. You're an absolutely lovely girl."

She blinked and stared down at her chicken with Shawn's special spices. The ring sparkled against the blue tablecloth by the crockpot.

"I'm hurting you," Angus said.

She stood. "You've a good heart, Angus. Everyone at the station knows it." She hesitated, as if about to say more.

Angus wheeled around the table, and took her hand. "I don't have a good heart," he said. "And I'm being a coward, waiting for you to do my dirty work."

She stared down at their clasped hands.

"I was wrong," he said. "I believed if I was kind enough, I could give you happiness."

She touched his hair. "I believed it, too. I believed I would be happy just to be with you, and if I loved you enough, you'd love me back."

"I'm sorry, Claire," he said.

She gave a quick sniff, and harsh, staccato laugh. "I've been an eejit. You were right—when you said.... I never should have settled...but I hoped...in time...." She drew her hand under her nose with another sniff, and said, "I guess, maybe I'll see you at work when you get back." She stood abruptly, and hurried down the hall.

Angus didn't remind her he might not be back. He didn't want to remind himself. He wheeled after her, past the palms and bisque crucifix. "I'm sorry, Claire."

"Aye, as am I." She leaned down to kiss him on the cheek, and held his hand, silent for a moment. "You told me how it was." She blinked, hard, twice. "I enjoyed our time together." She spun and hurried out into the night.

He stared at the door, relief and self-loathing mixing, a potent and awful brew, in his gut.

But it was time to heed his father's words. He wheeled to the front room and opened his laptop. Thomas Ritchie, director of the psychiatric unit in Bannockburn, had answered him.

London, England, 1318

His face dark, Edward strode into Simon's rooms in Edward's palace in London. Simon rose, bowing. He'd been immediately upgraded from a chamber in the Tower when the physician confirmed the queen was, as he'd said, with child.

"Your majesty." Simon bowed. His insides trembled. But surely this second thing would come to pass, too. He rubbed at his shoulder—sore from the attack in Inglewood.

Edward threw himself into a chair at the table, and lifted the jug there, pouring wine. "What magic do you possess?" he demanded.

Simon dropped his hand from his shoulder. "Deydras has appeared at Beaumont Palace." Relief washed through him, though he had to force himself to state it, rather than ask.

"I was given to understand by my men that the wound received in Inglewood healed with unnatural speed."

"Herbs taught me by the old woman," Simon assured him. "Ones I will gladly share with your army for their speedy recoveries in battle. Deydras?"

Edward waved his hand at the other chair. Simon seated himself. "As you said. Missing an ear. Saying I am the carter's son."

Simon held back a smile. "Your troubles will pass," he said. "If you allow me to advise you."

"If?" Edward took a long drink and slammed the chalice down on the table. "If you can predict the future, do you not already know what happens? Do you not already know if I am deposed over the word of a madman?"

"I know what can be and may be," Simon replied. The books of Amy's time said the man was tried, at which point he claimed his cat, speaking for the Devil, caused him to tell the tale. Deydras and his cat were both hanged. "And that means I can best advise you."

"The future can be changed, then?" Edward picked the choicest piece of cheese from the basket beside the wine.

"The future is not *entirely* set." Simon leaned forward. "And I am the man who can see that it changes, if need be, to your benefit." He raised his cup, lifting his eyebrows in request.

Edward indicated the jug of wine, granting permission.

Simon seethed, even as he lifted it and poured a careful ruby stream into his cup. Was it not his own wine, given to him in his own chambers? And he, Lord of Claverock, must await this foolish man's permission to drink?

"What more can you tell me of my enemies?" Edward asked.

"They are many, and their plots thick. You will fare well against them—with my foreknowledge."

Edward swallowed his cheese, and rose promptly, going to the window to gaze down on the snowy London streets.

Simon rose. He waited, his heart thumping. The queen was pregnant—as the books had said. Deydras had appeared with his one ear—as the books had said. There was no reason for Edward to doubt him. Still, as the silence grew, he wondered if Edward might charge him with dark arts, in knowing these things. If so, he would blame the old crone of his story, he decided—any old woman would do. If he, a great lord, chose an old woman and said she'd nursed him—and cursed him—he would be believed. She would be hanged or burned, and he would go home to Claverock, the victim of dark arts mercifully saved.

Edward set down his cup on the window ledge and turned. He studied Simon for a long moment, and finally spoke. "You shall be my chief counselor."

Simon bowed his head. "Your majesty, I shall serve England well." His smile slipped out. England, indeed, he would serve. Edward—perhaps not.

Midwest America, Present

Gray morning light poured into his mother's breakfast nook. In summer, it would be a sunny place for morning meals. Now, it showed streaks of pink reflecting on the lake outside, and glancing off fresh snow, as Shawn shuffled in wearing drawstring pants and a t-shirt. Carol, in a plaid flannel robe, poured coffee and set it on the table. "You have to leave soon?" she asked.

Shawn glanced at the clock on the microwave. "An hour."

"Are you accounting for the snow? It came down pretty heavily last night."

"The plows will have been through."

"Are these talk shows paying off?" she asked. "You don't have the backing of an orchestra anymore."

He gulped his coffee black. "Taking an interest in people always pays off. The Laird said so."

"The Laird...? Shawn, this has to...."

"Besides, I made the orchestra what it is. I can certainly do it for myself."

"I suppose you can," Carol agreed. "You never lacked confidence."

Shawn rose from the table, and dropped two slices of bread in the toaster. "It's not confidence. It's competence. I deliver what I promise."

"Yes, you always do. I sort of wanted to talk to you about that." She leaned against the counter, arms across her chest.

Shawn threw open the refrigerator, searching for butter. He turned in surprise. "What? Are you saying I don't?"

"Oh, exactly that you do." She took the butter from him. "Sit down. I'll do it."

"I'm beginning to hate these sit down discussions," he muttered, but he sat, staring into his coffee. It had to be about Amy, although what his mother expected him to do, he had no idea. He heard the drawer slide open, the clink of silverware, and the soft scrape of a knife on toast. "Is it Amy?" he asked.

"No." She slid the toast in front of him, the plate scraping against the soft honey wood of the table. "It's Clarence."

"Clarence." He repeated the word with the solid thunk of a hammer chunking against a nail head.

"It scares me when you say it in that tone," she said.

"What tone?" He looked up as she lowered herself into the chair opposite him. Morning sun picked out amber threads in her soft brown hair. "I didn't use any tone. I just said Clarence."

"So flat," Carol replied. "Like something's dead inside you. I don't know what you're thinking. Neither of you has said what happened when you saw him on Christmas Eve."

Shawn shrugged. "You know, I'm not prone to talking about—some things." He found even that much difficult to say.

"You should try sometime. Have you at least talked to Amy about it?"

"Well, you know, funny story. She sort of dumped me," he reminded her. "I'm not sure it's appropriate."

"How are you coping with him being out?"

Shawn sighed, and lifted his eyes to the sparkling glass of Carol's windows, the morning blues of the lake, and the sun casting shades of pink and orange over the fresh snow. "It's weird."

She waited.

"Weird," he repeated. "I don't know what else to say. You know the hate was so deep and so strong for so long, and then it was just gone."

"Is that what happened?" she breathed softly.

"And yet, he's out. It's sort of like the Clarence I knew then and the one I see now co-exist side by side, and I just don't really know what to make of it or feel about it. What do you want me to say?"

"Just the truth," she said. "It would just relieve me to know you're coping with it."

"I'm not drinking." He lifted a piece of toast to his mouth. "I don't know. It's like not knowing which side of a dream you're on. I don't know which is the nightmare and which is reality. The life I'm seeing or the life I thought I knew. I feel disoriented and disconnected." He jammed the toast into his mouth, focusing his eyes on the lake outside. Beyond it, a deer appeared out of the wood, lifting cautious ears before stepping delicately to the water's edge to drink. A fawn followed. Shawn's heart lurched back to the many nights he'd slept under the stars with Niall, and seen roe deer in the great Caledonian forest. A smile played on his lips. *A fine meal,* Allene would have said of the beautiful creature. He'd developed quite a taste for them himself, especially at the end of a long winter lacking in the plentiful meat to which he was accustomed.

The smile withered. Niall was living through the Great European Famine, fighting for king and country on a stomach that was most likely never full. And now, he faced a horrible fate, and didn't know what was waiting for him around the corner. Shawn dropped the toast to his plate. It tasted bad in his mouth.

"Sometimes I feel that way myself," Carol said. "But mostly, I see the boy who came to our door that first day, and the hope in his eyes. And I see the look in his face every time his mother and a new boyfriend decided he'd go live with them again. Especially that last time." She stared out the window, speaking softly. "There was such utter despair in his eyes, and I couldn't do a thing except hug him. I couldn't even promise it would be all right, because I *knew* it wouldn't be."

"No, it wasn't," Shawn agreed. "Is that what you wanted to talk to me about?" He stared out the window at the deer. Niall and Allene and James and the infant William he'd never met filled his mind. He wanted to be in their solar, arguing theology with Brother David, or playing chess with Hugh. He wanted to know it would be all right for Niall. So far, the parchments remained unchanged, saying it wasn't—wouldn't be—hadn't been. At her silence, he lifted his head from the half-eaten toast on the white plate.

"You made me a promise, when I flew over."

Shawn's eyebrows knit. "Yeah? So what is it?"

She spoke so softly, staring into her coffee, that he barely heard her. "I wanted to ask if you can find it in your heart to help him."

A fist seemed to crush his sternum. "Help him how?"

She lifted her eyes. "A job. A place to live. Anything. He's having trouble finding a job. He's living with a friend who, honestly, is not the best influence. All he has is friends from prison."

"A job." Shawn pushed his chair from the table and re-filled his coffee. "You want me to offer him a job? The hatred just disappeared that day you left me with him. I don't know how or why. But working with him? I don't know."

"I don't want him going back to what he was because life shuts off any chance for him to do better," Carol spoke quickly. "You can understand people are reluctant to hire him."

Shawn snorted. "Yeah. I of all people can understand that." His own words, *we're brothers again* filled his mind. They had slid out that day, as if someone else spoke them. But he'd *meant* them. The Clarence who had talked in the night, who confided about a girl he liked who wouldn't look twice at him, stood side by side with the angry boy in the courtroom, throwing daggers with his eyes, as sharp and full of hate as the blade Shawn himself had driven into Duncan MacDougall's gut.

"Just think about it, Shawn," Carol said. "Even if you could talk to someone you know about giving him a chance." She picked up her phone, poked at the screen, and looked up. "I've texted you his number," she said. "I understand if you feel you can't do it, and I won't ask again."

It's what your father would have wanted.

He started at the words. But when he looked at her, she sat with her head bowed once more over her coffee. He re-filled his own mug, sipping slowly, pondering the words, pondering Duncan MacDougall, who had done less than kill. He wouldn't want the man near Christina. But then, Clarence had shown remorse, even if he could never undo his actions.

Duncan hadn't. Duncan would have hurt Christina again. Maybe he had. Maybe his last act, before the battle in the courtyard, had resulted in her pregnancy. His jaw tightened. It was too much to think about. "I have to get ready," he said. "I threw my clothes in the dryer last night."

"Those medieval breeks?" Carol sighed.

"You don't like them?" Shawn grinned at her as she looked up.

"It seems half the girls in this country have taken to wearing them, along with the shirt and leather boots. It's not about liking or disliking, it's about not understanding and finding the whole thing a little disturbing. I worry about you."

"Well, don't. I'm fine." He glanced at the clock. "I better hurry. I better hope those plows cleaned the roads well. I've cut it too close."

Dumbarton, Scotland, 1318

"Wine."

Niall tore his gaze from the River Clyde, sparkling under a full moon.

Hugh handed him a skin, and leaned his arms on the parapet wall, his own wine skin dangling from his hand. "'Twould be warmer inside." He lifted his skin and drank deeply.

"Aye. 'Twould." Niall took a draft.

"'Twould taste better from a goblet before the hearth with good company."

"Aye. 'Twould." Niall stared at the water. His thoughts turned to other times standing on castle ramparts gazing out over water: The night so long ago when Iohn had come up to Glenmirril's tower and they'd spoken in comfortable quiet. Iohn had been plotting against him the whole time. *He tried to save you,* Shawn had said.

Aye, but he'd not tried so hard as to decline MacDougall's bribe.

He had climbed to Glenmirril's tower just days after his marriage, Allene at his side, to see Shawn there, drunk, sobbing to find himself still caught in Niall's time, his head on Allene's shoulder.

Months later, he had ascended to Tioram's ramparts and stood with Shawn, gazing out over Loch Moidart. *Your songs have become sad of late,* he had said.

My life has become sad of late.

As will everyone's, for a season. Niall remembered his answer clearly.

Spare me the lecture. At least you're heading home to Allene. Niall could almost hear his words, a ghostly echo of what had been, from a man not yet born.

At least you're alive. His own words now seemed flippant, even callous.

"We're alive." Hugh echoed the memory. "While one lives, there is yet hope. Does the good Scripture not say so?"

Niall smiled. "It does. And I myself said it to Shawn."

"We've ended MacDougall's threat to us."

"We don't know that," Niall pointed out. He drank deeply from the skin.

Hugh lifted his own skin, indicating Niall's. "What man does not drink the wine given him?"

Niall sighed. Starlight danced off the waters of the Clyde. "My mind shall be at ease when someone tells me they have seen the body."

"Have we not reached Dumbarton?" Hugh persisted. "And arranged for the arms the Bruce has requested? Does it not give you hope that we will see, in *our* time, the days of peace you saw in Shawn's?" When Niall said nothing, he added, "Edward *cannot* hold out against us forever."

"Edward, no," Niall agreed. "But perhaps Claverock can."

"You saw a future of peace," Hugh reminded him.

"I also saw one where Scotland lost at Bannockburn." Niall took another drink. "If Shawn and I changed that, Claverock can change things, too."

"He was struck in Inglewood," said Hugh. "He may even now be dead."

Niall frowned, staring at the River Clyde below. With the arrangements made, more weapons would come from the continent. He had seen miracles in the last three and a half years. He, of all men, should walk confidently in faith. "Yet look what our lives are," he said. "My earliest memory is carrying rocks to the parapets with Alexander. The whole of my life has been war, killing, fighting. Death at every corner. All my brothers gone, Iohn, William. Where is God in all this?"

Hugh gave a nod of his head toward the dark water, dappled with starlight. "I see Him there in the the stars dancing on the river. I feel Him in the peace of the night around us. I hear Him in your music. I see Him in our king riding at our head." He took a long drink of his wine. "And in a child born on a battlefield under the very eyes of an enemy who did not strike, though they outnumbered us."

Niall drank.

"Sure we're walking in the Valley of Death," Hugh said. "And we've a promise about that, have we not?"

Niall smiled. "We have."

"I thought we were done with this doubt," Hugh added. "Perhaps you'll come back to the great hall and see if they've a harp to play."

"My heart is not in it," Niall said.

"Perhaps it will be, if you but start."

"Perhaps," Niall said. But he made no move to go.

"We leave on the morrow for Glendochart." Hugh straightened. "Sure they'll have a harp and you'll find your heart in it there."

Niall smiled. "Perhaps."

"I am not asking, *Sir* Niall!" Hugh laughed, thumped him on the back, and departed, leaving Niall alone with the stars on the river and the peace of the night around him.

CHAPTER EIGHTEEN

Midwest America, Present

The talk show flowed smoothly. Shawn took his seat on a black couch, a mocha on the desk at his elbow, while Jen, the hostess—fresh out of college with the ink on her diploma still needing sanding—smiled behind the desk, and a small local audience applauded with enough enthusiasm to put a Manhattan crowd to shame. They chatted easily about his recent tour and upcoming album.

"I did some asking around before the interview," the hostess finally said. "Tell me about college. I think our audience will be surprised."

"College?" Shawn didn't think his partying days there would surprise anyone.

"What did you major in?"

"Music and marketing." Those days were an eternity ago. Seven centuries, in fact; a world away from riding hard with Douglas into English villages and setting them ablaze, pillaging churches, the screams of women and children.

"I was shocked when one of your classmates told me that!" Jen's eyes opened wide. "Excuse me for saying so, but it wasn't your, uh, *academic* prowess you were known for."

Shawn coughed. He thought of Brian, dead from driving drunk, and the clarinet player who had lost his house, of Caroline and Amy. Wouldn't they and their families all just find her comment hysterical, he thought with irritation. He put on a bright smile. "Well! Surprise!"

"You're known for surprises! Speaking of which, I was also surprised to hear you speak several languages. What are they?"

"French, Gaelic, and Latin."

"Latin!" Her eyes opened wider still. "Say something."

Shawn was beginning to enjoy himself. "*Sunt item, quae appellantur alces,*" he quoted. "*Harum est consimilis capris figura et varietas pellium.*"

Jen smiled at him blankly, blinking her eyes. "What does it *mean*?"

"*There are also animals which are called elk. The shape of these, and the varied color of their skins, is much like roes.*"

She continued to smile like a Maybelline commercial, head tilted. "Elk? Why?"

"It's from *The Gallic Wars,*" he explained. "Julius Caesar." Allene had made him memorize it. "*A tant vienent devant la roine,*" he added, switching to the story of Lancelot arriving to secretly visit his queen. It was another piece Allene had made him repeat over and over while drilling him in French. He closed his eyes as he recited, feeling sunlight pouring in the stone solar, feeling as if he would open his eyes any minute to see Allene's fiery red hair and Christina's shining black tresses, covered by the barbette tucked under her chin. She would be hiding a smile at some slight mispronunciation on his part.

The phone shrilled. He jolted. His eyes flew open to a modern studio with glaring electric lights and cameras and women in the audience wearing what seemed almost nothing to his medieval senses.

Jen pushed a button on her desk. "Hello, Caller," she said.

"I don't know about the Latin," said a male voice, "but he's speaking medieval French."

"*Medieval* French?" Jen's eyes shouldn't have been able to open any wider. But they did. "Are you sure?"

"I'm a professor of French," the man snapped. "I teach both modern and medieval, so yeah, I'm sure."

Jen looked to Shawn. "Is it?"

He cleared his throat. "Well, I guess. I never thought about it."

"How can you not know you're learning *medieval* French?" the voice demanded from the phone. "Didn't you notice you couldn't speak to anyone in French?"

"I spoke just fine with everyone I knew," Shawn said.

"Was the Latin medieval, too?" Jen asked.

He thought about denying it. But there just might be someone watching out there who knew the difference. "Yeah," he said. "I think so."

"You don't know?"

He gave a sudden grin. Best to laugh it off. "Yeah, it's medieval Latin. And I speak both modern and medieval Gaelic, too."

"How, why?" She looked up and down his distinctly medieval outfit.

"Just something I do. For fun."

"Not the kind of fun..." she began.

He interrupted. "Let's take the next caller." He was relieved when she nodded agreement. Still, his stomach knotted when the phone rang. But it was positive, as it used to be. *We love your albums. So glad you've breathed some life into some of those great pieces.* "I agree!" Shawn leaned forward, his bell-sleeved arms resting on his woven breeks. "Glenn Miller! Amazing musician!"

You've inspired my son to get back to playing, the next caller gushed. *It's changed his life.*

"Thank you," Shawn said with feeling. "It means a lot to me. Music has brought me the best experiences of my life." He thought of Niall in the window niche at Glenmirril, playing *The A Train.*

"Tell us about some of your favorites." The hostess leaned forward.

"There was the time...." Shawn stopped. There was the time in the Bat Cave, the first time he played for the Laird. There was the time around the campfire in Hugh's camp, playing for dozens of bearded mountain men preparing to lay down their lives for their country, their wives, their children. There was the night before battle with the greatest king Scotland ever had, watching.

"Uh, well." He cleared his throat, feeling a lot like Rob. "Lincoln Center was pretty great." It didn't compare to the acoustics in the Bat Cave—or the look on Christina's face the first time she heard the sackbut sing and echo off those high walls.

"Tell us about that," the hostess encouraged. "Didn't you play with Wynton Marsalis?"

"Yeah," Shawn said. "Yeah, he's pretty incredible. Very cool guy. It was an honor to meet someone so accomplished." He wanted with a force so great it nearly tore him off the couch, to race from the studio, fly to Scotland, and throw himself into that tower, hoping to be back with Niall, playing the recorder and harp together—dragging him back from MacDougall's trap. "Meeting him was a high point," he added.

It was almost with a sigh of relief that the girl announced there was no more time for callers. "But I do have one more question for you," she said.

Shawn's insides tightened, and nerves shot up his arms, as they had waiting in a medieval forest, knowing Arundel was near, though he could neither see nor hear him.

The hostess smiled. "You probably expect this by now."

"Where was I for two years?" Shawn forced a smile.

The hostess frowned. "You were gone a year."

"That's what I meant," he said. "I was having the time of my life."

"With an old couple in the west of Scotland?"

"Yeah. They're great people. Very relaxing year." Shawn smiled like a Cheshire cat. "We had fun learning medieval French."

"Not your usual pursuit." The hostess smiled back, and the tension in Shawn's stomach grew. "That's not what I was going to ask," she continued. "I don't think anyone expects the truth from you on that, by now." She turned to the screen hanging on the wall behind her, and it lit up with a YouTube video.

At the first glimpse of himself on screen, Shawn resisted the urge to close his eyes in humiliation. Never let them see it, he reminded himself. He plastered a smile on his face, unable to stop the video playing, as he watched a night scene of himself prowling the dark streets of Berwick.

Niall! Came his own voice from the sound system in the small studio. *Niall, it's a trap!* A part of him wanted to die. They must all think he was crazy. Insane. Stark, raving mad, and then some. A part of him churned with slow-rising anger. Niall was about to die and to them it was a big joke.

Niall! came his voice again. *Stay away from MacDougall! It's a trap!* A part of him wanted to rip off the microphone, throw it on the floor and race to the nearest airport, commandeer the fastest plane and get back to Scotland, throw himself against the walls of Berwick, shouting even louder. *Niall! It's a trap! Don't go!*

He stared at the video of himself looking ridiculous. Five million views, he noted. Five million, three hundred thousand two hundred and thirty eight, to be exact. Great.

He tried to console himself that any publicity sold albums. But what the hell was he selling albums for, he wondered, as the humiliating video played on, of himself creeping like a prowler around a corner, cupping his hands and hissing. He was earning piles of money to eat all he wanted in a house with central heat while Niall starved in the chilly winters and rainy springs and raced to a hideous death, trying to save Allene.

He scrambled desperately for a story to explain his behavior in the video. Since Amy's mention of the last video, he hadn't yet thought one up. Nobody had asked to his face, and he'd convinced himself he could skate past this, too.

The video came to an end with one last shout of *It's a trap!*

Silence descended on the studio. The audience didn't laugh. They didn't applaud. They stared at him with questions in their eyes, as if trying to decide where the joke was.

He was tired of the lies. They were exhausting. Trying to laugh off two years of his life as if it meant nothing—it was exhausting. His smile slipped, even as the girl's brightened. "You called yourself Niall Campbell at your last concert before you disappeared. Now you're shouting *to* someone named Niall. I can't wait to hear your explanation," she said.

Shawn looked at her sadly. "Have you ever had a friend die?" he asked.

She looked confused. "Well, no, but that's not...."

"I've watched men die in the last two years," Shawn said. "Good men. Men who were my friends. And I've learned that God exists and he has a terrible sense of humor, because I used to lie to a lot of people. And now when I finally tell the truth, it's so insane nobody believes it. And I'm tired of lying. I don't find it entertaining or challenging or fun anymore. I find it exhausting."

"Okay." She looked doubtful. "Who is Niall?"

The audience leaned forward.

"He's the best friend I ever had, and I don't want him to die. Not like that."

"Like what?"

"Hanged, drawn, and quartered."

She leaned across her desk, blinking, the smile unwavering, as if waiting for a punchline.

"Do you understand what that means?" he demanded. "They draw you—that means they slice you open. Sternum to pelvis." He drew a hand down his own body, demonstrating. "They tear out your bowels—while you're still alive, while you watch your own innards lifted up for the crowd to jeer at."

Jen blanched, though she forced a smile. "Where's the punchline?" she asked. But her words were weak.

"There is none." Shawn unclipped the microphone from his linen shirt, held it awkwardly for a moment, and set it down on the desk.

"You're leaving?" she asked.

"Yeah, we're done, aren't we?"

"Where were you for a year?"

"Scotland."

"Scotland in the modern day, right? Not medieval Scotland."

"I was there from June of 1314 to June of 1316. With some of the best men who ever lived." He lifted the latte. "Thanks for the mocha. It tastes good after two years of ale."

To silence, he left the stage. As he reached the edge, a few faint claps sounded from the audience. More joined them, but they sounded more like a question than anything else. He disappeared into the wings. A camera man backed away. "You okay, Mr. Kleiner?" he asked.

"Fine."

Onstage, Jen give a little laugh. "Well, he certainly sounds like he means it. And he's got the languages and clothing down. That was amazing! Thanks, Mr. Kleiner." She clapped and the audience now gave a burst of enthusiasm.

Ben would be furious. Shawn threw his head back, gulping the last of the mocha. James and Amy ripped at one half of his heart. Niall tore at the other. He had to do something. He couldn't keep living like this.

Bannockburn, Scotland, Present

"Thomas Ritchie." The man in the white coat held out his hand, giving Angus a firm shake before seating himself. "So the police have taken an interest in our unusual visitor." He shook his head. "Can't say I'm surprised."

"Tell me about him." Angus propped his notebook open in one big hand, his pencil poised, excited to learn what he could after the long wait before the man could clear room for an appointment. He could almost forget for a moment the look in Claire's eyes, or that he was confined to a wheelchair. In the last few days, he'd even felt cheerful again, as he managed ever more without Ken's help. He'd become enamored of the sweet tones he could play on the flute. He smiled, the notebook on his knee.

"He was angry when he arrived." Thomas Ritchie's words pulled him back to the present. Or was that the past, he thought sardonically. "Fighting everyone, violent. He'd recovered physically, but he spoke gibberish, and he clearly wasn't safe to be on his own."

"He eventually calmed down?" Angus asked.

"Oh, aye." The director nodded. "It took several days, but he seemed to finally understand he was goin' nowhere. He stopped fighting, just sat and watched. Now and again, he'd speak to us, slowly, as if we hadn't our wits about us."

"What did he say?" Angus asked.

Thomas shook his head. "That's just it. His language was like none we'd ever heard. I'd the feeling he was repeating himself to us in *several* languages, none of which were intelligible."

"You learned what he was speaking?" Angus kept his voice professional, clinical, almost uninterested.

"Aye, the director back at the hospital—Alec—his assistant has an uncle, a rather renowned historian. He asked to come and see our patient. You see, this man had a sword, and mail, which were of interest to the uncle. So Alec brought him, and he knew instantly what this man was speaking."

"He was able to communicate with him?" Angus clarified.

"Oh, aye, he came many times. 'Twas something, the two of them going on in a language no one else understood!"

"He learned your man's name?"

"Simon Beaumont."

Angus nodded, not surprised. "Did Simon Beaumont say where he came from or tell you anything—through his interpreter—about himself?"

"Oh, he began to speak proper English," Thomas assured him. "He was a professor. Though we could never find any missing professor called Simon Beaumont."

"I imagine not," Angus murmured.

"We wondered if he might be from Germany or France. Or America."

Angus asked, only because he must appear to ask all the right questions. "Did you try to find out?"

"We tried." He spread his hands. "But the world's a large place."

"When did he leave?" Angus asked. "He was deemed healthy?"

"He made great progress, once he had someone who could speak with him. I've seen it before, you know, a person entirely losing their native language. But his English came back, and though he seemed a bit odd, still, we couldn't hold him. There were no grounds, really."

"Do you know where he went?"

"Brian signed him out and took him home."

Angus lifted his eyes. "Brian?"

"The interpreter. Alec's assistant's uncle. The historian."

Fear rose, swallowing what excitement Angus had felt. "What is Brian's surname?"

"Let me get that, now," Thomas turned to his computer, clicking, and wiggling the mouse. "Here it is," he said. "He's the curator over at...."

Angus finished the sentence. "Creagsmalan. Aye, he's a mate of mine."

London, 1318

The wooden doors opened, admitting Simon to Edward's chambers. The king and his commanders sat at a table near the hearth in fur-lined robes. Edward waved a careless hand at a chair. "Sit," he ordered.

Geoffrey of Dunville and William de Crecy watched silently as Simon laid his scrolls on the table. In his new and well-appointed room in Edward's castle, he had worked for days and late into the night, eyes straining in the flickering light of a dozen candles, copying diagrams from his treasured notebook onto the vellum these men would not question. He was pleased with the results.

Simon's eyes flickered over Edward's commanders as he took his seat. He was glad Isabella was not there. She was a woman of sharp wits—and sharper cruelty when angered. The d'Aunay brothers had been executed shortly before Bannockburn—one of his last memories of his own time before waking in Shawn's. "Your Majesty," he said.

Edward was silent a moment before saying, "What is it you propose to do about Deydras?"

"Deydras is a small matter," Simon replied. "Easily dealt with. Might we speak, rather, of greater things?"

"I would deal with it immediately," Edward replied.

"All things in their time," Simon countered.

"And yet," William said, leaning across the table, "we must know we can trust you. Show us this in the matter of Deydras."

"All things must work *together*, must they not?" Simon looked from one to the other of Edward's advisers. "Have I not shown you I see the future? I see that which you do not, which determines the time to address Deydras."

Geoffrey dropped his gaze to his hands on the table. William averted his eyes.

Satisfied, Simon turned to Edward. "The barons are against you," he reminded him. "There's trouble with Lancaster."

"The cursed Ordinances!" Edward pushed away from the table, and marched to the window between the hearth and a great tapestry: Eve in the garden, an apple of glistening red in her hand, woven with thread of gold, and the serpent curled around a tree limb.

He was tall, like his father, and as fair as Longshanks had been in his youth, Simon thought. But he was no Longshanks, and that was the trouble. Simon smiled—or rather, it was the opportunity. A man more interested in thatching roofs and shoeing horses than running a kingdom suited Simon quite well. "You will have to at least appear to give way a bit," he advised.

Geoffrey nodded, almost as if unaware he did so, before staring again at his hands on the table.

Simon turned his attention back to Edward. He suspected the Ordinances were still such a grievance as a result of the Gaveston affair. He himself had been glad to see the upstart go. It left him space to guide a king more comfortable with peasants than with princes and popes, who desperately wanted someone to step in and tell him what to do. Simon was happy to oblige. He poured wine into a jeweled goblet, satisfied with his rise in the world in this short time. "Say you'll abide by the ordinances, and let us address the real matters."

"Mm." Edward stared out the leaded glass to a city white with frost.

"What are the other matters of which you wish to speak?" Geoffrey asked.

Simon met his eyes. "The Scots are surrounding Berwick."

William tilted his head, speaking as if to a child. "The walls are impregnable."

Edward turned from the window. "They've tried many a time. Berwick has always held."

"I've told you this time they will succeed." Simon bit back his irritation, reminding himself it was exactly this sort of foolishness that served him. "In the meantime, they are *attacking*, when they ought be *defending*."

Edward studied him a moment before nodding slowly. "So be it. You say you can stop the Scots and take the cursed country back in hand." He returned to his chair and poured himself wine. "What must you have, in order to do so?"

"Give me ten thousand troops. I shall first take control of our north, stopping them at the border. I will sway Lancaster to work with me."

"How will you do that?" Edward tipped his chalice back, a swift gulp, before setting it down with a sharp rap. "My dear cousin is not wont to do aught that would help me."

"I shall give him what pretty words he pleases and make what promises I must," Simon replied, "and with his help, take back what your father held. Two years, and the Scots will be subdued. Three, and they will be annihilated."

Edward studied him a long moment before asking, "How do you propose to do so when the last eleven years have been fraught with failure?"

Simon held his scathing thought—*I propose to do it by being more like your father than like you*—in favor of more diplomatic words. "I have seen the future, your Majesty. I have seen their *weapons*, the likes of which will spread terror to the Scots and subdue them with ease."

"What kinds of weapons?" Edward asked.

"Instruments which shoot balls of lead—large and small—that fell men, and crumble walls."

Edward leaned forward, his eyes alight. "As a bow shoots an arrow? Do we not already have catapults?"

"See for yourself, your Majesty." Simon smiled, a grim tightening of the corners of his mouth, anticipating the moment of reveal. He lifted the first of his scrolls—his right shoulder gave a twinge at the motion—and unrolled it across the large table. William retrieved lead weights from a chest to hold it down. Simon stood back, admiring his work, as he rubbed the aching arm.

In the morning sun, Edward studied the drawings. He traced the outlines of the *gun,* of its working parts carefully depicted. "What size is it?" he asked.

Simon held up his hands, demonstrating, and saying, "Do not be fooled by its size, your Majesty. It shoots fire with a great roar." He'd seen them on the *tellies* of Shawn's time.

"Sea fire?" Geoffrey asked.

"Sea fire does not roar." William, who had fought gallantly since the days of Longshanks, leaned in, his coarse gray beard brushing the table as he studied the intricate design. "And this weapon appears to have but a small bore. Explain how these mechanisms work."

Simon did so, pointing to various illustrations, and looking up as he finished. William nodded, frowning. He was an intelligent man. He'd been known to try his hand at inventions, himself. But he had no way of knowing the parts Simon had deliberately left off the diagram.

Simon smiled. It would not serve to let too many know how to construct his weapons. He already had plans to separate the stages of production so that no one man would have the knowledge to make his own. "The real secret," he said, pointing to another picture sketched in one corner, "is the *bullets*—small balls of lead which will pierce armor."

"How can something so small pierce armor?" Edward lifted his head to meet Simon's eyes.

"The force with which the fire propels it." Simon moved his finger to another picture. "This one fires much larger balls. These ones will cut through entire armies, even knock down castle walls, such is their power."

Edward and his commanders moved around the table, studying the diagrams from different angles, asking questions. Simon pulled out another scroll, showing them the ingredients for *gunpowder*—thought not the measurements—and a list of where such things might be acquired.

Finally, Edward seated himself. Simon and the commanders followed suit, waiting. Edward beamed. "These weapons," he announced, "will turn the tide against Scotland. What must you have to proceed?"

"Give me men of learning, give me smiths," Simon forced himself to speak calmly, though elation jolted through him. The pieces were falling into place!

William leaned forward. "How soon can they be ready?"

"As soon as I meet with them and teach them." Simon turned from William back to Edward. "Within the month, I hope to have the first dozen *cannon.*" He had no idea how long it would take to build one, but it sounded about right, from the drawings he'd seen. They would need to cast molds of the heavy equipment, but a skilled smith should be able to do so. Failure to meet the deadline would be blamed on the workmen. "I need men to find the ingredients and bring them to me in great quantity. I need space to store them and to work."

"It shall be done," Edward said. He stared at the tapestry, at Eve with her apple, his eyebrows drawn together long enough that the nerves began to rise in Simon's neck and arms, and in his shoulder. Finally, the king spoke. "We still must, of course, plan our campaign wisely."

Simon unrolled a second parchment, depicting a map of Britain, Scotland, and Wales. "With a united force, we secure our own north." He pointed as he spoke. "We then move into Scotland, from east and west simultaneously. The Scots will crumble in the face of our united forces and new weapons." He looked from Edward to William to Geoffrey, and back to the king.

"Your plan is solid," William said. He spoke to Edward. "The first step is peace with the barons."

Geoffrey nodded.

Edward's lips tightened.

"Your majesty," Simon said. "I have foreseen unity. It will happen." It would happen, he thought, without Edward. He himself would lead the barons. He'd already sent a missive to Lancaster. But he would get there through Edward. "I ask that you allow me to meet them as your envoy. With promises— and with the understanding of the new weapons you possess—they *will* stop their clamoring and join you."

Edward rose from his seat, returning to the window. Snow trickled down the leaded panes, white flakes striking, sticking, and sliding down the glass. Simon's nerves stretched. It all depended on Edward agreeing. The tapestry stirred in a draft, Eve, and the serpent faintly undulating.

"Yes." Edward turned abruptly from the window. "Begin production immediately." He turned to Geoffrey. "Take the list of ingredients and begin— today—to organize expeditions. Find a place outside the city walls to make *gunpowder.*" To William, he said, "Search out smiths who can make the parts for these *guns.* Have answers for me at tomorrow's dinner."

William and Geoffrey rose, bowing, and hurried out.

Edward turned to Simon. "Choose buildings on the Thames. Pick your commanders. Send a summons to the barons. We will meet again in a week."

Midwest America, Present

In his music room, at the table under the floor to ceiling leaded glass windows where he had written so many scores, Shawn spread out the last parchment, reading it for the third time, feeling sicker than he had the first two times. Nothing had changed. If he didn't figure out something else, this horrible thing would happen.

He skimmed Christina's letter. Her words hurt in a different way.

Ka-ching! His phone announced Ben's call. Fourth one in the twenty-four hours since the interview. Second one since he'd sent Ben a text apologizing and promising to call. He slid his finger across the screen, sending Ben to voice mail and lifted his head to stare out the window, across the lawn, covered in a fresh layer of snow. He had loved the wide, open stretch the moment he'd pulled into the driveway with the realtor, four and a half years ago—three and a half, he reminded himself. *He couldn't go on like this!* He let go of the parchments abruptly, letting them curl up around each other, and strode to his hall closet, grabbing a heavy jacket.

In moments, he was out on his circular drive, staring at the place he'd intended to put a fountain. It was so clear in his mind. He wanted it. He *wanted* Amy and James there, James leaning in, trailing his chubby hand in the water—maybe with koi—and laughing his cheerful laugh. He could make that happen, he could save Niall, and he could prevent the future Brother Eamonn had predicted. He *would!*

Shawn turned away, hands in pockets, and rounded the house, out onto his twenty acres, heading for the stream. Soon, he could hear it burbling, a happy chatter in the quiet day. He'd loved this property before he'd even seen it, reading the descriptions on the realty website of the wooded copse, the stream, the small waterfall. He fell into step beside the brook. Sunlight glinted silver off its running surface. He'd walked here dozens of times with Amy. He smiled, remembering taking the kittens, in their basket, into the woods, delighted with the surprise and joy he had known he'd see on her face. *Something in you needed her,* Dana had said.

The smile slipped. He hadn't spoken to Dana since he'd come back. She was yet another person he'd hurt. He thought of walking these woods with Celine—the one time. He shoved his hands deeper in his pockets, shoulders hunching. The cold bit through his jacket. He hadn't found a way to make it up to her. Maybe Aaron was right, maybe there really was nothing he could do, except—Niall's voice spoke in his mind, in a heavy Scottish brogue—*Pray.*

A bird chirped, high up on a snowy branch, as he passed into the woods, his own private woods. He snorted. Niall hadn't needed his own private woods. He'd had the whole of Scotland. The creek babbled between its banks of ice, between trees reaching bare, brown fingers to the sky, carrying Shawn back to riding along Scottish streams with Niall.

He wanted to give James riding lessons, ride one day with his son. He wanted to ride with Amy. He smiled, remembering leading her along this same path, two years ago, promising next time he'd buy her a horse.

He couldn't abandon her. Couldn't abandon James. Except...he couldn't abandon Niall, either.

Pray.

He was no good at prayer! Shawn heaved a sigh as he came to the small waterfall, flowing from the stream, with fir trees towering on either side. Amy had loved it. They would be happy here together. She'd see Angus wasn't changing his mind, and she'd join him, and they'd be a family. He sank down on the boulder at the waterfall's edge—a boulder much like the one behind Glenmirril. *There was Niall.*

Pray!

Shawn bowed his head over clasped hands. No words came. Only the distress, the knowledge he couldn't abandon Amy and James—and he couldn't abandon Niall. *I've tried!* His thoughts echoed in his own head, as he lifted his eyes to the blue winter skies overhead, glaring at an unseen God. "I've tried!" He spoke the words aloud this time. Tree limbs swayed overhead, and he shouted up past them, "What do You *want*? There's nothing else I can do!"

The bird chirped, a musical dance of joy. Shawn pursed his lips. He loved his home. He loved the stages. He loved the music he was writing. He loved the swell of the orchestra behind him.

There was Niall, who would die.

There was the orchestra, his publishers, the musicians he hired, his commitments to symphonies, people who earned their living off his work and success. He shook his head, faintly. No, that wasn't right. Zach was here. These were things Zach could take over, and Amy.

Shawn stared at the water spilling over the rock for another two minutes. There *was* one thing he could do. He pulled his phone from his pocket, and pulled up the text messages from his mother. He studied the number she had sent, the wind sighing in the bare branches overhead, the bird chirping, the water burbling. He could do that one thing. He touched the number—and immediately pulled his finger away, bowed over the phone, feeling the tremors shake his body.

London, Present

"Congratulations," MacDougall said dryly. "Chief Counselor, is it? High adviser? Commander of armies?"

"Did you find the brigands as I ordered?" Simon asked, ignoring MacDougall's caustic words.

"Ordered?" MacDougall raised an eyebrow. "I was not aware I had sworn fealty to you."

Simon merely watched him.

"No," MacDougall said. "My men searched all the way to Michael Scott's tower. Apart from a rumored farmer and rumored gypsies, neither of whom we ever found, no one saw hordes of brigands."

"Meaning?" Simon asked.

"Meaning we were lied to."

"*Someone* shot me," Simon snapped. "You've found nothing?"

"After you left," MacDougall said, "someone poisoned my finest stallion and attempted to poison me. I believe it was he who spread the false story. He has gone to Canterbury. I am on my way there now."

"Bring him back to me. He will wish he was never born."

MacDougall raised an eyebrow. "Is it now your place to order executions?"

Simon grinned, as he poured smooth streams of ruby wine, first into MacDougall's goblet and then into his own. He lifted it in salute. "My rise is your rise. Did we not agree to an alliance when last we met at Linstock?"

"We did." MacDougall lifted his cup and leaned forward to touch the edge to Simon's. "However, your reputation precedes you. I'd as soon trust a viper."

Simon laughed. "A man is wiser to befriend a viper than to antagonize it."

"One does not, however, invite a viper into his bed." MacDougall watched Simon drink. "Your swift healing is spoken of with suspicion." He stared into his own wine for a moment before raising his eyes. "I've no wish to be part of any dark arts."

"The talk of small minds!" Simon waved a hand. Light flashed off a ruby ring. "Did not our Lord Himself heal by the power of God? What of Hildegard von Bingen?"

"What of her?" MacDougall asked.

"She wrote extensively of healing herbs. An abbess and two popes have pressed for her canonization."

"Was it her potions you used?" MacDougall asked.

"Now I wonder." Simon rose. Hands clasped behind his back, he sauntered to the window. "Were it your own arm, would you concern yourself with the wagging tongues of uneducated fools?" He turned, meeting Alexander's eyes.

MacDougall said nothing.

Simon took a step closer. "Suppose the potions in my book—the book your man Roger took from my chambers without my permission—could have saved your son?"

Alexander's jaw pulsed. "It is as you say. The uneducated speak foolishly." He glanced at his wine goblet, still full. "You now have troops and power. What of our previous discussion?"

"We want much the same thing." Simon seated himself. "You want Niall Campbell and the man called Shawn. I want Shawn, and I want his son. The three are tied together...by fate. We stand the best chance of getting all we desire by joining forces."

MacDougall twisted his goblet between his fingers. "You have not yet told me where Shawn is."

"I *have* told you." Simon dropped back into his chair. "He is beyond reach." He re-filled his goblet and took a long draught.

MacDougall glanced at MacDougall's drained goblet and took a small sip from his own. "From a man like you," he said, "vague words do not suffice How so?"

"You would not believe me if I told you!" Simon smiled, his eyes alight.

"How then do you propose to deliver him to me?"

"It is what you must do. Leave messages at Creagsmalan, at Monadhliath."

"Monadhliath?" MacDougal repeated. "The monastery in the Highlands?"

"The very one," Simon affirmed. "Write multiple missives, telling Shawn that Niall will die. Tell him when and where it is to happen. Do you recall I asked you what the two most powerful forces are?"

"I recall," Alexander said tightly.

"What does Shawn *love?*" Simon asked.

"Christina," MacDougall answered. "My son's wife, Christina."

Simon smiled. "Then you shall let him know in your messages that she is also in danger."

"And this man who is beyond my reach is going to look at Creagsmalan— he will walk right into my son's home—for my messages?"

"Creagsmalan will be confiscated by the Bruce," Simon said. "It is only a question of how soon your wife will be driven out and sent to join you in England. Yes, Creagsmalan is one place Shawn may look."

MacDougall poured more wine and lifted his gaze to Simon. "Forgive me for being a wee bit skeptical. Speak plainly. Where *is* he and how will he know I've left these messages?"

"Do as I say, and he will come. You will have not only Niall, and Shawn, but Shawn's son."

"How?" MacDougall demanded.

"That is to be arranged," Simon said. "Let us start with Niall." He thought of Angus stumbling on the stoop outside his house, and of the satisfying crash and shout that had come from his house when the mantle had fallen. He wondered if slicing his gear had had its impact yet. Killing was easy. No, it was playing with a man first that brought the real joy to it. "What does Niall *hate* the most?" he asked.

MacDougall grinned. "Water."

Glendochart, Scotland, 1318

Niall bowed his head over folded hands in the little chapel at Glendochart. He'd come as soon as he'd cared for his pony, grateful to be in from the long, cold ride. He had not forgotten his king's command to pray. *You do not always know my plans,* Bruce had said. *Yet you follow me.* Still, he felt empty and alone. He reached for elusive gratitude, reminding himself of all that was good.

They'd reached Glendochart safely and would have more than a week of beds and roaring fires in hearths. Even now, the cooks were preparing a hot meal in the priory's kitchens. He had a reprieve from the mind-numbing boredom of a siege. That alone was a blessing.

He'd had time—if only a little—with Allene and his sons.

They had procured the weapons from Dumbarton, which greatly improved their chances at Wark. Success there brought them closer to cajoling Edward into peace and that meant being home—for months, even years on end—with Allene and his boys and the new bairn.

He bowed his head lower still. There *was* much for which to be grateful. Yet the unease remained. They had injured Simon—perhaps even killed him—and a number of MacDougall's men. And here the agitation spiked. He would not feel at ease until he knew MacDougall himself was dead.

He closed his eyes, shutting out the chapel, the crucifix, the candles, and felt himself back in the St. Michael Chapel at Stirling, with Shawn spinning the Rosary; how his irritable hand had fallen on it, stopping the spinning. *We ought pray and fast,* Brother David had said. *I'm on it!* Shawn had jabbed a hasty sign of the cross on his forehead, chest, and shoulders.

It's a trap.

His eyes flew open at Shawn's voice. But the words were weak, this time, a thin gray ribbon of thought through his mind. He jumped to his feet, hand to knife, looking around the small chapel. He was alone in the gray stone walls, with the crucifix and candles and smell of wax and old incense.

"Shawn?"

He listened intently. But all was still, quiet. He'd imagined it.

But he hadn't imagined it those other times—had he?

Stress, strain, he told himself; life under constant threat from England, from MacDougall—of course it led to fear of traps. He was imagining it.

He eased himself to his knees again, trying to settle his mind to his Father in heaven and the next day's work of planning a chapel.

London, 1318

Simon lifted his goblet in a salute. "Then water it shall be. I believe I know just the place. The trick is to lure Shawn in, correct? We have, then, but to see that Shawn knows. I tell you—for love and for hate, a man will act. Shawn and Niall *will* act for those they love, and then we shall have them both." He leaned forward. *"You* shall have them both."

MacDougall twisted his goblet between his fingers. "I know you. You will want aught in return."

"Indeed." Simon smiled, as he plucked a piece of meat from the carcass on the table before them. "Have no fear. What I want will cost you nothing. You are kin to John of Lorne."

"He's dead," MacDougall said.

"He commanded a fleet," Simon said. "I want it. As his kin, you might influence Edward in choosing a commander for the ships that were his."

"You've no naval background," MacDougall said. "Why do you want ships?"

Simon's smile grew. "I have seen visions. Not only will it cost you nothing, but you shall profit beyond all you've ever dreamed. Get me his ships, and I will show you wonders you cannot imagine. You will have more land than Edward could ever give you."

"You have been evasive on many questions," MacDougall replied. "Why would I believe this?"

"I have told you the truth is more fantastic than you can believe." Simon smiled. "I have seen the future. And I have seen a great land, so vast it dwarfs our England to nothingness. Get me his ships and you and I will rule it."

"A land so vast it dwarfs England?" MacDougall rose, wandering to the window.

"A land vaster than all Europe combined," Simon said.

MacDougall stood still at the window, watching the snow drift down. After several moments, he turned. "I will recommend you to Edward as commander of his fleet after we have Shawn and Niall," MacDougall said. "I want troops there. To be sure he's not escaping."

"I have troops from Claverock," Simon assured him. "And more from Edward."

"Enough to be sure we get through the North of England? You were gone more than three years. Campbell has risen in the esteem of the Bruce and James Douglas. They will almost certainly attack us to take him back."

Simon's jaw tightened. "I am aware of this." He didn't need lectures from this fool who couldn't even take a small holding like Glenmirril. "I will see that he ends up in England."

"A public execution of the enemy." MacDougall returned to the table, selecting a grape from a bowl. "A little show of force and kingliness can only help Edward now, with more and more people whispering that maybe Powderham's story is true." MacDougall snorted as he poured a red stream of wine. "Just last week, he was seen standing by a creek, watching a man fish. His father never would have done such a thing."

"I'll bring the troops," Simon assured him. "When and where are we to meet?"

"There's a farm directly across from the island. Send twenty of your men across, and wait there with the rest. Be ready before dawn that day. I'll take them both, as planned, and bring them across, and we'll start for London."

"I've given you a plan." Simon's lips pursed in a tight smile. "Now send word to Edward recommending me as commander of his western fleets, leave the messages I spoke of, and go north for the bait. When Edward grants me the fleet, *then* will I bring the troops and join you."

MacDougall glared at him. "I seem to be making all the first offerings."

Simon smiled. "Do not forget I've also seen the weapons of the future. Talk to Edward and I'll bring the troops."

Midwestern America, Present

Standing under snowy trees, by the stream on his property, Shawn glanced at his watch. It was mid-afternoon in Scotland. He studied the number his mother had sent, and shook his head. In a minute. After he'd talked to Amy. He hit her picture on his phone, wanting to hear her voice.

"Shawn!" She came on the line, a little breathless.

He smiled. Did he imagine she sounded happy, saying his name? "How are you?" he asked. His smile grew. It was a stupid question. A make-talk question. A question to hear her speak. Something fluttered inside him, dancing between her voice and Christina's.

"I'm good," she said.

He felt warm inside, despite the winter chill. "Are you writing music?" he asked. "Playing violin? What is James doing?"

As he entered the path through the copse, she fell into easy conversation, laughing, joking, her voice alight with joy as she relayed James's new words, and sparkling as she told him of her quartet arrangement of *Der Erlkönig*.

"The Elf King." Shawn thought of the three days he'd spent in the Eildon Hills with Niall, hoping for an Elf Queen. What insanity had possessed him now, after all he'd gone through, after all Niall and the rest of them had gone through, to get him back?

"The Goethe poem, Schubert," Amy clarified.

"Yes, the father assuring the son it's nothing, just the wind, just the trees." He knew the translation by heart. *He reaches his courtyard with toil and with dread. The child in his arms finds he motionless, dead.*

There was Simon. But he'd seen no more of Simon on the parchments. And hey, they were all still here, America still existed, right? It had to have been the ramblings of a drunk old man. This was about Niall.

"It'll premiere in Inverness," she said. "A big fundraiser for the arts." Her words fell and rose, seconds, thirds, an occasional fourth, a melody that echoed on his heart, a flute to Christina's clarinet.

"You?" she asked. "You're good? Is it good to be home for a bit?"

Shawn looked across his land, at his house rising far away. "It's good," he said. "You know I love this house." He thought of his fireplace, his white leather couches, and the kitchen with its granite counters, re-stocked with spices, with the saffron and the grains of paradise Glenmirril's cooks had loved. Yes, he loved his house. He loved the memories it held. He loved the future he hoped it held. He cleared his throat. "So. Angus."

There was a moment's silence, the breeze lifting his hair, a bird swooping overhead, before she spoke, her voice dropping a diminished fifth. "What about him?"

Shawn turned, heading back toward the house, envisioning the room that had sprung up in his imagination, the room he could make happen, the room that would change everything. "You know what about him," he said. "Is he coming around?"

"No." The word snapped out, quick and sharp and high as a rim shot on a snare.

"I'm sorry." The words slipped out, unplanned.

She said nothing.

"I want your happiness," he said. He climbed the flight of stairs that led to the deck, remembering the first time he grilled there for her.

"Even if that means I'm with Angus?" she asked.

He scanned the deck, remembering the way she'd laughed, that night, as she told a story about her first memory of a barbecue, with Miss Rose. Something about a cat. He wanted that laugh to be her life. A pain shot through his heart at the thought. But he meant it. He wanted her happiness. He wanted her laughter to always, forever, be that happy, carefree laugh he'd first seen the day he went to audition for the orchestra, before he even knew her name. "Even if it means you're with Angus," he said.

"You mean it?" She sounded incredulous.

"I mean it." He thought of Christina, telling him to go, to leave. He knew, in a rush, what she had felt, what it had been for her to say those words. "I mean it with all my heart. I hope your happiness will be with me." He could find another way to solve these problems, he told himself. The scrolls might yet change. "But I want it for you, no matter what."

She was silent.

"I'm not going to try to turn you against him," he said. "Or sway you. I'm here for you, Amy, to be your husband if that's what you want, to be your friend if that's what you need." He stared over the grass, seeing the new solarium.

"He's engaged to Claire," she said softly.

Shawn's heart leapt in joy and twisted in pain all at once. "I'm sorry," he said. And he meant it. But the new room would be great for parties. Because surely he'd find another way of doing what needed to be done. He could see them—himself and Amy, welcoming friends, musicians. A chamber orchestra playing while people drank and laughed. Potted ferns. Glass panels curving up to form the ceiling, letting in starlight and the soft yellow glow of the moon. And James would dart among the guests, and Shawn would teach him to swim, and he might play an eighth size violin with some of the chamber musicians. "Would you be happy?" he asked. "Married to me?"

"Yes," she said.

His heart tripped. He almost blurted out, *Then why not!* But an invisible finger seemed to touch his lips, silence him.

"And I'd be happy married to him," she said. "And I miss him every single day."

THIRD MOVEMENT

CHAPTER NINETEEN

London, 1318

"I have one request," Simon said, as the door closed behind Edward's advisers. The second meeting had gone smoothly. Smiths were making molds for the *guns* and *bullets*. Men were ranging across the countryside gathering the ingredients for *gunpowder.* Commanders had been chosen for Simon's troops. A great warehouse on the waterfront had been outfitted for the production and storage of the new weapons.

"Does it regard the missive Alexander MacDougall left me on his way to Canterbury?" At the window, Edward rubbed his hands together. Snow trickled down the leaded pane. Flames crackled in the hearth.

"It does," Simon said.

Edward strode to the table, re-filling his goblet with red wine. "A fleet of ships is a great deal to ask."

"I shall give far more in return." Simon slid a sheet of parchment from behind the others. "I served your father well and have shown myself eager to do as much, and more, for you."

Edward regarded him a moment before saying, "In a few short weeks, you have acquired command of an army, and risen to the highest power in the land, short of my own. You have a campaign outlined to recover Scotland. *By land.* What do you wish with ships?"

Simon dropped his hand on the map and abruptly closed his eyes, intoning in a deep voice, as he had long ago heard Thomas of Erceldoune do, *"Far across the ocean grand, lies a great and glorious land! Far across the great water, a land shall be ruled by king and daughter!"* He opened his eyes, pleased with his rhymes, but kept his face stern, his eyes boring into Edward's, daring him to question.

"A...land?" Edward dropped his eyes quickly and leaned forward, studying the inked drawings of water, mountains, and islands.

"A land greater than all Europe." Simon held back a smile, well satisfied with the king's reaction. He pointed to a continent on the west of the map. "It has vast mountains, rivers so long they would stretch from the north of Scotland to the south of England! Wide plains that would feed our people by the millions, with new foods and animals we have never seen."

Edward leaned in, peering at the tiny depiction of a shaggy creature with massive shoulders.

Simon wondered, himself, whether the illustrations in the books of Shawn's time were real or fantasy. *Buffalo.* The depictions of them had called to mind the mythical beast of the labyrinth on Crete—though on four legs instead of two. The captions had suggested they were real. In truth, it didn't matter. He had read something even harder to believe that had been—would be —used to draw men to the new continent. He leaned forward, dropping his voice. "The cities have streets paved with gold! It is a vast, uninhabited world, your majesty."

"Who built these cities?" Edward lifted his eyes from the map. "If the land is uninhabited?"

"Men who have left." Simon was pleased with his quick wit. *Helen of Malley, how you underestimated me!* He found he enjoyed making up stories. "It is yours for the taking. Give me command of John of Lorne's fleets, your Majesty. I shall have a team of master shipwrights outfit them for a great voyage—one that will be spoken of till the end of time—and when I am done with Scotland, I will secure this land. There will be no greater king on earth than you."

Edward sipped his wine before saying, "I've just authorized new and untested weapons. You showed me how such things could be. We can test them. But a land across an ocean. Entire fleets. Sending my men into an unexplored ocean on your word alone...."

Simon carefully crafted an expression of sorrow. "Your majesty, was I not the most loyal of knights to your father? Did my kin Sir Henry not lead one of your cavalry divisions at Bannockburn? Did I not fight beside you there, myself?"

Edward gave an unmajestic snort. "As did others I would not necessarily trust."

Simon straightened. "Did I not predict Deydras's arrival with his one ear?"

"There are those who say you put him up to it to appear to prophesy."

"It would be possible to do such a thing." Simon's eyes bored into Edward's. "But certainly your majesty, I could not have arranged the queen's pregnancy."

Edward's eyes narrowed. "Certainly not," he responded dryly.

Simon leaned back in his chair, satisfied. "Deydras is no threat to you." He put only the smallest disdain into his words. "You will see."

"It may be that I *will* see," Edward said, "but I have *not* seen *yet.*"

"What you *have* seen," Simon countered, letting his impatience show, "is that you have problems with Scotland, your barons—which I have shown you how to solve. Problems with Lancaster. You know he is not to be trusted."

"He would say the same of you," Edward commented.

Simon's mind flickered to Isabella. It was she who would put such questions into the king's ear. "My loyalties," he said smoothly, "are assuredly not with Lancaster. He is hungry for the throne, your majesty. Several of your own knights have named him traitor. In the matter of his wife's abduction, he has accused you..."

Edward rose from his seat, anger flashing across his face. "Lies!"

"Falsely, of course!" Simon held up a hand, forestalling him. "I say merely that Deydras is of no consequence. It is Lancaster who is the danger, stirring your barons against you."

"It does not take a seer to see that," Edward snapped.

"You doubt me?" Simon's temper flared. He must remain calm, he reminded himself. He drew a breath. "Know this: Your daughter will be born on June 18." He pulled a small leather-bound book from his pocket, and carefully tore a leaf of vellum from it. He handed it to Edward. "Sign your name, Sire, and I will write below what your daughter's name will be."

Edward hesitated a moment, then took it to his desk, where he dipped a quill in ink and signed. He took vellum and quill to Simon, watching as Simon carefully wrote a name under his, then sanded it.

"There is also this," Edward said, when Simon looked up. "If Deydras convinces enough people of this outrageous story, I may not be here to see my daughter's...." He stopped, glanced at the paper, and corrected himself. "My *child's* birth. Or this great land you say is across an ocean. When do you plan on dealing with *Deydras*?"

Simon smiled. "Your Majesty, I will sway Deydras. But in truth, it is what *you* must do."

"Which is?"

"You must treat it as a jest, your Majesty."

"A jest!" Edward's hands slammed down on the maps. "Do you not understand the seriousness of this claim? The people may turn on us! On Isabella, on me, on our children! What then?"

"It will not come to that," Simon said, "if you do as I say."

Edward spoke over him. "Already people had questioned if I am a changeling!"

Simon closed his eyes suddenly, and pressed a hand to his forehead. He let his eyes fly open and his gaze snap to the painted ceiling, as if he saw beyond. *Why, Simon, an imagination!* That hideous woman knew nothing of him.

"What is it?" Edward leaned forward. "What do you see?"

"*Before the end of this dreadful year,*" Simon intoned, "*end the lies of the man with one ear!*" He opened his eyes wide and lowered his voice. "Heed my words. If you yourself treat it as a jest, to be laughed at and scorned, so will others. You *must* do as I say. Was it not for my advice that you made me your councilor?"

Edward turned to stare out the window, his jaw tight.

"Your Majesty," Simon said firmly, "ask the queen what name she wishes for the child. You will know then that I see the future."

Edward sank back into his seat. He opened the parchment, scanned the name, and looked up. "I will ask her."

"And when she speaks the name I have written, you will give me a fleet?"

Edward nodded. "If she does not, your wheel of fortune may spin down as quickly as it spun up."

Simon reigned back the hatred that boiled up at the king's threat. He rose, smiling. "You are all-wise, your Majesty. Gramercy." He bowed low.

Glendochart, Scotland, 1318

"You are unsettled." The old abbot spoke as his garron ambled beside Niall's, alongside the River Fillan.

"Has Hugh been speaking to you?" Niall asked.

Dunod shook his gray tonsured head. "Nay. I see it in your face."

Niall gave an easy grin, thinking wryly that he was unconsciously doing exactly as Shawn would, making light of it. "'Tis an uneasy time."

"Sure an' ye've seen unsettling things."

Niall's turned away, his eyes on the gray winter waters rippling beside them. The image of the dead woman in Ireland flashed before his eyes, the dog gnawing on her leg, the dead priest on the altar. Though it was hearing Shawn's voice, the warning, that really unnerved him. "Who has not, these past twenty years?" he asked.

"There are those who have been more sheltered," the abbot replied. "Our bairns, our women we protect as well as we can." He stopped at the river's edge, near a chapel ruin, and slid off his horse. "Even I have seen little of the destruction and violence, and we have managed well enough, these years of famine." He patted his lean stomach, and said wryly, "Though we've none of us grown fat, now, have we?"

Niall, too, slid off his pony and looped its reins onto a low-hanging tree limb, beside the bishop's. "We've much to do."

"Aye." Dunod walked with him to the ruins. "'Tis a shame, sure, the very stones of God's house taken."

"Who took them?"

"The men in the hamlet—to build houses, stables."

In the crisp air, they walked the length of the crumbling wall. It rose from the height of his knees at one end, to reach above his head at the other, jutting against a sky hovering between pearl gray and pale blue in the early morning.

"The restoration," the bishop said, "should be modeled on this, St. Fillan's own. See, here was the nave, small though it was. And the confessional there."

"Bruce would like the new one to be ten fas." Niall ran his hand over the lower edge of a window in a wall still largely intact. "Imagine, fifteen, twenty such windows stretching the length of the church when it's done." For a moment, the chilly morning seemed to slip away. He felt the hope and excitement of a new day as he once had, as he almost saw Bruce's new church rising proud against a summer sky.

"A wee bit bigger, aye." Dunod beamed. "We're no so many now, but I understand Bruce foresees us growing."

"He does," Niall agreed, and looked up to the empty window. "Stained glass! 'Twill be a fair sight!"

"A great honor to our Fillan," the abbot said. "Our king is good."

"He does not think he is." Niall wandered to the far side of the ruins, envisioning the new transept.

"Ah, the Greyfriars incident gives him no rest?" Dunod asked.

"It does not," Niall said. "You met him just before Dalrigh. Sure you knew of Greyfriars?"

The old bishop sat down on a low wall that had once formed part of the confessional. "He did not kill Comyn. Another did."

"True," Niall said. "But he struck the blow, and on sacred ground. The Pope ex-communicated him. Yet you blessed him. Why?"

Dunod stared up to where a black bird perched on the chapel wall. "A story is never simple."

Niall gave a faint smile, thinking of Shawn and his time in the Scotland of centuries hence. "'True enough," he agreed.

"As I saw Bruce's men ride over the hill, that day, 'twas a sore ragged lot they were. Tired, hungry, desperate. Weary of soul. Yet I fancied I saw a light around them and I heard our Lord bid me listen."

Niall wandered to where the altar had stood, gazing up through the broken wall rising behind it. He turned back to Dunod. "Did he tell you his story?"

"He did." Dunod grinned. "A benefit of my calling, aye? We hear stories. Indeed we hear stories!" He chuckled softly before saying, "And *those* stories, I cannot repeat. But we both know Scotland sorely needed Bruce and Comyn to work together. They agreed do so."

"And instead Comyn betrayed him to Edward."

"A most gruesome death 'twould ha' been," Dunod said, "had he not been warned and fled. Think you, that Comyn would have done better for Scotland?"

"I think," Niall said, "his betrayal tells us we'd have had Edward's man on our throne. No, I don't believe Comyn had Scotland's interests at heart."

"Nor do I." Dunod lifted his gaze to the sky, brightening to the steel gray of a winter day, staring a moment into the past, before speaking. "None but the Bruce—and I—perhaps, will ever know what transpired at Greyfriars that day. What words were spoken." He stared at the trees above. "What threats, were made." He looked to Niall. "We simply do not know all we think we know."

"I imagine the Pope heard a very different version than you heard."

"Aye." Dunod smiled. "He'd have heard a version from Comyn's kin and Edward's supporters—none of whom were there." He rose from his seat, rubbing his hands together briskly. "What we choose to believe of others, without adequate evidence, has grave consequences, indeed. I gave him my blessing. Who else would free Scotland? But more—I saw his heart, as he rode over the hill. 'Twas pure."

"'Twas you warned him of the MacDougalls and MacNabs in the area?"

"It was." Coming up beside Niall, the bishop pointed to a door in the wall to the side of the altar. "There was the night passage from the cloisters. I've come here of a night and felt the holiness of those monks, of Fillan himself, still filling this place. As if their prayers blessed it for all time."

He walked several steps, trailing his hand along the half-wall. "Mayhap Bruce felt it, too, for I found him here, the night before they left, on his knees. It was then he begged my blessing. I gave it, my hands on his head."

Once again, he seemed to stare into the past. "What is it to know your wife and daughter are imprisoned—your child of nine years of age?" he asked. "To know the woman who rode through the night to crown you kings of Scots, lives in a cage, humiliated daily, for her faith in you?" Dunod lifted his gaze to the empty rose window, where the rising sun burst through. "He has suffered physical hardship, emotional devastation. *Yet he thanks the saints for assisting him.*"

"It seems his confidence was well placed in our St. Fillan," Niall remarked. He gazed around the ruins. "'Tis a thing to be envied," he added, "to live here in peace among these holy walls." He sat down on the stair to the altar. "I look forward to the day I am *building* instead of destroying."

Dunod turned to him, his eyebrows drawn together. "But you *are* building. You are building the future. You are building a new Scotland. You are building a land where your sons can live free."

Midwest America, Present

On the deck, Shawn ended the call with Amy, his heart heavy—*my heart is torn.* He'd meant it when he said he wanted Amy's happiness—and his heart ached for Christina's broken words. And there was James, his own flesh and blood, who kissed his cheek with gleeful abandon and laughed as he hit piano keys. And there was Niall and there was Simon. *Maybe* there was Simon.

His gaze traveled once more across the wide lawn below the deck. He was seeing the long, meandering stream of a pool—it would have to be deep at one end—very, very deep—and the walls. They would be medieval stone on the right, and gradually become glass, over there on the left, arching up into a glass ceiling overhead, letting in starlight.

He turned abruptly. There might be no reason for it. He charged back through the breakfast nook, across the great room and hall to his suite, to the oilskin. In moments, the parchments were spread out across his desk. He scanned the one that mattered.

Nothing had changed.

His heart pounding, he rolled them up, slid them carefully back inside, and returned to the deck, seeing it all. The space under the garage would become a living suite. His chest felt tight. The suite—the pool—everything he had in mind—depended on making that phone call. The one he'd been avoiding.

He could do it, he told himself. He pulled his phone from his pocket and stared at it. Hadn't he called record producers and publishers?

His guts twisted much the way they had as he'd stood after Skaithmuir, staring at a man's brains spilling out across bloody snow. *It's just a phone call,* he reprimanded himself. *To a nobody!* But his hand trembled.

Hadn't he called Ben, as a brash young 20-year-old, and glibly told the up and coming agent, already representing some big names, that he could make them both rich with his trombone and his music? Surely he could do this.

He jabbed angrily at the number. Of *course* he could call a nobody. It didn't commit him to anything. It bought him time, left him with the option—if he couldn't think of another way. He hit *dial.* It rang...rang...rang. And suddenly, *I'm unable to come to the phone,* spoke the once-familiar voice.

Memories assailed him, of laughter, of hate, of an unexpected swell of peace. The return of agitation. He could still hang up. *Leave a message.* Shawn hesitated. Amy would come around. *I'll get back to you.*

He gave his head a sharp shake. No, he *had* to have the option. He *had* to. It didn't obligate him. Voice mail beeped, waiting for his message.

He stabbed *end,* studying the deck, and the land stretching around it. There was the community center—but that would raise too many questions. And it didn't suit *all* his needs. He paced the deck, studying the land, the lawn where he'd held parties, the place he'd planned to put in a garden.

He had twenty acres. He had more money than he knew what to do with, and more pouring in by the hour. He lifted his phone, punching in a search term, and moments later dialed. "I need a job done," he said, and outlined it. "I want it done in a week. Whatever it takes. I need it done."

That other call—he'd get to that when the job was done. He had time.

Glasgow, Scotland, Present

I hurry through Glasgow's drizzle, my violin tucked up against my chest, safe under the umbrella. My thoughts swirl around Shawn—the Shawn I once believed I knew. I smile, thinking of his question, of his voice warm over the phone, telling me he wants my happiness. I find I want his in a deeper way than I've ever known before.

I'm on the other side now—I used to always be on the defensive, off balance. Would I be happy married to him? Yes. But I would always miss

Angus, too. But he's engaged. It's over. I can't see this going well for either of them. But it's what he's chosen and what she wanted.

I reach my car as the drizzle grows. My phone rings as I get in. I shake the umbrella once, snap it shut, toss it on the passenger seat, and slide my finger across the phone, under the unfamiliar number.

"Amy?"

"Clive?" I ask in surprise.

"Aye, 'tis me! Now I've a very interesting story for you!"

I listen, sitting in the parking spot with the rain spattering the windshield, as he hits me with a verbal explosion. I miss half the words, but he seems to be talking about a fiddler's rally. "Clive," I finally interrupt, "Why are you telling me this? It was last month. Are you saying I should take part next year?" The rain drops dance and bounce on the windshield, a frantic Bolero.

"Oh, aye," he says in a rush, but my gut tells me he thought no such thing. "You should! Angus has been busy, now."

My heart jumps at the sound of his name!

"He's at physical therapy every day and I'm just after driving him to an appointment and Mairi's given him a flute that he's taken an interest in."

I want to stop him, tell him it hurts to even hear it when he's with someone else. A car flies by, sending spray across my window.

"And it seems he and Claire are no longer engaged. He'd not speak on it; nor would she, but you see, maybe you'd like to look into that fiddler's rally. 'Tis a grand chance to play, and you played on a CD in America, aye? Now, sure, I'd love to buy that CD when it's out."

His words fly, a verbal volley, as the rain lightens. He gives me no chance to answer, finishing with, "It's grand talking to you, Amy! Grand! It's grand to hear all you've been up to. And I'm that sorry, I have to run, or I'll be late."

"Bye," I murmur, even as the dial tone hits my ear. A last triplet of raindrops bounces off the windshield, and a ray of sun smashes through the clouds, brightening the city. I smile, my head leaned back against the headrest. My stomach dances a trembling, tentative jig. Angus is not engaged after all.

London, 1318

Edward found Isabella at her desk, a quill idle in her hand as she gazed out the window to clouds drifting across a blue sky. A blanket of royal azure lay over the back of the divan. Edward picked it up, noting the damask and threads of gold woven through the blue.

"From my brother," Isabella said. "For the new child."

"Beautiful." Edward held it up to the light. "Woad, perhaps."

Isabella sighed. "Have we not weightier concerns? Lord Claverock...."

"It was of the child I wished to speak." Edward laid the blanket down.

"I don't trust Claverock," she said.

"It is a peculiar situation," Edward answered. "But he has outlined a fine campaign for re-taking Scotland." He glanced at her stomach, barely rounded. His heart hammered. But he had to know. "Have you thought on names?"

"Certainly this is not...."

"Please." He sat on the couch, taking the brilliant blue damask in his hands. "I should like to know what you have in mind."

"Joan," she said.

Edward felt his breath sigh out in relief. *Not Eleanor.* Simon's story didn't sit right with him, either, and Isabella had invariably proven herself astute. Yet Claverock's prophecies had unfolded exactly as he'd given them. He had shown weapons Edward believed would work. He realized he wanted evidence to prove Simon a fraud. And yet—though she named Joan, something was wrong with her one word. He studied her. "If it's a boy?"

"My heart tells me it's a girl." She pushed herself from her chair, a hand to her back, to stand at the window. A frown flitted across her delicate features. "No."

"No?"

She turned from the window, and crossed the room to join him. Her fingers closed around the blanket, kneading it. She rubbed it against her cheek, smiling. "No. As much as I like the name Joan, I wish to honor your mother." She smiled, taking her husband's hand and placing it on her stomach. The child kicked. "We will call her Eleanor."

Inverness, Scotland, Present

Angus sat alone in his front room. Why hadn't Brian mentioned Simon? The question had plagued him since his visit to Thomas Ritchie. But his days had been filled with physical therapy, helping Hamish with flute and chess, visits from Ken and Mairi and colleagues from the station and neighbors. Mrs. MacGonagle had come with dinner and taken the opportunity to chat. That fanciful Kathleen! The neighbors with their new baby!

For once, he had a moment's silence. His father had only just left him off after therapy and he'd ignored the ring of the doorbell, wanting nothing more than a moment, finally, to think through his conversation with Thomas Ritchie.

Waiting for the hearth to take off the winter chill, he tried to remember the last time he'd talked with Brian. After a moment, he dug in his pocket for his cell phone, pulled up his call history and scrolled back. It was nowhere to be seen. He tried to remember. He'd been up at Castle Dollar with Amy, James, and Carol. Late March. So long? He'd called Brian asking for information. As he recalled their conversation, another piece of the puzzle fell into place. Brian had talked about meeting Amy at Melrose.

And Simon had been living with him.

And Simon had shown up at Melrose.

An ugly fear curled and stretched inside Angus's stomach. There had been unusual deaths in Bannockburn and Inverness—both places Simon had been. He was jumping to conclusions, he told himself sternly.

He touched Brian's number. It rang twice before clicking. A tinny voice informed him the number had been disconnected. No further information.

Brian had wanted to tell him something. The memory grew in clarity. He'd wondered at the time why Brian didn't just tell him over the phone, but he'd been with Amy, thinking about the hike to Monadhliath. Fighting back rising suspicion, he hit Brian's work number. A woman answered. "Creagsmalan Archives. Can I help you?"

"Is Brian there?" He heard his own voice come out in a rush.

"Brian?" She sounded perplexed.

"The curator, Brian." Angus locked his eyes on the flames snapping in the hearth, trying to keep his voice calm. But the nightmare was raising its head.

"Oh, *Brian!*" Her voice dropped. "I'm sorry, he's gone these last months."

"Gone?" Angus asked. "Where?" Maybe Brian had only gone to give some talks in America or Australia. Maybe he'd taken a sabbatical.

The woman cleared her throat. "Perhaps I can help you?" she said. "I know the museum quite well. I've taken over the post of curator, since Brian's death."

London, 1318

"And now," Isabella said, "now I have told you my choice, let us speak of Claverock. What has he to say in your meetings? What of his campaign?"

Edward told her. The barons. The weapons.

"You seem ill at ease," Isabella remarked. "Is this not all good tidings?"

"But for such weapons in the hands of one we know to be both brutal and thirsty for power." He rose from the couch. "Now he wants command of John of Lorne's fleet. He says there's a land to be had across the ocean."

"You trust him no more than I." She watched him.

He shook his head. "I don't trust him, yet he has great weapons. I cannot alienate him."

Isabella's lips pursed. "Surely you and I will not be outwitted by such a brute." She turned to the blue sky, shining over snowy fields.

Edward came to her, a hand on her shoulder. He was about to speak, when she laid her hand over his and said, "You and I have both changed a great deal, these eleven years."

"You were but a child," Edward said softly.

"You have changed, too," she countered. "And that suggests to me an answer. Give Claverock the fleet, but attach conditions he cannot refuse."

"What conditions?"

She smiled. "Conditions which may soften his edges. And failing that— give us a close eye on him."

Midwest America, Present

Shawn stood in his breakfast room, watching the work in the new solarium below. On his right were walls of heavy stone, like those that built Glenmirril. They had risen practically overnight, surrounding the dry, tiled pool—twenty feet at its deepest. At the middle of the room, glass topped the stone walls, the panes becoming steadily longer until, on his left, the whole wall was glass, arching up into a glass ceiling. Exactly as he'd envisioned it. Two men laid slate decking while a team of electricians wired the sound system and lighting. A box of black iron sconces sat near the stone wall, waiting to be installed.

He smiled, imagining Amy in the finished space. He'd pushed the designers on acoustics. He wanted her to have music. He wanted it to be her oasis. She would need one, if he followed through on his plan. And if he didn't, there would be parties here, there would be....

The doorbell rang. Tearing his eyes from the emerging creation below, he passed back through the great room, into the foyer. He opened the door to a delivery man propping a tall, thin carton against the house. His eyes flickered over Shawn's trews and linen shirt. "Shawn Kleiner?"

"You need to ask?" Shawn raised one eyebrow as he reached for the stylus and signed his name.

"If I want to keep my job, yes." The man took his board back, hesitating for just a moment, as if about to speak.

"Ancient Egypt." Shawn reached for the box, confirming the sender's name. "I was directing the building of the pyramids."

"Well done," the man said. "I was just talking with Marco Polo about that." He turned and left.

Shawn grinned. Most people were a little more cowed by him. He hefted the box and took it back out to the deck, down the stairs into the solarium, and toward the glass wall, maneuvering it past the workmen. On his left, a door led into the new suite built under the garage.

There, he laid the box down and knelt, drawing his *sgian dubh* from his boot to slice it open. He stared at the contents, nestled in bubble wrap, and ran his fingers down the smooth wood of two slender bows. They were perfect— just what he'd ordered. Of course, he still had to try them out, make sure they were as good as he'd been promised.

His smile fled. Amy might yet come home with him. They could have a beautiful life together, with James, with more children. He didn't have to follow through, he reminded himself. He was only keeping his options open. Of course, there was still the matter of that phone call—and the promise to his mother. She wouldn't like his plan for fulfilling it.

"Medieval?"

Shawn looked up.

The project manager stood in the doorway with a clipboard. "They're beauties," he said.

"You shoot?" Shawn rose to his feet, reaching for the clipboard.

The man offered his hand. "Sir William de Clare, archer general of Calontir."

"Calontir?" Shawn gave a firm shake with his free hand. The name on the man's shirt said Drew.

"Society for Creative Anachronism," Drew explained. "I do medieval archery when I'm not building additions."

"You pretty good?" Shawn asked.

"The best." The man grinned. "That's not vanity talking, it's the competition scores."

"Isn't that damned lucky for both of us?" Shawn smiled, thinking of Niall. He would say it wasn't luck—especially not *damned* luck—but the holiest of help, dropping everything into place. "Do you have room in your schedule to make me even better than you?"

"Trombone, composing, arranging," the man said. "I'll make room in my schedule." His eyes flickered over Shawn's outfit. "But why is it so important to be that good at archery, on top of everything else you do?"

"I'm an eccentric multi-millionaire." Shawn gave a broad smile. "Does there need to be any other reason?" He glanced over the items on the sheet, half of them checked off, and looked up. "I want to start with two hours a day. I'll make it well worth your while, financially, to make me a priority."

With Drew's promise, he returned to his office up in his house, his mind skimming through his priorities. He had an arrangement to finish for Ben, and a quick trip north—he wanted to deliver the music to the church personally. He opened a browser on his computer, found the name of the music director, and dashed off an e-mail requesting an appointment.

When he returned, the shooting and swimming would begin in earnest.

Glendochart, Scotland, 1318

As the night sky lightened to gray, Niall led Hugh, Conal, Lachlan, and Owen into the chapel for their last mass before they set out for Melrose. Shawn had hated daily mass. He'd scratched his ears and stared slack-jawed at the women. They'd argued over it and Niall had made him go as Brother Andrew. *At least sully the name of a man who does not actually exist!* he'd raged.

But now he smiled at the memory, as he dropped to his knees beside Hugh, sinking into the soothing balm of incense. He missed Shawn. He missed the easy humor between them. He missed having a *brother.* He lowered his head, his brows touching his fingertips, as Abbot Dunod filed in with his acolytes behind him. Inept and oafish as Shawn may have been—he missed him.

And he envied him. And he wondered why he had imagined hearing his voice these past months.

He bowed his head lower still as incantations flowed over him, the prayers that had been the warp and woof of his daily life at Glenmirril. They wove a comforting cloak of peace, gentle as apple blossoms, on his shoulders. Dunod's haunting chant washed away his cares. He had envied Shawn? Shawn was surely having no such mornings of peace.

The stress of the last months melted away into the incense, floating away on the drifting melody. He felt refreshed, and ready to set out for Melrose. And a word came to him. *Trust.*

For just a moment, with Hugh, Conal, Lachlan, and Owen around him, it felt easy to trust, to forget the fears, for himself, Allene, his sons, that hounded his every step.

Cum transieris per aquas tecum ero et flumina non operient te cum ambulaveris in igne non conbureris.... The abbot's words floated through his heart. *When you pass through the waters, I will be with you.*

Beside him, Hugh rubbed his nose. *When you pass through the rivers, they will not sweep over you.* The rhythm of the Latin words washed over him, taking him back to the time when he'd always felt peace and joy and hope. Allene's time was near. He'd be home one day. Things would come aright. They always had.

Inverness, present

I wait in the lobby, two coffees by my side, wearing his favorite of my sweaters, in royal blue. My hair, brushed to a high sheen, hangs free—the way he likes it best. I even spent the morning with my face in a clay mask and manicuring my short nails. I feel myself glowing with the care—but even more with excitement at seeing him again.

Angus wheels out, his hair damp from therapy in the pool. He wears jeans. A black t-shirt stretches across his chest. His jacket is draped across his lap. His cheeks are ruddy and his hair long enough to show the curls he hated. I love them. The flutter grows in my stomach. I want to take his hand, I want him to look at me as he did the first night he said I love you. I want....

He glances at me. Color flushes his ruddy cheeks and he turns away.

It's what I expected. What I steeled myself for. Still, I had hoped.

Pain shoots through my heart. Tamping it down, I stand, offering a coffee. "Your father couldn't come. He asked me to pick you up."

He leans into the wheels of his chair, rolling through the sliding glass doors. At the curb, he pulls out his cell phone.

I stand in the lobby, watching his back, wrestling hurt and anger and knowing I must be patient. He struggles into his jack and sits under the awning, staring at the circular driveway, his back stiff, until a taxi pulls up. He slides himself from the chair into the back seat, and stares straight ahead, while the driver stows his wheelchair in the trunk...while I struggle with my hurt.

The taxi pulls away.

It's about what I expected. I know his pride. I know he thinks I came back out of pity. It's going to take more than this to show him that's not true. And slowly, the sun comes out again. I smile, glad to have seen him, and go to my car for the long drive home.

Glendochart, Scotland, 1318

"'Twas good to have daily Mass for a time," Lachlan said. The day had passed with few words as they pushed their mounts through snowy hills, and down the shore of Loch Lomond. "D' ye think Father chose his words for us?"

"His words?" Owen asked.

"The reading from the Epistle."

Niall glanced at the blue waters flashing sparks of light under the winter sun. *When you pass through the waters, I will be with you.* He had felt comfort in the words. He wondered if Lachlan was referring to his—*dislike*—he refused to even think the word fear—of water. Anger stirred in his heart. Except, he chided himself, Lachlan had never done such a thing. In truth, none at Glenmirril had ever acknowledged what they all knew.

He cleared his throat, striving for nonchalance, "What has water to do with us?"

"Water?" Conal asked. "I heard naught of water."

Niall glanced at him. "Isaiah."

Hugh's brow furrowed. "'Twas not Isaiah read in chapel this morning." He turned to Conal. "Was it Isaiah?"

"*When you pass through the rivers, they'll not sweep o'er you,*" Niall said.

Conal shook his head. "No, 'twas not what was read."

"I heard it clear as day," Niall insisted.

"Clear as a day covered in clouds." Hugh snorted, giving Niall a grin.

Niall pulled his cloak more tightly around himself. No clever response sprang to mind.

"'Twas Deuteronomy," Lachlan said. "*The Lord your God fights for you against your enemies to give you victory.*"

"No mention of water," Hugh guffawed. "Days of rest have addled your wits, Niall laddie!"

"I misheard," Niall muttered. "My mind drifted." He thought of Shawn, swimming in the river by night, and in the loch behind Glenirril, teaching Hugh to swim. He saw it in his mind, as he had so many times, the way his arms swung over his head. *Sing,* he had said. Niall pushed the thought from his mind. They would soon be in Melrose.

CHAPTER TWENTY

London, 1318

"I've good tidings!" In his throne room, Edward beamed at Simon.

Simon bowed, relief washing over him. Edward might yet turn on him. But these weren't the words of a suspicious man. He raised his head. "My Lord, what tidings are these?"

"I have decided you shall wed the daughter of John of Lorne."

"*Wed*, your highness?" Relief washed away in the memory of Brother Eamonn's words. *Your son shall be your undoing.* He'd had no desire to wed, even before the old monk's words.

"Indeed." Edward beamed. "John commanded my fleets in the west. They are Alice's dowry."

"Your highness...." Simon faltered. He couldn't refuse a king's command. "I know not what to say." He would simply ignore the girl, he decided.

"I expect her to be treated well." The smile slipped from his Edward's face. "I wish to hear no complaints of her being neglected."

Annoyance crept up in Simon. "I will often be away on your business," he reminded the king. "In truth, I've little time for a wife."

"Apart from her dowry—the ships you desire—she is intelligent, and comely," Edward replied. "She shall make an excellent wife in every way." His eyes became hard. "I'm sure you'll find time."

Simon gave a cursory bow. "I'm sure I shall."

Edward smiled. "Her guardian has agreed. They are waiting in the chapel."

"Now?"

"Yes, now. John was loyal to me." Edward rose, gesturing toward the door. "I can't leave his daughter without a husband, and your aspirations are a fine match with her family. She has lived her life around ships. You will, in fact, find her knowledge a great asset."

"Your Highness, it seems abrupt."

Edward's smile faded, and for a moment, Simon saw Longshanks looking out through his eyes. "*You* want ships. *I* want Alice well cared for."

"Yes, your highness." Simon bowed. "Gramercy, your Majesty."

Midwest America, Present

Behind his desk, Dennis Martin, the music director, studied Shawn's scores —a full set of the psalms, many set to the very music Shawn had played for the monks of Monadhliath in 1315. Shawn smiled. They were as authentic as they could get. Those he hadn't played in his month at the monastery, he'd arranged based on his own research into extant manuscripts or, when he couldn't find original music, copying the style of that which he had played.

He glanced around Martin's office. It's dark wood paneling reminded him of Brother William. He held back a laugh, remembering Niall's praise of the man and his own sarcastic response. Still, he thought, Brother William *had* been a good man. His mind drifted back to Eamonn's story of Simon Beaumont at the monastery, and his alarm over Brother William's fate.

Dennis, a more than middle-aged man in a gray sweater over a white shirt, looked up from his perusal, and lowered small glasses off his nose, setting them on the desk. "Beautiful," he said. "The plain chant, everything. I'm no expert— which is exactly the problem—but I know enough to be sure this is exactly what we wanted."

"It's as authentic as you can get," Shawn assured him.

Dennis's eyes flickered over Shawn's trews and bell-sleeved shirt. "You'll let us use this royalty free? We can of course offer a stipend...."

Shawn waved him off. "It's something I can give." On some level, he thought, it wasn't *giving* so much as a plea bargain: *Hey, God, I gave them music! Now will you save Niall and give me Amy?* "Don't worry about it," he said to Dennis.

The man looked up, past Shawn's shoulder, and jumped from his seat. "Pastor Justin! This is amazing! He's given this to us to use!"

Shawn turned to greet Pastor Justin. He was young, barely older than Shawn. He studied Shawn quizzically, his eyes traveling over his medieval clothing.

"Pastor Justin," Dennis enthused, "I'd like you to meet Shawn Kleiner!"

Shawn offered his hand.

Pastor Justin's hands stayed by his side.

Dennis's smile fell. "Is something wrong?"

"He's giving us this music?" Justin didn't look at Shawn.

Dennis nodded.

"Yes, I'm giving it to you," Shawn said.

"Where did he get it from?" the pastor asked.

As Dennis started to speak, Shawn interrupted. "I arranged it based on original sources."

"Do you know who this is?" Pastor Justin asked Dennis.

"Hey," Shawn said sharply, "do you think you could quit talking about me like I'm a piece of furniture? I'm right here."

"What's wrong?" Dennis persisted.

"You've never heard of Shawn Kleiner?" Justin demanded. "Drinker, gambler, philanderer. He makes Hugh Hefner look like an altar boy."

"They're properly called acolytes," Shawn corrected, his annoyance turning to anger. "And thank you, but that's all long past. You're about seven hundred years behind the news. And the last person to ignore me like this was a jackass in Glenmirril's stables."

Justin glared at him. "Are you calling me....?"

"Technically, it was a garron," Shawn amended. "So, no. What does it have to do with the music, anyway?" He gestured at the manuscripts on Dennis's desk. "Everything you wanted, right there!"

"What does it have to do with the music?" Justin repeated. "*Everything*. Your antics have been all over social media." He turned to Dennis. "You didn't hear about that talk show?"

Shawn felt heat climb up his face, infused with anger. Niall was going to die and this idiot....

Justin turned back to him, "Everyone knows who you are."

"Was," Shawn said. "Doesn't change...?"

Justin waved his hand, dismissing it. "What does it say to my congregation if I accept this music from you?"

"Um, that you have good financial and artistic sense?" Shawn snapped. "Despite behaving like a *garron.*"

"It says I approve of your behavior!" Justin corrected.

"*Past* behavior," Shawn corrected in return. "And nobody's asking you to approve anything!"

"Please! *Please!*" Dennis hurried around his desk, waving his hands as if to ward off a fight.

Justin snatched the papers up off Dennis's desk. "I can forgive you..."

"Forgive me!" Shawn scoffed. "Who are you to 'forgive' what wasn't done to you? The people I actually hurt—you know, *they've* forgiven me."

"But I *cannot* do business with you," Justin finished as if Shawn hadn't spoken, the parts gripped in his fist.

"Business?" Shawn demanded. "I'm *giving* you the damn things so you can have a happy fucking merry Easter!"

"This is exactly what I'm talking about!" The pastor thrust the papers angrily in Shawn's face. "Take them and go!"

"Now let's all calm down!" Dennis fluttered.

Shawn's hands went to his hips, thinking of Niall, and Christina. She would be horrified, not only by the language, but even more by linking it to the Resurrection. "I'm sorry," he snapped, staring at the floor. "I shouldn't have used that word. I'm just saying, I'm *giving* it to you. It's not even business."

Justin drew a breath, glaring for a moment out the door. "I can't appear to condone your behavior," he said, with less force.

"Pastor Justin," Dennis pleaded, "this is exactly what we prayed for! He's apologized, he's changed, he's trying to make amends!"

"Don't say where you got it." Shawn felt his own anger descrescendoing.

Justin pushed the parts at him. "I want nothing to do with you. Take them and go."

"No." Shawn shook his head. "No, *you* take the parts and do with them what you will." He tossed the pages on Dennis's desk and walked out the door.

London, 1318

The girl waited, silently, at the side of the priest, no more than seventeen, Simon guessed. A man of roughly Simon's own age stood at her side. He touched her back and murmured to her. She lifted her eyes once, a hasty glance at Simon, dipped a low curtsy, and dropped her gaze again.

She was as fair as Edward had promised. Hair the color of honey fell down her back in thick waves. Her eyes, in that fleeting glimpse, had been a bright blue. Her dusty rose gown made her skin pale, almost luminescent, giving her a fragile quality. A gold belt emphasized a slender waist.

Having a wife might not be bad after all, he decided. He exchanged small bows with her guardian, and the priest began. Promises were made. Her name was Alice. Edward handed him a ring, which he slipped onto her finger. Her voice was soft as she spoke her promise, in turn.

Simon wondered, for a moment, when she had been informed of this arrangement and if her initial reaction had been like his own. Surely, though, she'd been told of the fine home she would have at Claverock, of his renown as a seer, and his reputation on the battlefield. She was in awe, certainly, afraid to lift her eyes to such a man.

The promises concluded, her guardian grasped Simon's hand.

Alice stared at the ring on her finger.

"As soon as Lent is over," Edward promised, "we shall have the wedding and the feast," and the men clasped hands.

Simon hesitated. He could now rightly take the girl to his rooms. Her beauty and the old monk's words warred against one another in his heart. *Your son will be your undoing.*

Edward laid a hand on his shoulder, as he addressed Alice. "He will be a most attentive and loving husband." He squeezed Simon's shoulder, a firm reminder of his previous words. "After dinner, your chambers await. Alice will join you."

"Yes, Sire." Simon offered his hand. When she laid her thin, pale hand in his, his lifted it to his lips, kissing it, and smiled. She *was* beautiful. And there were ways to make sure a child wasn't born. Maybe a wife would be pleasant, after all. He would watch carefully for pregnancy. He smiled, a small, tight smile. The old monk was a fool.

Inverness, Scotland, Present

Angus drew in a deep breath and let it out slowly into the flute. The deep breathing, the gentle sound, all soothed his nerves while he waited for Clive, and the upcoming talk with Faith MacKenzie at the Creagsmalan Archives. It was hard to say if that or Amy's unexpected appearance had him the more rattled.

He'd been shocked to see her. And happy. He'd wanted to drink in the hair he loved and her cobalt eyes, the shape of her cheekbones, her ears. He'd wanted to hear her laugh.

He played a second long tone. He'd been happy—and angry. He'd reacted without thought, calling the taxi.

He drew breath and let out a higher tone. It was easier than bagpipes—and likely easier on the neighbors' ears. He could play it in his chair. He went up the B flat scale, slow with all those fingers moving between C and D. He tried the two notes a few more times.

Mairi might be transparent—Hamish still had no flute—but she was also wise. He loved bagpipes, aye. But why had he not thought, himself, that it wasn't strictly about any one instrument, so much as the need to make music at all, to breathe in deep and sigh it out into a flutter of notes dancing in the air around him, creating from nothing what hadn't existed before—and seeing people smile in response. He *could* still do that. He wanted to tell....

He lowered the flute to his lap. No, he *didn't* want to tell Amy. He didn't want to *see* Amy. Not like this, not in a wheelchair. Maybe if he were walking again. Maybe if—

No! *No*, he told himself sternly. She had chosen Shawn and really, that's where she belonged. That's where *James* belonged.

He glanced at the clock and started taking the flute apart. He'd barely clicked the case shut when the bell rang and the door opened. "Cheers, Angus!" Clive stuck his head in the front room. "You ready to go then?"

They were soon in Clive's car, Angus swallowing shame at the help he needed to slide from his chair into the vehicle, and setting out on the long drive to Oban. Clive chatted about work, weather, the upcoming music festival. As they drove past Laggan, he talked about a nephew entering his first Highland games, and at the first sign for Glencoe, a play at Eden Court. "I'll be picking you up for that Thursday. Sure an' you've no plans, aye?"

Angus sighed.

"You *do* have plans then?" Clive glanced over.

"I do not," Angus clarified. "And you surely knew that."

"Then what's the sigh for, Mate?"

"Seeing as you're doing me a favor," Angus replied, "'twould be courtesy to not mention it, but we've been mates a good long time now."

Clive grinned. "So we can dispense with courtesy, aye? I'll still buy you a pint on the way home."

"I suspect I'll need it," Angus said. But he smiled at Clive's humor, less irritated than he'd been the moment before. "Amy showed up to bring me home from therapy."

"She did now?" Clive guided the car onto the Ballachulish Bridge, stretching high across the narrows of Loch Leven.

"You'd not make it as an actor," Angus said. "Now I know Amy, and she'd not be showing up thinking I'm engaged. So it seems someone—and that would likely be someone in this car—has had a chat with her."

Clive glanced over his shoulder, and at Angus. "You told her, then?"

Angus snorted. "I did not."

"Your mam? Your da? Ken? Mai...."

"Today's perhaps the first day ever you've *not* brought up Amy," Angus interrupted.

"Such a suspicious mind!" Clive shook his head, grinning. "She took you home, then? Did you make her dinner?"

Angus said nothing. He wasn't especially proud of ignoring her. But then, he wasn't especially happy about her showing up, either. And he couldn't encourage her when James had to be with Shawn. He leaned back against the seat, staring at the waters of Loch Linnhe, rolling along now on their right.

"Were you glad to see her?" Clive asked eventually.

"I was not," Angus snapped. He held back adding, *You'd no business telling her!* Clive was, after all, taking his entire day to drive him to Oban.

Besides, he'd just told perhaps the first lie of his life. His heart had leapt at sight of her. He missed her terribly.

Glenmirril, 1318

The pains came suddenly, a sharp ache in the lower back, like vice grips crushing the spine. Allene rose from the divan in her solar, her hand to her back, drawing a deep breath.

Christina looked up from her easel.

"My lady?" Margaret laid down her sewing.

Allene nodded. "It is time." A laugh came over her—she would see her long-awaited child at last!—mixed with the ever present fear. Would she live through childbirth? They took Shawn's advice. Everything was boiled and well-cleaned. It gave her a better than average chance and many women gave birth to a dozen or more children.

The pain crushed her spine again, and she gasped, eyes wide.

James looked up from where he played on the fur rug, with the horses and soldiers her father had carved. "*Mather?*" His eyebrows furrowed. He looked so like Niall, her heart almost stopped with missing him, wanting him, desperate to see him again.

Christina was at her side, ushering her to her chambers as Margaret scooped James up off the floor and Bessie ran from the room, shouting for the Laird and the midwife.

"Tell them to boil everything," Allene gasped. "Shawn said."

"Aye." Pain flashed across Christina's face. "We'll boil everything."

"If I dinna live," Allene whispered, "tell Niall...."

"You'll live," Christina said firmly, helping her into her bed. She squeezed her hand. "Ye'll soon have a braw laddie or bonnie lassie! A deep breath now."

Canterbury, south of England, 1318

"No." Roger slid into a dark corner booth, across from MacDougall, in the tavern under their lodgings. "I have not found him." He nodded at the men at the next table. "They have not found him. My lord, Canterbury is a big place. We are hunting for a lad among thousands, in a city with a constant stream of pilgrims swelling its numbers."

"He poisoned Pelops." Alexander glared over his dinner. "He tried to poison me."

"We don't *know* this kitchen lad did either of those things," Roger argued. "My lord, have we not bigger concerns? We have scoured the city for days. Even if he did these things, even if he's here, we could be one one side of the cathedral while he's on another, and never see him."

He lifted his hand and a tavern maid appeared with a thick wooden goblet and a jug of ale. "Meat pie," he said, not looking at her, and to Alexander, "My lord, we've described him to every vendor, every merchant, every priest, every goodwife. Every waif running in the streets!"

He watched the tavern maid poured his ale. When she left, he said, "No one has seen such a lad. We've found a dozen who have led us to a Jackin, and none are the man you seek."

"We were attacked in the forest, my horse was poisoned...."

"Pelops may have simply had an infection as the wise woman said."

"My own life has been threatened."

"We don't *know* the wine was poisoned."

"Are we to do naught?" Alexander jabbed a piece of bread angrily into his meat pie, scooping up rich dark sauce, and pushed it into his mouth.

"My lord," Roger said quietly, "let us return to London and wait on the lands from Edward. Bring your lady to join you. Enjoy life a wee bit."

MacDougall swallowed his bread. "Am I to live forever at Edward's whims and mercy?" He glanced around the tavern. "Even now there is question whether he or Powderham is the real son of Longshanks."

Roger gave a sharp shake of his head. "My lord, pay no heed to rumors."

"Niall killed my son!" MacDougall hissed. "No, I will not *enjoy* life until I have land—which Claverock has promised me—and until Niall is dead."

Roger stared into his ale.

"We will follow through with the plans we made with Lord Claverock," MacDougall took a long drink of cool ale, slamming the goblet to the scarred wooden table. "Was the lady Allene not with child?"

Roger nodded, looking none too happy. "Her time is soon."

Alexander smiled. "She will be back to her visits to the poor outside the castle walls. It is time to head north, Roger." He looked to his men, at the nearby table. "Eat your fill and get rest," he told them. "We leave at dawn."

Midwest America, Present

Shawn sat in his Jaguar in the snowy driveway of his mother's house—the house he had bought her. Her Great Dane, the one he'd given her, barked at the front window. It wouldn't be long before she looked out and saw him there.

He wished he'd driven home. But it was too far. He wished he'd taken a room at a hotel. But she would ask questions if he did. He was only grateful he hadn't told her where he was going.

He wasn't prone to blushing, but heat suffused his face at the recollection of Justin's words. He wondered what details Justin really knew, and was bothered to know he didn't like to think about some of those details, or the whole world knowing them. There was a time he would have laughed about it. It no longer seemed funny. They'd been private, he defended himself, and it was only Amy any of it concerned; only Amy to whom he owed apologies.

He stared at the manila envelope on the passenger seat, that had held the parts he'd worked on for weeks—parts left on Dennis's desk—parts rejected by this *ja—jaaaaa—jacka—*

He stopped his thoughts, thinking guiltily of Christina, of how he'd already referred—with language which would appall her, make her gasp, even—to Easter, *our Lord's resurrection,* as she would likely call it.

Our Lord. As if she knew Him. As if He knew her. For a moment, Shawn envied her certainty. For a moment, he remembered his sense, months earlier, of her in the car beside him. He reached his hand across the gear shift, and felt instantly the foolishness of the action. She wasn't there to take his hand. She wasn't there to talk to. He didn't have her serenity, her soft acceptance of life, to turn to. He would have to just cope with it himself.

The dog barked again at the window, a deep *woof,* its big head thrust forward, insistent.

Jerk! Shawn amended his choice of words. *Justin the Jerk.* But he wasn't sure even this milder epithet was entirely justified. Justin was right: He *had* influenced people. The clarinet player had followed in his footsteps. Brian had, too, and ended up dead. He'd hurt Dana. He'd hurt Celine. Shawn stared at his hands on the steering wheel, hating—*hating* with a passion that bordered on rage, on the unbridled violence of James Douglas on the battlefield—that he could not *deny* his behavior had impacted others.

Had *hurt* others.

Carol appeared beside the dog, peering out into the dark, and suddenly smiled, waving. He waved back, though he doubted she could see it in the dark. She disappeared from the window, and a moment later, stood in the light of the open front door. Shawn watched her, thinking about his plan for fulfilling his promise to her. He doubted she would approve. He wasn't sure, himself, it was fair to ask of anyone. He sighed, and climbed out of the car, preparing to meet his mother with a smile.

Glenmirril, 1318

Evening sun cast through the arched windows of Allene's bedchamber when Niall's mother stuck her head in the door, bearing a bundle in swaddling clothes. "He's been fed," she said. "And how do you fare?"

Allene rested against her bed, accepting the baby into her arms. She pulled back the blanket, admiring the face of her third son. Like James, he had thick red curls. His eyelids rested peacefully against his cheeks. He was perfect. She smiled, looking up at her mother-in-law. "I'm blessed."

The woman smiled. "Aye, that you are. Three braw laddies."

Allene gazed down at her son. Her mother-in-law had had seven sons, and lost all but Niall. Her heart clenched, dreading the same fate. But the Bruce would lead his country out of these dark days. And Niall was helping him— building a world where their sons would be safe. She had to believe it.

Midwest America, Present

"Are you okay?" Carol asked, as Shawn shrugged off his heavy jacket in her foyer. Cain woofed, nudging his hip.

"I'm great!" He beamed, swallowing the shame at Justin's disdain. He wanted a drink. He scratched the dog's ear, saying, "Hey, how are you? You been ignored?"

"Come here, Cain," Carol reprimanded, pulling his collar.

A pain stabbed Shawn's heart. The dog had been named for one of his father's favorite songwriters of the 80s. It wasn't exactly what he wanted to be reminded of at the moment. "Big dumb dog," he joked, swallowing the emotion. "Broken any vases lately?"

Cain woofed in response.

"He's calmed down a lot," Carol said.

"Yeah, that little operation will do that to a guy."

"Where have you been?" she asked.

"Out and about," he said cheerfully. "Proposing music here and there." He pushed past the dog, toward the kitchen. "Any good food here? You use the spices I told you about?"

"Saffron is pricey," she replied.

"And worth it." He found a pot of spaghetti steaming on the stove, filled his plate with noodles and meatballs, and seated himself at the table in her breakfast nook, where the windows looked out into the star-filled rural sky, across the dark lawn and into the woods beyond.

She served her own dish and joined him. Cain snorted, and flopped down on the floor at their feet, looking up at Shawn with sad, dark eyes.

"No food," Shawn told him.

The dog sighed, head on paws.

"You can have all you want," Shawn said to his mother.

"Thanks. Considering I made it." She smiled.

They ate in silence. Shawn's mind spun around her request about Clarence, and his struggle to fulfill it—because of who Clarence had been.

"It always bothers me," she said at last, "when you don't want to say where you've been. When you're so overly cheerful for no apparent reason."

Shawn set his fork down, the words coming out before he could stop them. "I was at a church, giving them music they wanted. The pastor refused to have anything to do with me. Or my music."

Carol's lips tightened. She touched his hand. "I'm sorry, Shawn. Why?"

"Why?" he exploded. Cain lifted his head, letting out a whimper. "Because of who I was! I know I screwed up. I know that. I know things have happened because of me that I can never fix."

His mother watched him, saying nothing.

He rushed to fill the silence. "But I could have done this. I could have given them something good. They don't have to mention my name."

Cain rose, sniffing his hand, and rubbed his head against his hip, with a soft whimper.

"Say something," Shawn said.

"Yes," Carol agreed. "They could have given you a chance to show you're a new man."

Shawn sighed, and sank back into his chair. "How is I feel you've just given me a lecture when all you've done is agree with me?"

She smiled, touching the back of his hand. "Because you knew it already."

"Karma." Shawn shoved a fork through his spaghetti.

"Or God has a sense of humor," she replied.

Shawn grunted. "You and Niall."

Carol frowned. "Shawn, I wish you wouldn't....Niall didn't...."

"Yeah, of course not," Shawn said. "All the same, he's on your side, and I guess between his non-existent self and God's supposed sense of humor—or karma as I call it—you and Clarence have a couple of allies."

Scotland's West Coast, Present

Faith MacKenzie was a polished woman of middle years. She wore a pencil skirt, white blouse, and low heels. A smooth bob of blonde hair tucked itself into a polite curl behind one ear. She held her hand out to Angus, as she opened the door to what had been Brian's office. "I'm so sorry to be the one to give you this news," she said, for the second time. She glanced around her office, and blinked sharply, with a quick intake of breath. "He was your mate. I'd not thought it might be hard to see...." She beckoned him back out. "Will we walk...." She stopped, flustered, trying to cover her glance at his wheelchair. "We could...go through the gardens, will we?"

"Aye." Angus wheeled alongside her numbly as she led him from the office, out to the castle courtyard Brian had loved and been so much a part of. Going to the gardens, Angus saw through hazy, numbed thoughts, wasn't much better. They had been as much a part of Brian as his office.

"He was your friend?" she asked, as they passed through the cool shade inside the gatehouse.

"Aye. We met years ago, volunteering together at Stirling. We've been mates ever since." Weeks and months had flown ever more quickly, though, with each passing year, Angus thought, and their communication had dwindled, such that he'd hardly noticed how long it had been since Brian had contacted him. He cleared his throat. He couldn't let emotion cloud his thoughts. He had a job to do, and finding out what had happened to Brian—though he already had a guess—would certainly be a more worthy remembrance to his friend than delving into sorrow. "What do you know of his disappearance?"

"Not much," she admitted. "Only that he simply didn't show up to work one day."

"Which day?"

"'Twas near the end of March. Let me think now." She pulled out a slender cell phone, touching its screen. "Here." She held it out to him, pointing.

It was the day after Simon had met Amy at Melrose.

"Was he not scheduled to work the day before?"

Faith shook her head. "No. He'd planned on working from home. When he didn't arrive, we made calls over the next few days. He didn't answer."

"Someone went to his home?" The bright afternoon sun hit them as they rounded the castle walls and emerged on the west side.

Faith seated herself carefully on a rock wall, overlooking the place where Angus had eaten a picnic lunch, months ago, with Amy and Brian. Now they were both gone. "Aye." She smoothed her skirt. "The police went. It was empty, no sign of violence, or forced entry."

"His car?"

"It was found abandoned on the side of the road."

Angus turned to the sparkling blue waters. "The road to Melrose?"

When she didn't answer, he turned back to her. She stared at him, her head tilted to one side. "How did you know?" she asked.

"I'm guessing he was with a man named Simon Beaumont." He cleared his throat. "They found his car. And...?"

She looked away, staring into the east, where dark clouds hovered on the horizon. Then she turned back to him. Her hand touched his shoulder softly. "Aye, they found him."

Angus squeezed his eyes shut, his shoulders huddled. He felt her hand slide down to cover his.

"I'm sorry," she said.

CHAPTER TWENTY-ONE

Midwest America, Present

Shawn hunched at a table in a local McDonald's, largely empty in the lull after lunch. It was a good place to grab a burger before the long drive to his own home. *It ain't exactly McDonald's,* he thought wryly. He'd once said it, complaining of the food, to Niall.

Of course not, Shawn could almost hear Niall snapping disparagingly in his Scottish brogue. *MacDonald's cooks are back at Glenmirril.* Shawn smiled into his Pepsi. Niall was going to die the worst possible death. And he was sitting here drinking Pepsi and watching snow come down.

Irritation gnawed at his guts—at Justin. *I can forgive you. Didn't you hear about that talk show?* The irritation veered back to Jen. *It wasn't your, uh, academic prowess you were known for.* He muttered her words under his breath, his face twisted in angry imitation. One of the employees at the counter glanced at him and looked away quickly when Shawn scowled back. Let Ben hear about that—intimidating the local teens working at McDonald's. Tessa, Caroline, Jen, Justin, Rob: all of them left him grating with irritation.

Had there been any point coming back? Amy might never come back. He'd hardly ever see his son with this schedule of touring. His own mother thought he was mentally unstable, along with five million three hundred thousand two hundred thirty eight YouTube viewers.

In the quiet of the empty McDonald's, the squeak of the door opening and stamping of feet sounded. Shawn looked up, and stared in disbelief.

Clarence hesitated in the doorway. Snow dusted the floor around his scuffed boots. He wore jeans faded at the knees, and a brown jacket with stains on cuffs and sleeves. His hair had grown. Shawn rose from the table. Their eyes met across the empty space. The boy at the counter looked from one to the other, as Clarence stood in the half open door.

Clarence spoke first. "I'll leave."

"No." Shawn fired the word like an arrow from a bow. He made himself say it more quietly. "No, you don't have to." He hadn't seen him since Christmas Eve in the prison. His stomach twisted on itself, not sure he wanted to now. But the coincidence hit him full force. "Sit down. I'll buy you lunch."

"I'm fine," Clarence said stiffly. "I can buy my own lunch."

Shawn tried not to let his eyes roam over the rough clothes and the hair growing out to the corkscrew curls Clarence hated. "Yeah, okay."

He tried to ignore the irony that he'd been trying to call Clarence for a week. Christina and Allene would call it God's hand. He called it damn funny. His eyebrows drew together at just how funny. And the thought flashed through his mind—maybe there *had* been plenty to come back for. To set things right. Besides, it was only in coming back that he had the knowledge to save Niall. And he needed help to do it.

But he couldn't talk about it here. "Get your food, and I'll give you a lift wherever you were going."

Clarence's shoulders tensed.

"Oh, come on," Shawn said, louder than he'd intended. "It's fifteen below. You're not planning on sitting here all day. Let me give you a ride." He strode to the counter, beckoning.

After Clarence ordered off the value menu and pushed an anemic handful of coins across the counter, they stared at one another. Shawn resisted the urge to yank out his credit card and order a full meal. "Cold day," he said, instead.

Clarence shrugged. "Well, it *is* winter in Minnesota. Not bad, considering."

"Lots of snow, though." Shawn wanted to kick himself. It was exactly the kind of small talk he would once have ridiculed, had he overheard someone else saying it. *Yeah, nine or ten inches. I hadn't noticed. Thanks for pointing it out,* he would have said.

"Yeah. More coming tonight." Clarence snatched up the bag with one value menu chicken sandwich as the teenage boy placed it on the counter.

Shawn jangled his keys, and they headed out into the snow globe world.

"Do you want to tell me why you're so determined to drive me somewhere?" Clarence asked.

"How about I tell you in the car?" Shawn hit *unlock* on his fob.

"When it's too late for me to run?" But he opened the passenger door.

They climbed in, as dim as a bear's cave inside, the windows covered by a thick layer of snow.

Nerves hummed up and down Shawn's arm, at being so near the boy. In his mind, he saw the corkscrew curls he'd had at the trial, the angry face sneering out at the world from beneath the black fringe of hair. His hands tightened on the steering wheel.

"I'd spend a lifetime saying I'm sorry." Clarence pulled on his seat belt. "I know in the end it doesn't bring him back."

Shawn turned the key in the engine, let the wipers clear the window, and without speaking, guided his Jaguar out of the parking lot.

"I'm committed to spending my life at least putting something good back in the world, and maybe there will be fewer people like me in it."

"Mm." Shawn grunted as the Jaguar slid like a shark onto the empty street. "My mom seems to think the world needs fewer people like your mom and her boyfriends."

"And fewer people like *their* parents, or whoever turned them into what they were," Clarence said. "Generous of you, considering, but the buck stops somewhere. I could have walked away. I didn't. But maybe I can do at least a little of the good your dad would have done."

Shawn guided the car through the suburban streets. "First order of business is a job. How are you doing on that?" He glanced over.

Clarence stared out the window at the passing traffic and stores, his jaw tight. "Why are you doing this?" he asked. "I can walk, I can take a bus, I can call other people if I really need a ride."

People, Shawn thought, who his mother worried would drag him back the wrong way. He answered with his own question. "Why did you get in, then?"

"Partly curiosity. Mostly, though, I figure I owe it to you to listen to whatever you have to say."

Shawn eased into the right lane, pulling into Applebee's. He killed the engine. "Come on. I'm buying you a real lunch." Clarence sighed, but didn't object.

Shawn didn't broach the topic until their meals had come. "My mom's worried about your future."

Clarence nodded. "Yep. Not everyone's racing to hire a convicted killer." The anger that had been on his face gave way, revealing the despair behind it.

"What are your plans, then?" Shawn asked.

"Continue the ministry I started. Build it up. Live with friends till I get on my feet."

Shawn stared into his quesadillas a moment before looking up. "These friends of yours. Were they in prison?"

"Come on, Shawn," Clarence said, "who else have I known since I was seventeen? It's not like the cheerleaders from good old MGHS are pounding down my door."

"They've gotten jobs?" Shawn ignored the sarcasm. "They're able to move ahead with that in their background?"

Clarence pushed a nacho through artichoke dip, making no move to eat it. Finally, he shrugged. "Some of them. Some aren't doing so well and will end up right back there."

"I've found something since coming back." Shawn sought the best words. "People aren't always willing to accept me as I am. They hate me for what I was and won't forget that. Or they *like* what I was and hate me for being different now."

Clarence laughed harshly. "I don't think anyone is going to hate me for being different. At least you were fun. For some people."

"Too much fun." Shawn sipped his Pepsi. He thought of Brian, his high school friend killed while driving drunk. It wasn't fair to Clarence, he thought, hiding his own misdeeds. They were on more equal footing than Clarence knew, and Shawn couldn't bring himself to tell the whole truth. But then, it wasn't quite the same, either.

"Where are you going with this?" Clarence gulped the nacho chip. He closed his eyes, a smile flashing across his face. "I've missed this." He picked up another one, scooping up dip.

Shawn studied him. A tremor of irritation washed through him. His father wouldn't be having artichoke dip again. But he'd promised his mother. *Anything you want,* he'd said. God—Niall's God—seemed determined he keep that promise. And he'd made his decision and brought Clarence here for a reason. "My mom wants me to give you a job."

"No." The chip in Clarence's hand slapped down onto his plate. "I don't want handouts. Worse yet, from you and your mother."

"Forget handouts," Shawn said. "What if there's something I really need?"

Clarence narrowed his eyes. "Such as?"

"Trust, for one," Shawn said. "There's something I'll pay you—and pay you very well—to do for a few months at least. Something you'll even enjoy, I think. There's a suite ready for you, fully furnished, included in the job. I want you to come down to my house, give it a try. A week or two. Leave if you don't like it."

"What is it?" Clarence asked again.

"No questions. A home and a job. You can walk away any time."

"I'm not doing anything illegal or immoral," Clarence said.

"It is neither," Shawn assured him. "It is for the most moral, noble purpose there is. Are you in?"

"I need to think about it."

Shawn's heart sank. He needed Clarence, he realized. He *needed* him.

Melrose, Scotland, 1318

As his garron trotted along, Niall lifted grateful eyes to the town of Melrose, floating in a sea of evening mist. *When you pass through the waters, I shall be with you.* The words filled his mind at the sight. It should disturb him, he thought, that no one else had heard that reading. Yet he felt peace as they rode into the town, hooves clopping on cobblestone.

Before nightfall, they were seated with William, abbot of Melrose, at his head table, warmed by the blazing fires, with sweetmeats and roasts and sauces streaming in from the kitchens, the likes of which they had not tasted since leaving Glenmirril. "Best partridge pie there is," William boasted, as a man set it on the table before him.

Niall bit back a smile. Had it been partridge pie he'd ordered at the Two-Eyed Traitor with Amy? Or woodcock? The shock on her face amused him.

"You like partridge pie, then?" the abbot asked.

"'Tis among my favorites," Niall assured him. A boy set a trencher before him, filled with boar meat.

"As is fresh boar," Hugh said. "He's particularly fond of fresh boar."

Niall laughed. "So I am."

"Now funny you mention boar," the abbot said. "Ye ken we're just back from journeys ourselves—I and brothers Rowan and Benedict." He gestured to two monks on his left. "We've just come from Holm Coultram."

"Have you?" Niall's smile fled. He exchanged a glance with Hugh.

"Had you not heard the good abbot Robert died?" William asked.

Niall nodded. "We did. You were there for the election of the new abbot?"

Abbot William nodded. "Quite the stramash." He took a drink of wine, before explaining. "We'd set out from the abbey but hours before, when we were met by Alexander MacDougall's men coming the other way, asking had we seen a horde of bandits and warning us."

Hugh leaned forward, speaking across Niall. "Had you?"

William shook his head. "We had not. Nor did we later. Though it gave us a fearful trip, for the story was, there was a boar of immense size in Inglewood, and while MacDougall and Lord Claverock hunted it, they were attacked from all sides."

"A strange tale, Claverock," Conal said. "He disappeared at Bannockburn, they say and has now turned up again." He glanced at Niall, who kept his eyes on the abbot.

"So he did." William dug his knife into the partridge pie. Steam escaped its crust, wafting out rich scents of meat. "Moreover, he claims a gift of prophesy."

"Has he foretold aught to prove such a claim?" Hugh asked.

"Edward's queen is with child." William lifted eyebrows. "And Claverock knew before Edward himself."

"How do you suppose he knew?" Niall asked.

"Dark arts?" William shook his head with a sigh. "'Tis evil times."

Mid-evil. Niall thought of Shawn's word for his time.

"And if the *Butcher of Berwick* sees the future," William added, "sure 'tis not the voice of *God* speaking to a man who would murder a woman in the very act of birth."

Down the table, Owen bowed his head and crossed himself.

Brother Rowan leaned forward, talking across the abbot. "He predicted, as well, the arrival of a man called Deydras, with one ear, claiming to be the true King of England."

"Where did you hear these tidings?" Hugh asked.

"We stayed with the bishop of Linstock but several nights hence."

"What had they to say of the bandits in Inglewood?" Conal asked. "Were any killed?"

"Three of MacDougall's men, and several injured, Claverock among them. He made a dramatic recovery, however, and had left for London with Edward."

Niall and Hugh once more exchanged glances.

"But he *was* with MacDougall?" Niall clarified, and at William's nod, asked, "For what reason?"

"That I could not say." William gestured to a boy, who poured wine into Niall's goblet. "MacDougall seems not overly fond of Lord Claverock."

"Seems?" Conal asked. "You met with him, then?"

"Aye." William nodded heartily as he gulped a mouthful of the partridge pie. "He's staying on at Linstock."

Niall's heart hammered. The abbot spoke in the present tense. "He himself was not injured in the attack, then?" He lifted the freshly filled goblet, trying to drink slowly, trying to remember what he—a man who supposedly hadn't been there—should or shouldn't not know.

William shook his head vigorously. "He was not. Though his horse collapsed the next morning as he prepared to set out after the bandits."

Niall lowered the goblet, waiting.

"By the time we arrived, the beast was much better, though still lame. MacDougall now...."

Niall's breath stopped. Hugh leaned forward.

"MacDougal now, he's a different story."

Conal set down his chalice.

Lachlan and Owen stared down the table, waiting.

"He's in a fine fettle. In a fury, lashing out at all the staff...."

Niall felt his limbs go cold. MacDougall was alive.

Midwest America, Present

"Decision time." In his Jaguar, in the lot of a run-down apartment complex, Shawn stared at Clarence. "I can see the clouds of meth rolling out half the windows. I can smell the pot. I bet cat pee is the nicest thing that's been on the couch you're sleeping on."

Clarence stared out the front window. Dark shame flushed his face.

Realization hit Shawn. His brash ways had gotten him far. But maybe now was not the time. "That wasn't a criticism. I didn't mean...."

"Yeah, forget it." Clarence's hand went to the door handle.

Shawn hit the lock button. "My mom thinks the world of you. She really does."

"Why are you really offering me this?" Clarence asked.

"I told you I need something."

"You're not going to kill me in my bed?"

Shawn snorted and rolled his eyes. "Ah, for Go...." Christina would hate to hear him take her Lord's name in vain. "No!" He scowled at Clarence. "Don't you think if I wanted to kill you I could have just driven you somewhere and done that by now?"

Clarence shrugged. The color receded a bit in his cheeks.

"You have a phone. Tell my mother you're with me. Believe me, if you go missing, she's not going to cover for me."

Clarence said nothing.

"I'll buy a new lock for your suite when we get there," Shawn said. "Change it yourself so you know I don't have an extra key."

Still Clarence stared straight ahead, silent. From a second floor apartment came the sound of a woman shouting words that almost made even Shawn blush.

"You're leading this ministry," Shawn said. "You're all into God and all. Have you prayed about it?"

"No," Clarence admitted.

"Fifteen minutes." Shawn hit the unlock button. "My hubcaps and probably a tire or two are likely to be stolen by then. Use the time to get your stuff and get back out here. Or don't. In fifteen minutes, I leave and I won't bother you again."

Clarence opened the door.

"I'm offering a chance!" Shawn spoke quickly, aware he was giving away his desperation. "In good faith. I need your help and I mean it for the good of everyone."

"Yeah." Clarence slid out into the swirling snow.

Shawn watched him disappear through it, into the apartment building. He set the timer on his watch and leaned back, staring at the ceiling of the car, as the snow gently covered his windshield. Maybe he should try praying, too. *Come on, God,* he thought. *Tell him not to be an eejit.* Minutes passed. He drummed his fingers on the wheel, flicked the wipers to clear the window and watched the unmoving door. His mouth tightened. *Come on, God, you know I'm right about this!* The door remained closed while the woman continued her tirade on the second floor.

Melrose, Scotland, 1318

William turned to Rowan. "What was milord MacDougall on about?"

"A baker. Was it Hob?"

"No, no!" Abbot William waved a hand. "Hob was the head cook. 'Twas a lad. Jocelin?"

"Jacob." Benedict spoke decisively. "'Twas a youth called Jacob. Hob and all the cooks were quite insistent on his name, now, and the tall lad—he said Jacob had come up from Canterbury, and was going back anon."

"Jacob? Canterbury?" Lachlan asked incredulously.

Niall glanced at him with a small shake of his head.

"Has MacDougall plans to follow the youth to Canterbury?" Hugh asked.

William took a drink of his wine, before saying, "I suspect so. He blames the lad for the horse, you see, and was in a rage over the wine. He threw out a great quantity and insisted on testers for all the remaining batches."

"Wine, horses." Niall feigned ignorance. "Why should a lame horse send him into a rage over wine?"

"The boy brought him wine, see," Benedict explained. "MacDougall believed 'twas poisoned."

"Did any of the testers fall ill?" Lachlan asked from down the table.

Niall heard the urgency in his voice. He glanced at him, giving another shake of his head.

"Nay, not a one," Rowan answered. "He'd thrown out the jug the lad brought. The servants believe 'twas but another of his rages."

Niall felt his shoulders relax. He resisted the urge to look again to Lachlan, but could guess his friend felt the same relief. He got through the rest of the meal with feigned smiles, making conversation on things for which he cared naught. His thoughts stuck on MacDougall through the courses, through a juggler brought in for the feast, a fire-eater, a traveling minstrel who sang of the victory at Bannockburn. A lithe young woman played the lute, her hair spilling in thick black tresses to her waist. His mind turned to Amy, to her son James, his own sons and Allene, who must certainly have been delivered of child by now, to Christina, and the fear of MacDougall on his way to Glenmirril to do them harm. It was consolation MacDougall was away in Canterbury.

It was with relief they were finally able to leave. They were barely in their room, the five of them in a dormitory, before Niall exploded. "We have to find him! We have to stop him!"

"We haven't time," Hugh said sternly. "The Bruce expects us back anon."

"He'll kill Allene!" Niall all but shouted.

Conal glanced at the door, a finger flying to his lips. "Hush, Niall! No one must know we were there!"

"We must follow him to Canterbury."

Hugh laid a giant hand on his shoulder. "We will not, Niall. 'Twould be suicide to go so far into England."

"They'd not expect...."

"And we'll not disobey our king's orders," Hugh finished.

Niall sank down on one of the narrow cots.

Hugh dropped to another, facing him. "We will have faith, Niall. When you pass through the waters, will He not be with you?"

Niall lifted his head. "You said this was not what was read."

"'Twas what you heard. And that, then, shall be your answer. Have faith He is with us."

Niall raised his eyes to the timbered dormitory ceiling.

"And with Allene and your bairns," Hugh added sternly. "We leave on the morrow for Berwick. Mayhap we'll have news of Allene by then."

Midwest America, Present

Shawn glanced at the timer on his watch. *9:12.* His heart sank, thinking of Niall. He would do it himself, he decided. But he *knew* MacDougall's plan. He *couldn't* do it himself. And he had been so sure this was right. How could God let him down?

Did you ask? The words flowed through his mind, a gentle swell in a lake.

You know I'm right! he thought angrily at the disembodied voice. He watched the door, thinking of Niall. The woman's shouts had been replaced by a child crying and a man yelling to *shut the fuck up.* How could Clarence possibly choose this, and a filthy couch, and no job, over all he was offering?

Pride. Don't take a man's pride.

Shawn closed his eyes. It was easy to hear the Laird saying that. He hadn't meant to. It was just the way he was, he defended himself. He looked at his watch. *11:28.* He hit the wipers, clearing snow from the windshield. The door to the building remained closed. What more could he do? He had promised to leave Clarence alone. He couldn't go barging in and begging. Wouldn't *that* be treating Clarence like less than a man? He'd given his terms.

But he *needed* Clarence's help. He stared at the building. In truth, he realized, it *hurt* to think of Clarence trapped in that awful place. And no...he hadn't *asked.* He had demanded. Would Niall die for his arrogance?

He dropped his head to the steering wheel. *Okay, look, I'm sorry for demanding. Please. Can you please convince him I mean him no harm?*

He hadn't been fully honest with Clarence, he thought. He was envisioning a world where Clarence would just agree to this. He had to have a real choice in the matter, didn't he? *Tell me what to do,* Shawn thought to Christina's God. *I mean to do the right thing. If I'm screwing it up—well, I'm trying. Doesn't that count for something?*

He stared at his watch. *13:52.* He watched the numbers tick away, his hopes sinking. *Christina—you're up there somewhere. You must be. Maybe you could tell God I mean it? You knew me. Maybe he'll listen if you ask?*

Something grazed the side of his car. He jolted from his seat—*the hubcaps!*—bursting out into the snow, knife in hand.

Clarence stood on the other side of the car, a small duffel bag slung over his shoulder. "No cat pee on your couch?" He glanced at the knife. "Do I have to let your mom know you're already pulling a knife on me?"

The adrenaline drained out of Shawn. He laughed. "Bad news. You're going to have to sleep on a real bed." He shoved the knife back in his boot. "Please don't tell my mother." He glanced up at the sky, thinking of Christina. A ray of sun dropped through the clouds and hit the hood of his Jaguar. He stared at it quizzically for just a moment before saying, "Get in."

Clarence tossed his duffel bag in the back seat, and the Jaguar rolled them smoothly away from the crying child and the smell of drugs and hopelessness.

Inverness, Scotland, Present

Amy's heart pounded as she watched Angus in his jeans and fisherman's sweater on the shore of the River Ness. A cane rested on the pebbles beside him. Hamish and Gavin played on the shore in woolen sweaters much like Angus's, stacking stones and pebbles into what must be a castle. Two coffees in Styrofoam cups warmed her hands, preventing her edging the collar of her leather coat higher. Hamish looked up and saw her. His face lit. He scrambled to his feet, ran up the shore, and slammed into her, wrapping his arms around her waist. "Amy! You're back!"

She hugged him awkwardly, holding the coffees. "For a bit at least. I've missed you. How's Angus?"

Angus turned. His eyes rested on her.

"He's grand!" Hamish said. "He's teaching me chess! He's loads of time now!" He tugged at her wrist, pulling her down the shore.

Amy's smile slipped as Angus stared at her, not speaking. She was expecting too much too soon, she reminded herself. His pride had taken a blow that would heal more slowly than his legs.

"Amy's here, Uncle Angus!" Hamish announced needlessly. "She's brought you coffee."

"So she did." Angus turned away, staring out at the silver ribbon of river.

Amy lowered herself carefully with the two Styrofoam cups. Her insides trembled. "Mocha with hazelnut."

"What are you doing here?" Angus didn't look at her.

"Visiting a friend."

He didn't smile.

Amy held out the coffee.

He took it, and sipped, but his eyes remained locked on the silver-blue river. A fish jumped and landed with a plop, reminding her of the night on the bridge that crossed the river just down from where they now sat. She wondered if the same thought passed through his mind. He offered no clue.

Gavin wiggled onto her lap. "D' you see our castle, Amy?" He pointed to the stones, stacked to form the four curtain walls of a castle courtyard. "There was once a castle over the river there. King Robert broke it down so the English couldna use it against us."

Amy smiled. "And there was once a boy named Niall, only a bit older than you, who stood right on this shore, maybe right where we are now, watching King Robert's men tear it down."

Hamish scooted into the space between Amy and Angus, and pulled

Angus's arm around his shoulder. "Uncle Angus told us about Sir Niall, did you not, Uncle Angus?"

Amy dared a look up.

Faint color tinged Angus's cheeks. His fingers curled over the cane at his side. He raised it half an inch, set it down again, and looked away.

Amy bit her lip, understanding more deeply how hard his helplessness had hit him. "Do you need to go?" she asked.

"Soon." Angus tossed a stone in the river. It landed halfway across.

She didn't know if he'd be more upset if she assumed Mairi had to come for him, perhaps bringing his chair, or if she assumed, perhaps wrongly, that he could walk home with his cane. She guessed he didn't want her to see him struggle, either way.

Gavin pulled on his arm, squinting up in the sun at his uncle. "You promised us a walk on the shore."

So he could walk that much, Amy thought.

"In a bit." He sipped his coffee. "You two go play a wee bit, now."

They scampered off, chasing one another down the rocky shore. Angus sat silently, his hands wrapped around the coffee.

Finally, Amy checked her watch. "Ina and James will be waiting for me."

Angus watched a seagull hop on the pebbles. "How is he?"

"Good." She hesitated, considering whether to say more. If he wanted to know, she decided, he could ask. The last thing he wanted was coddling or sympathy. "How are you? What've you been up to?"

He turned, studying her face. His was inscrutable. She wondered if he was going to simply not answer. Then he said, "Cooking. Chess with the lads."

"That's great!" The worry that had weighed her heart lifted a little, knowing he was doing things he enjoyed.

"Mairi gave me a flute. I always liked James Galway."

Amy's smile grew. "I didn't know you liked the flute."

"Aye, well, don't tell the other coppers." He receded into himself, pulling back from the burst of conversation.

"Shawn's a parent to James," she said. "Nothing more."

Angus turned to her, his eyes narrowed, and turned away, throwing another rock in the river. It skipped three-quarters of the way to the other shore. "Only because you're being foolish. The reality of living with a man who can't walk willna please you."

"There's no wheelchair here. You promised Hamish and Gavin a walk."

He shot her a glare. "You know there's a reason I'm not walking in front of you."

"I suspect whatever happened between you and Claire had nothing to do with how much you can or can't walk."

"That's different."

Her anger rose. "Maybe you give me far too little credit. You've never been able to completely believe I don't need Shawn's flash and glamour. Maybe you're right and we don't belong together, but it's because you insist on believing I'm that shallow."

He dropped his head, staring at the pebbles between his feet. "Who told you about Claire?"

"Who do you think?"

"Nosy bastard," he muttered, and a minute later, "Go home, Amy."

His cold voice shot through her like a lance. She laid a hand on his arm. He didn't react. "You made me a promise at Monadhliath."

His jaw tightened. "If I tell you no often enough, you'll go back to him. You never stopped being torn."

"What's the statute of limitations?" she asked.

He took a gulp of his coffee. "What does that mean?"

"How long do I have to sit by myself, being rejected by you? A year after your last no? Two years, before you finally believe me? What if I finally go back to Shawn ten years from now? Do you still get to say *I told you so?*"

"You wasted your time," he said. "James needs his father."

She picked up her coffee, and left for her three hour drive home. He'd said far more than she'd hoped for, after the last attempt.

London, 1318

Isabella waited in her chambers, seated on her couch near a table piled high with food. Her gown rose over her rounded abdomen. She looked Simon up and down and gestured a pale, slender hand at the chair opposite the table. When he'd bowed and seated himself, she spoke. "It is highly unusual, my Lord Claverock, to seek out a lady during her confinement."

"These are highly unusual times, your Majesty," Simon replied. "May I?" He indicated the jug of wine.

"Please," she said.

He poured ruby wine into her goblet and handed it to her, before pouring his own. "You are aware of the growing discontent in our fair kingdom."

"I believe you have spoken to my husband of these things," she said.

Simon stared just for a moment into the blood-red liquid in his goblet. Her answer gave him a moment's hesitation. The history books said there was discontent in their marriage. Far worse than discontent. He had not mistaken that. He lifted his eyes.

"Your Majesty, Edward does not heed my words. Indeed, he aggravates the problems. He laughs about Deydras, as discontent grows, as revolts fester—in Wales, among his own barons. He does not heed my warnings regarding Damory, Audley, and Despenser, but continues to lavish gifts on them. Your Majesty, I have seen the future. These men will be his downfall."

"So have I told him. One hardly needs the gift of prophecy to see that."

She watched him with steady eyes that reminded him uncomfortably of Amy, the day he'd pressed the gift on her, tried to help her carry her groceries home.

"We must separate Edward from them. Have you authority, your Majesty, to send them away, or perhaps give them aught—lands in France, a mission—that might draw them away for a time?"

"I will think on it," she replied noncommittally. "Why do you bring this concern to me?"

"Concern for England, my queen." He held her gaze fast. A woman, especially one with child, was easy to dispatch, should she make it necessary. Even now, there was but one guard, outside, such did they trust him. "Surely you share that concern." A smile curved his lips. "Despite your French origin." She would see the implied threat.

"What is it you propose?" One slender eyebrow arched upward. "That need be spoken to my ears alone?"

Her manner gave no hint. Yet she was not ending the meeting. And the history books had *assured* him her hatred of Edward would grow murderous. "I propose," he said, "merely that you think on these issues, that you might best advise Edward. The situation in Wales...."

"Llewelyn Bren surrendered in November," she replied. Her interruption rankled Simon's nerves. "While your whereabouts were still unknown—an issue that continues to interest people."

"I have given a full account." Simon swished his wine and took a quick drink. He took her meaning, in turn. He should not be surprised at her wiliness, he thought. Had she not spent her early years in the court of Philip the Fair—hardly a man to be trifled with? "I have shown my loyalty to England, bringing great weapons to Edward. Still, it is best to prevent revolts altogether, is it not?"

"Without Llewelyn," she said, "his people will not fight."

"Be warned." Simon's goblet hit the table with a sudden clatter, and he leaned forward, his eyes boring into hers. "Llewelyn's people love him. Prophecy says...." He let his gaze rise to the ceiling, his eyelids flutter closed as if going into another world, he dropped his voice, and intoned, *"The day Llewelyn dies, then shall Wales arise."* He let the words hang, held himself still several moments, and slowly opened his eyes. He held back a smile. No imagination, indeed! That foolish Helen O'Malley underestimated him.

The young queen gazed steadily at him. "What prophecy is that?"

"The one that was given me." He held his voice firm, though her attitude infuriated him. "I know it to be true, your Majesty. Not only will his death rouse a fury from Wales such as you have never seen, but it will cause division among our own. Llewelyn commands the respect of the greatest men in England." He paused only a moment for emphasis before saying, "Roger Mortimer, for one. He is a great man—a *wise* man. He alone can bring peace between the barons and Edward. Otherwise, there will be revolt, wars." He gave a quick but unmistakable glance at her abdomen. "Attacks on the royal palace.

She said nothing—but Simon thought he saw fear flicker through her eyes.

He smiled. "Urge Edward to listen to Mortimer. If Edward will not, you must."

"Roger Mortimer is presently in Ireland." Isabella pushed herself up from the chair, a hand to her belly. "Was that all?"

"Only this," Simon said. "I have told him these things. If you see the wisdom in my words, your Majesty, you must speak them to Edward as your own, such that he hears it from both of us." He, too, rose, and lifted her hand, touching his lips to her fingertips. He headed for the door and stopped, turning. "He was heard but this morning making light of the Deydras situation again. I am on my way to find him now. I was told he's been down at the river since early morning, watching men fish."

Isabella's lips tightened. "Thank you, Claverock."

Simon sketched a courtly bow and turned for the door, his smile coming out. She hated his commoner's habits more than anything.

Midwest America, Present

"Swimming?" From the deck, Clarence looked down on the new pool, stretching across seventy-five yards of what had been Shawn's lawn, enclosed by stone and glass walls.

Shawn beamed, gazing over the blue water. The last workmen had swept their way out as Shawn entered his home, leaving him master of his brand new watery domain. The last of the day's sun shone through the glass roof.

Clarence rounded on Shawn. "You're paying me to *swim*? I don't need make-work." His face darkened with an angry flush. "I'd rather go to a shelter than be patronized like this!"

"Stow it," Shawn snapped. "I'm not paying you to *swim*. I'm paying you to *improve* your swimming."

"Big difference!"

"I'm paying you to...."

"I don't need...."

"You need a *job*," Shawn snapped, "and I *need* a strong swimmer. I told you part of what I'm paying you for is to *do* it, no questions asked."

Clarence's eyes narrowed. "I'm not doing anything illegal or immoral."

Shawn's lips tightened. "What do you take me for?"

Clarence gestured at the pool. "Unexplained job—no questions. Usually means one thing."

Shawn shook his head sharply. "I'm asking just the opposite. *If* I ask it at all. But I want the option. That means I'm paying you to be prepared."

"I don't want a sympathy job." Clarence glared at the pool, his hands on his hips.

"It's anything but," Shawn said softly. "It's the most important job in the world. Can you please just trust me? At worst, you get paid good money to do what other people go on vacation to do."

Clarence heaved a sigh, staring up at the ceiling.

"You got a better offer?"

"No." Clarence shook his head. His shoulders fell. "Is that it? Swimming?"

"A bit more. Running, climbing, studying language." Shawn started down the stairs, into his new wonderland. "But mostly," he pointed to the right, as they reached the hot tub at the bottom of the stairs.

Clarence peered across the long room, to half a dozen targets. A row of bows hung on the wall beside them. "Archery?" he asked in disbelief. He walked the length of the room, to examine the bows. His eyes, and his hand, fell on the replica English long bow.

"Your own private suite." Shawn indicated the far end of the new solarium. "Ten an hour. A dollar for every bull's eye. Twelve for ten in a row, fifty for twenty-five in a row. Three hundred when you make two minutes underwater. I have archery and swimming coaches lined up."

Clarence turned to him, studying him. "Why?"

"No questions," Shawn reminded him. "Are you in?" His chest tightened. Everything rode on this. *Everything.* Still, he forced the words out. "I'll buy you the bus ticket home if you change your mind." He bit his tongue not to remind Clarence of the couch waiting at the end of that bus ride.

Clarence stared at the water. A muscle twitched in his jaw.

Shawn's throat tightened. He could find someone else, he assured himself. Anyone would jump at this chance. But his entire body tensed. Clarence was *right* for this. Clarence wouldn't expose him to the press. Someone else might. Besides, it was an opportunity for Clarence.

"You're crazy," Clarence said.

"You in?" Shawn asked.

"I can back out any time?"

Shawn nodded. He'd have to convince him not to. He needed him.

"When do I start?" Clarence held out his hand.

Shawn grinned. His tension melted away. "As soon as we put our things away, we start."

"We?"

"I'll be swimming and shooting with you about three hours a day." He led Clarence the length of the pool to the new suite—a bedroom with a small sitting room, bathroom, and kitchenette—tucked under the garage, beside the deep end of the pool and the glass walls.

Clarence dropped his dusty duffel bag on the fresh beige carpet beside a taupe suede couch.

"You can start now. Trunks are in your dresser. The faster you swim, the more I pay you. The longer you stay under, the more I pay you. Next week, I have a trainer coming to work on rescue skills."

Clarence's eyes narrowed. "What is this about, Shawn? What's going on?"

"The fewer questions you ask, the more I pay you. I have a few other things I need to do every day." He tossed the keys to the suite to Clarence. "Start swimming while I get dinner in the oven."

London, 1318

Simon found Edward under a tree down at the river's edge, talking with a peasant holding a string of fish. "Look at these fish, Claverock," the king called with excitement. "Do you see the size of them?"

Simon forced his mouth upwards in a tight smile. "A fine morning's work." It suited his purpose to take the appearance of interesting himself in the king's peculiar fascinations. Because of doing so, Edward had grown fond of him, finally, and trusted him all the more.

"Did you know," said the king, " his line is made from hemp-nettle?"

"I did not, your Majesty," Simon replied.

"How do you think," Edward asked, "it first occurred to a man to spin nettles into line?" They watched silently while the man dropped his fish into a big wooden bucket, and threw his line back in the water. "To think, this will be our dinner tonight!" Edward enthused. "How few people *think* what it takes to feed us each night, and what goes into a single meal?"

"Indeed," Simon replied. He had things to discuss. But his mind flickered to his attempts to prepare meals on the *stoves* of Shawn's time. He much preferred the food of his own and wasn't particularly interested in what it took to get it on the table before him.

"Far away in Russia," Edward continued, his eyes still on the man with the pole, "they tell a story of a talking fish that grants wishes, when the fisherman spares its life and throws it back."

"Fascinating." Simon fought back his impatience.

Abruptly, Edward turned to him, clapping his hands together with a huge grin. "The morning air is bracing! It's good to be out in it! I presume you had something you wished to speak of."

Simon gave a slight bow of his head. "Several things, your Majesty. Might we?" He nodded his head farther down the flowing river, and Edward turned to walk its shore, away from prying ears. Simon fell into step beside him. "Your troops are even now gathering to ride north. I leave as soon as we finish speaking. I shall return with the Scottish traitor. But there are things you must know. I saw them in visions last night."

Edward stopped, turning to him. All excitement over the fish fled. "What things?"

"Three, your majesty. First." He ticked off a finger. "Llewelyn must die."

"No." Edward shook his head. "I vowed to Mortimer to protect his life."

"And yet," Simon said, "if he lives, he will send word to his people, calling them to rise again. I saw the future, your Majesty. I saw the Welsh rising as they have never risen before." He lifted his eyes to the pearly pink clouds over London, as if seeing a vision still, lowered his voice, and recited a new rhyme. *"If Llewelyn lives, death he gives."*

He let the words hang, the brilliant dawn burning his retina for a moment, before meeting the king's eyes. "The vision was horrid, my liege. Death, destruction, even to London's door. They will join with the Irish and Scots, and Llewelyn shall lead them."

Edward looked troubled. "Mortimer has pledged his protection. I *cannot* in honor kill him."

"Which is why it mustn't be by your hand," Simon replied.

Edward turned to look out over the rippling waters, agitation in his face. "I can have no part of this. There must be another way."

"Hugh Despenser," Simon said. "The second vision I saw was that Despenser, Audley, and Damory are your staunchest allies. You are wise to trust them. I will command Despenser to take care of Llewelyn, should it be your wish. You cannot then be held to account." He glanced around. The banks of the Thames were empty for a long stretch. "There is no one here but us to know of this conversation, and, your Majesty, is not the good of England what matters —more than one man's life? Think of the lives of your subjects, your court." He gave the briefest pause, and lowered his voice. "Your own wife heavy with child, the young prince Edward, prince John."

A vein pulsed in Edward's neck.

Satisfaction filled Simon. He pressed harder. "Your Majesty, do you wish them to be caught—to *die*—in this hellish vision I have seen? Will you let it become reality?" He paused long enough to let Edward imagine it. "Or will I save their lives?"

"Save them," Edward whispered.

"Despenser, Audley, and Damory are your hope and salvation," Simon said. "It is they who have your best interests—and those of England—at heart. They will stand by your side. Treat them well. Give them gifts. Secure their loyalty at any cost. Look with suspicion on any who oppose them."

Edward nodded, still not meeting Simon's eyes. "The third thing?"

"Mortimer. Roger Mortimer."

Edward turned at last, to Simon. "My trusted commander in Ireland."

"To the contrary," Simon said, "you must *not* trust him." No rhyme sprang to mind. It might be overplaying his hand, anyway, he thought. "Is it not he, after all, who swore protection of Llewelyn, an enemy of our realm? He will try to seize power from you. Above all, you must not trust Mortimer. Nor must you trust any who vouch for him, advance him, or recommend him." He cleared his throat, emphasizing, *"Any.* Not even those closest to you. They are either blinded, or traitorous, your Majesty."

Edward began walking again, turning up a path back toward the palace. "I will heed your words."

"One thing more, your Majesty. As I dressed, I was overcome with a chill, and I saw the queen's face."

Edward stopped again, alarm on his features. "Is she unwell?"

"There are different kinds of ill-being," Simon said. "She must rest these next weeks, for the sake of the daughter, Eleanor, whom she carries." An idea sprang to him. "A child who shall play a great role in the rise of England. Do not disturb her with these things. Leave her to her rest that the child might be well."

Edward nodded. "You said there are different kinds of illness."

"There are illnesses of the heart and mind. I fear the queen distrusts me. She is influenced by your enemies, perhaps by Mortimer himself. You must leave her to rest, but when the child is born...."

"Which is not for weeks yet."

Simon nodded agreement. "Yes, you must leave her alone at least until then. But if she wishes to speak of matters of state when she is out of her confinement, you must say *you* are decided on these things, that they are *your* choices—if indeed you see the wisdom of my words."

"I do," Edward said.

"Then stand by them as your choice and do not speak my name." Simon pierced the king's eyes with his own.

Edward nodded.

"If I have your leave," Simon said, "I will head north to bring you the Scottish traitor, and, by the grace of God, James Douglas."

CHAPTER TWENTY-TWO

Midwest America, Present

The days fell into a grueling rhythm. *No pun intended,* Shawn thought wryly. Up at six, Skype Amy, talk with James through the monitor, swim with Clarence—three days a week with a coach. Breakfast at eight, work on arrangements of *Laughing Brass,* the *Cillcurran* score, other great music lost to the world through his actions, while Clarence practiced archery. Lunch at twelve-thirty.

Four afternoons a week, they worked with Sir William de Clare, archer general of Calontir. On the other three, while Clarence studied Gaelic, ran, and lifted weights, Shawn got commitments from musicians, arranged with Conrad to record *Cillcurran,* and played his trombone, pushing himself to learn more and more music—big band hits, 80s rock, orchestral greats—memorizing melodies and lyrics he didn't already know.

He and Clarence swam again for an hour before eating dinner together just past six. "Venison again?" his housekeeper asked. "Boar? How am I supposed to know how to prepare boar?"

"Cook it over a fire," Shawn said. "It doesn't matter."

She did so, leaving them charred meat several nights in a row. When it brought no reaction, she returned to her normal home cooking on which she prided herself.

Just he and Clarence alone in this huge house, Shawn thought with irony, as they ate dinner at his long mahogany table night after night. It should be full of people, Amy, James, friends, parties for the orchestra, rehearsals. None of that would happen anymore—if he couldn't find another answer.

After dinner, Shawn drilled Clarence in medieval Gaelic, recording every lesson. "No questions is part of what I'm paying you for," he reminded him again and again. "Just do it!" They finished the day with more archery, Clarence's accuracy swiftly improving.

Shawn's dissatisfaction came from missing Amy at the same time he knew it was better to stay away from her—to encourage her to be with Angus.

It was several weeks into their routine, while he sat at his score table jotting out the melody of *Soul to Soul*—another of the *Lost Pieces*—a shout rose from the solarium.

"Shawn!" Sir William bellowed.

Shawn bolted from his chair, racing across the great room and onto the deck. "Is someone hurt?"

Below him, Clarence stood near the stone wall at one end of the solarium, a wide smile across his face. "Take a look," he called up.

Shawn took the stairs two at a time. The targets were now placed outside, so that they shot through the glass door fifty meters away by the deep end of the pool, and out across Shawn's land. "How far?" he asked.

"A thousand feet," Sir William announced proudly. "Go ahead, Clarence. Do it again."

Clarence fit arrows in his bow, firing them one after another; grabbed a handful from his quiver, and shot—again and again, as Shawn and Sir William watched. As the last arrow struck home, Shawn whooped. "A thousand feet! Ten in a row." He danced in a circle, shouting, "You did it! You *did* it!"

Clarence lowered the long bow, grinning.

"You *win*, Clarence, you *win!*" Shawn charged the length of the solarium, Clarence at his side, through the door in the glass wall and across his land. "Now we go for a thousand thirty-five."

"When are you going to tell me what it's about?" Clarence asked.

Shawn laughed, the breeze lifting his hair. "Do you care? Are you having fun? Do you have money in the bank?" He picked up the target and paced it off.

"I have money in the bank," Clarence acknowledged. "And a great home." He looked, suddenly, abashed. "I mean—you know—a place to live. I didn't mean...."

"It's your home." Shawn planted the target thirty-six feet past where it had been. "For as long as you want. Come on, let's nail this."

They jogged together back to the far end of the pool, where Sir William had arrows laid out. Clarence fit four in his draw hand, turned, and let them fly, one after another, and six more, across the solarium, through the glass door, two-hundred-ninety-five meters across damp winter grass. He lowered the bow, smiling.

Shawn grinned. "You nailed it, didn't you?"

Clarence nodded, beaming.

"You beat the sixteenth century record by a foot. Now do it again." His smile slipped, thinking of Niall. "And again." He looked to the undulating length of pool. "Another twenty shots, then join me in the pool."

As Clarence fit another arrow, Shawn's phone rang, number unknown. He answered it to a deep voice. "Hey, Shawn! Mike here. I hope you don't mind— Ben gave me your number. I'm hoping you can help me with something." Shawn wandered up the stairs, his emotions raging as he listened to Mike's request, while watching Clarence and Sir William shooting below.

It would bring Amy home. She would ask questions. She should stay in Scotland with Angus. But it was too good an opportunity for her. He couldn't deny her that. *I would do anything for your happiness.* She'd never seen the note. That didn't change his promise.

"You think you could?" Mike asked.

"Yeah." Shawn made his decision, though it through his emotions into tumult. "I'll do what I can." He ended the call, unsettled. He watched Clarence and Sir William for several more minutes, thinking about Niall. There was nothing for it, he decided, as he headed back down the stairs, but to be prepared. The music forgotten, he stripped off his linen shirt and pulled off his leather boots. From the wall, he took a dummy, a heavy burlap sack tied around its waist and falling to its feet, and threw it in the water. His heart clenched for just a moment, then he tightened his jaw and dove in after it.

Bannockburn, Scotland, Present

I should be thrilled, I think, as I read the e-mail again. James sits on my knee at my desk. How many people spend their lives aspiring to such a chance? And it's fallen in my lap. Am I actually considering throwing this away for someone who won't even speak to me? Yet another choice I don't want to face.

The smell of cooking drifts up the stairs—Shawn's favorite spices. Just before he left, he used them on venison. He's been partial to venison, lately. It's understandable, right? There's no reason to worry about the upswing in venison. But hasn't that been my problem in life? Not trusting my gut?

Before I can pursue the thought, Skype comes to life with Shawn's face. "Hey, James!" He couldn't look happier if he'd just been given the world's greatest trombone. "How are you?"

"Da!" James says, and Shawn grins even more broadly. His eyes cut to me. "How are you?"

I smile. "Making lamb with your favorite spices."

He grins. "Wish I was there." His eyes scan my face and he frowns. "Is everything okay?"

I nod. "It's fine. I just got an e-mail."

"You don't look happy," he presses.

"Mike Mansfield asked me again to join him on his tour," I explain. "He's offered more money. He'll pay someone to travel with us to help with James, so he can stay with me. It's centered in the Midwest, so I'd be...." I pause at the word. I'm torn in two. But there is no other word. "Home. Midweek."

"Perfect!" Shawn beams. As suddenly, his eyebrows furrow. "So why aren't you happy?"

"Well," I fumble for a reason that doesn't involve Angus. "To just pick up and go—it's not that easy. I don't have an apartment there anymore."

"There's loads of room in my house." Shawn holds up a hand immediately.

"You know I won't..."

"I'm not saying you would," I break in. *"Look, it's everything! I have students here."*

"It's only four weeks."

"James," I say. *"It's not that easy to pack up a baby."* Something niggles at the back of my mind. I feel I've just missed something. A wrong note.

"Jay! Jay!" James laughs and touches the screen, his father's face. *"Da?"*

Shawn smiles as broadly as if he's just won a Grammy, his eyes on his son, and says, *"Yes, Dad!"* Then to me, *"A few bottles, a few diapers."* He waves away my objection. *"He's got everything he needs in my house."*

"It's a long flight." What is it that feels out of place? Am I still this suspicious from the old days?

"You flew alone with him to...." He stops.

To get to Angus. We both know it.

"Yeah, I know it's a long flight," he says. *"But Amy, this is big."*

I have no answer. He's right.

"Don't throw away a major career opportunity, Amy."

I stare at the keys of the laptop my eyes averted from him. We both know it's about Angus.

"Think about it." Shawn leans forward, his face filling the screen. *"When does he need an answer?"*

"Next week." I lift my eyes, meeting Shawn's—the warm brown eyes with the golden flecks. I'm taken aback by the rush of emotion, remembering the first time I noticed his eyes, fell in love with his eyes. My stomach tumbles and flutters. In his eyes, I see his soul shining out, love for me, genuine concern about this opportunity.

"Week!" James laughs and wiggles off my lap. Our gaze breaks as I turn to see him toddling down the hall, step by step, holding the spindles of the rail.

"Is...um..." Shawn pauses just a second, tilts his head quizzically and asks, *"Is Angus still not speaking to you? When are you going up to see him again?"*

"A day or two." James is sliding down the stairs on his belly, laughing the whole way. *"He's heading downstairs,"* I say.

"Better make sure he doesn't dump the wrong spices in that lamb," Shawn says with a grin, as I hear James bump down to the bottom of the stairs. A shadow falls across his eyes for just a moment, and he says softly, *"Amy, you can't pass this up. It's only four weeks. Let me know, okay? I'll put you in first class. Whatever you need."* He signs off.

I head down the hall, down the stairs. Maybe I am overly suspicious. But my gut is ringing. Is it the timing of his Skype, so soon after Mike's e-mail? Irritation swells. Yes, it's an opportunity—an incredible one—but what message does it send to Angus if I disappear after two attempts, as soon as something better comes along?

As I touch the bottom stair, it hits me: How did he know it was four weeks?

London, 1318

Simon entered the small chamber that housed John of Powderham, his last stop before heading north. It was sparse, but not a dungeon. John rose to his full height, as tall as Edward and holding himself with a regal bearing that left Simon resisting the habitual bow to nobility.

He understood, on seeing the man, why he had been put in a chamber rather than the dungeon that would normally be the home of a pretender to the throne. This man bore himself more like a son of Longshanks than did Edward. His jailers didn't want to be on his bad side if it turned out a man who liked thatching roofs and swimming with commoners was in truth the impostor John Deydras called him.

John gestured to a chair and seated himself as Simon did. Clever, Simon thought.

"Wine?" John smiled, indicating the jug on the table. "The King of England has little to offer at the moment, but what I have, I gladly share."

Simon noted the irony. This pretender showed him more graciousness and respect than had Edward. He considered whether it might not be to his benefit to back Deydras' story and install him on the throne.

But he knew how one path ended; he didn't know about the other, and thus could take no chances.

He lifted the jug. "Thank you." He sized up the pretender as he took a long draft. It wasn't the best wine; neither was it the poorest. Yes, they were being cautious. So would he be, Simon decided. He set the goblet down, wiping at his lips. The man bore an uncanny resemblance to Edward. Shawn and Niall flickered across his mind.

"You are most welcome," John said pleasantly. "What did you wish to discuss, Lord Beaumont?"

"Your fate," Simon said.

"Ah, yes, my fate." John smiled. "It seems a great many people see the truth in my words, or I'd be in a less pleasant place right now. I wish him no harm, of course. It's none of his doing. But I believe the country is suffering, for a carter's son on the throne, struggling to do what my father did so easily and so well."

"It is not at all assured who your father *is*," Simon reminded him.

"You can see I am the spitting image of him." John smiled again, and indeed, he did look very like Edward the elder had, as he sat in the castle at Berwick, so long ago, beaming with his victory, with the long line of Scots nobles waiting to bend their knee and do him homage. "You do see it," John confirmed.

Simon nodded. "Yes, I fought beside your...." He cleared his throat, surprised at how quickly he'd just followed this man's lead. "I fought with Edward. The problem is, Deydras, you are making a very bold claim with no evidence." He glanced at the door, and lowered his voice. "There are a great many who would fain see you on the throne. You could hardly do worse by our fair country." He stopped short of naming Edward. "So it is said in the streets," he added. There would be no direct words of sedition or disloyalty from his own lips. "Do you know my reputation?"

"I've heard you whispered of by the guards." John poured his own wine. "They say you see the future. If you are a seer, then you know I speak the truth."

"What I know," said Simon, neither knowing nor caring who was the real king, "is that most men who claim to be king end up dead. I offer you this wisdom. When it appears your word will not be accepted, tell them Satan came to you in the guise of your cat and harried you day and night to make this claim."

The smile slipped from John Deydras's face. "You surely do not expect me to claim witchcraft or trafficking with Satan? There was no such thing. A careless moment by a servant, no more."

Simon narrowed his eyes, fixing Deydras with a dark gaze, and dropped his voice, intoning, "A man is king by another name and on the word of the cat his fate shall hang."

"The cat?" John scoffed. "I'm supposed to say a *cat* told me to say this? It was the servant...."

"If you wish to live, do it," Simon snapped, and rose to leave.

His spirits rose. Edward's troops waited, armed and mounted. He would make a stop to check on his fleet and meet up with Mortimer, and from there, he would meet MacDougall and together they would bring the Scottish traitor in—and by God's grace, James Douglas, too.

Inverness, Scotland, Present

Amy waited in the empty lobby, reading to James from one of Hamish's old Gaelic books, two coffees on the table at her side, until Angus limped in on his cane, wearing well-fitting jeans and a black t-shirt, his hair damp. He started at seeing her. His face flushed, he sat down in the first available chair and picked up a magazine.

"Aigoo!" James shouted. He squirmed down off Amy's lap, and, grasping chairs, began working his way toward Angus. Angus kept his eyes focused on the magazine.

"Really?" Amy said at last. "None of this is his fault."

"So let's not encourage him." Angus turned a page, not lifting his eyes.

"Mairi couldn't make it," Amy said. "She asked if I could pick you up. So if you're going to ignore me, I guess you better get your phone out and call your taxi again."

James reached for a table, his eyes locked on Angus, shouting, "Angs!"

Angus sighed, and slapped the magazine onto his lap. He stared at the ceiling. "Now we both know my father could have come."

"He was busy," Amy said.

James launched himself away from the table, catching Angus's knee.

Angus's mouth tightened. "As were you. Do you not have rehearsals, then? Students?"

James reached for his hand, patting it, and babbling, "Aigoo, Angs!"

"That was yesterday," Amy said.

"It's an awfully long drive to pick me up from an appointment."

"I was here anyway." Amy smiled. "Visiting a friend."

Angus looked up, startled. Then a slow smile crept across his face. He glanced at the coffees on the table.

"Your favorite," she said. "Mocha with hazelnut." She didn't offer to bring it to him.

James slapped his knee. "Angs! Up!"

Angus touched his soft black hair absently, then suddenly seemed aware of what he was doing, and drew his hand back. "I'm sorry you came all this way for nothing," he said. "I'll call a taxi."

"You're being ridiculous." Amy rose, stuffed the book in her purse, and scooped James up. He let out a shriek, leaning so suddenly for Angus that she nearly dropped him. Angus reached to catch him, and James clutched him around the neck, babbling, and planting wet attempts at kisses on his cheek.

Angus smiled, laughing, and hugged him back, his face buried in his hair. "How are you, then? Determined, aren't you?"

"Aigoo book," James replied.

"All right, you convinced me." Angus glanced up, questioning. Amy pulled the book from her purse, handing it to him. She sat down, a few seats away, watching and listening while Angus read. He reached the end of the book, and shut it.

On his lap, James tried to open it again. "Just once this time," Angus said, and handed the book back to Amy.

"Dinner?" She reached for James, settling him on her hip.

"Did you not say you've a friend to visit?" Angus asked.

The corner of Amy's mouth quirked up. "He can wait."

Inverness, Scotland, Present

Angus is already seated when I reach the restaurant, having left James with Mairi. My gut rings as a waitress escorts me to his table. He's not inviting me to his home. But he chose the River House, a table looking out on the River Ness, and his eyes are soft, watching me approach. His dark hair isn't quite so shorn, revealing small curls.

My heart jolts in a flood of memories—that first day he offered me coffee behind the Heritage Centre, our kiss on the windswept shore behind Glenmirril; his head on my shoulder as we stopped on our hike up to Monadhliath and he spoke of watching a child die. I feel swept on a river of every single thing I love about him—his gentle humor, his archaeology jokes— who would think there's such a thing—his deep, gruff voice, his smile, the light in his eyes when he looked at me. Surely he's remembering our time together, too, the way his eyes are shining now.

'They're harmless,' I remember him saying as we drove a small road in the Highlands. 'The cows, I mean, though I'm sure your parents are, too.' I smile as I seat myself.

He doesn't return the smile. Doubt returns. Maybe he's not remembering.

"I've ordered oysters and wine," he says. And just when I think he's not happy to see me, he reaches across the table. I put my hand in his, the shock of his touch electric on my skin. I almost gasp, and bite back a quick smile.

His lips curve, and I guess he felt it, too. He tamps it down and pulls his hand back. "Thank you for coming," he says.

"So formal," I say. My heart flutters.

He smiles. "Aye, well."

"Aye, well," I tease.

"Things have been maybe awkward between us," he defends himself.

I raise my eyebrows wordlessly.

He acknowledges my unspoken words. "Aye, 'tis all on me. I walked away when he came back. I told you to leave the hospital. I've ignored you when you've come to pick me up."

"You have done all of that," I agree.

The waiter arrives with a bottle of wine. My eyes fall on the label—Merlot. I'm yanked back to the night in the tower, the night that started all this, and Shawn pulling out Merlot. 'I listened this time,' he said proudly.

As the waiter trickles a sample into Angus's glass, I look out the window to the River Ness, sparkling as the Loch did that night, my mind caught between two men, two Merlots, two uncomfortable moments. The river is blue under a bright moon. The loch was black that night, and shrouded in mist. Lights shine off the lacy white bridge. It was there I told Angus, under an earlier shining moon, that I was pregnant, and he walked away.

And walked back, I remind myself.

"Very good," he says.

I turn back to our table, where he sets down his wine, the approval ritual complete, and the waiter fills my glass. I thank him. My insides knot; I suspect this isn't the night I hoped for, the night where he accepts that I can still love him, that it is him I choose to be with.

Angus lifts his glass in salute.

I raise mine warily, and we clink rims. "What are we toasting?" I ask.

"Your success."

"You heard?"

He nods.

"How?"

He smiles. "Now, that would take all the fun out of it."

The motifs weave together: The timing. Angus relenting just now. Shawn knowing it was four weeks and asking when I'd see Angus again. Irritation hums through me. "Shawn told you."

Angus grins, neither admitting nor denying. "Or it's on your fan page."

"I have a fan page?"

He smiles, in lieu of an answer.

I frown. "Did you set it up? Did Shawn?"

"You'll be going, sure," he says.

"Shawn called you." The irritation grows. "No, I'm not sure...."

The waiter returns with a plate of oysters, fitting it onto our small table.

"Thank you," Angus says, with the most cheer I've heard from him in months, and before I can resume my sentence, says, " 'Tis a great opportunity, Amy. This can make your whole career."

"No," I protest. "There will be other offers."

"Twill open the door, too, for your arrangements. Orchestras will be begging to work with you."

"What do you know about the workings of a symphony orchestra?" The words snap out more irritably than I intend. A woman at the next table glances over and hastily averts her eyes. He's trying to get rid of me.

"I know a wee bit about Smetana," he reminds me with a grin.

"Right, because I told you." I snatch up the small fork, ready to stab an oyster. "And Shawn told you this."

His hand flashes across the table, taking mine. He is no longer smiling. "Amy! You've a great gift—and a great opportunity. You must take it!"

"Conveniently for you." I yank my hand back and stare out the window. The river reflects the moon in rippled silver patches. A boy and girl run past the window hand in hand, laughing. Was there ever a time life was that simple for me? Oh, right—up until the second before I met Shawn.

"Amy, please," he says softly.

A moment passes. I feel his touch on the back of my hand. I let the tiny oyster fork fall, and curl my fingers into his, their touch soft in my palm.

"I've possibly—likely—done everything wrong," he says. "But is there any part of your heart that can see where I'm coming from?"

I stare at our fingers wrapped together, thinking of the way I kissed Shawn with wild abandon in the tower, the morning he returned. I nod. I understand with my whole heart. But I felt the same about Angus. "Is there any part of your heart," I ask in return, "that can grasp the position I found myself in? I love you, and I can't do any better than that."

"You've a great opportunity in America." He doesn't address my words.

"You haven't the same opportunity here."

"I chose you. I chose you! Why couldn't that have been good enough?" We're speaking past one another. I ignore his words as he ignores mine. But I want him to hear me.

"Go," he says. "I'm asking you to please go and take this opportunity."

I lift my glass with my free hand and sip the Merlot. Unable to meet his gaze, I once again look to the river, deep royal blue as the night grows darker. I miss him. I'm angry about being called here for this. I understand why he feels as he does. I can't fix it for him. "Are you asking me to go away forever?"

"I'm telling you," he says softly, "that you've a chance I canna give you." He leans forward abruptly. "I canna do all you need!"

"You can do plenty!" The intensity of his words baffles me. "You can walk, you can cook, make music, read to James, play chess...."

"I can't rescue," he says. I canna do...."

"Didn't you once tell me music rescues souls?"

"'Tis not...."

"Just for a bit," I say, "let someone take care of you. Maybe this once you can stop being the rescuer."

"Claire said the same thing."

I paraphrase his own words of long ago. "There you go. Claire's smart."

He stares out the window, his jaw tense. "Amy, you must go. 'Twould be foolish to pass this up."

"Love is never foolish," I say.

"Perhaps sometimes it is," Angus counters.

"To let me love you," I argue, "is not foolish. All that's happened....."

He drops his gaze to his plate. "Sometimes, we only see in hindsight that a choice, that hurt at the time, was the greatest love there could be."

"No." I tighten my fingers on his. "It's just another kind of pride, always having to be the strong one, always being the one who's needed."

Angus stares at our clasped hands. "You and James need Shawn." His words are strong and firm. He raises his gaze to mine.

I shake my head. For just a moment, he seems on the brink of giving in. Then he takes a hasty sip of his wine, sets it down, and stares out at the moonlight on the white lacy bridge. Something in him becomes steel. "You were with him. In America. Before—I fell."

"When I heard you were...."

He cuts my protest short. "You must really look at the lives you think you're choosing between. In a year, you'll be tired of this dull life."" He bows his head momentarily, and adds, "James deserves that." He squeezes my hand. "I need that."

I meet his eyes, grateful for the warmth of his hand on mine, gratified he reached for my hand a second time. His words reach me on a level I didn't understand before. He needs me to do this.

"I need to know I didn't hold you back from this," he says. *"You must go."*

His words are a hammer to my heart. I scramble, seeking a way around the hard, high wall of his firm announcement. *"Okay,"* I say. *"But what if I go —and I have to stay there. For business. But I still want to be with you. Would you join me? Come to live with me in the States?"*

He stares at the plate full of oysters. He's squeezing my hand harder. I think he doesn't know he's doing it. *"There are immigration laws,"* he says at last. *"I'm no longer fit to work."*

"Marriage grants a green card," I say, but even as the words slip out, I know they're the worst kind of wrong. They're exactly the words Angus doesn't want to hear.

"I don't want a pass," he says, confirming it. *"I earn my way in life."*

The words I'm sorry hover on my tongue, but irritation swallows them back. *"You're walking,"* I remind him. *"Your mind is alive and well. You may not rescue again, but you're hardly helpless."*

"That's well and good." A muscle twitches in his jaw. *"But I've no skills or education other than police work and rescue."*

"Then start getting some," I snap. I lift the glass and gulp.

A corner of his mouth quirks up, and he says, *"Careful now. I've seen how alcohol affects you."*

"That was brandy." I set the glass down, clattering, on the table, and add, *"and obviously you're not interested anyway, so I don't think it'll be a problem."*

His smile slips. I realize his fingers still clench mine. *"Go back to America,"* he says. *"Do your composing. Do this tour. Spend time with Shawn. I'll learn something other than police work. In six months, we'll see where things are, aye?"*

I feel tears sting the corners of my eyes. It's not what I wanted from this evening. I had hoped, despite my suspicion, that Shawn had nothing to do with it; that Angus relented because he wanted to see me.

"Please, Amy," he says softly. And now his other hand wraps around mine. *"Do this for me, and I'll do my own thinking. I know I've too much pride. I know that. Do this for me, and I'll do that for you."*

I nod. I brush blindly at my eyes, and stand up, knowing this isn't proper social etiquette, to walk out on dinner like this. *"Yes."* I nod, trying to smile, trying to Not Make a Scene. *"Okay, I'll go, and you do that. Promise?"* I move, half-blinded by the tears, to the door, out into the cool evening, to the sound of the River Ness lapping at the shore, and I walk, trying to think straight. I've just promised him I'll go. I know he's right, but I don't want to. He needs this. But I don't want to. What if he falls in love with Claire after all, while I'm gone? Or marries her anyway? Or meets someone else?

Tears spill down my face, chilling my skin, as I try to remember where I left my car.

Glenmirril, 1318

Allene stood in the second floor hall, the child in swaddling clothes in her arms. She and Christina gazed down into the courtyard, dotted with snow, to the group of travelers mounting up. Her father handed up a packet of letters and a bag of gold to their leader. Allene smiled, her eyes flickering down to Alexander in her arms, and back to the men below, as they touched heels to their ponies and headed out through the gatehouse.

"Sure an' Niall will be pleased to hear he's another bonny son," Christina said softly.

"I once told him," Allene replied, "how hard it is to sit home waiting. But he waits, too."

As her father turned to climb the stairs to the great hall, a tall, thin rider passed from the southern bailey into the northern and fell into line behind the travelers. He wore a long brown cloak, the hood pulled up.

"Was it not ten who arrived?" Christina asked. "There are eleven leaving."

"I've not been to dine in the great hall," Allene reminded her. "I can't imagine any of our people would leave with them."

CHAPTER TWENTY-THREE

Midwest America, Present

In a whirlwind rush, I accepted Mike's offer, packed myself and James, and raced for a hastily booked flight. Now, with James settled upstairs, I blink down, speechless, into Shawn's new addition. I'm half-dazed from the long trip, the suddenness of it. In my exhaustion, I almost wonder if I'm dreaming.

He mentioned it as we entered his foyer. "By the way, I put in a pool." A pool in the yard, I thought, a small pool or a hot tub on the deck. But this—this is beyond what anyone would conceive from, "by the way, I put in a pool."

I stand on the terrace where Shawn loved to barbecue. Black velvet sky shows overhead. Starlight shines down—but now it pours through a glass ceiling. I catch my breath as I take it in—on my right, a room like a medieval castle vault; stone walls with Gothic arches at intervals. The arches alternate between windows of leaded glass and stone niches framing....

"Sconces," I breathe. "You put in sconces."

"They're electric," he says defensively. And then, with the child-like joy I loved, the innocent joy that made me believe his public self was the facade: "You like it?"

"It's incredible." The brand new medieval walls enclose a long, narrow pool, rippling blue in the soft yellow light; Olympic length, but twisting like a stream. It connects to a smaller pool, a foot higher, that stretches back to the house, below the terrace.

"That's the hot tub." Shawn's voice bursts with pride—and a plea for approval. I've never understood why Shawn, holding the adulation of the whole world, craves my approval.

"It's beautiful," I say. Moving down the stone stairs, my hand on a wrought iron rail, I see the trickle of water flowing down the terrace wall to the hot tub below. My eyes travel over the potted plants scattered around the slate deck. At the bottom of the stairs, I see the far wall. I stare, speechless.

Here, Shawn's worlds collide. Medieval stone meets soaring glass, letting in night sky and thousands of twinkling stars. A bridge—a stone bridge!— arches over the pool to a veranda with a white wrought iron bench and potted plants.

"You can have breakfast there!" Eagerness fills his voice.

"Breakfast?" My thoughts jumble. I don't live here.

"I want this to be yours," he says.

"But..." Is he asking me to marry him? I was having dinner with Angus twenty-four hours ago. I'm still not sure I should have agreed to this. I walk along the pool, toward the glass wall, the bridge and veranda.

Shawn trails an inch behind. "You can read there, or play your violin!" he says. "Can you imagine filming in here? The acoustics are incredible!"

I nod. Yes, his voice resonates, his bass voice that always touched me deep inside. I stop at the bridge, running my fingertips over the rough white stones of the stanchion, and turn to him. "Why?"

"To swim." He grins. "Why else would you build a pool?"

Of course. But—why so elaborate?

Because he's Shawn, I tell myself. Because he can.

Because his mind creates music so beautiful it sweeps the world. His physical creation could be no less. I have no doubt the vision was his, not an architect's, and for just a moment, amidst the trickling waterfall and mist rising off the hot tub, the strong stone walls and airy glass, with starlight shining in, I'm swept into his soul, and I know I was right, in what I saw in him in that first year. A craving for beauty fills his every cell and bursts forth from him, and I feel I'm sinking into that beauty, falling more deeply in love than I've ever been, even as I wrestle with leaving Angus.

But my brain protests! There's something I'm missing. And suddenly, I know. It's the speed with which it happened. My eyebrows furrow. Why this instead of the fountain he wanted out front? The ground is still half-frozen—it must have cost ten times what it had to, to do it now, to do it so fast.

"You don't like it?" He sounds crestfallen.

"I love it!" I turn to him, eyes alight. "It's incredible, but...why the rush?" My eyes fall on a door, in the wall that was the back of his garage.

He laughs, takes my hands, pulling me into a spin, and suddenly, we're swaying, his cheek pressed to my hair. He's humming Moonlight Serenade, his humming reverberating in my own chest. He's just evaded—and distracted me from whatever is in that room.

And it's working.

He's singing now, singing softly, with the ghosts of a dozen big bands playing behind him and the ghosts of a thousand couples swaying beside us, in love. I'm tired of questions. Beside the sparkling pool, under the glittering stars, to the melody of his voice and the harmony of the trickling waterfall, I melt into him, letting him embrace me and sing to me. He and Angus are pushing and pulling me between them. I'm tired. Just for tonight, I quit fighting it all and melt in his arms.

Inverness, Scotland, Present

"Have you heard from her?" Clive talked even as he swam in the lane beside Angus, pulling one arm over another.

Angus grunted, dipping his head back in the water and focusing on his legs. They felt stronger than last time. He lifted his head for a breath.

"No, you didn't say you want to talk about it," Clive said.

"Would you ever just swim?" Angus ducked his head in again, pulling with his arms. He was keeping up with Clive today. There had been a time he could do three laps for every one of Clive's. Not that it did anything about Amy, or for Brian. The memory of his talk with Faith MacKenzie still left him ill. *Knife wounds.* And to Brian, of all people—innocent as a child—Brian who had never done harm to anyone. He smiled, thinking of Amy's irritation with him for defending the MacDougalls.

He lifted his head, saw the side of the pool, and grabbed it, heaving himself up to the edge. Clive, too, lifted himself out. He climbed to his feet and grabbed two towels, tossing one to Angus. "You should call her, you know."

"No, I should not." Angus rubbed at his hair and chest. "She'll call if she wants to." But she had to stay with Shawn. James needed *Shawn.*

Clive handed him his cane and held out a hand. Between the cane, the rail, and Clive's help, Angus was able to get to his feet with a wee bit more grace than he'd managed the week before, though the embarrassment of needing so much help stung as greatly.

"I'm not sure she will," Clive said. "You keep sending her away."

They fell into step, Clive matching Angus's slow gait, a hand on Angus's arm to steady him. The wet decking made Angus willing to suffer the indignity. "She had an opportunity," he said. "She has a gift. I canna take that from her."

"Aye, well, there's always a cost. For both of you." Clive opened the door to the locker room, holding it for Angus and they made their way to the sauna.

"We had a wee talk before she left," Angus said, as he took his seat. "I've my own things to look to."

Clive poured water over the stones, sending up a sizzle and a plume of steam. "Speaking of which—have you heard from the chief?"

Angus shook his head. "I've not. But I suppose they'll need to make decisions soon on a more permanent basis."

"You may be off the rescue team for another month or two...."

Angus snorted, hating the image of Shawn it conjured up. "A wee bit beyond that, I'm thinking."

Clive grinned as he joined Angus on the bench. "A year or two? Angus, there were days we thought you'd not *live* and weeks we thought you'd not walk again, ever. Count your miracles, would you ever?"

Angus gave a half-hearted smile. "You're right, Mate." So much seemed to be going wrong since Shawn's return—losing Amy, the fall, Brian's death. And yet, Clive, was right. "I *am* beating the odds, am I not?"

Clive slapped his back. "That sounds more like the Angus I know. The chief will be calling you."

"About?"

"Giving talks. Teaching." He paused. "Working on some cases."

"Cases?" Angus raised his eyebrows.

"There may be some limitations on your mobility at the moment," Clive said, "but your mind is quite well. No one snaps pieces together like you do. Aye, there's one they've managed to keep off the telly."

Simon Beaumont sprang to mind. Angus didn't know why

"A man was found in a rubbish bin—in an alley."

"Knife wounds." The words came out unbidden.

Clive stared at him. "How did you know?"

Angus shook his head, as if to dislodge a buzzing. "I don't know."

"That gut instinct of yours. Aye, he was in a dumpster behind a shop, you know the run down area that's had trouble keeping anyone? Most of the shops had been empty for months. A new tenant let one and went back do some cleaning up." He shook his head, leaving it to Angus's imagination.

"How many?" Angus asked.

"Five."

Angus wiped at his forehead as the heat rose. "First thing," he said, "tell the chief look for the same thing in Bannockburn and around Oban. All at least three months ago."

"Bannockburn again." Clive glanced at the door of the sauna, and lowered his voice. "I've no idea how you *knew* 'twas stab wounds, but when the chief told me, my first thought *was* that a certain Lord Claverock might be familiar with a knife."

"Familiar with a knife and in need of money—easily found in alleys," Angus replied.

Clive stared at the door, his eyebrows furrowed. "Poor buggers," he said. After a few moments of silence, he turned to Angus. "If you're right, what do you tell the chief? How do we explain it to the media?"

Angus heaved a sigh. "I suppose if I'm right, it won't happen anymore. If I'm wrong, we need to know, aye? What we say—well, some cases are never solved, are they? In the meantime, tell him to start putting out pictures of Simon and looking for anyone who saw him."

London, 1318

Edward lolled in a chair, long legs stretched before him and eyes closed, listening to his Italian musicians, lost in the wonder of the intertwining notes, when the doors of his chambers crashed wide. His eyes flew open. He leapt to his feet as the music screeched to a halt, one last drag of a bow across a protesting viol string.

The guards scrambled at the intruder and abruptly stopped.

"What do you mean," Isabella demanded, "by ignoring my messages for days on end?" She pressed one hand to her lower back, the other to her protruding stomach. The men in the room turned their eyes away.

"Isabella!" Edward looked from her to his musicians and back, and gave a wave of his hand. They lifted bows hastily to viols, fingers to lutes, and music drifted back into the room. Edward waved his hand at the door, and a guard closed it and sprang back to attention. Edward guided his young wife to the window. "You must rest," he said softly. "Your time is near."

"I could rest the more easily," she said tartly, "had you come at any of my summons, such that I had no need of leaving my chambers to seek you out."

"I did not come," Edward said, "*because* you are in need of your rest."

"Was I not quite well but days ago when last we spoke?" she asked. "You've not seen me since to know whether I need rest or not."

"My dear queen." He glanced at the guards. "Let us not give a show for the servants to whisper of. Let me escort you back to your rooms."

She shook her head. "And have you walk out when we've much to discuss? No, we'll stay here and speak. Or your guards and musicians shall witness the scene you fear."

Edward's eyes took on a hard cast. "My lady, you are with child. *My* child, a princess...." He cleared his throat, still irritated with himself for taking Simon's word as truth. It had not been proven. "Or a prince of the realm. This child's well-being—as yours—must be our first concern."

"*England* must be our first concern," she countered.

He glanced at his guards. "You may rest well, my lady, and not agitate yourself or endanger your health. I am dealing with England."

She rolled her eyes and flung out a disdainful hand, indicating the four musicians at the far end of the room. "It seems to me you are listening to music. How, pray tell, does this benefit England? This morning—they told me when I sought you—you were out in the country building a wall. *A wall!* With a farmer beyond the city gates. Why, Edward, is the king of England *building a wall?*"

"It was in disrepair. He had lost his pigs as a result."

"A *farmer!*" Isabella emphasized.

"I learned a great deal from him," Edward said. "He has fascinating views on soil and ways to improve the crops."

"You are *king of England*, may I remind you? You may lose your *people* if you do not attend to matters of the kingdom."

"Is he not one of the people of England?" Edward asked. "My advisers are seeing to these other things."

"By advisers, you mean that unholy trio."

Edward stiffened. "They are my friends. They are loyal and shall be the salvation of England. I trust them."

"*I* do *not*," Isabella snapped. "And you know your lavishing of gifts...."

"Stop," Edward commanded.

"...upon them is...."

"Stop!" His face grew dark.

Her lips pursed tightly. She stared at him defiantly, but fell silent.

He turned to the musicians and guards. "Go. Leave us."

The music stopped again. The musicians hurried through the great door and the guards followed them out, shutting the door again. Isabella and Edward stood for a moment in silence, before he laid a hand on her rounded belly and said, "Isabella, I am concerned for your health and for the child."

She lifted her eyes to his, reading his sincerity. Her anger dropped. "Why now? Was I not quite well through Edward and John? Have I not been well throughout this pregnancy?"

"Lord Claverock had a vision. He said you must rest."

Her eyebrows dipped.

"Come." Edward laid his hand on her shoulder, steering her to the table laden with cheese and bread. "Eat and drink." He poured her a goblet of wine as she eased herself down into the hard-backed chair.

"I don't trust Claverock." She lifted the cup to sip.

"He predicted Deydras." Edward poured a second goblet of wine, and drank deeply. "He told me, before you knew yourself, that you were with child. He has given me weapons and knowledge no man has ever had."

"Yes, he's shown you how to produce these weapons." Isabella glanced at the bread and cheese. "But so far anything that has been made is in *his* hands, not yours. How do you know he'll not turn them on you? How do you know he's told you everything? Perhaps he has stronger weapons to use *against* you."

"How do we know any man will not eventually turn on us?" Edward replied. "Could he not have taken this information to Bruce or France?"

"I hear he is quite close with Alexander MacDougall, not so long ago a subject of our difficult neighbor."

Edward waved a hand. "Alexander is loyal to me. I've no concern there."

"Claverock also had a conversation with Deydras."

Edward studied her momentarily, a frown creasing his forehead. "And so? Surely he was advising him to drop this foolish story, just as he advised me how to respond to the ridiculous claims."

Isabella's face darkened. "And that is exactly a point on which I don't trust him. There are those who say your response, to laugh him off, only proves you are perhaps not your father's son. Your father would have challenged him to combat and killed him. Or ordered him hanged for such insolence."

"I am not my father," Edward said tersely.

Irritation flickered across Isabella's face. Her lips parted, but the child kicked, making her gasp.

Edward's mouth pursed. "He prophesied and he was correct," he said. "I will follow his advice regarding Powderham."

A servant entered the room, bringing a plate of grapes and thinly-cut meat. Isabella waited until he left before saying, "He was telling Deydras how to save himself." She helped herself to a thin slice of meat, adding, "It seems to me that one of you wins and one loses and if Deydras saves himself—as Claverock advised him how to do—then it is perhaps not *you* who wins."

Edward chose a grape, studying it and holding it up to the light.

"Thinking about the soil that would produce such a grape?" Isabella asked. "We've bigger concerns, *my Lord.*"

Edward stuck the grape in his mouth, ignoring her taunt. "How do you know of Claverock's conversation with Deydras?" he asked.

"Because while you were once again watching men fish..."

"It helps me think," Edward said tersely.

"...I was seeing to things. His guards report to me daily. Despite Claverock's attempts at keeping his voice low, they heard bits."

"I see." Edward smeared herbed butter on a thick slice of white bread and took it, with his flagon of wine, to the window. He stared out, as he ate.

"You see," Isabella said, finally, "but what will you *do?*"

"What is there to do?" Edward asked. "We do not know what he really said or what his intentions were, from 'bits' of conversation. I shall speak to Deydras again, myself. In the meantime, I am concerned about your health and wish you to retire to your rooms."

"I will retire when we have discussed Claverock. Is it he, perchance, who has you so concerned for my health, as of the day he left?"

Edward sipped his wine before turning and saying, "It is. He had a vision, that you would lose the child were you not to rest from these matters."

"I think," Isabella said, "it is time you tell me the visions and prophecies he has given you. And I ought likewise tell you what he has told me."

Inverness, Scotland, Present

Angus woke early after another restless night—half night, he corrected himself, glancing at the dark sky outside his window. That last dinner with Amy filled his mind. He'd immediately regretted agreeing to dinner, fearing she would see him hobble along, barely able to walk. She'd bowed her head at his insistence and left the restaurant. His heart had hurt, watching her go.

But he was right. He knew he was right, though even a stiff drink each evening did nothing to stop the dreams of Simon in dark alleys, Simon pursuing him through choppy waters, Simon pounding and slashing through a bedroom door, trying to reach James—Simon towering over a helpless Brian.

The dreams had increased since she'd walked out, turning to nightmares of James fighting Simon, and himself watching helplessly, having failed to teach James anything but playing a flute. No amount of swimming, not even his two trips into the station to talk to the chief, had worn him out enough to give him the deep sleep he craved. He forced his mind to the dreams, to Simon.

He could piece bits together. What if Simon, waking up in the wrong century, had spoken of Bannockburn? Brian would laugh in his child-like way, and mention Amy and her questions about Niall. Simon was wily. He'd known. He'd wanted to meet Amy, to learn how to cross back to his own time. Learning where she would be, he must have insisted Brian drive him to Melrose.

Angus stared up at the ceiling, his head pillowed in his hands, following the chain of presumed events. Brian had pulled off the road, and run up into the hills. There was no other reason he'd have been there, on the other side of the ridge. He'd been found, Faith said, just outside a tiny cave, one that little-known lore said had once been used by Edward's men on their marches through Scotland. Angus frowned at the shadow of the tree, swaying on the ceiling as a breeze blew outside. It had been exactly the wrong place to hide from a medieval warrior, one of the few men outside of medieval enthusiasts to know of its existence.

Angus rolled his head to stare out the window at the dark sky. Brian had realized—somehow—that Simon was a danger. And he had given his life to keep him from Amy—and James. Another pain stabbed at Angus's heart. Heroism came in the most unlikely guises.

He rolled over, lowering his legs to the floor and feeling for his cane. He thought again how much had gone wrong since Shawn had come back, even to the string of accidents in his house. He pushed himself to his feet, stretching his back as he did. It had a dull ache most of the time, that grew worse at night. His therapist couldn't promise it would ever go away. His father was right, he supposed, that he ought thank God not only to be alive, but to be walking at all. He inched slowly down the hall, stepping carefully over the torn carpet at the top of the stairs. He'd tripped on it the week before his fall off the cliff—one of the many accidents he'd been having. Stress, Mairi had said.

He glanced back at it, as he descended. Stress didn't loosen a carpet. Stress had only left him too preoccupied to notice it—or the broken cement on the back stoop, he thought as he reached the kitchen. And it *hadn't* been loose, now he thought about it.

He made coffee, and lowered himself gingerly to one of his chairs. It had broken under him, a month before his fall. He frowned, wondering now at so many incidents—and more importantly, how had he not noticed at the time?

He took his coffee and cane and hobbled to the front room. Setting down the mug, he inspected the mantle that had fallen. But it had been solidly re-installed. There was nothing to see now. And who would have been in his home messing with the mantle piece anyway?

He sank down in his armchair, thoughts flashing through his mind: The party, here in this front room. Mrs. MacGonagle's comment to Kathleen. Kathleen's insistence there was a ghost next door. Simon waiting for him near the station. He pushed himself painfully back out of the chair, knowing he shouldn't, and knowing he would.

It was a slow walk down the short hall, but finally he stood on his back stoop, contemplating the low wall. In the gray before dawn, nobody was about to see, as he sat down on it and pulled his legs over, one by one. On the matching back stoop on the other side, he twisted the handle. The door swung open—it didn't even surprise him—letting him into a kitchen the mirror image of his own, and still dark in the early morning. The odor of rotting meat came to him. Kathleen had been right, at least in part. Someone *had* been here.

But it had been no ghost.

He stood, listening. No sound came from the house. He inched painfully, down the dark hall, to the front room. He glanced around, staying clear of the window. It was empty.

After listening again, assuring himself of silence, he climbed the stairs, grimacing at every squeak. He should have called Clive. It had been foolish to come alone, in case the 'ghost' was not who he thought. But only silence came to him from above.

It took almost ten minutes, at his slow gait, to climb the stairs and search the three rooms there. One room held a pile of blankets that must have served as his bed. Opening the closet, he found a dozen hats and a pile of clothing. He poked his cane at a pair of black pants, jeans, several long-sleeved shirts in dark colors, and a heavy knit jumper with an Aran stitch on the front.

He glanced at the window, lightening as the sun struggled from sleep. He didn't need his neighbors asking what he was doing in this house, should any of them come to their back gardens for a morning cuppa.

He inched down the stairs. As he entered the kitchen, now bathed with light, he stopped in shock. On the walls, shoulder to shoulder, someone had drawn rough, life-size outlines of men.

Every one of them was gouged at head, heart, and guts, plaster spilling out.

Angus stared in shock. Kathleen had been the only one to see it, and she had been called fanciful. Although, Angus thought wryly, if Simon were back in the 1300s, he was—technically speaking—a ghost *now.*

A shiver went up his aching spine, as if Simon's evil still lurked here. *The Butcher of Berwick.* He had learned that of Simon just before his fall. He was even more lucky to be alive than he had understood. He was fortunate Simon had decided to play, a cat with a mouse, instead of going straight for the kill.

He let himself out with only the swiftest of glances at the house on the other side, eager to be away from the stench of spoiled food and the heavier stench of evil, and made his way awkwardly over the wall. A nerve pinched in his back, making him cringe. But he got into his own home with none the wiser.

As he did, his phone jangled the station's ringtone. He grabbed it, his heart already racing for the rescue ahead, before he remembered. He would not be rescuing. Nobody should be calling him from the station. "Hello," he said.

The chief's voice boomed. "MacLean, how do you know these things?"

"What things?" Angus sank into his chair, pressing his hand to his lower back, as pieces began to fit together. Simon had certainly watched him. He had laid out his gear. Simon had been in his home.

"'Tis as you said—two similar deaths in Oban, six in Bannockburn. All in alleys, all with knives, most found days or weeks later. How did you know?"

Angus grimaced with more than pain this time. "A guess," he lied. "A man who uses a knife once—sure he'd do it again, aye? Did you find anyone who's...."

"That's why I'm calling!" Alexander's deep voice boomed over the phone with the same excitement Angus used to feel when on a trail. "There was a message this morning! A girl down in Bannockburn swears she talked to him! Clive will take you down."

London, 1318

Anger flushed Edward's face as Isabella concluded her recitation. "The exact opposite of what he told me," he said. "Yet there are these prophecies and visions. He could have put Deydras up to it. But knowing you were with child...." He could not bring himself to mention the name she had chosen. "It is *impossible* for him to have known such a thing when you yourself did not."

"My dear Edward." Isabella pushed herself up off the chair, a hand to her back. He reached, helping her, and she laid her hands on his shoulders. "I know not how he knew. But there be evil sources as well as good, that give a man foreknowledge. He has tried to pit us against one another, and that is not the work of a man to be trusted. Surely you know Claverock's reputation."

"The Butcher of Berwick," Edward laid a hand on her abdomen. He spoke softly. "I overheard my father speak of it." He had been not quite twelve, coming into the stables after training, to hear his father whispering darkly to a knight. He didn't elaborate. He didn't want the image of the woman giving birth, being attacked by Claverock, in Isabella's mind. "Is it not safer to have such a man on my side?"

Isabella laid her hand atop her husband's. "You cannot trust such a man not to turn on you. Is such a man ever on the side of any but himself?"

Edward nodded. He looked again out the window, to the courtyard below

"My Lord," Isabella said softly, "I fear for you, for myself. For our *children.*"

Edward tilted the flagon back, downing his wine as a soft breeze whispered through the open window, and turned back, shouting at the door as he did, for his guard. "Gather my men," he said, when the door swung open. "Follow the Lord of Claverock. Tell the commander to arrest him, take command of his troops as well as my own, and bring Claverock back for trial."

Bannockburn, Scotland, Present

Angus hobbled carefully on his cane, as Clive ushered the girl, chewing gum and carrying a toddler on her hip, to a table at the back of a small restaurant. She wore a black leather mini skirt and ripped fishnet stockings. Three hoops hung in one ear and a ring clipped through the septum of her nose. Black hair matched black nails and lips. It had been a lot of work, a lot of talking—all of which kept his mind off Amy and James whom he missed terrible—and a touch of luck, to convince her to meet them at all.

"You know," she said, as she slid into her chair, settling the baby on her lap, "I've not done anything. If you're judging by the clothes...."

Angus shook his head as he pulled out his own chair, forestalling Clive's move to assist him. "We need your help." He held back a sigh, fearing that even now she'd bolt. The waitress appeared, taking their order for coffee.

"I shouldn't have called," she said.

Seating himself, Clive reached into a leather folder he carried and pulled out both the artist's rendering of Simon and a copy of the photograph from when he'd been booked for attacking Brother Eamonn.

She glanced at them and shook her head. "If he did something—you know, some of my mates 'ave 'ad trouble with the law, but that's nothing to do with me." The boy on her lap picked up a spoon, and set to sucking loudly on it, his big blue eyes on Angus.

Angus thought of James. He lifted his eyes back to the mother. "Please," he said. "Will you look a wee bit more closely? We're not here to accuse you of anything. There've been some murders and we think he might be responsible."

She glanced from him to Clive, and lowered her eyes to the picture. After a moment, she looked up. "I did see him." She cracked her gum as she bounced the baby on her knee. "He was asking about—about some woman, now I don't remember the name."

"Amy?" Angus asked.

She snapped her fingers. "That's it! Aye, he was on about how she was..." She drawled the next word. "'Unwed.'" She rolled her eyes. "Who talks like that in the twenty-first century?"

In another situation, Angus would have smiled at the irony. The string of murders left it not all that funny. "You never saw him again?" he asked.

On her lap, the baby fussed. She shook her head, and then almost as immediately, her eyes opened wide. "Wait! *I* didn't. But me friend—hold on, now." Setting the child on the chair beside her, she took a phone from a small black leather purse.

Angus and Clive glanced at each other as she punched at a name on the screen and a moment later, burst into a rapid-fire explosion of speech so fast Angus couldn't understand a thing. A volley of words shot back. The girl listened, her eyes alight. As the waitress set out three coffees and a basket of bread, she sobered. "Maybe you should tell the inspector yourself."

She handed the phone to Angus. He listened to the girl's story, his excitement growing as he pulled out his notebook and scratched out directions. He ended the call and turned to Clive with a broad grin. "Up for another drive?"

"You think it's him? Yer man? He's the one who's been killin' people?" the girl asked. Her face looked white even under the pale foundation she wore. "He could have killed *me* that day."

"Aye." Angus's grin disappeared. "He could have done." His eyes flickered over her son. "You were fortunate."

"Let's go." Clive handed him his cane.

"I mouthed off to him," the girl said. "He was in a rage. I could see it in his eyes. And I laughed in his face, just to take the piss out of him even more. What if he comes looking for me?"

"He's long gone." Angus climbed awkwardly to his feet, his emotions tumbling over one another. Perhaps he'd been fortunate, too, he thought, as the realization swept through him: he was doing police work again. There was nothing he could do for Brian, but he was doing police work again and he could at least reassure this girl, with her child, as defenseless as Amy. "He's dead. He'll not hurt you."

"You're sure?" She grabbed her boy onto her hip as she, too, rose. "How do you know?"

"He knows," Clive said. "Angus, let's go."

CHAPTER TWENTY-FOUR

Northern England, 1318

As the sun slid into the western hills, Simon led Edward's troops up the rise to the monastery nestled at the top. Alerted by those he had sent ahead, the gates stood wide and the abbot waited in his brown robe to give a deep nod, almost a bow, as Simon rode through the gates.

"As soon as your horses are cared for," he said, "we've dinner in the hall."

"Very good, Brother! I've looked forward to your meal all day!"

"You're in high spirits," said one of Edward's commanders as they swung down off their horses.

"I am indeed," Simon said, and realized it was true. His spirits had risen as they'd left London, as he thought of Claverock and—he realized with a jolt—Alice.

"Good to go home to your young bride, eh?" The man gave a wink as he waved for a stable boy. A youth of twelve or thirteen came running, in a brown woolen robe like the monks, and took the reins of the animal, leading it away.

Simon grinned. It felt odd and light on his face. Smiles, in his experience, had always been a deliberate tensing of muscles. This time, his features moved on their own, without his will, and his heart lifted, too.

"Cat's got your tongue," the man laughed. "She must be a fair delight to the eyes!" He slapped his horse on the rump and melted into the crowd, leaving Simon alone and feeling foolish in the midst of the courtyard, as more men poured through the gates.

Horses bustled around him, boys running to help with the work, soldiers pulling off helmets, drinking deeply of wine skins that monks, gliding in their brown robes in their midst, offered them. He patted his horse absently, thinking of Alice sitting in the window, her hair blowing in the sea breeze, and the wooden flute to her lips. He liked hearing her play. He liked the way her sleeves billowed and the way she looked up at him and lowered the instrument, smiling shyly when he came in.

"She is indeed fair."

He thought at first it was his own words drifting through his mind.

"You could find great happiness with her."

Simon's eyes slowly focused. A tall, thin monk stood on the other side of his horse, drawing a brush over its coat, his head bent to the task and covered by his cowl.

"I *am* happy with her." Simon almost jumped to hear his own words, soft as they were, spoken aloud. Shame flooded him abruptly, to be saying such things to a stranger. He was hardly behaving like the knight trusted by the great Longshanks, himself. "How do you know if she's fair?" he demanded.

"You know your fate," said the monk. "Should it comes to pass, you shan't have long to be happy with her."

Simon's hands stilled on his horse's shoulder. There had been a letter, back in Shawn's time, penned to Douglas, telling of his death at the hands of Amy's son. The horse twitched under his touch. "I make my own fate," he said. "I know James is coming. I am ready for him. And so shall that letter change."

Around him, the sounds of men and horses took on an oddly hollow sound.

"No." The monk shook his head. "You'll not escape your fate by killing James."

"My fate," Simon ground out, leaning over the horse's back, "is to rule England, to reach this land across the ocean and rule it, too."

"Your fate," replied the monk, reaching for his cowl, "is to die at James's hand, and the only way to escape that fate is to stop trying." He pushed the cowl back.

Simon stared at Brother Eamonn, at his lined face and bright blue eyes, the few strands of white hair over his bald pate.

"Give yourself up to the love you feel," the old monk whispered. "Let yourself be who you once were and could be again." He leaned across the horse's back, his eyes bright. "Be happy with Alice and give up this hatred, or you will surely die!"

The horse twitched its tail.

"I will have the power I came for," Simon hissed, "and I will *live!*"

"I'll be at Monadhliath," Eamonn said.

"No, you'll die today!" Simon started around the animal.

Sounds of merriment hit his ears. One of Edward's men bumped into him, laughing back over his shoulder at something someone said, and spun, apologizing, hand to Simon's shoulder. "My Lord Claverock, my apologies!"

"Unhand me!" Simon snapped, trying to push pass, to round the horse and reach Eamonn. He would kill him!

"Give yourself up to the love you feel," the old monk whispered. His bright blue eyes burned into Simon's.

A stable boy pulled a horse in front of him, a large animal draped with its knight's colors, yanking back on its reins and snorting. Simon pushed at its rump, pushed at his own mount's head, shoving his way finally to the other side. Eamonn was gone!

Simon spun, took a step, turned, took another. There were monks everywhere, tall monks, thin monks, short monks, fat monks, moving among Edward's men, among the horses, and boys in brown robes darting in and out.

Simon shoved boys aside, snapping at men who got in his way, pushing through the crowd shouting for the abbot. "Brother Eamonn!" he demanded. "Where is he? I want him!"

The abbot tilted his head. "My Lord? We have no Brother Eamonn."

"Tall, thin, elderly. Bright blue eyes."

The abbot shook his head. "I'm sorry." He looked with gentle worry on Simon. "We have no such monk. Please...come in to dinner."

Midwest America, Present

Shawn had no trouble getting Amy solo spots with their own orchestra. When she wasn't traveling with Mike Mansfield, he pushed her out the door for rehearsals so he and Clarence could swim and shoot without questions. When she wasn't rehearsing, he encouraged her to visit friends, run errands, take James to parks, drive up to spend the weekend with his mother.

"It's like you don't actually want me here," Amy said, laughing, when he suggested she fly to New York to visit her mother.

Shawn laughed in return, wrapping his arms around her and swaying. "Of course I want you here. I just don't want you to get bored. Or miss out on opportunities. You're back to Scotland in a few weeks, aye? Easier to visit your parents now."

"I'm not sure I can," she said.

"Not even Miss Rose?"

She shook her head. "Mike scheduled two extra concerts."

Shawn hid his sigh of relief. More time for Clarence to swim and shoot.

He skimmed the parchments regularly, his eyes falling on Christina's words despite his reprimands to himself that he shouldn't. It hurt, in a different way, to read them, to think of her married to Hugh.

And nothing had changed for Niall.

He cut back on trombone to swim more—deeper—farther—pushing himself to stay under longer. He increased the weight of the dummy. In trying to consider every possible problem, he replaced her burlap skirt with a full medieval gown and added a long wig, hair curling to its waist.

"Are you crazy?" Clarence demanded, after his third aborted attempt to drag it up from the bottom. "This stupid thing's hair is getting in my eyes and blinding me!"

"And we figure out how to deal with that," Shawn said, "so we can get this *stupid thing,* which is a human being, to the surface and all the way to the other end of the pool!"

"Reality check," Clarence shot back. "This is *not* a human being."

"It clearly represents one, and I'm paying you to get this stupid thing and its stupid hair out of the water, not to complain! Was there anything in the contract about complaining?"

"There *was* no contract," Clarence reminded him. "What's the plan? For me to become your household lifeguard? Is this meant to be Amy's hair? Are you expecting her to routinely have problems with drowning? In long dresses? Wouldn't it make more sense to buy her a swimming suit and give her swimming lessons?"

"Just help me get her out," Shawn snapped.

With Clarence in the water and Shawn wielding the pool hook, they managed to drag her out, and drop her, dripping, on the deck. Shawn lifted a stopwatch, clicked it, and shoved the woman back in the water, shouting, "Go!"

Clarence flipped under, legs slicing. But the seconds, more than a minute, ticked by, and he surfaced alone, flinging water from his hair and eyes, gasping for air.

Shawn tried, himself, until his body shook with the effort of the deep dives, the weight of the medieval gown, the hair in his eyes blinding him, and the long skirt tangling around his legs. He couldn't get her to the surface and the length of the long pool.

"It's time for a break," Clarence finally announced.

Shawn looked at the clock on the wall and shook his head. "We used up our break time." Charging past the hot tub to the bows hanging on the stone wall, he thrust one at Clarence. "Twenty in a row."

With a sigh, Clarence nocked the bow, three arrows in hand, and shot, one after another, out the open door. They slammed into the bulls eye, *one, two, three*. And six. And nine.

Shawn watched, squinting at the distant target, seeing a future—a past— that had to happen. The arrows must be straight and true. His tension climbed as the arrows flew. The eighteenth struck the bull's eye. He saw only Niall— and MacDougall. The nineteenth struck. And Allene in danger. The twentieth hit the edge of the second ring from the center.

"You missed!" Shawn exploded with a curse, even as one more arrow flew. "You *missed!*"

The last arrow struck dead center amidst twenty others. "Twenty out of twenty-one," Clarence said tersely.

"It has to be *every time!*" Shawn raged. "You *cannot afford* to miss!"

Clarence threw down the bow. "I *get* it! I *get* that you think it's important to hit the bull's eye without fail. But it would really help if you'd give me a clue —any idea at all—why I'm doing this!" His hands went to his hips. "You're nearly drowning me trying to rescue an idiot...excuse me, a *dummy*...in a wig. Now this. Tell me what this is about."

"It's about hitting the damn target!" Shawn shouted. "How hard is that?"

Clarence shook his head, slung the bow on its hook, and walked away.

"Get back here!" Shawn shouted.

Halfway down the length of the pool, Clarence turned. "I'm taking a break. A real break. And it lasts until you tell me what this is about."

Bannockburn, Scotland, Present

"Commandeering ferries." Clive looked around the cavernous warehouse, exactly where the girl on the phone had described meeting a man of Simon's description. "Breaking and entering."

"We broke nothing," Angus countered. "Simon left the door unlocked."

"Simon Beaumont, Butcher of Berwick," Clive breathed softly. "'Tis too mad to think."

"Yet you've accepted 'tis true." Angus scanned the empty space. It might have once held large shelves for products waiting to be shipped. Or it might have been a mechanic's workshop. There were no windows. He looked at the wall beside him. A life-size outline of a man stared blankly back at him. His eyes fell to the chest. The plasterboard gaped there, torn out. He touched it—it had to be a knife. He noted the vertical length of the deeper holes and guessed a sword, too.

Clive whistled. "He's mad," he said. "Do you see them? They're everywhere?"

"*He's* mad?" Angus asked. "Who's everywhere?"

"These—men. Targets."

Angus turned slowly. Hundreds of silhouettes lined the walls, as they had in Simon's kitchen, shoulder to shoulder, a blank-faced army surrounding them, each heart and most guts torn—presumably by Simon's sword. Each had a harsh slash across the left arm. Angus drew a deep breath, thinking of the letter Helen had shown him. Simon knew. Simon was preparing.

"If I'd doubted," Clive said, "this would convince me."

"You see why I didn't want to get a warrant and have to explain what we might find here?" Angus pressed a hand to his back, beginning to ache again.

"Aye," Clive agreed. "I'm seeing very uncomfortably what Kleiner was warning us of in that labyrinth under Glenmirril. You're sure he's gone?" He glanced around. "I've no wish to meet the man who did this."

And I'd be little help, Angus thought grimly. But he only said, "He's gone. And ye ken we've had no deaths in back alleys since that night."

Angus hobbled slowly to another of the outlines, looking down the line of formless men. Many of them had their throats slashed. A chill crawled up his spine, thinking of Carol alone in the house with him, that night. "Thank God Shawn got to his mum on time," he said. "To my knowledge, Beaumont didn't have his sword with him when he disappeared at Glenmirril. It wasn't found at Brian's house. So my guess is it's here."

"There's a room over there." Clive pointed.

"Go look," Angus said. "I'll be right there."

Clive jogged across the room. Angus was grateful. They both knew it humiliated him to have someone trail beside him, as if helping a doddering old man. He was as happy if they both pretended there was some other reason for Clive to go ahead. He followed slowly, watching the floor for unevenness or debris that might snare him in his unsteady gait.

A stench hit him when he reached the room at the far side. He moved cautiously over the transition strip from the concrete floor of the warehouse to a hardwood floor. A quick glance suggested it had been some sort of break room. A kitchenette stood along the narrow wall with cupboards, sink, and a stove. A long table occupied the center. The refrigerator hung open, as did the cupboard doors and drawers. A ratty mattress occupied one corner. The walls here, too, were lined with Simon's featureless, defeated army.

Angus saw the pile of apples rotting on the counter. "Quite a stench for fruit," he said.

"Meat in the refrigerator," Clive announced, turning from his search.

"It all fits," Angus said. "My guess is he saw Brian put food in the refrigerator and did, too."

Clive grimaced, his hand to his nose. "He didn't know it had to be plugged in."

"I found he likes to aim for the throat," Angus said—as if Clive believed for a moment that's what had taken him so long. "What about you?"

"He wasn't much of a cook.

"He'd not have had to before," Angus replied. "I suppose he did well not to have burned the place down. Anything else?"

Clive waved a hand at the doors and cupboards. "Utensils, pots, the last of his food—presumably his, unless someone else moved in."

"Did you look under the mattress?" Angus began the slow trek across the room, watching the floor.

Clive was there in several quick steps, dropping to his knees and lifting it up. "Nothing," he announced, leaning down to peer under.

"Anything inside it?" Angus took another cautious step, moving his cane ahead of him. It caught suddenly on something. He reached for the table to steady himself, even as Clive jumped, a hand out to catch him. "I'm fine!" Angus's words snapped out more forcefully than he'd intended. Heat climbed up his face.

"I'm sorry," Clive mumbled. "'Twas just...."

Angus dropped his gaze to the floor. "No, I was wrong." He was silent for a moment, staring at the wooden boards. It was the edge of one that had caught the cane. "Clive," he said. "See that? Down there?"

Clive dropped once again to his knees, working at the loose board, as Angus made his way to the kitchenette. He returned, carrying a spatula with a long, flat handle, and a knife, and dropped awkwardly to a chair, leaning to help

Clive pry at the board. It creaked up grudgingly, revealing a long, narrow space beneath, in which lay a long leather case, with a notebook on top of it.

Clive handed the notebook to Angus and eased the case out.

"A sword?" Angus asked.

"I'm thinking so," Clive returned. "Though 'tis still hard to believe." He handed the notebook to Angus, and reached into the case. A heavy medieval sword slid out, glinting in the light of the bulb dangling overhead. Angus touched the blade reverently. "Seven hundred years ago 'twas made," he said.

"Aye, will we have to turn it over to the authorities or a museum?"

James. The name flashed through Angus's mind. He shook his head. "We've no way of explaining it, have we?" He would give it to Shawn, for James. "I think his notebook will prove even more interesting." Angus flipped through it. Modern blue ink covered page after page in distinctly medieval script.

"He'd a great deal to say," Clive remarked.

"Aye." Angus turned back to the first page and leaned in.

"You can read it?" Clive asked after several moments.

"Much of it." With one hand pressed to his aching lower back, Angus lost himself in the words, well-known in the days he'd volunteered at Stirling with Brian. But these were not history. They were Simon's future—his plans for taking power from Edward, his notes on Edward's life after 1317.

"I couldna do it," Clive said.

Angus looked up, disoriented as Clive's words yanked him from the medieval script. "Do what?"

"Read that. You know a great deal."

"Aye, well, 'tis my passion, is it not?"

"What does it say?"

Angus turned the page, wading through several more paragraphs of the jarring medieval script in modern ink, before saying, "He's plans to destroy Scotland. He spent his time here well—studying weapons, in particular. I'm guessing from what he says here that this was not the only notebook and that he has quite a set of diagrams of guns and canon. Furthermore, he plans on being the one to find America and settle it, taking even more power."

"Well, sure he failed," Clive said, "or we'd know of him."

Angus frowned, staring at the words, and shook his head slowly. "Niall and Shawn changed history when Niall went back. If they did, so can Simon."

"There's not much you can do about that," Clive said.

"No," Angus agreed faintly. Still, he couldn't do nothing. Only one person came to mind, and he still wanted nothing to do with him.

"What will we do with this?" Clive asked. "We can hardly explain it to the Chief."

Angus pulled out his phone.

But everything in him rebelled at talking to Shawn.

Midwest America, Present

Shawn found Clarence at a nearby pancake house and slid into the booth opposite him.

Clarence looked up from an omelette. "Ready to tell me what's going on?"

After putting in an order himself, Shawn spoke carefully. "What if you could have a completely fresh start?" he asked.

"You said that before. You and I both know there are no fairy tales."

"Princesses, knights, castles." Shawn swirled his coffee. "Streaming banners, noble deeds."

"The song you play," Clarence said. *"He's gone with streaming banners where noble deeds are done."*

"You know it?" Shawn asked in surprise.

"You think I never saw you on TV? I looked up the words once. But it's not real. There are no streaming banners." He jabbed a bite of omelette into his mouth and said bitterly, "and very few noble deeds. Certainly not from me."

"But there *could* be," Shawn said. "What if you *could* go there, start over, none of this hanging over you?"

"Sure I would," Clarence said. "You buying me the ticket to Neverland?"

Shawn leaned forward. "Yeah. I'll buy you the ticket."

Clarence stared at him. "Why do you say these things, Shawn? What are you, trying to make a fool of me? Why? You know, I *get* that I'm the lowest of the low—you know, through all this shooting and swimming—and you've treated me like a brother—I never forget that. I *know* I don't deserve all you've given me. I don't deserve *anything*." He stared at the omelette a moment before looking up, his eyebrows knit. "But why not just—just kill me or something? Why this elaborate game?"

Shawn's shoulders sagged. "I'm not trying to make a fool of you. I swear to you it's win-win. I need your help, and I'm offering you a place where you can be who you've become, none of this following you."

"That's nice, but I gave up drugs a long time ago." Clarence wiped his hands on his napkin. "Look, maybe it's time for me to move on." He stood up.

"Clarence, I'm serious." It was the voice that commanded an orchestra. "Sit down."

Clarence sat.

"I'm considering going back," Shawn said.

Clarence gave his head a sharp shake. "Going back?" he echoed.

"A friend of mine needs my help," Shawn said. "Help me save him. And give yourself a new life."

Clarence stared a moment before asking, "What kind of help?"

"He's going to be attacked," Shawn said.

"Going to be? You have a crystal ball now?"

"No, someone told me."

"Someone involved in the attack? So warn your friend."

"It's not that easy," Shawn said. "There's no way to contact him."

"Phone, facebook, e-mail," Clarence ventured.

Shawn's eyes blazed. "*Look* at me, Clarence. I'm dressed in breeks and a medieval shirt and leather boots."

"So are half your fan girls. So what?"

"I got outed speaking medieval French on a talk show. *Medieval* French. Do you understand it's *medieval* Gaelic you've been learning? What do you know about the day I got back?"

"Your mother said you showed up in armor. That's ridiculous, of course. I couldn't figure out if she was speaking metaphorically, if...."

"It wasn't a *suit of armor.*" Shawn snorted. "It was *chain mail.* Big difference. Much easier to fight in."

"We're back to your favorite story." Clarence fixed his eyes on the exit sign, his jaw tight.

"*Story?*" Shawn stood abruptly, jarring the table. He lifted his shirt high, revealing the vicious red scar circling his body. "Does this look like a *story?*"

A crash sounded from the kitchen doors. Shawn's head shot up. The waitress stared at him, the black tray tilted in her hand, a pile of plates on the floor. The manager raced up behind her. "Sir! Would you mind...?"

"Oh, yeah. Sorry." Shawn dropped the shirt, unabashed. He turned to Clarence. "Battle of Bannockburn, June 1314. I can show you the scars from a wolf, a knife." He indicated his leg.

Clarence glanced at the waitress, gathering plates off the floor. "I think you better keep your pants on," he said dryly.

"I was *not* going to drop trou in a public place," Shawn snapped.

"Well, you just practically took your shirt off."

"I'm telling you something important!" Shawn held out his palm, showing the fine white line from Allene's knife. "She stabbed me when I hit on her."

"High time somebody did!"

"I can show you a medieval crucifix, the ring of Robert the Bruce."

"Who's that?"

"Who's that? Are you kidding me?" Shawn stared in disbelief. "Scotland's greatest king. He saved his country at Bannockburn. I mean, Niall and I helped."

Clarence stood, too. "You have delusions of grandeur! Wasn't having trading cards made of yourself enough? Now you're saving entire countries?"

"I can show you the chain mail. I can show you the sword." Shawn snapped his fingers, remembering. "I can show you Niall and Allene and Hugh's letters to me."

"And who are they?"

"They're who I was with. They lived at Glenmirril in the early 1300's."

"You could fake those."

"I didn't...."

"You really believe this," Clarence said. "I didn't get that, back there in the prison. "Have you considered talking to someone?" He backed away.

"Talking to who?" Shawn stared blankly, till the meaning of Clarence's words hit him, and he slammed his hand on the table, clattering the silverware, sloshing the coffee. "After all this time, you think I'm *crazy?*"

The waitress's head shot up. She scrambled to her feet with her tray.

The manager came over. "Is everything okay, sir?"

"Fine!" Shawn shouted. He took a deep breath, his jaw tight. "Yes, I'm fine. Sorry." When the manager left, looking back over his shoulder, Shawn hissed, "I do *not* need mental help." He rose, glancing at the waitress, who turned and hastened into the kitchen. He looked back to Clarence. "It's time we made a trip to Scotland."

He took his phone from his pocket, and headed out into the evening sun.

Claverock Castle, 1318

Two dozen ships bobbed in the harbor, a glorious sight under the March skies—and they were all his. Simon smiled. He'd enjoyed hosting Edward's troops when they'd arrived several nights hence. They saw the expanse of the Claverock lands, its fertile fields and wide shores and hunting lands rich with game. He liked the envy in the young knights' eyes at his lovely young bride on his arm. He had been filled with joy to return to Alice, and her very pleasant way of welcoming him home. And now—he was enjoying the sight of his newly outfitted ship almost as much.

A fleet and a wife! Yes, it had been good having a wife, after all! She had managed the home and servants well in his absence—and was a pleasant distraction. His smile grew. She was more than pleasant—and more than a mere distraction. She spoke to him shyly at night, listened to his stories of his youth, of learning to fight, his first battle. She enthused about Claverock—and the place had gained warmth as she hung tapestries and strewed the floor with scented rush mats and brought in wild flowers.

He'd loved Claverock. He'd been proud of it. But he saw it with new eyes as a magical transformation settled over it—and over his life. He liked it. He looked forward to seeing her in her chair at his head table when he returned for his evening meal after hunting with Edward's men. He'd even begun to understand what King Alexander had been thinking, rushing home to his bride so many years ago. He felt peace in her presence, a wonder at her way of seeing the world, that he'd never felt before.

"My Lord."

Simon jolted from his reverie.

Rolf, the ship master stood beside him, grinning broadly behind his beard. "Marriage becomes you!" he said, with a great laugh.

"The ships!" Simon snapped.

Rolf's grin grew. But he only said, "I believe we have outfitted the first to your satisfaction. If so, we'll proceed with the rest."

"Show me." Simon swept his cloak behind, striding alongside the ship master, up onto the deck of the first ship, one that had been commanded just two years ago by Alice's father. But his use for these ships far outstripped anything Lorne had dreamed of. He led the way down the deck, to find the holes cut into the rails for the cannon, and the iron reinforcements laid in the floors, just as he'd ordered.

"What is it for?" Rolf asked.

Simon smiled, as pleased as a mother with her new child. "Surely you've heard there's work at the foundry?"

"They're tight-lipped," Rolf replied.

Simon nodded in satisfaction. "Else they'd hang. You will soon see the greatest weapon ever dreamed by man. Six to be mounted on each boat. The Scots will no longer hinder us. Nor will anyone else."

Rolf drew a deep breath, and gave a bow. "I am honored, my Lord."

"You, too, shall be tight-lipped." Simon led the way further, stopping to inspect the tall mast, the oars that lay waiting, and the storage areas below, peppering Rolf with questions. "How far can she sail? She can stand up to the worst storms? How much food can she hold?"

"My Lord, more than enough for any stretch of English coast."

"What about leaving England's coast for weeks at a time?" Simon asked.

Rolf looked perplexed. "Why would we do that?"

Simon smiled. "Never mind why. *Can* we?"

The ship master, his eyebrows furrowed, heaved a sigh. He walked the length of the deck again, peering into the cargo areas, touching the mast as if to assure its strength, running his fingers over the sails rolled up and waiting. "You'd want to be well-supplied," he said. "Extra sails, extra lumber for repairs, plenty of fresh water. You'd want to travel with a fleet, of course." Suddenly, his eyes cleared as he looked around the two dozen ships bobbing under the blue skies. He laughed out loud. "My Lord! What are you planning?"

Simon grinned broadly. "I've had a vision. You, Rolf, shall sail into history with me. I have a campaign in Scotland. I expect to be done by next spring. I wish to leave, then, with these two dozen—all outfitted as you have done with this one. By the following year, I want a much larger ship to sail with us."

Rolf nodded, excitement on his face.

"We'll take her on a maiden voyage," Simon said, "across the channel. I have business with Roger Mortimer before we begin the Scottish campaign." Simon slapped Rolf on the shoulder, and they commenced their tour of the rest of the fleet, Simon noting all he wanted done. They would make a crossing that would so far outshine Columbus, that that man would never even try. And the land across the ocean would be his to rule.

Inverness, Scotland, Present

Angus sank into his armchair in his front room, the hated cane on the floor beside him and a bowl of popcorn on the table at his side. A Guinness stood beside it, tall and cold and sweating. It had taken some doing, getting it all to the front room, but it felt good to be able to relax in the big chair with a beer.

And now—he had to make the call. He'd put it off for hours. It would be evening in America. She was with him, maybe in his mansion, maybe feeding James in the sunny kitchen he'd once seen in an interview on the telly; maybe taking a swim in the new pool he'd built. He'd overheard Judy and Pete talking about it—before they saw him and fell silent.

He found Shawn's name and hit call. It was over with Amy. She had advantages in America he couldn't give her. More importantly, James *needed* Shawn. And for that, Shawn needed to know about this notebook. As the phone rang, he steeled himself for the angry response he'd gotten on a previous call.

But Shawn, picking up immediately, said, "Angus!" A pause and then, "What's wrong? Has something happened?" No animosity.

It made it harder to hate him, Angus thought. Shame rose in his throat. He was failing to live up to his own beliefs. This was business. He cleared his throat. "I found Simon's sword. And a notebook he kept."

"Just a minute."

Angus waited a painfully slow minute, through a slow sip of the cold beer, dreading and hoping for the sound of Amy's voice in the background, before Shawn said, "Yeah, what was in it?"

"His plans to take power, destroy the Scots." Angus gave a brief summary, before adding, "He clearly had another book with diagrams, with weapons. He's outlined his plans very carefully—including backup plans with Lancaster."

A brief silence greeted the end of his story, before Shawn spoke, his voice heavy. "Brother Eamonn told me. I've tried to convince myself..."

As his words dropped off, Angus filled in the rest of the thought. "You tried to believe 'twas the ramblings of a crazy old man."

"Yeah." The word came out short, a sound of defeat.

"I know there's nothing we can do about it from here," Angus said. "I just felt—you needed to know. I couldn't *not* tell you."

"Actually," Shawn said, "there *is* something. Maybe."

Angus listened to Shawn's request. "How is that going to help?" he asked.

"Can it hurt?" Shawn countered. "I might have an answer."

"Well, aye, 'tis easy enough," Angus said. "I'll arrange it and text you."

"Thank you, Angus," Shawn said, and hung up.

Angus squeezed his eyes shut. Did he *have* to say thank you? It made it that much harder to dislike him. And he missed Amy. He missed her terribly.

Midwest America, Present

Shawn ended the call as he let himself into his house, and pushed the phone into his pocket. In moments, he had the parchments unrolled on his bed. It took only seconds to see the message from Brother David had not changed. There were no further letters, nothing to tell him if Simon was on the path to succeeding. He stared at them a long time, lost in Angus's words, before realizing his room had grown dim. He rolled them back up carefully, sliding them into the oilskin, and, hands in pockets, headed across his dark, silent home to the French doors and down to the solarium.

At the bottom of the stairs, he stood, staring at the moonlight shining down from the arched glass wall onto the rippling water of the long pool. He had really convinced himself, he realized. He had convinced himself Eamonn was half mad.

Hell, he'd convinced himself Eamonn was completely mad. That Simon had no such plans; that he didn't have the ability to carry out such plans anyway. Nothing but the ramblings of an old man. He would reach Niall across time, and stay here, with Amy. Even as he tried to convince Clarence, he had clung to the fairy tale ending, in which he had it all.

The notebooks, showing that Simon had taken diagrams, studied and learned, were clear evidence he intended to do exactly as Eamonn said he did.

Who was he kidding? Shawn crossed the bridge—the stone bridge that had been a ridiculous extravagance, almost foolish, even, in leading to the small patio. But Amy loved it. He had stood many times just outside the breakfast nook's French windows, on what had been his back deck, watching as she crossed the bridge to play her violin there in the sun pouring through the glass walls, or curl up on the lacy wrought iron bench with a book.

He sank down on the bench, with the moonlight grazing over his shoulder to land in a splash of soft silver light on the pool. He stared at the water for long moments, before he became aware he was being watched, and lifted his head.

Clarence stood on the far end of the bridge. "Need company?" he asked.

Shawn glanced up at the deck, remembered Amy was in Omaha with Mike Mansfield, and gave an almost imperceptible nod. His eyes went back to the water, his mind to countless hours of swimming and diving with Clarence.

Clarence dropped into one of the chairs, setting a bottle and two glass tumblers on the table. "I wasn't trying to spy. I saw you at the bridge. You had a need-bourbon kind of droop to the shoulders." He poured without waiting for confirmation.

A corner of Shawn's mouth quirked up. "Yeah, I think you read that right." He reached for the glass and drank deeply, before glancing at the label. "That's a hundred dollars a bottle!"

Clarence grinned. "I've been hitting a lot of targets. I thought it could be a gift—you know, you've done a lot for me—but it seems now's a good time.

Bad news?"

Shawn shrugged. "It shouldn't be. Who was I kidding?"

"Back up, explain," Clarence said as he re-filled Shawn's glass and poured one for himself.

"It was...." He stopped. He hadn't exactly convinced Clarence, back at the restaurant. He sighed. "Let's say I was told the whole world is in danger."

Clarence raised his eyebrows, but said nothing.

"But I told myself it was just the ramblings of a crazy old man. Then there was proof it wasn't."

Clarence sipped his bourbon. "What crazy old man?"

Shawn heaved himself up off the bench, pushing a hand through his hair. His fingers snagged in the leather thong that held it in a ponytail. He planted his hands on his hips, staring out the glass wall.

"Okay, so I was a little skeptical, back there at IHOP," Clarence said.

"Yeah, you were. So let's say I'm writing a book."

"Sure," Clarence agreed. "In between swimming, archery, arranging and writing music, pumping out recordings, and doing the odd concert, you were writing a book. Or you're asking for a friend."

Shawn lifted the tumbler and finished the bourbon. "I guess I *am* asking for a friend." He sat down at the table and poured another before saying, "Okay, so this guy meets this old monk—Brother Eamonn." He told the story, concluding with, "Angus just found this notebook, where Simon spells it out. Now there's proof he intends to do exactly what Eamonn said he would."

Clarence stared at him, his head tilted and a quizzical look on his face.

"I sound even crazier than I did before, don't I?"

Clarence's eyes flickered over Shawn's trews and bell-sleeved shirt, and down the length of the long pool.

"Let's say just for a minute it's all true," Shawn said. "What would you say —you know, with all your great hard-earned wisdom and all...."

Clarence's lips tightened.

"Sorry." Shawn glared out the window, irritated with himself, with Clarence, with everything. "I'm sorry, you did nothing to deserve that. What would you say *if* it were true?"

"I'd say," Clarence said, "it brings it home. But what does it change?"

Shawn stared up into the night sky, visible through the glass ceiling. He loved his new solarium. He loved the extended days, the evenings, and sometimes even the nights, with Amy and James. It could be his life, for the next forty, fifty, even sixty years.

"You know," Clarence said slowly, "*if* all this is true—and I've been hoping the drugs I used to do didn't leave me messed up enough that you could ever convince me it is—but if it is, you were never going to let Niall die. You always knew that. No matter how you tried to spin it to yourself. If those parchments don't change, you're going back."

Shawn stared into his bourbon, and after a moment took a long, slow drink. "You're right," he said. "I've told myself something will change, I won't have to. But Simon—that isn't going to change by shouting a warning for Niall."

And, he thought, if Eamonn was right about Simon, he was also right about James contracting an infection.

"Obviously whatever is going to happen...." Clarence cleared his throat and gestured at the water. "Whatever you *believe* is going to happen, involves swimming. Rescuing. Possibly staying under for a time." He glanced at his suite where the bows hung. "And shooting. Like our lives depend on it."

"Like Niall's life depends on it." Shawn stared at the rippling water. "Do you believe me now, at least, that this was not a make-work job?"

Clarence nodded. "Yeah, I believe that. The rest...." He shrugged. "You can understand...."

Shawn's eyebrows quivered together. "Look at the letters. Talk to Amy. You can talk to two members of the orchestra who met Niall."

"Wait—you said Niall lived in the 1300's."

"He was here for a couple of weeks."

"You expect me to *believe* this? And a madman is trying to conquer the world!"

"There's a cop in Inverness who believes it," Shawn pressed. "A couple of cops. Come to Scotland and talk to them. I'll show you a picture of Amy with a pair of earrings and the ring she wears. The Glenmirril archives will tell you it's been there for seventy years."

"That's impossible."

"Don't take my word for it. Come and see it." He pulled out his phone. "I'll show you something they call the Glenmirril Hoax—drawings of airplanes and the Foshay Tower. The staff there will tell you they carbon date to medieval times. I drew them."

Clarence stared out the window for a long moment before turning back to Shawn. He took a drink of his bourbon and set down the glass. "Since I left the restaurant—I've been trying to make sense of this. You obviously *believe* this."

Shawn's phone beeped. He pulled it from his pocket, scanning the text from Angus, and smiled. "Start packing," he said. "I'm buying you that ticket."

Claverock Castle, Northumbria, 1318

Simon reached Claverock, joyful with his tour of the newly-outfitted ship, as the sun slid into the western sea behind him. There, he climbed to the chambers he shared with Alice, followed by a servant who removed his boots, damp with a day on the shore and docks.

"Where's Lady Alice?" he asked.

"Begging your pardon, my Lord, there was food from the feast—it was going to spoil."

"And?" prodded Simon.

"Begging your pardon, she's taken it to a family in the village, my Lord," the man answered. "I ask your patience. She's young and does not...."

Simon waved a hand. "We'll not miss food that would have gone to the pigs anyway."

The man's eyes widened. His mouth gaped for just a second, as he stood frozen, one damp boot in his hands.

"Why do you gape?" Simon asked.

The man snapped his mouth shut. "My apologies, my Lord. You'll not mind, then, she's ordered more rushes for the dining hall and glass for the windows in the upper hall?"

Simon beamed. "Give my lady what she wishes. It's a bit magical what she's doing, is it not? Now quit gawking and get my dry boots!"

The man nodded hastily, scrambling to the wardrobe for the fresh pair.

"What else has she planned for Claverock?" Simon glanced at the flowers Alice had cut and left in a bowl on the window sill. Such a small thing, and so cheering. He breathed in their sweet scent, wondering that he had never before noticed the scents and smells of Claverock.

As the servant dropped to one knee, working the dry boot onto Simon's foot, he said, "Lady Beaumont has asked the carpenter in the village to make a cradle, my Lord, and has the weavers making tapestries for the room she has chosen as a nursery."

Simon cocked his head. Of course it had been on his mind. Especially after the recent nights they'd enjoyed. But it was too early to know such a thing. He wiggled his toes into the boot as the man tightened the laces. The monk's words whispered on his mind. *Give yourself up to the love you feel.* "Send for the physician," he said.

"Certainly, my Lord," the man answered. He helped Simon on with the second boot, and hastened out, at Simon's wave, leaving him to tie the boot himself.

When he had done so, Simon arose, poured wine from the jug on the table, and stood at the window, looking through leaded panes, down into the gardens below. He sipped his wine, imagining the blue and pink and yellow blooms that would soon line the paths and fill the beds. The gardens had lain fallow since his mother's death, years before. A frivolous waste, he'd thought. But now an image flashed through his mind, of Alice and a child, laughing in those gardens, happy there. Maybe the minstrels would practice in the small house at the far end. He found the fantasy soothing to the soul.

Why, Simon, a soul! You do continue to surprise me! He almost heard Helen's words, though she wouldn't be born for hundreds of years. His lip lifted, anger surging through him. Of *course* he had a soul! Hadn't he poured it into serving Longshanks and fighting for England! He had a soul as surely as he had an imagination! Did he not have the vision to do more for England than even the king himself?

His jaw tensed. And to do that, there must be no child to undermine him.

Give yourself up to the love you feel!

The words brushed softly on his mind, a cat twisting around his leg. He *would* give himself up to that love—just as soon as he held power in England!

A knock sounded on the door.

Simon seated himself as a servant swung the door open, admitting the elderly physician. His eyes flickered from Simon's face to the floor, as he gave a hasty bow. "My Lord, you summoned?"

"Indeed. I'm told you have great medical skills."

The man gave a deep bow. "I am honored if people think so, my lord."

"What of pregnancy and childbirth?" Simon asked. "How soon will you know when the Lady Alice is with child?"

"Within three months." The physician dared a smile—a tentative, fearful smile. "Are we hoping for such a blessing, my lord?"

"Is she pregnant now?" Simon asked, in lieu of answer. Twin images filled his head: The garden, Alice, a child, music among the flower beds and herbs— and a child who would be his undoing.

"No, my lord, she is not," the man replied.

Simon turned to the window, staring out into the gardens, to the gazebo at the far end where the musicians might play.

Give yourself up to the love you feel.

He would. As soon as he dealt with the Scots and found this new world. The weapons were being made. Edward had given him troops, who waited even now in his own castle for the plot to bring in Niall Campbell and, by the grace of God, James Douglas. He would appear a fool to back out now.

More importantly, he did not wish to. Power was within his grasp.

He turned back to the physician. "Send word the moment it happens," he said. "No matter where I am. You are to see to it that the old healer woman makes sure there is no child, do you understand?"

The elderly man paled. "Yes, my lord."

The servant glanced between the two nervously.

"Good, then," Simon said. "Do not fail me on this or you shall pay the price." To the servant, he said, "I'll be making a short trip across the channel, and then on to Berwick. Let my men and the cooks know. Begin the preparations at once." He waved his hand at the two of them.

The servant bowed and hastened out, followed by the physician. Simon strode to the window, staring out into the gardens.

CHAPTER TWENTY-FIVE

Midwest America, Present

The plane hummed as it lifted smoothly into the sky. A palindrome, Shawn thought, his mind relaxed with the free red wine of first class. *Lulu,* a musical palindrome. His life was spooling out in themes and sequences, flying *from* Scotland, *to* Scotland, in unspooling and repeating themes and motifs.

He was once again racing across the skies, the theme developed now, with Amy and James just across the big armrest. A violin solo would join in, and a trumpet, for James, playing the same theme in high, short, sharp, laughing staccatos of joy. Maybe life was playing with the rhythm a bit—but it was the same melody—flying to Scotland, to Niall, torn between two worlds, between his best friend, a place where he'd mattered—and wanting his family.

He took a sip of red wine. The timing of Angus's phone call did not go over his head. Damned funny, he'd call it. God at work, Niall would say. What did Niall know? He lived in a world of medieval superstition.

James, buckled in with Amy, clapped his hands, saying, "Da-da-da!"

Shawn rolled his head, smiling, to look at his son. His black hair was thick, like Amy's. Time to tie it back, Shawn thought. He reached a hand across the wide armrest. "You liking this?" he asked.

James grabbed his hand, laughing. "Da! Buh!" He planted his mouth on Shawn's hand, drooling, and looking up with big eyes, brown with gold flecks like Shawn's own.

"Are *you* liking it?" Amy asked. "You didn't seem thrilled that I wanted to come along."

"It's only that I'll be busy," Shawn said. "Just a quick trip." Of course he didn't want her along. He'd switched Clarence's seat into her name, bought another seat three rows back, and booked a second room—on a different floor —at the hotel. He'd have to get a separate taxi to take Clarence to the hotel. Easy enough. It wasn't like he was going to need all this money for anything else, soon, if those parchments didn't change.

"I've never prevented you working before," Amy reminded him.

"No, I was being stupid." Shawn pulled his hand from James's mouth, grinning, and ruffled his son's hair. "Yes, I'm glad you came, yes, I'm liking it."

But it wasn't just Clarence, he had to admit to himself. It was the realization, in her eagerness to go, that her heart had not let go of Angus at all. She'd said nothing, these past weeks. But she was still thinking of him. It hurt, even as he reminded himself—he was steadily laying his plans to abandon her. He should be grateful there was someone there for her—he should *hope* Angus would be there for her.

But who had ever accused the human heart of being rational? It hurt, knowing he wasn't alone at the center of her heart, as he once had been. He should have valued that when he had it.

"Buh!"

James's insistent voice broke through his thoughts.

"Book?" Amy asked. She reached for the bag at her feet, as the plane climbed, and offered him a book with a lamb on the cover.

James shook his head sharply, though his grin remained.

Amy pulled out another, with a truck. Again, James shook his head, shouting, "Buh! Buh! Nigh. Nigh."

"Night night?" Shawn asked.

"Knight. Knight." Amy dug a third time in the bag.

"That's what I...." His eyes fell on the book she held up. A medieval warrior held his helmet in one arm, his other hand on his horse's nose. The man's auburn hair blew free around his face.

James let out a short, sharp squeal of joy, and grabbed at the book, patting it, and saying, "Nigh! Nigh!"

"Where did that come from?" Shawn asked. His stomach turned.

"It was in a thrift store," Amy said. "He practically jumped from his stroller, grabbing it, and wouldn't let go."

Shawn turned away, staring out the window across the aisle. Clouds soared past as the plane climbed. It meant nothing. He would stop this happening, no matter what it cost him. He thought of Clarence, three rows behind. He had insisted on keeping first class for him. So far, he'd managed to prevent Amy meeting Clarence at the house. She hadn't noticed him, sitting across the waiting area, his face half hidden by a book, or following them onto the plane.

Small flames of guilt licked at Shawn's heart. He was lying to her again. But wasn't it for a good purpose, this time? To spare her anxiety and worry—over things he couldn't help, this time?

Beside him, her voice rose and fell softly, reading to James about the knight and his horse. Riding the green hills. Saving the princess. Rescuing....

"Are you ready to place your dinner order, Mr. Kleiner?"

Shawn pulled his eyes from the clouds, and his mind from Glenmirril. A young flight attendant stood before him, clutching a small clip board, her skirt impossibly short. He snatched up his menu, averting his gaze, surprised at how quickly medieval views wrapped their tentacles back around his mind.

"Chicken," he said.

"What dressing on the salad?"

Her legs seemed so bare! Allene would be horrified to have her legs show like that. "French," he said.

"We don't have French, we have..."

"Honey mustard."

"We don't have..."

"This is first class!" Shawn met her eyes, escaping the sight of her legs, and ignoring the passengers who turned to stare at his outburst. "Just bring me something expensive! And more red wine."

"Yes, Mr. Kleiner." Her voice became frosty.

"Please," he said. "Thank you. I'm sorry. I'm just...hungry."

"Yes, Mr. Kleiner." She moved to the next passenger.

"Bring me something expensive." Amy shook her head. James snored on her lap now, his head lolling against her arm. "Did you really just say that?"

"Yeah, I think I did," Shawn muttered.

"You've been agitated all day." Amy reached for his hand. "Why the sudden rush to Scotland? What is it you're going to be so busy with?"

He let her fingers twine their way, warm, into his. A small jolt shot up his arm. Her touch still had that effect on him. He smiled, squeezing her hand. His options wrestled in his mind. To tell, not to tell, how much to tell? He wanted to be with her. Forever.

"Why, Shawn?" she asked softly.

"I'm going to an archives."

A sixteenth beat of a rest, and she asked, "Which one?"

"Um." He took a swift sip of wine. "Glenmirril."

"Why?"

"Just—to see a few things."

"Why?"

He shrugged. "I just want to see them again. You—um—you'll see your students in Bannockburn, visit Ina, maybe?" He knew he sounded like an idiot.

She lowered her eyes. "Maybe we're not very good at this honesty thing yet. Am I right in thinking Angus is getting you into the archives?"

"Maybe." He drew the word out, avoiding her eyes.

She laughed. "Really, Shawn?"

"Okay, yes," he snapped. "What, did he call you and tell you right away?"

"No," she said. "He didn't. And I'm not sure I appreciate you two keeping secrets behind my back." In her arm, James drew a deep shuddering breath. She stroked his brow. He smiled in his sleep, and curled into her shoulder. She lifted her eyes to Shawn. "Did he call *you* right away and tell you I let him know I'd be in Inverness?"

"No." Shawn spoke through a tensed jaw. "No, he did not."

She said nothing.

The flight attendant appeared, a tray of damp, steaming towels in one hand. With a pair of silver tongs, she held out a towel.

"Is it very expensive?" Amy asked her.

The attendant's mouth quirked up. "No, ma'am, I'm afraid it is not."

"Will you be able to make do, Shawn?" Amy asked, "or shall we have a better one sent up from the Marriot below us?"

"Very funny." He snatched the towel from the tongs and looked at the attendant. "I said I'm sorry."

The flight attendant smiled, the frost of their last encounter gone. "Thank you, Mr. Kleiner."

He pulled his hand from Amy's, wiping his fingers, even as an attendant in the other aisle handed her a matching towel. "Do you have any idea what it's like," he hissed at Amy, "to see all these *legs* after two years where it was complete taboo? Sometimes—you know, go ahead and laugh—but sometimes it's like my mind is still there, okay?"

She bit her lip and looked down at James. "Okay," she said. "I'm sorry. I shouldn't have made fun of you."

He sighed. "No, I mean—she's forgiven me, right? She's not mad anymore. It's all good." Shawn dropped his towel on the tray. "So—yeah. Angus. He didn't call either of us to tell on the other." He gave her a grin, but it slipped immediately. "You're going to see him, then?"

"He didn't answer my text." Cradling James in one arm, she worked to wipe her hands.

Shawn leaned over, helping her. "I'm sorry."

"I only texted last night. Maybe he hasn't had a chance."

Shawn said nothing. He'd have answered if he'd wanted to.

"I guess I'll see him at the archives."

Shawn frowned. "No."

"No?" she asked. "You're forbidding me to go to the archives?"

"I just think—you know, I just think you'd be bored."

Amy laughed. "Shawn, are you forgetting I know you better than anyone ever has? No, you don't think I'd be bored there. So what is it you don't want me to see?"

Shawn heaved a sigh, dropping his head against the seat and staring up at the overhead bins. Clarence. She couldn't see Clarence. How was he going to explain his presence, or even who he was, to her? He craned his neck, twisting around to see him three rows back. Clarence tilted his head, questioning. Shawn grinned and gave him a thumbs up. Clarence shook his head and rolled his eyes.

"I assume Angus knows," Amy said.

"Knows what?" Shawn asked sharply.

"What it is you want to see there."

"Just the Glenmirril Lady, the Glennirril Hoax," Shawn said. "You've seen them. Really, it's nothing."

An attendant emerged from the front of the plane with his dinner, and set it before him. "Do you want me to take James?" he asked, as Amy's dinner appeared, too, on her tray.

She shook her head. "We'd wake him up trying. I'll be fine."

They fell quiet, eating. When she finished eating, he laid his arm on the tray between them, palm up, his eyebrows raised in question. She smiled, and placed her hand in his, lacing her fingers with his. He closed his eyes, reveling in the feel of her hand. There might not be much more time to feel it.

Castle Trim, Ireland, 1318

Simon gave a cursory bow to Roger Mortimer as he entered a small private chamber at Trim Castle.

"Claverock." Mortimer gestured to the table, where bread and cheese sat on a board. "How was the journey? Your new fleet?"

"A joy." Simon dropped into his chair. "They are the fairest of ships." Though it was Alice at his side, leaning into the curve of his arm, her face lifted to the wind, that left him smiling. She waited even now in his rooms. His insides stirred at the thought of joining her.

Roger grinned. "Lorne's daughter is quite fair, too."

"The seas were rough," Simon replied, unwilling to talk about Alice.

"Hardly a surprise this time of year. A good draft of wine can but help." Roger lifted the jug, pouring a ruby flow into Simon's cup and his own.

As Mortimer lifted his goblet, Simon studied him, remembering all he'd read, in the books of Shawn's time, of Mortimer and Isabella. "You've been in contact with Edward?"

Roger downed the wine in one long gulp and set the cup down with a sigh. "Every week," he snapped. "Rather, Edward is in contact with me."

"And Isabella?" The history, centuries from now, told of the affair between the queen and Mortimer, starting with their meeting in 1325. But that was what history *had* been. He could arrange an earlier meeting, and they would work to depose Edward the sooner. "Has she spoken with you?"

"Isabella?" Roger reached for a thick slice of meat. "Why would she speak with me?"

"Lovely lass," Simon replied. "Very lovely. Frustrated with Edward, I believe."

Roger's eyebrows lifted. "Who among us is not?"

"Things have been difficult?" Simon inquired.

"One missive after another. He makes me his lieutenant in Ireland and then hounds my every step."

"Surely he's worried, what with his problems at home."

"As well he ought be." Roger reached for a knife. "Did he learn *nothing* from the Gaveston affair? And now Damory."

"Hugh Audley is also among his favorites, as is the younger Despenser. I believe he was just given some large property of other."

"Something must change." Roger shook his head as he sliced an end off the cheese. "His barons are in revolt, trouble with the Scots, trouble in Wales." He pushed the cheese into his mouth.

"It was a noble deed you did," Simon said, "intervening for Llywelyn. He sought but to speak for his people!" He let indignation show in his voice.

Roger's eyes opened wide "Claverock, I'd not have thought you'd a heart for the Welsh."

Why Simon, an imagination! Why Simon, a soul! Simon gave a tight-lipped smile. "Now why should you not think so?" He knew the date and manner of Roger's death. That date could also be moved forward—when Roger was of no more use.

"How do you know about that, anyway? You were missing those years."

"I've caught up on news." Simon sliced his own cheese, washing it down with wine before saying, "Llywelyn is a great man. A true leader who cares for his people. Not all do." He named no names. Let others do the naming and risk charges of treason. "The younger Despenser will move against Llywelyn."

"Will he?" Roger laid down his knife. "Edward promised his safety."

"Are you his favorite, or is Despenser?" Simon lifted an eyebrow. "To which of you will he pay heed? How long before there are more uprisings?"

Roger held his gaze a long moment before saying, "I wish to hear no word of treason."

"I speak no treason," Simon replied evenly, "but of the good of England. How can seeking the good of one's country be treason? Is the good of the country not a king's sole purpose?"

"Indeed," Roger said.

"Are not a husband and wife one?" Simon asked.

"Meaning?"

"Our fair queen shares our concerns for England."

Roger stared at him. Then he abruptly rose from his seat. "Good to see you back, Claverock. If you'll pardon me, I've a great many things to attend to."

"Indeed." Simon watched with narrowed eyes as Mortimer left the small chamber. He rose, going to the window to look out into the hills, as the sky faded to dark velvet blue. Mortimer had left no clue as to whether he was in agreement with Simon, or would report his words as traitorous. These were dangerous times.

Suddenly, a pounding erupted at the door. "Enter," Simon called, turning.

The door burst open. Herbard, one of his men stumbled in, dropping to his knees, a hand on the table to steady himself.

Simon jumped to his feet in surprise. "Were you not with the troops?"

"I came with the next tide," Herbard gasped. "My Lord, we must flee."

Glenmirril Archives, Scotland, Present

"Clarence, Angus, Angus, Clarence." In Glenmirril's archives, Shawn made the introductions. Angus, leaning heavily on a cane, glanced from Shawn to Clarence, before offering a hand.

Clarence shook firmly. "Good to meet you." He looked to Amy, standing apart with James sleeping on her shoulder.

Shawn saw the hesitation cross his face. He could guess: Clarence assumed Amy knew who he was. It made him wary again. He was right, he thought, in offering Clarence a place where his history would no longer matter.

"Amy," Shawn said, "this is Clarence."

Amy offered her hand, looking as nervous as Clarence. "Nice to meet you," she said, but she stepped back quickly, looking at Angus, who avoided her eyes.

"Where's this Hoax, and picture?" Clarence asked.

"There." Angus gestured to the long black table.

Silence settled on the room, but for James's deep breathing, as the three of them watched Clarence. Amy laid James on a blanket on a floor, where he snuggled into his rabbit and sighed.

Clarence studied the drawings of airplanes and the Foshay tower, and of the Glenmirril Lady, looking from it to Amy. She removed one of her earrings, and held it out to him. He took it, comparing it to the drawing.

"The ring," Shawn said.

She wiggled it off her finger and handed it to Clarence, who once again compared it. He turned to the three of them. "You all claim to believe this."

They nodded as one.

"As does my sister and my partner on the force," Angus said.

Clarence looked from one to the other. "I want to talk to the rest of the staff like Shawn promised me."

Trim Castle, Ireland, 1318

"Mortimer can do naught," Simon said

"Not Mortimer," Herbard gasped. "Edward."

Simon's eyebrows drew together. "Explain."

"Edward sent a messenger to recall his troops." The man straightened, still catching his breath. "Dumont has orders to return you to London."

Simon's face grew dark. "For what reason?"

"My Lord, I know not. Edward's man rode hard into Claverock, shouting for Dumont. They conferred in his room. When they came out, Dumont called your men together, demanding to know where you'd gone."

"How does this mean Edward wants me back in London?"

"I listened at the Laird's Lug" Herbard replied. "King Edward wants you brought to the Tower. They didn't say why. When they turned to other matters, I took a horse and rode hard."

"Well done," Simon said.

"Someone may tell them where you've gone," the man added. "Given the messenger's haste, I fear they'll send word to Mortimer, and I believe they'll wait, hoping to catch you unawares at Claverock."

Simon returned to the window, looking out. The first stars showed in the east. He had no choice. He turned to Herbard. "Rouse the men. We leave as silently as possible." His mind spun. He had just lost all of Edward's men and some of his own. He needed troops.

A slow smile came to him. He knew where to get troops. "I want them to start slipping out one by one. Meet in the forest. Ride ahead; tell Rolf to have the ship ready. We land north of Claverock and ride for Wendale.

Glenmirril Archives, Present

As the door closed behind Marjory and Charlie, Angus looked from the wiry man with the cropped black hair to Shawn, in his trews and *leine*.

He'd watched in sympathy and excitement, as Clarence asked the same questions he himself had; studied the Hoax, the Lady, Amy's ring and earrings, and asked endless questions of Marjory, Charlie, Amy, and Angus himself. He'd spoken to Clive and Mairi on the phone. He stared now at the Glenmirril Lady, looking shaken.

Amy looked from Shawn to Clarence to Angus, James now playing on the floor with his book about the knight, and started toward Clarence. Shawn laid his hand on her arm. Angus felt irritation rise in his throat at the gesture.

"Leave him." Shawn said softly. "Remember the shock you felt."

They watched, to the soundtrack of James babbling, "Knight! Knight!" at his book, punctuated by occasional squeals of glee.

Clarence lifted his head, looking as if the air had been sucked from his body, his eyes round, his face gaunt. "If this is a joke, it's rock solid. I can't find a single chink anywhere. Are you trying to make a fool of me, convincing me of this?"

Shawn touched his abdomen. "Would I really put myself through that? You saw the difference in how the scars healed. You can *see* which one got modern medical care."

Clarence nodded, but he looked as though he did so unwillingly, a puppet on someone else's strings.

Angus's heart beat faster, waiting for him to speak. Shawn had refused to say why Clarence was there.

Clarence pushed a hand through his hair, and pierced Shawn with a stare. "If making a fool of me by convincing me of this insane story is what you want, you win." He walked out the door, as if in a daze. "I'll be at the castle." The door shut behind him.

Angus stepped closer. "Who is he?"

"Just a friend."

"Is he who I think he is?" Amy pressed, and when Shawn wandered to the table, staring at the Glenmirril Lady without answering, asked. "Why is it so important to make him believe this?"

Shawn turned, studied her till she took a step back, her hand flying to the crucifix. He glanced at James sat on the blanket, and headed for the door.

"What's going on?" Amy swooped down on the boy, snatched him up, his book falling to the floor, and followed Shawn. The door swung shut in her face. She spun on her heel. James twined his hand in her hair. She lifted her eyes to Angus. "Why did he do that? Why did he look at James like that?"

"I don't know, Amy. Bring him here." He dropped into a chair.

She brought him, settling him in Angus's lap. He climbed to his feet, grabbing Angus's ear, planting a slobbery kiss on his cheek, shouting, "Angs!"

Amy turned her back, her arms clutched around herself.

"Who is Clarence?" Angus asked.

Her shoulders tensed. "I think it's the boy—I mean, obviously he's a man now, but Shawn never called him anything but *the boy*—who killed Shawn's father."

"Oh." Angus drew in a breath. "He's out of prison?"

"He was seventeen at the time. He was released in January. Carol was worried how Shawn would take it."

"I imagine," Angus murmured. James planted another kiss on his cheek. "And how he's taken it is to bring him here to try to convince him of this story? Why would he do that?"

"That's why I wonder if I'm wrong!" Her words rang with frustration. "It makes no sense. And why did he ignore James?"

"Go ask him," Angus advised.

"You know he won't tell me."

"Then I'll go." Angus lifted James, still reaching for his ear, back to Amy, and hobbled out of the room.

The Hills above Glenmirril, Scotland, 1318

"My lord, how long will we wait?" Roger, MacDougall's steward, spoke from among the dozen men sitting around the small fire high in the hills overlooking Glenmirril. He warmed his hands over the flames. "She gave birth only weeks ago. She'll not be out."

MacDougall shook his head. "She'll be back to her routine soon." He'd watched it often enough from the southern tower, Christina and Allene going out with their single escort to bring food to the widows and poor who lived near Glenmirril. It hadn't taken her long, after the birth of the second child, to be back to it. "'Tis a simple matter of patience." He rose, pacing to the edge of the small clearing, his fur cloak keeping him almost warm in the chill of the March evening. From here, he could see the candles burning in the windows of the north tower, where Niall—no, the man called Shawn—had killed Duncan. Hatred burned deep inside him, glowing red hot.

"Surely there must be another way." Roger appeared at his side. He hunched in his tartan, his face pinched with cold. "Suppose they've already taken Berwick and return home before she comes out?"

"'Tis the beauty of it," MacDougall said. "Wherever he is, Niall will come for her." He didn't *just* want Niall. He wanted to be sure Niall also knew the pain of losing a loved one, before he died himself. Allene would do.

Glenmirril, Present

With dusk falling, Angus found Shawn staring out over the water gate to the small patch of shore behind Glenmirril. He made his way slowly down the path. Shawn turned, when he was halfway there, but didn't walk away. Angus limped up beside him, and together, they watched Clarence sit on the shore.

Finally, Angus spoke. "I think James should have Simon's sword."

Shawn raised an eyebrow. "Do you?"

"After an exorcism and blessing, perhaps."

Shawn turned back to Clarence on the shore. "You have his notebook?"

"Aye." Angus reached inside his jacket for a school composition book. "A very detailed plan to take power and destroy the Scots."

Shawn opened it, scanning page after page of Simon's medieval script: Copious notes on Edward's personal and political life, the events of 1318, Lancaster, Llewellyn's death, Isabella and Mortimer. Set apart on one page were the words: *Deydras, John of Powederham, Beaumont Palace, one ear. Isabella, daughter, Eleanor, 18th June.*

"He'll use this information to pretend to prophesy," Angus said, when Shawn closed the book.

"Brother Eamonn told me." Shawn spoke, his voice heavy.

Angus stood silently beside him for another minute in the cool evening breeze. Shawn still didn't volunteer who Clarence was. Angus didn't ask. "Amy's upset you ignored your son," he said, instead.

Shawn stared out to sea. "Yeah, well, she needs to see I'm no good for him anyway."

"You've a great deal to teach him." When Shawn said nothing, he spoke again. "Helen O'Malley has a letter. It tells of a man who can only be James charging over a hill to kill Simon."

"Eamonn told me," Shawn stared at Clarence on the shore below them.

"So you understand how vitally he needs you?"

"Yeah, I know," Shawn said. "I'm working on it."

He and Angus fell silent, watching as Clarence stood on the shore, throwing pebbles into the loch. Angus turned it over in his mind, but no answer presented itself. Neither did any avenue of getting the information. He asked, "What's happening with Niall?"

"Niall?" Shawn turned to him.

"Sure you're still reading the scrolls. What happened?" Angus noted the heightened pitch of Shawn's voice, the tightening of his shoulders. He'd guessed right. "'Tis bad, aye?"

Shawn's shoulders, under the home-sewn shirt, sagged, and Angus knew.

"How bad?" he asked. An ordinary death, death in old age, even death in battle, wouldn't upset Shawn, not after two years in medieval warfare.

"If you have to ask," Shawn said, "I think you know."

"How does Clarence come into it?" Angus expected Shawn to tense, deny, pretend he didn't know what Angus meant.

Shawn turned to him, eyes dry and red in the wind. "Let me handle this."

"Handle what?" Angus added the words to the sparse information he had.

"I'm taking care of it." Shawn opened the gate and headed down to the shore, leaving Angus in the courtyard.

Berwick, 1318

The letters reached Niall within days of his return to Berwick, delivered by a shivering band of travelers riding into camp astride ponies with shaggy winter coats.

"You made it through the mountains," Niall observed, giving them a handful of coins. He glanced at the bundle of letters they handed him, fearful of the news they bore. Had Allene survived childbirth? Had MacDougall done them more harm?

"Barely," spoke their leader. "Your good-father gave us shelter. 'Twas the least we could do."

Niall stuffed the packet of letters into his tunic before the gray sky could open up and dampen it. A siege was hardly the time to read, though his fingers itched to tear them open. "My wife is well?" he asked.

The men on the ponies exchanged looks and shrugged. "We were there but a night," said the first.

"MacDonald seemed in good spirits?"

"Niall! The trebuchet!"

A hail of arrows and rocks erupted from Berwick's walls. The travelers wheeled their ponies and spurred them hard out of shot of the enemy. Niall threw his shield up over his head, ducking under it. Arrows scattered around him, piercing the earth muddy with March rains. Somewhere, a man screamed. Arrows quivered like flowers, their feathers a garden of death.

He shouted for his men at the trebuchet. They'd already put up a wall of shields, guarding those who rolled another boulder into the machine's giant palm. A dozen men jumped back, letting the arm fling the rock through the drizzle just starting. It struck the wall too low, doing little damage, and slid into the moat. A shower of smaller stones and a second flurry of arrows came flying back. Men shouted and ducked.

"Another one!" Niall yelled. "Keep the shields up!" Already, Lachlan and Owen had a massive stone rolling down the incline, and Taran and a half dozen others were hauling on the ropes, pulling the arm down against the massive counterweight. In his weeks with the army, Taran had gained bulk. He shouted in a voice deepening into a man's. "Pull harder, lads!"

A spray of arrows spattered around them, piercing shields and the soft mud. "Let fly!" Hugh roared, and the second boulder spun out against the gray sky. They backed up as the archers on the wall took aim again. Their stone struck the wall, digging out chunks of masonry. His men cheered, even as they dragged the mangonel back, just beyond reach of the arrows.

"Messages from home?" Hugh asked. He and Niall tugged the war machine through the mud.

Niall nodded, as he strained on the ropes. The drizzle grew to great, splattering raindrops. Taran slipped in the mud, shouting as he went down. Lachlan pulled him to his feet, and they hauled the trebuchet another two yards, before Niall gave the signal, and they once again rolled a boulder into it. A last hail of arrows shot through the air, every one landing a yard short, as they loosed the rock toward Berwick's walls.

"Pull!" Lachlan shouted, and half a dozen men heaved on the ropes, tugging the great arm down for its next load.

Niall helped Hugh roll the boulder. "A whole bundle. There'll be summat for you, too."

"You'll be anxious for news of Allene."

"Very." They shoved the rock into the waiting hand. "And of Christina."

Hugh grunted with the exertion. "They'll be okay."

The word slipped by Niall at first. Then he looked up in surprise.

Hugh grinned, and raised his voice. "Let it go, lads!"

The stone arced skyward, crunching into the town walls. Faint cries came to them.

Bruce rode up on his pony. "What news?" he asked.

"Little success, your Grace," Niall replied. "They are on us constantly with arrows and their own trebuchets. They seem able to defend all walls at once."

The Bruce looked out over the drizzly landscape to the walls that had so long defied him. "We must have Berwick," he said. "Keep working. We'll have it this time, one way or another."

The day dragged on, boulder by boulder, arrow-storm by arrow-storm, into the darkening evening, till James Douglas came around calling off the assault. "We can do no more tonight," he said. "Sir Niall, your men will guard the southern edge of camp."

"Aye, my lord." Bidding Douglas farewell, Niall looked around his weary men, at Taran looking like the boy he still was. "See to them," Niall told Hugh. "I'll take the first watch."

Hugh shook his head. "Read your letters. Get some sleep. I'll take the first watch."

Niall hesitated.

"You've waited long for this news," Hugh said softly. "Read your letters."

"Thank you." Niall clapped Hugh on the shoulder. After swallowing hard bannocks that passed for dinner, he settled himself by the campfire and pulled the packet from his tunic, grateful, at least, that the rain had stopped. He wished for one of Shawn's *electric* lights, though he feared reading the worst.

He sorted through the missives, sliding one addressed to Hugh to the back of the small pile. His own name stood out on the second one in the Laird's bold handwriting. He stared at it, wondering if that meant Allene was no longer alive and able to write to him herself. He closed his eyes, wandering the halls of his memories, treasuring each one, before sliding that letter to the back.

His gaze landed on her neat, light script. He touched the dried ink, not reassured. She might have written him while still with child. He might be holding, right now, her last words to him on this earth.

He bowed his head, praying about events that had passed weeks ago. Shawn had insisted women rarely died in childbirth in his time. *Wash your hands,* he'd told them on any occasion requiring a physician. *Boil the water.* Niall had insisted they follow Shawn's rules in these matters, regardless of what they thought of other points of Shawn's time. It gave Allene a better than average chance.

But that meant little. She was either dead or alive.

Father, he prayed, *keep me strong, whatever the news. Thy will, not mine.*

He chose the Laird's letter. If the news was bad, he wanted to know, reading Allene's words, that they were her last. Even apart from life or death, he thought, he wanted her words for last.

He unfolded the stiff parchment and leaned close to the low flames, squinting to make out the words. *You will be anxious for news of Allene,* his good-father had started. *She is delivered of another son. She and the bairn are in good health.*

The breath rushed out of Niall in a great smile. She was alive! And well! She was *okay!* He had another son! He wanted to jump from the log, shout the news to the men around him, but they slept, exhausted by the day's siege. He closed his eyes and breathed his joy and thanks to God.

He is called Alexander after your brother whom you loved. We believe you will be happy with the name.

Niall re-folded the letter, and opened the one from Allene, knowing she and the child were well. He read through it, through her words of love and her description of the new child. Finally, he folded the letter, half the words lost, and pushed it, with the others, back into his tunic, smiling. Life was good. There was no word of MacDougall, Christina was safe, he had a new son, and Allene was well. He settled into his tartan, praying until sleep came.

Glenmirril Castle, Present

The notebook in hand, Shawn pushed through the water gate and headed down to the shore. A cold wind blew. Clarence looked up from the rock where he sat under the lone stunted tree. "You believe," Shawn said.

Clarence nodded. Shawn sat down on the rock beside him, shoulder to shoulder.

"Niall is in trouble and you want me to go back with you to save him."

"He's going to be hanged, drawn, and quartered."

"Going to be?" Clarence squinted into the breeze lifting his hair. "He lives —I mean lived—in the 1300's. That's was. *Was* hanged, drawn, and quartered."

"In the chronology of my life," Shawn argued, "he's *going to be*. If I don't find a way to stop it. And Beaumont...."

"But if he was, then he was," Clarence broke in. "There's no stopping it."

"History said the Scots lost at Bannockburn," Shawn said. "Niall changed the battle. He changed history." He leaned forward, his eyes lit with excitement. "*I can change this, too!*"

Clarence's eyebrows puckered with an unasked question.

"What?" Shawn demanded. "I mean apart from thinking I'm crazy."

"If you can change history, why not change the history that matters." He met Shawn's eyes. "Why not go back and change what I did? Change my mother and her boyfriends? Go back and change that I was born at all?" His voice rose. "Change your parents letting me into your lives."

Shawn sagged, where he sat on the rock. The ache hit him, slicing deep into his heart like a medieval flail, missing his father. Why not? He shook his head. "I don't think I can. It happened twice in the tower. In Scotland. And I just went to his time and came back. I had no control over it."

"What good is it," Clarence asked, "if you can't use it to change what matters?"

The memory of that day, of the words, *your father is dead*, hit Shawn harder than they had in years. But he had no way of going back to his own earlier life. And besides.... "Bannockburn *did* matter," he said. "*Niall's life* matters. And I *can* change that."

"He's seven hundred years ago. What difference does that make now?"

"It makes all the difference," Shawn said. "He's the best friend I ever had. I can't let him die. Not like that. They're setting a trap for him."

"A trap." Clarence's eyes narrowed. "Yeah, I saw the YouTube clip."

"You and five million other people. Thanks to that ditz talk show host, ten million by now." Shawn pulled out the oilskin, sliding the parchments out. "Take a look."

With the castle rising strong behind them in the dusk, Clarence leaned over the scrolls, scanning. He stopped now and again to ask about a word. As the gloam rose, Shawn pulled out his phone to shine light on it. After the last scroll, with the stars coming out above, he stared at Brother David's script on the last parchment. "I feel like I know them," he said. "They're real."

"Doesn't seem so irrational when you know them, does it?" Shawn asked.

"But how are you expecting him to hear you if he lived in the 1300's?" Clarence asked.

"Angus came to Glenmirril to tell me when James was born. He appeared like a ghost. We could see him and hear him—every word—even though he couldn't see us."

"Okay, so maybe Niall has heard you, too."

"The scrolls haven't changed," Shawn said.

"Okay." Clarence stared at the water, the wind lifting his hair. "Can you just jump back and forth at will?"

"I don't think so," Shawn said. "Brother Eamonn hinted it's forever this time, if I go. I don't know how it's happening." He didn't mention the crucifix. He didn't believe it.

Clarence watched the ebb and flow of the loch against the shore, finally looking up. "You're talking about abandoning your mother again. And Amy."

"Amy's life will get easier when I leave," Shawn said.

"Your son."

Shawn dropped his gaze. "He won't remember me. Angus will be a good father to him."

"You know none of that matters," Clarence said. "Every boy wants his father. And you're going to just disappear on him? Don't you care about him? What about your mother?"

"Of *course* I care about them," Shawn snapped. "But it comes down to this: They'll live without me. They'll live *well*. Niall will die. And die *horribly*. How can I stand by and let that happen?"

"You may not be able to stop it," Clarence pointed out. "*Any* of it! And there's Simon. What if you get there and fail and they kill you, too? Do *you* want to die like that?"

Shawn remembered all too well the searing pain of his injury on first waking up. He shook his head, but he understood a deeper truth. "I'd rather die like that at Niall's side, trying to save him, than live as I do, knowing I did nothing when I *know* what's coming."

"Okay," Clarence said. "That's noble, but what makes you think you *can* just hop back? You said you can't control it."

"The key seems to be being in the same place as Niall on the same date, same time. I know where it happens and when—the exact time. I have to *try*."

Clarence was silent for a time before asking, "How would we do it?"

"It happens at dawn on the causeway out to Lindisfarne. There will be archers. MacDougall in a boat."

"And what am *I* supposed to do?"

"Take out the archers. You can fight."

Clarence snorted. "Yep, sure can, thanks to prison."

Shawn could feel the excitement burning in his eyes. "I'm offering you a world where you can use that for good. You'd be welcomed there."

Clarence drew a deep breath. "You want me to go back to medieval Scotland to live out my life?"

"I'm doing it myself," Shawn said. "I'm offering you a fresh start."

Clarence tossed a pebble into the water. "You're sure you're going to save Niall and not to run away from all your own problems?"

Shawn rose from the rock, going to stand by the water. He smiled, thinking of Hugh in the loch, learning to swim. *No wonder they're so immoral in your time if they've time to embroider trees on braies,* Niall had said.

The smile faded. He wished Niall had come in that night, or any of the nights he'd swum in the river down in England during the siege. It sickened him, thinking of MacDougall's plan. He turned back to Clarence. "I wondered myself," he said. Christina flashed across his mind. "But I never considered it —not seriously—until I read the scrolls Niall left me."

"It doesn't kill you to think of walking away from your son and your mother?"

Shawn nodded. "Yeah. But it kills me *more* to think of leaving Niall to die that way."

Clarence stared at the loch.

"It's not just Niall." Shawn's heart took up a steady pounding. He *needed* Clarence. "There's Simon. I intend to stop him, so James doesn't have to. I'm doing this *for* James!" The seconds ticked by, till he finally asked, "Will you do it?"

Clarence lifted his eyes to Shawn's. "You're asking me to give up everything, my whole life."

"Give up what?" Shawn asked—immediately wishing he hadn't said such words. "I'm sorry," he said. "I didn't mean...."

But Clarence nodded. "You're right. I have no family. No friends, no home. Am I really giving anything up?" He met Shawn's eyes. "But—to live in medieval times?"

"You don't want to?" Shawn's heart sank. But he said, "I understand if...."

"It's not that," Clarence said. "It's that I need to know one last thing."

"I've told you everything. You've seen the evidence, read their letters."

Clarence bowed his head onto clasped fists. His voice, when he finally spoke, came out low. "Are you hoping I get gutted and quartered when we go back? That something awful happens to me?"

"No!" Shock hit Shawn. "No, this is all on the up and up, it's about saving Niall and giving you a chance you don't have here, at least in the States. Why would you even think that?"

"There's this little issue of me killing your father. You hated me for years. Why the change?"

"Oh." Relief and deflation mingled in Shawn. "I don't know." He searched his mind, but he could no more answer the question now for Clarence than he had been able to weeks ago for his mother. "What about you?" he asked. "Do you hate your mom, those step-dads?"

Clarence shook his head, his eyes locked on the water. Moonlight reflected on its ripples. "No, I pray for them. I pity them. I hope you don't pity me."

"No." Shawn shook his head, wondering what the difference was. He saw it clearly almost before his mind finished asking the question. "They're still locked in their messed up lives, aren't they?"

"Yeah." Clarence sighed. "Still drinking, drugging, sleeping around, in and out of prison, miserable, loud, unhappy."

"You've tried to do something better," Shawn said. "That's not pitiable. It's admirable."

"But your father is still dead. How can you possibly forgive?"

"I don't know," Shawn said. "It just happened."

"That's God healing something," Clarence stared at the water another moment before asking, "When does it happen?"

"A couple weeks."

"We'll go home and practice more?"

"That and I have things to wrap up." He waited.

Clarence rose. He walked the short length of the shore, stared up at the dark ramparts, and turned back to the water. Sweat broke out across Shawn's brow. He didn't think he could save Niall without Clarence's help. He definitely couldn't save Niall and Allene both and would likely be captured himself if he tried it alone.

Clarence turned at last.

Shawn stood up, waiting.

"You're offering me more than a fresh start," Clarence said at last. "You're offering me redemption. I can never, ever bring your father back, but I can save your friend. Or die trying. I'm in."

Shawn gripped Clarence's shoulder. He felt his eyes sting. He blinked. "Thank...." His voice went hoarse. He looked up to the high walls of Glenmirril rising dark above them, to the second floor where Niall had lived. For just a second, he thought he saw light flicker in the room that had been his. Then it was gone, a trick of the mind. He turned back to Clarence, clearing his throat. "Thank you, Clarence."

CHAPTER TWENTY-SIX

Crossing the Atlantic, Present

Lulu—life playing forward and reverse—the plane lifted off again, carrying him, with Amy and James, back to the States. The same motif played—Amy beside him, James's happy laughter—but it had shifted from the flutes and violins to the dark, low brasses, alto clarinet, and basses, with more minor chords that stabbed at his heart every time he looked at James.

Shawn waited through the first free wine and wine with dinner before launching his next offensive. "You have great opportunities in the States. With Peter retiring, you know they'll hire you as concert mistress."

"Shawn...." she started.

"Apart from Angus," he said hastily, "why would you turn down such a good opportunity?"

"I've just found it a little..." She hesitated before saying, "strained. At rehearsals." She turned away, staring out the window, and refused to say more, despite his pressing.

Finally, he cleared his throat. "You know, Angus could move to the States with you."

"Well," she replied, as the flight attendant cleared her tray, "there are a few issues with that and it's got nothing to do with taking Peter's place, anyway." She wiped James's face clean, and pulled out another of his books. This one featured a castle on the cover.

"Another one he picked out?" Shawn asked.

"Mmhm." She began reading softly pointing to pictures, and James listening intently, saying, "Cass. Cass-uh."

Shawn sighed, relaxing against the headrest, wondering what she meant by *strained at rehearsals*. And suddenly—he knew. And he'd created that problem, too. He sighed, staring at the overhead bins. If he wanted her fully free to seize her opportunities, he had to fix this, too.

He glanced over his shoulder to where Clarence sat across the aisle and three rows back. Clarence lifted a glass of red wine in toast. Shawn grinned, and turned back, to the sound of Amy's reading. Yes, there was an answer. She wouldn't like it—but he had to try. After all, hadn't she told him the same thing? And she had been right.

Glenmirril Castle, on the Shores of Loch Ness, 1318

From the divan in the solar, Allene helped James move his small wooden soldiers about the landscape of the fur rug at her feet. William, too, lifted the pieces her father had carved, trying to name them in his limited speech. Alexander slept in his swaddling clothes in a cradle in her room, while Christina sat at her easel, James Angus in his own cradle at her side.

A knock landed on the door and on calling *Enter,* Gil appeared. "All is ready, my Lady. The horses are saddled and waiting in the courtyard, and the baskets of food packed."

Allene rose from the divan, smoothing her skirt. "You've firewood for the widow?"

"Enough to last until summer," he said.

"Flour?" she asked Christina. "Another blanket?"

Christina nodded. In the splash of sunlight under the window where Shawn had so often played Niall's harp, her hand rested on James Angus's cradle. "And wee brogans for the youngest boys."

"Are you sure you'll not come?"

Christina rocked James Angus's cradle, gazing at him. "Quite certain. And I wish you'd not go either."

"I'm eager to ride again!" Allene strode to the window to look out on hills and forest. The first spring colors were just beginning to push through the last winter snow. Fat white clouds hung high in a blue sky. She turned back to Christina, full of joy at the prospect of being free of the walls that had closed her in for so many months. "MacDougall is far away in England," she said. "He's been exiled by Bruce and his lands confiscated. He's no reason to return."

"Vengeance," Christina warned her. "He is a man who will always return for vengeance, and he'll not rest until he feels he has it."

"Gil rides with me," Allene said. "I'm only going to the widow's and the village. I'm quite safe."

"What does your father say?"

"He also believes he is in England or near Berwick," Gil answered.

Allene pulled her gloves on. "'Tis Niall he'll seek vengeance upon. I'm far more concerned about him than myself."

"I fear you and father underestimate MacDougall," Christina said softly. In the cradle, James Angus stirred and opened his eyes, blinking in the sun.

"Pray for Niall." Allene gathered her cloak, and followed Gil to the stables.

Soon they had crossed the drawbridge. The worst of the snow had melted, leaving only scattered patches of white in the forest, and mud on the path. Their horses climbed the hill, past the great entwined fir trees, and descended the other side. "I wish I could set Christina's mind at ease about MacDougall," she said to Gil as they rode out into open fields.

"She is wise to worry." He led a horse bearing firewood, and scanned the countryside as he spoke.

"Sure he's in England looking to his own affairs or at Berwick. He'll not be here."

Gil shrugged. "One cannot know with a man like him. Your father, too, has forbidden you to go beyond the village for the near future." He scanned the fields stretching north toward Urquhart, watching for danger, his sword at his side, until they reached the widow Muirne's small cottage.

Gil slid off his horse, and looped its reins over a limb before helping Allene off her mare. "I'll bring the firewood round to the back," he said, as she took her basket to the ramshackle hut.

Allene pushed open the small wooden door, peering into the first of the widow's two small rooms. "Muirne?" she called into the dark interior. Maybe she was in the fields. But her smallest son had been ill. Perhaps she was in the back room with him. Allene stepped into the front room. The door swung to, leaving the room dim.

A motion behind her caught her attention. Before she could spin, a hand clamped over her mouth. Something hit her head, and the weak light in the room faded to black.

Midwest America, Present

"We need to talk about something." Shawn laid his knife down on his cutting board. They'd barely reached home before she'd had to leave again for two days playing with Mike. It was the first chance he'd had to bring it up.

Amy sliced the onion, even as she glanced up at Shawn and the moon shining through the windows on either side of his fireplace. "More surprises?"

"A party from me is hardly a surprise," he said. "I used to throw them all the time."

"Used to," she emphasized.

"Is it a bad thing?" he asked.

She shook her head. "No, it's great, throwing a party for Peter's retirement."

"And for your gig with Mike Mansfield." His eyes shone warmly.

She smiled, but said, "I meant the surprise about Clarence moving in."

He shrugged. "He needed a place to stay."

"You can understand...."

"Mm." Shawn grunted acknowledgment as he slid the venison flanks into a marinade. The aroma of spices filling the kitchen. "Yeah, I know it's weird. But —well that's kind of what I wanted to talk about." He didn't exactly want to bring the name up.

Amy set the knife down. "What?" She scanned his face.

"This forgiveness thing." He ripped a sheet of foil off the roll, wondering what Bessie would be using in the kitchens below Glenmirril for the same task.

"Clarence?" she asked. "It has to feel—freeing. To not...."

"Not Clarence." He coughed and mumbled, "Dana." He busied himself crimping the foil over the casserole dish, trying to pretend the silence wasn't stretching out.

She cleared her throat. "Dana." The word snapped across his eardrums like the release of a bowstring.

He opened the refrigerator, sliding the venison in, scratched his ear and dared a look at her.

Her hands were on her hips, the onion forgotten. "Dana?" she said again. "So you figured out what was making rehearsals strained. Well done. But you thought the problem was my *lack of forgiveness.*"

His irritation swelled and burst into anger. Dana had been his friend, a *good* friend—a *friend*—whatever Amy thought. He slammed the refrigerator on the unsuspecting venison. "Murder!" he spun on her. "I get what Dana did hurt you...."

"It didn't just *hurt,* it was *wrong....*"

"...but, you know—she doesn't see these things the same. It was nothing but a game of tennis to her. It didn't *mean* anything. She didn't mean to *hurt* you."

"And a drunk driver never *means* to hurt anyone." Amy wiped her hands, and threw down the towel. "So I suggest we get rid of that pesky little manslaughter thing they get hit with. No pun intended." She crossed the breakfast nook, through the French doors into the solarium. He followed, watching from the terrace as she threaded among the potted plants below, and over the stone bridge to the veranda. She sat on the iron bench, her feet pulled up on the seat, arms around her knees, staring out the glass wall into the night.

Shawn hesitated, to the melodious trickle of the waterfall, unsure. Push the issue? Leave her alone? Apologize? *Was* he wrong? Wasn't forgiveness good?

He pushed a button on wall, tapped the keypad, and *Moonlight Serenade* crooned softly from the speakers around the solarium.

Berwick, 1318

"Another day." Niall lifted his harp to his knee, at the edge of the campfire. Berwick's walls rose high and strong, just beyond reach of Scottish arrows and catapults, while the Scots settled in for another night just out of reach of Berwick's arrows and catapults.

Fires glowed up and down the ridge. Niall touched the strings of his harp, and let a slow ballad drift out. He was tempted to play one of Shawn's pieces, the one about the pillow and the *phone*. Shawn had seemed to think it was a bad idea, mixing the future with Niall's time. *It might disrupt history,* he'd said, and Niall, in his rare moments of uncertainty, had never known how serious he was. Regardless, pillows and *phones* would take explanations he didn't have the energy to think up. His fingers stirred through the ballad and moved on to an old song about a victory of Kenneth MacAlpin.

He felt, for the time, as if he and Shawn were much closer than seven hundred years. Somewhere, in time and in the world, Shawn was no doubt playing, too, and even though his audience would be different, Niall felt they both, in their own times, were putting joy into the hearts around them. A few men sang the old piece along with him, and the tension of the day melted into the late March mist. Men laughed around his campfire. Men spoke of home. Men smiled. It wasn't so at every fire.

He played a love song, a dance, and another ballad, and, as the fire burned lower, stowed the harp back in its oilskin. The tension of the long day had seeped away in the music, leaving him smiling himself, and hopeful of seeing Allene, James, William, and the new bairn, Alexander. Maybe even soon. He pulled his tartan close and stared across the no-man's land to Berwick. Three times, Bruce and Douglas had tried to take it back from the English. Three times, they'd failed.

"Remembering the last time we were here?" Lachlan dropped onto the log beside him.

Niall turned to him, studying a face that still seemed young. It had been Shawn at the last attack, with Lachlan. "Aye," Niall said. "A terrible blow, Douglas's injury." He hoped Lachlan wouldn't discuss anything Shawn hadn't thought to mention. He would have to claim forgetfulness.

"This time, we'll take it." Lachlan sat silently at his side, arms folded across his chest, watching the town as if he already owned it. They could see the dark shapes of men, small at this distance, walking the walls.

Niall wasn't so sure. Berwick was strong. But he nodded. "Aye, this time, we take it back." A moment later, he added, "You'd best get some sleep. I believe you're on the next watch."

Lachlan clapped him on the back, and headed to his spot by the fire. Rising from his makeshift bench, Niall wandered the length of the Scots camp, till he reached the end, and headed off by himself. He sat on the hill near a small copse, watching the fires burn on the walls across the empty stretch of land, watching the English on the ramparts, looking back out at the Scottish fires. He crossed himself, praying for the time of peace he had seen in Shawn's time.

A scuffle of stones made him leap to his feet, his dirk to hand, his heart pounding.

"Peace," said a voice in the stillness. An *English* voice. "I offer help."

Niall's heart pounded. He was alone and the English were not known for *help*. Shawn's warning, *It's a trap,* shot through his mind.

Midwest America, Present

Feet pulled up on the lacy white bench, arms wrapped around knees, I stare into the black night, to stars twinkling like the shimmering notes of the triangle. I wish I could talk to Angus. I crave his steady wisdom. Isn't Shawn right? Isn't forgiveness always right? I've failed in the very thing I pushed him to do.

But it's different, isn't it? Carol tells me of Clarence's repentance. Dana rips the wound open at every rehearsal, with caustic comments and haughty looks down her nose. I rest my head against the glass, and shiver as the cool air touches my cheek, my shoulder.

The soft croon of the saxophones comes over the sound system, caressing my ear; the lilt of muted trumpets harmonizes. I smile, thinking of dancing under these stars with Shawn, the first night I arrived. I hate how torn I am, how susceptible to both of them. Angus was right to tell me to come back. I have to let one of them slide into the past.

I become slowly aware of Shawn's reflection superimposed over the dark trees outside. He's standing beside me. I never heard him cross the bridge, his leather boots softer than the clarinet now playing.

I swing my feet to the floor, making room on the bench, and he sits beside me, his hands clasped between his knees. "You forgave me," he says. "Why not her?"

"You dare to sit in judgment?" I ask softly. "You were part of it."

"It's only that I don't understand," he protests. "Yeah, I was part of it, and you forgave me."

I study his eyes, deep brown with the gold flecks. "You apologized," I say. "I saw real remorse. I believe—no, I know—you'll never do it again."

"Dana?" he asks.

I stare at the water rippling in the stream. Nat King Cole begins singing Stardust, to the sigh of violins. "She said I'm not even trying to see her point of view." My words tumble out as anger rears. "She said I'm just a little bit outdated, and it didn't deserve all this." I turn to him, my eyes blazing. "Do you see the difference, Shawn? She's not only not sorry, she actually blames me. She would do it again in a heartbeat." My heart races. "Could you have forgiven Clarence—if he took that attitude?"

He stares at his clasped hands. "No," he says. But then he lifts his eyes, his brows furrowed. "But—I had to stop hating."

It's my turn to drop my gaze. "I don't know if I hate her, Shawn. I just know it hurts to see her at every rehearsal. To see her look at me—like I did something wrong."

"She was your best friend," he says. "I just want to fix what I broke."

We sit silently for a moment in the muted light, amidst the strength of the stone walls, and the ripple of the pool, before I say, "So you tell me to forgive her. Have you told her a real apology is needed?"

He shakes his head.

"I wonder why not," I say. I hate the bitter acid in my words. 'Bitter' is not who I want to be. But I know why not.

He clears his throat, and says what we both know. "Dana doesn't give an inch."

"So you thought I should," I say. "Even though I'm not the one who caused any of this."

"I just hoped," he says, and clears his throat, speaking more loudly, "you two could be okay again."

"Because it'll ease your own guilt if Dana and I are like we used to be?"

He stares at the pool, and he's silent long enough I know I'm right.

"You can't unring some bells," I say.

"I'm sorry," he says. "For every single thing. I just want to fix it for you."

He holds out his hand.

Our eyes meet, and after a moment, I place my hand in his, our palms grazing, then clasping. "I want your happiness," he says. "I want your peace, more than anything in the world."

I'm shocked at the intensity in his voice, his eyes.

He wraps an arm around me, and I drop my head on his shoulder. I can feel my love for him exploding. We sit quietly, while Nat King Cole sings Unforgettable.

Outside Berwick, 1318

A man emerged from the scant cover of trees, silhouetted by the moonlight. He wore the armor of the English. He held up his hands.

Niall took two steps back—closer to his own people, if he needed to call for help. "You're English." He spoke bluntly. "Why would you help us?"

"I am called Peter Spalding. My wife is kin to Sir Robert Keith, your Marischal." The man's lip lifted in a scowl. "Edward's governor has ridiculed and shamed her, though she has done him no wrong."

"You'd turn over a town for this?"

"Take me to Keith," the man insisted. He lifted his hands higher, showing them empty.

"You could be hiding weapons in your coat or boot," Niall said.

"I'll walk ahead of you. Send for Sir Robert. He knows me."

Niall made his mind up in an instant. It was dangerous. But failing to bring such an offer to Keith carried a larger danger of missing their best—perhaps their only—chance. He jerked his head, indicating Peter should fall in line before him, and ushered him, murmuring directions, through the maze of fires and sleeping men, to the commanders' tents. A man stood by Keith's tent flap. His eyes locked on the unknown Peter Spalding, Niall spoke briefly with him, and he disappeared into the tent. Two minutes later, in which Niall and the Englishman eyed one another, each full of questions, Keith pushed through the flap, shoving his *leine* into his trews. He studied the Englishman.

"I am married to your kin, Elizabeth," Spalding said.

"Aye, I mind Elizabeth. How fares she?"

"Poorly, milord!" The man's agitation burst out. "It is why I came! She is treated shamefully by the English soldiers."

"Come in." Keith stepped back, and with a glance at Niall, indicated he should join them.

A thick candle burned on a small table, casting soft light on the tent walls, a cot, table, and two chairs. Shadows fell across Peter Spalding's face, giving rise to Niall's fancy of deception. He'd sung too many poetic songs, he thought, for the shadows swayed equally across Keith's heavy beard and hooded eyes.

"It's a disgrace," Peter Spalding said, "a woman of her birth, treated so by common soldiers. It has gone on for months, and naught I do stops it. She was accosted today, going to the shops."

"You love her?" Robert asked.

"Deeply!" the man cried. "And the actions of my countrymen shame me. When I see the bearing and patience of my good wife amid these bespawling dalcops, I'd rather be a Scot." His voice rose. "I beg to be taken to Bruce. I can help you take the town!"

"What is it you offer?"

"I am on watch the night of April 1. The western wall. I'll meet them in Duns Park and let them scale the walls. Sir." He bowed low before Keith. "I beg that I might be allowed to return with all haste, ere any notice I've been gone. If you've questions, I'll answer them. Otherwise, that is what I can and will do for you."

Sir Keith inclined his head slightly to the man. "Wait here." He beckoned his man in from outside, and said to Niall, "Come with me."

It's a trap, Niall. Niall started at the faint whisper. But he had no choice. He rose, following his commander with a backward glance at Spalding.

Midwest America, Present

Dana's eyes lit up when she saw him standing in her door. "Shawn!" She reached to grip his linen shirt, pull him in, kiss him, as she once would have done.

He shook his head, taking her wrists and removing her hands from his shirt, reluctantly. He *had* missed her. Her ginger and cinnamon hair fell straight today. Without her normal moussed-up spikes, she looked younger, fragile. Her white button-up shirt hung to mid-thigh over black leggings. The top three buttons were undone, revealing a delicate silver pendant that drew the eyes down. "You look great, Dana," he said.

She grinned, flicking the leather lace of his bell-sleeved shirt. "Likewise."

He smiled. She *had* been his friend. A real friend. He missed the friendship more than anything. But he'd made himself, and Amy, promises. He cleared his throat. "Can I come in?"

"Mm." She purred and stepped back, holding the door wide. "You know I love it when you do." She winked.

Shawn drew a deep breath as he followed her into her front room, a large airy space of windows and whites. Her French horn rested in its stand beside the grand piano. She loved her home. He remembered the day she bought it, showed it off to him, the most joy he'd ever seen on her face, when their salaries had skyrocketed with albums and tours and trading cards.

Guilt gnawed at his heart, thinking of the calls he'd made that morning.

"Did you come with a personal invitation to Peter's retirement party? I've heard the new pool is something else." She gestured to the couch, white leather like his own, facing a stone fireplace. They were well-acquainted with that couch.

He looked away from it, shame stirring in him. "Maybe we could talk in the kitchen?"

"Sure." In the kitchen, looking out on a wooded yard, she stretched up on tiptoe to reach for two low-balls, her shirt riding high to expose her derriere covered only by the thin leggings. He knew her well enough to know she was fully aware of the impact on him. "What can I get you?" She beamed as she set the glasses on the table. "I still have your favorite bourbon."

He dropped into a kitchen chair. "Water's fine."

The smile slipped from her face. She filled his glass from the refrigerator. "Okay, so I think I'm the last to get it. I'm being visited by the same old *new* Shawn." She set the glass in front of him, and leaned against the white marble counter, arms across her chest. "I take it I'm still on Amy's shit list, then? So why are you here?"

"For Amy." He swirled the water in the glass, but didn't drink.

"Right. Not like you and I were friends. Not like it could be for me."

"It *is* for you," he said. "For us. I want things to be okay between us again. I want things to be okay between you and Amy."

She raised her eyebrows and cocked her head. "There's not much I can do about that."

"Did you apologize to her?" he asked.

"Yeah, I apologized. Over and over." Dana shrugged. "I guess she had a script I was supposed to follow, or I was supposed to grovel or beg or beat myself or something."

Shawn glanced out the white lace curtains. A cardinal perched on a limb, tilting its head at him quizzically. Amy was right. Clarence had shown nothing but remorse. He turned back to Dana, picking words cautiously. "Did your apology include the word *but*?"

"But?" she repeated. "What do you mean?"

"'I'm sorry *but*—it didn't mean anything. I'm sorry *but*—you know what I'm like. I'm sorry *but*....*"

"Yeah, I get it." She cut him off. "Of *course* I tried to explain to her. It's not like you were ever going to leave her. I always sent you back."

"That's not really the point. She's hurt, by the lies, the deception. She feels you're insulting her and brushing it off. She thinks you would do it again."

"This is *ridiculous,* Shawn!" Dana rolled her eyes. "I'm sorry she's stuck in the fifteenth century, but that's not my fault. Maybe she should open her mind, try on some new ideas. You know—join the twenty-first century!"

"Maybe you and I should have opened our minds," he countered.

"She's not the only one who's been hurt." Dana tilted the bourbon bottle over her own glass, and raised it in cheers to Shawn. "But hey, what does it matter if I'm hurt? A rumor here and there, bitchy gossip, my personal life discussed in the string section, ugly looks as I pass. I guess I don't matter." She drained the glass in one long draft, and poured more.

"I know I hurt you," Shawn said. "I know I have, and I *am* sorry and I want to make things right."

She drank half of the second glass and set it sharply on the counter. "But you're not going to change the way you've been treating me, are you?"

"You and I hurt her," he said. "We lied to her. We did something wrong."

She stared out the window, to a small creek running across her property. "Why are you here, Shawn?"

"If you could *really* apologize to her," he said, "maybe you two could be friends again."

She turned back to him, eyebrows raised. "She knew from the start who I was." She snorted. "She knew that about you, too. The only reason we had to lie was because she insists on staying in her little world. Besides...." She drank the rest of her bourbon, and poured a third. "She *wanted* to be blind. She knew. How could anyone *not* have known?" Dana laughed and gulped the drink.

"Maybe you should lay off the bourbon," Shawn said.

"Maybe you should join me and you'd loosen up. Wherever you were hasn't done you any favors, and neither has staying with Amy."

His jaw tightened. "You need to leave," he said.

"Leave?" Her eyes widened. "It's my house. *You* leave."

"You need to leave this town."

She laughed. "Well, look who's back."

Anger flashed through him, and an urge to push her up against the wall. He held himself still, only saying, "You wanted the old me back. Here I am."

"Only the good parts." She sneered. "Such a saint now, aren't you?" Her empty low ball clattered to the counter and with a swing of her hips, she straddled his lap, lips to his ear, whispering, "The sinners have *much* more fun. You know that."

He stood abruptly, pushing her off, and knocking the chair backward. It landed with a crash on the kitchen's slate tiles, as Dana stumbled backward. "Beloit has an opening for French horn," he said. "I've already talked to them."

"Be-*loit*?" She spit out the name with disdain.

"You can also teach at the college."

"Where the hell is *Beloit*?"

"Wisconsin. They can't wait to have you."

"You just went ahead and arranged my life?" She laughed, a sharp report. "You're insane."

His jaw tensed. "I hear Amy's not the only woman in the orchestra having problems with you," he said. "And things are pretty strained between you and Jack. Little uncomfortable, considering you can't get too far from the other horn players."

She scoffed. "That's his problem. I don't care. Send *him* to Beloit. I'm not going to *Beloit*."

His anger flared. "Then don't." The words *But you're not staying here* formed in his mind.

She backed away, anger and pain on her face.

He thought of Christina. Of Amy. Did they really want him playing that game? He thought of Joan, the washerwoman used and discarded by Duncan MacDougall, and compassion welled up in him. He was no better than Duncan. His hands went to his hips. He drew a deep breath, calming himself. "Then don't. But I'm offering you a clean start."

"And if I don't take it," she demanded. "Are you going to drive me out like you did those others?"

Shawn stared at the floor. Shame flooded him. "Fifty thousand." He lifted his eyes.

They watched each other warily. Her eyeliner slowly smudged under her eyes. Her lips tightened. "You can put a price on me," she whispered.

"Dana," he said softly, "are you happy here, with all this?"

She shook her head, then abruptly glared at him. "I couldn't care less about them. They're small-minded, narrow...."

He closed the distance between them in two quick steps, grasping her shoulders. "Dana, we were friends. I *know* you. I knew you better than anyone."

She turned her head, drawing her hand swiftly under her nose.

"I'm sorry," he said. "I'm sorry for my part in this." He pulled her into an embrace. For long moments, she sagged against his chest. He felt her trembling.

Then suddenly she yanked back, pushing him away. "A hundred and fifty," she said.

"A hundred."

She glared at him. "A hundred and buy my house to save me the hassle of selling."

"Done," he said. "Have it written up and to me by noon tomorrow."

They stared at each other. "Okay, she snapped. "You got your way. Just like you always do. Get out."

"I don't want it to end like this," he said.

She swiped a hand under her eye again and spun, grabbing a cloth and scrubbing at the spotless counters. "Yeah, well, check in with your girlfriend if you're allowed to talk to me. But I'm not sure how you expect it to end any differently when you just told me to get the hell out." Her hand grazed the glass, sending it crashing to the slate floor. It shattered, shards flying. Shawn jumped back, as Dana swore. She flung the cloth into the sink and snatched a broom from the pantry, sweeping viciously, her back to him.

He tried to take the broom. "Let me help."

"You've done enough!" She shoved his hand away.

Shame, rage, and guilt all vied. "I hoped...." he said.

"You hoped I would apologize for being who I am!" She pushed the shards into a dustpan and flung them into the garbage.

"We hurt her."

"She hurt herself." Dana threw the broom in the closet, slammed the door, and faced him, her arms thumped across her chest, defying him to disagree.

"I want Amy's happiness," he said.

"Good. Have the contract and a check for me tomorrow and it's done."

"But I want yours, too."

"You can go now," she said.

"Dana," he pleaded, "we were friends. I'm sorry for what I've done to you, too. I was wrong."

She drew a hand across her eyes. "Get out."

"There are five pieces waiting for you in Beloit. You'll be their featured soloist."

Blinking furiously, she picked up an egg timer. "I told you to leave."

"I'm sorry," he said. "I'm sorry I hurt you. I'm sorry for every..."

"I *loved* you!" she screamed. "Didn't any of it *mean* anything to you?" She hurled the timer. "*Get out!*"

Glenmirril Castle, 1318

Christina entered the great hall for the evening meal. Knights lined the tables, though many were gone at Berwick. Minstrels played in the gallery. Niall's mother sat beside her daughter Finola, two seats down from Brother Eamonn. Christina glanced at him curiously, realizing she hadn't seen him in some time. He smiled at her.

Christina glided across the hall, skirting the hounds that lay before the fire, and took her seat between Niall's mother and the old monk. "Has Gil returned?" she asked.

Finola leaned across her mother. "He'd work to do after escorting Allene to the widow's. He'll be here soon. Was there aught ye needed?"

Christina shook her head. She needed only to set her nerves at ease, but it would be foolish to raise an alarm over Allene and Gil simply not being early to dinner. Even the Laird had not yet arrived.

In the gallery, the recorder lilted a sweet tenor melody, with the harp rippling chords beneath it. She lowered her eyes to her hands, clasped in her lap, wondering what Shawn was doing. He'd described his home to her, something small enough to fit in half the northern bailey. Yet he'd laughed when she'd imagined it like the widow's hut.

"More like one of the really nice houses the rich merchants have in Inverness," he'd said, "except it stands all by itself in the middle of a huge field and forest, a mile from the town."

Amy had her own home, an *apartment,* he called it, one of many in a building together, though he insisted it was nothing like their chambers in the castle. She wondered if he'd married Amy, if they'd had more children, if Amy had forgiven him. She hoped—she prayed—for his sake that she had. Hugh would return soon and she would marry him and they would be at peace. She must stop thinking of Shawn.

"Has Allene not come for dinner?"

Christina jolted from her thoughts at MacDonald's voice. Her head shot up. It took no time to see that all the seats, but for Allene's and Gil's, had filled. The nerves took up their steady humming in her arms again. "I've not seen her since she left for the widow's," she said.

The Laird's mouth tightened behind the red and silver beard. The scar on his cheek turned white. He looked around the hall, and called, "Has anyone seen Allene or Gil?"

Christina looked to Finola. Her face had become pale. She reached for her mother's hand. Around the hall, heads shook.

"Men of Glenmirril," MacDonald boomed, already striding the length of the hall to the door, "don your mail, get your weapons and horses, we ride for the Widow Muirne's!"

A great rustle arose, the woofing of hounds and rush of feet as men hastened for the door. Beside her, Eamonn rose. Niall's mother reached for Christina's hand. "Mary, Joseph, and Jesus," she whispered hoarsely. "God protect them."

Midwest America, Present

"Dana's leaving."

In Shawn's music room, Amy stopped mid-Pachelbel, the last sixteenth note dangling.

"She got a better job in Wisconsin."

Amy laid her violin on the grand piano. "She got a better job?" Her eyes flickered over the bump on his temple. "Or you got a better job for her?"

Irritation flickered through him. He was trying to *help* her. He turned back across the foyer to his office, calling over his shoulder, "She was agreeable to it."

"She was my best friend, Shawn." Amy stood in the doorway of his office, watching him pull scores and parts from a shelf. "What made her so agreeable?"

Shawn laid the score in a box and set the parts on top. "It was too good a chance...."

She brushed her fingers over the bruise. "She's a good aim, isn't she?"

Shawn winced—at her touch and at her words. Three years ago—two years, he reminded himself—no, three, by his time—he would have told her a story, tried to convince her her instinct was wrong. He nodded. "Yeah, she's a good aim."

"She played baseball with her brother. I guess she never got around to mentioning that. To you."

"Thanks for reminding me." Shawn's lips tightened as he fit the lid on the box. "I'm trying to help you." He snatched the tape and ripped off a long piece, taping the box shut, then grabbed a permanent marker from the pen holder and, checking the yellow note stuck to the shelf over his desk, copied the address of the Beloit symphony onto the box. He was trying to make it right! He was....

It was quiet—too quiet. He lifted his eyes to the doorway.

Amy was gone.

The smell of wood smoke drifted to him. He went into the foyer, looking down the short hall into the great room. A fire crackled in the hearth. Moon and stars shone through the tall windows on either side.

A glass of amber liquid waited on the coffee table. Amy sat on the couch, her long hair draped over one shoulder. She patted the white leather. "I'm sorry," she said.

He rounded the couch, staring at the dancing flames. He saw Niall, Bessie in Glenmirril's kitchen's, the Laird before the hearth in his personal chambers. He thought of Christina in her blue gown, standing before the fireplace in the great hall. *What kind of a man were you?* Simon was there.

"Shawn?"

He blinked. She sat on the couch, her legs curled up under her, wearing a royal blue shirt that set off her black hair. "I was trying to help you," he said, and almost reluctantly, because he felt the guilt, "When do you stop reminding me of what I did? Are there enough apologies to make it stop?"

"Please." She patted the couch again.

He sank down beside her, and she curled into him, her head on his shoulder. He wrapped his arm around her.

"You're right," she said. "You've apologized. I didn't need to say that."

The fire crackled, shooting tiny red sparks up the flue. He pressed his lips to the top of her head. He wanted to be here, with her in his arms, forever.

She pulled back enough to touch the blackening bruise on his temple. He winced. "So what happened?" she asked.

"She's leaving," he said. "Do the details really matter?"

"Maybe not," she said. "But maybe being able to talk to someone does."

He picked up the drink, swirling it and watching the flames cast lights in the amber liquid. There were things he wouldn't say. Simon—his plans. James's part in it—how he must stop that. *Talking to someone* wasn't really a *thing* in the fourteenth century. Action *was*.

She kissed his temple. "You were right," she said.

He turned to her. "About what?"

"About me. Sometimes I have my own things to learn. To work out."

"Yeah, well." He took a quick gulp of the bourbon, thinking of the same bourbon in Dana's kitchen. "I saw what you meant. I tried to imagine Clarence right there in my face, at work every day, no remorse, no apology. Even sneering at me." No, he thought, Clarence was making the ultimate atonement —a stark contrast to Dana's attitude.

"But you're upset," she said. "About sending her away?"

"Yeah, I am." His voice rose in agitation. "I'm sorry if it hurts you, Amy, but I feel like crap for hurting her, too. I started it, I continued it, I thought...." He stopped.

"She was in love with you," Amy said.

Shawn stared at the fire. "That's about what she said as she threw the timer." He touched the swollen bruise with a grimace and said, "Okay, that's exactly what she said."

"What did you expect?" Amy asked softly. She took his hand.

"I expected," he said tersely, "that she was like me. I thought we understood each other."

"Of course she's not like you," Amy whispered. "She's a *woman.* Of *course* she felt something."

"I thought," he said, "that this would be win-win for everyone. A fresh start for her, with a high profile position, solos."

"You still thought you could shuffle people around your own personal chess board," Amy said. "But you don't always understand the rules each piece follows."

"So what should I have done," he asked. "How could I have made this right for everyone?"

"I don't think you could." She rested her head in the crook of his shoulder.

After a few moments, he cleared his throat and said, "Speaking of shuffling people like chess pieces, I guess—I arranged something for you. I hope you'll like it."

She lifted her head to look at him enquiringly.

"I'm having a short ad made to help promote your tour. And your work.

And anything else you do in the future."

"Shawn...thank you. What made you decide...?"

"You need to be at Eden Court in a couple of days, concert black." He grinned. "You'll see why!" As she started to press for more, he said, "You know how I like surprises! Don't blow this one for me, okay?"

"Okay," she agreed. "How can I argue with a trip to Scotland?" She settled back into his arm.

As the fire died, he contemplated how wrong he'd been, how he'd misjudged and hurt Dana, who had seemed to need and want nothing. He thought of her, even now packing her things in the house she loved, losing everything while he sat here with Amy. He'd managed to hurt anyone and everyone he'd ever cared about.

Did it make it up, in anyway, he wondered, that he was about to give up all he loved? Or was Clarence right? Was he really only running away from all the damage he could never undo, to a place where he never had to face these people again?

The flames sank down to embers. Stars twinkled in the floor to ceiling windows along the fireplace. Amy sighed heavily on his shoulder and he realized she was asleep. He tightened his arm around her, laying his cheek on top of her head. He wanted this moment to remain forever. He wanted a lifetime of her on his shoulder and James.

Clarence was also right. It could all go wrong. He could end up entertaining a medieval crowd with his bowels burning before his own eyes. He squeezed his eyes shut tight, fighting the image.

FOURTH MOVEMENT

CHAPTER TWENTY-SEVEN

Outside Glenmirrill, 1318

Darnley, Morrison, Brother David, Red, even the elderly Eamonn who had returned from his mysterious disappearance with vague explanations of prayer —the Laird looked over his score of men, as they climbed the hill outside Glenmirril. Half of them were old men, monks, and boys. It was no good, he thought, so many of his young, strong men gone at Berwick.

The old monk rode up alongside him. "Things will come aright."

"So you've said," Malcolm replied. Eamonn had followed them, insisting another hand against whatever they might find could not hurt. Indeed, he seemed to have grown stronger since his arrival. His stoop had disappeared, along with his incessant coughing. He had pressed to learn fighting, and Ronan had worked with him. He would never be a good warrior, but one never knew when an extra sword would make the difference. "Though I told Niall much the same," Malcolm said, "it seems there is little but trouble and death for my people. 'Twould be good to know *when* things will finally come aright."

"They will take Berwick soon," Eamonn replied softly.

"Careful." Malcolm glanced around at his men and asked softly, "You know this from your history?"

Eamonn nodded.

"Why did you not tell me?"

"'Twas of little consequence."

"Do you know of Niall, then?"

"He will lead them over the wall."

"And my daughter?"

Eamonn met his eyes. "All will be well," he said.

Ahead, a shout caught the Laird's attention. A pair of figures limped down the road, helping one another along.

"'Tis Gil and Ronan!" Red shouted, and kicked his garron, spurring it forward.

The Laird's eyebrows drew together in fury. "You said all is well! What has happened to them?" His voice rose in anger. "Where is Allene?"

"My Lord!" Darnley pulled up on the Laird's left, laying a hand on his shoulder. "'Tis no doing of his!"

MacDonald flung Darnley's hand away, roaring at the monk, "Why did you not stop her going!" He threw himself off his horse, and reached for the monk's robe, dragging him off his own mount. "Why did you not warn me?"

"I did not know when or how," Eamonn said. "'Twas not told in the letters left to me."

"My Lord." Morrison joined the group. "We must look to Gil and Ronan and be on guard lest there be still enemies near."

As the pair drew closer, they saw a blood-stained cloth bandaging Gil's head, and another around Ronan's upper arm.

MacDonald shot a glare at Eamonn. "If my daughter is harmed, your head shall be on a pike on Glenmirril's walls."

Eamonn met his eyes. "The path is strewn with many stones, but it takes us where we wish to be."

"My Lord!" Gil and Ronan hobbled up, and Gil collapsed at MacDonald's feet. "They've taken Allene! I beg your mercy, my Lord, I have failed you. They hit me from behind."

MacDonald pulled him to his feet. "Who?" He turned to Eamonn. "Where have they gone?"

"How would he know?" Darnley asked in confusion.

Eamonn lowered his eyes to the ground.

"I did not see them," Gil said.

"I saw only the man who attacked me as I went to bring in the wood," Ronan said. "I did not know him."

"'Tis MacDougall wishes us harm," MacDonald said. He seemed, suddenly, to notice Ronan's state, and asked, "The Widow Muirne! Have they harmed her?"

"She canna walk." Anger crossed Ronan's face. "My Lord, might I bring her and her bairns to Glenmirril for safety till we know who has done this?"

"Aye." MacDougall glanced among his men and sent three of them racing back to Glenmirril for more horses, before summoning the rest on to the widow's home.

There, they found the youngest children huddled against their mother's skirts. Bruises marked her face. "They accosted me in the fields, milord," she whispered.

"Was it MacDougall's men?" the Laird asked.

She shook her head. "I dinna ken. But the lad here...he's a good head on him, he has!"

The Laird turned to the boy, a child of perhaps five. He bowed his head, hiding behind his mother.

"G' wahn now, Nigel," she chided him. "Tell our Laird what a brave laddie ye were!" She lifted her eyes, one of them swollen shut, to the Laird, and said proudly, "He hid under the pallet, when they come in. He heard them talking."

"I couldna stop them taking the lady, milord." The boy's lip trembled. "I'm that sorry, I am."

The Laird dropped to one knee, touching Nigel's shoulder. "You could ha' done naught against armed men, lad. Ye see what they did to your mam and Ronan and Gil, aye?"

The boy nodded, though he stared at the floor.

"Ye're a brave lad," Ronan said.

"Ye did the wiser thing to listen," the Laird said. "Did they say where they're going?"

Nigel lifted his eyes. "To meet someone called Claverock, milord."

MacDonald looked from Eamonn to Brother David. "Why would he be meeting Beaumont?"

"I know naught of Beaumont," Brother David said, "but if 'twas MacDougall, we know his real purpose is Niall."

"Which means," said Malcolm, "he'll not harm Allene at the moment."

"Had he wished to harm her," Darnley said, "we'd have found her here. Harmed."

"Did they say where they'll meet Claverock?" MacDonald asked the boy.

"They spoke of an island, milord," the child said.

MacDonald lumbered to his feet, turning to Brother Eamonn. "Lindisfarne is the only island near Berwick. Near Niall. Is that his intention—to hold her there?"

"My Lord," Darnley spoke, "why do you continue to ask Brother Eamonn these things?"

"Let us not rush to conclusions," Morrison warned. "*Iona* was MacDougall's land. 'Twould be the natural place for him to go."

"We know not that it *was* MacDougall," Darnley reminded him.

"'Tis most likely," Gil pointed out. "We've no other enemies."

"Still, there *are* passing brigands who do mischief," Ronan suggested. "If we pursue MacDougall...."

Gil shook his head. "These are men who looked for the widow in the field, lay in wait for Ronan, and waited in the widow's home. 'Twas *planned.* That almost certainly means MacDougall. *Why* he's meeting Claverock is not of such import at the moment as pursuing him."

"We cannot leave until morning," Morrison said.

"Are we to leave my daughter in their hands overnight?" MacDonald countered. "We must take the chance."

They argued the point, squeezed in the darkening room, until the approach of hooves outside interrupted them. Darnley stepped outside, calling back, "They are here with the horses. Let us help Muirne."

Ronan and Gil joined hands to form a seat, and lifted Muirne together, her arms around their shoulders. They squeezed through the small door with her.

"The two of you, take her to Glenmirril," the Laird announced, following them. "Then gather weapons and food and follow us as quickly as you can. The rest of us will set out anon to see if we can find MacDougall on the road."

Midwest America, Present

With guests due to arrive in half an hour, Shawn hit *send* on the e-mail to the Philharmonic, assuring them Dana would be there in two weeks. The strains of Dvorak's e minor violin concerto, floated up from the solarium. In the music room across the foyer, James pounded on the piano, sending sprinklings of melody into the air, punctuated by his short, staccato laughs of glee.

All was right in his world, Shawn thought as he studied the hefty contract from Dana. He should be the happiest man on earth.

Except nothing was right. He'd hurt so many people and he could never repair all the damage. He rubbed the lump on his temple. It throbbed.

He scanned the contract. He'd been considering it for hours since it had arrived, through the final preparations for the party. She was asking too much. He stared at the number, but all he saw was the smudge of her eyeliner welling under her eyes. Heat flooded his face, thinking of that night after his first Christmas party, of the things he'd whispered in her ear, believing her a kindred spirit. The heat climbed, remembering her screamed words, full of anguish: *I loved you.*

He thought of her packing her things in the house she loved.

"Is that ready to go?"

Shawn blinked, and turned.

Clarence standing in the doorway. "The contract?" He indicated the thick sheaf of papers. "Ready to go?"

"Just about." Shawn looked again at the number. It was high, but not ridiculously so. He crossed it out, added ten thousand dollars, and wrote the check.

"Wanna talk about it?" Clarence dropped into the only other chair in the small office.

"I screwed up." Shawn pushed everything into a large envelope and sealed it. "Everything I do to try to fix things just make everything worse."

"Is Amy happy she's leaving?"

From the solarium came the sound of a phrase of Dvorak being repeated over and over, slowly.

"Yes. No?" Shawn shrugged. "Sure, the way you're happy to have a gangrenous arm cut off." He sighed, leaning back in his chair, hands behind his head. "That's not exactly happy, is it? Nothing can ever really make it right. It can only hurt a little bit less someday."

Clarence glanced out the door and lowered his voice, switching to medieval Gaelic. "Have you told her?"

Shawn shook his head, responding in the same tongue. "No. She's going to have these days of happiness. Tonight's a party she'll remember forever. You have everything ready?"

Clarence nodded, ticking off items on long, slender fingers. "Appointments with your publisher and your lawyer tomorrow, ten and noon. Amy and James booked on a seven am flight tomorrow morning, your mom's flying out of MSP tomorrow at eight. Cain is staying with neighbors."

"Cain!" James shouted from across the hall. A single key sounded on the piano.

Shawn glanced across the hall and back to Clarence. "You called my mom? What excuse did you give her?"

"I told her you have a place rented on Skye. She's wanted to go there and see where..." He dropped his eyes. "Where your dad's family came from."

The mention of his father hung between them for a moment. Shawn cleared his throat. "You got her a place there?"

"I booked a cottage and a car for a month."

"She'll need it," Shawn said grimly.

Clarence glanced down at a sheet of paper in his hand. "I got confirmation from the film director in Inverness. They're filming onstage at Eden Court."

Shawn smiled, gazing into a world only he could see.

"What's that all about?"

Shawn's grin broadened. "Well, we just better make sure we live long enough to get to Glenmirril and you'll see."

"Are you sure it's a good idea?" Clarence asked. "Bringing them?"

"Good? Maybe not." Shawn pushed himself away from his desk. He stared at his stockinged toes a long moment before looking up again to meet Clarence's eyes. "But necessary. There's something I need to show Amy there. I want her to have Angus. I want her and my mom to have each other." He stared at the floor again, and said softly, still in Gaelic. "And I'm selfish. I want to see her—them—up to the very last possible minute."

"I understand," Clarence said.

"What about the list I gave you? It's all ordered?"

"The books have been pouring in. I've been getting them on the shelves."

Shawn nodded approvingly. "And the rest?" From the solarium, strains of *Der Erlkonig* replaced Dvorak. "It's coming?"

Clarence nodded. "Spices? Yamaha solid silver flute? Gym equipment. Walk-in tub is on its way and I've arranged with Sir William to do the installation when it gets here." He lifted his eyebrows. "Which happens after we leave. It makes no sense to buy those things for Amy. What's going on?"

Shawn swiveled on the chair, turning his back to Clarence, and dropped his voice. "I want Angus to be happy here."

"Wooden swords, pint-sized bows," Clarence continued. "These are not for Angus."

The doorbell rang.

Shawn spun in his chair and climbed to his feet, breezing past Clarence. James stood in the doorway of the music room. Shawn scooped him up to his hip as he threw open the door to Rob on the doorstep with a six pack of beer in one hand and a vegetable tray in the other. Zach and Kristen and their girls stood beside him. Behind him, the sound of the violin stopped.

"Come on in!" Shawn beamed. "Let the party begin!" He slapped Zach on the shoulder and offered his hand to Kristen, shaking. "Hey, good to see you again! Go on in. Amy's down there. I'll be right down."

Clarence came out of the office after the group passed. "The bows?" he asked.

"I'll tell you about them later." Shawn took the thick envelope with the contract from his desk and handed it to Clarence. "Run it to the post office and join us."

Heat climbed up Clarence's face. He shook his head. "No, thanks. I'll see you in the morning."

Scotland, 1318

Allene swayed in a boat, rocking strangely. A jolt threw her to the side, hitting her head hard...it throbbed...thirsty, trying to call for water, around a mouth thick and dry, unable to get words out...dark water below, reaching.... But her wrists were caught, stuck, and the sea dark all around. A long journey— she couldn't remember why she was on a boat. And why was it so silent? Waves whispered around her, and a soft steady *clop clop clop*. A soft sough of wind. But no voices. Who was manning the boat, the oars?

She yanked at her wrists, trying to pull them from whatever held them. Something touched her shoulder. She jerked away, panicking, and the boat disappeared, leaving her awake, feeling the cloth over her eyes, the blackness beyond, the rock and sway of a pony under her. Something filled her mouth, while a cloth covered it, preventing her spitting it out. She fought panic, the fear of choking. She yanked again at the rope biting into her wrists, feeling numbness in her hands, and let out a grunt.

"Shh," murmured a voice. "Else they'll hit you again."

She could feel them now, riding close on either side. The widow Muirne's.... Who? Why?

The pony bumped along under her. She gripped the pommel, steadying herself. *Stay calm,* she thought. *Listen. Pay attention.* The trail was ascending. It told her only that they were likely still in the Highlands. She couldn't have been unconscious for long.

The men remained silent. She could only guess from the rustle of hooves that there were more than a score, fewer than two.

CHAPTER TWENTY-EIGHT

Midwest America, Present

Shawn waited, alone, in his lawyer's office, a place of sedate wealth, wood paneling, and fine art. Calm suffused him. He smiled, thinking of the party—like the ones he'd once thrown, but with less alcohol, and James thrilling to the electric atmosphere, food, and music. Zach's daughters had taken him by the hands, delighted with him, and he with they. Aaron had arrived with his son and daughter, and even Celine came, though she'd stayed clear of Shawn.

His smile grew. He would carry the memory with him. The only shadows on the evening were Clarence's refusal to leave his suite and Dana's absence—a constant reminder of his mistakes. Nobody remarked on it. But they all knew—and that hurt, too.

Amy had slept in his bed, wrapped in his arms, waking to look at him with love in her eyes in the morning, and kissed him at the airport when he'd dropped her and James off for the early morning flight. "I wish you would have booked us all together," she'd said.

He'd grinned. "A few things to finish up here. And you have that appointment at Eden Court to film that piece. I'll see you tomorrow night."

The heavy wood paneled door flew open. Shawn straightened, the businessman in him snapping his spine straight. He had work to do.

Charles Caruthers III entered, dropping a thick stack of papers on the table before him. "Everything you asked for." As he sat down, he patted the tall stack. "You're sure about this, Shawn?"

"You knew the answer before you asked," Shawn said dryly. "Why bother asking?"

"Because it's an extreme move."

"I just want things in order. Everyone should have their affairs in order."

"It's too detailed. This is not just putting your affairs in order." The older man's eyebrows, feathered and gray, dipped sharply. "I don't mind saying, I'm worried about you." He pierced Shawn with sharp gray eyes. "*Very* worried. There are only so many reasons someone does a thing like this."

"Settle your mind about that," Shawn snapped. "I don't pay you to worry. I pay you to draw up papers."

Charles smiled, undaunted. "I've known you since your first royalty payment." He removed his small round glasses, laying them on the mahogany table beside the papers. "Regardless of what you pay me for, regardless of being a lawyer, I am first and foremost a human being. I've spent four years trying to be just a little more than a lawyer—trying to maybe somehow, in my limited ability, step into that gap for a fatherless boy fighting to prove he's not affected by what happened."

Heat burned Shawn's throat at the unexpected words of kindness. He averted his eyes, snatching up the pen. "Where do I sign?" He hoped the crescendo of emotion in his voice wasn't as obvious to Charles as it was to him. He'd had no idea his lawyer felt this way.

"Tell me first," Charles persisted, "that you're not going to do anything drastic. You have a lot to live for."

Shawn slapped the pen down, his gaze going toward the ceiling even as his eyelids closed. Amy and James and Niall and Allene all swirled in his mind; his mother and Clarence and the Laird—the people of Glenmirril who needed Niall, the grief Niall's beautiful swan-necked mother would suffer if she lost this last of six sons—and in such a manner. *Simon.*

"I'm doing something very drastic," he said, unable to open his eyes. The people of Glenmirril danced behind his eyelids, as they had danced at Niall's wedding, thinking he was Niall. "But it's not what you think." In the silence, he opened his eyes.

Charles stared at him. "Please, Shawn, tell me what's going on. Trust someone, for once, to listen and help you."

Shawn stared at the pile of papers, fighting back emotion.

"If not me," Charles said, "talk to Ben. Your mother. Amy. Someone. Tell *someone* what's going on."

"I just need to sign these," Shawn muttered.

"Does your mother know?" Charles pressed. "Does Amy?"

Shawn shook his head sharply. "No, and you'll be fired if you tell them."

Charles laughed. "Um, Shawn, it looks like you're not planning on having assets left for me to manage, so you do realize that's an empty threat?"

Shawn's jaw tensed.

Charles sobered. "Look, I'm sorry. I shouldn't make light of it. But you understand my concern?"

"I just need to sign these," Shawn said tersely.

"The disappearance." His lawyer gestured at his clothes. "The new wardrobe, the sackbut, your new interest in archery...."

"It's not entirely new."

"A fantastically expensive new wing to swim in."

"Swimming is good for breath control."

"Hiring...." He hesitated. "Hiring *Clarence* to—to *swim*. Who hires someone to *swim*?"

"He's a good swimmer. He's coaching me."

"He's been in *prison* for ten years," Charles nearly exploded. "Shawn, this is me you're talking to. A guy who's been in prison since he was seventeen is *not* qualified to coach you in swimming. And you never hire or buy anything less than the best."

Shawn shrugged. "So I'm an eccentric multimillionaire." He leaned in to scan the documents, verifying his directions had been followed. His emotions screamed. There had to be a better way. "Twenty-five percent to my mother, twenty-five to Amy, twenty-five to Angus." He felt as if a steel hand squeezed the air from his lungs. He wanted to be the one with Amy. They'd been happy, these last weeks.

"Regardless of whether he marries her or not," Charles affirmed. "Yes, it's very clear."

Angus would have to be an idiot, if he refused to marry Amy, Shawn thought. He'd change his mind.

"Subsection D there, do you see?" Charles pointed it out. "I don't get why you want...."

"I don't pay you to get it. I pay you to do it." Pain sharpened Shawn's words, a slash of a whip through the air. He ran his finger under the next line. "Twenty-five to Clarence if he comes back...." He cleared his throat. "If he *comes* by August to claim it. Good, right."

"Comes back? From where? This is the specificity that tells me this isn't just putting your affairs in order. What are you *doing*, Shawn?"

Shawn's jaw tensed. "If he doesn't show up to claim it, his portion is split equally between the other three." He signed the bottom of the page with a sharp scrawl and flipped to the next, covering the hundred pieces of music he'd been able to re-write so far.

"What's the deal with those pieces," Charles asked. "Why isn't your name on them?"

"They remain anonymous," Shawn answered.

"But you wrote them?" Charles asked.

Shawn heaved a sigh, thinking of the irony. Lying had once come so easily to him. Now, when it was almost true, in a sense, that he'd written them, he couldn't say so. Because it wasn't *entirely* true. Someone else had written each of them, even though now, as things stood, none of those people had actually ever existed.

"You own the rights to them?" Charles pressed.

"I own the rights," Shawn said. "There is no one, living or dead, who has any claim whatsoever to the rights on that music. There is no one, living or dead, other than me, who wrote that music. But my name does not go on them. Is that clear?"

"Why are they separate from the rest? The pieces whose royalties go to

Amy and your mother?"

"What difference does it make?" Shawn asked. He wished he could lie as he once had—make something up, stop the questions.

Charles scratched his ear, with a troubled frown. "I suppose because people do the things they do for a reason. There's some *reason* you separated *these* pieces." When Shawn didn't answer immediately, he added, "I *know* you, Shawn. You never do anything without a reason."

"The reason is to fund a music scholarship and the kids I sponsor." He saw his mistake even as Charles arched an eyebrow.

"Which means you're not planning on being around to take care of them yourself."

Shawn threw up his hands, and leaned back in his chair, glaring at the ceiling. The best defense was a good offense. He met Charles' eyes again. "Look at me!" He indicated his own face with a theatrical gesture. "Do I *look* like someone who's about to commit suicide?"

Charles studied him a moment before saying, "No. I believe your friend Rob said it on TV just before you showed up again—you love life. But things have been rough. Not everyone can accept you as you are. Not everyone is willing to leave the past behind."

Shawn cocked an ironic eyebrow, and gave a *harrumph* worthy of the Laird. "Sometimes we *can't* leave the past behind." He picked up the pen, pulling the pile toward himself.

Charles' eyes narrowed. They flickered over Shawn's outfit. "My guess is you're planning on disappearing again. I'm guessing somehow..." he waved a hand over the thick pile of papers, "...you're going to be able to access your money. Live in hiding the rest of your life? Write more of this music?"

Shawn scanned the sheet under his hand, found his printed name, and signed.

"Shawn."

At the stern tone, Shawn lifted his eyes.

Charles had replaced his small, round glasses. "Those people, the ones who can't accept the new you—they don't matter. There are lots of us who care about you, who are cheering you on. Me, Ben, the musicians you work with."

Shawn thought of Tessa, Dana, Celine, Suzanne who wouldn't work with him. Pastor Justin. He pushed a page aside and scrawled his initials on an item.

"The audiences you reach, the people whose hearts you touch with your music, with your stage presence, your humor, your generosity."

Shawn dropped his eyes, and moved to the next page, signing again.

"Shawn."

He looked up once more.

Charles spoke softly. "Whatever it is you're planning, don't hurt Amy. For God's sake, don't hurt your mother again."

"Whatever it is I'm planning," Shawn said, rising to his feet, "you are my lawyer and if you breathe a word to them about these papers, I'll report you to every damned authority there is and see to it you lose your license and never practice law again."

"There's no need for threats." Charles, too, rose. His hand fell on Shawn's shoulder. "I just hope you've thought it through carefully, whatever it is."

"I think of nothing else, day and night," Shawn said. He hesitated, his heart full of Charles's words of concern, and suddenly, hugged the man. "Thank you," he said. "Thank you."

Northumbria, England, 1318

Beatrice's head shot up as the door burst open, yanking back from her husband's embrace at the window.

"Sir Kenrick, I...." The guard, Erol, stumbled to his knees, shoved by a man in chain mail, with a dark, curling beard.

"Lord Claverock!" Kenrick jumped forward, reaching a hand to help his man up off the floor, even as he said, "I'd heard you were back. Why did you not send word you were coming? I'd have met you in the hall."

Beatrice backed up a step, gripping her shift close as she studied her cousin. She hadn't seen him in years, not since he'd been a vile boy, dropping spiders in her hair. The malice in his eyes had not changed.

"I've no interest in formalities," Claverock snapped. He glanced at Beatrice, and stopped suddenly, a smile curving his lips. "Ah, fair cousin. How good to see you again. You're charming in the morning."

"What do you want, Simon?" Her hands clutched the neck of her night shift tighter, feeling his eyes rake over her.

At the same time, her husband said, "Are you staying? I'll order dinner and a hunt."

"I have my own hunt." Simon's eyes flickered over Beatrice and back to Kenrick. "I want two score of your men, ready to ride within the hour."

Kenrick's eyebrows drew down. "For what purpose?"

"Does it matter?" Simon demanded. "As lord of Wendale, you owe me fealty. I'm claiming your men."

"It matters a great deal what for," Kenrick replied. "I believe you've gotten yourself on the wrong side of Edward, and I'll not have my men take part in any traitorous uprising against my king."

Simon spat. "Your king is a weakling and a fool. He won't be your king much longer, and then you'll do well to have supported me. Send for your men."

"Meet me in the hall," Kenrick replied. "Give me a moment to at least pull on my breeks."

He turned, heading toward Beatrice and their bed chamber.

"Now!" Simon demanded. "I have a hard day's ride. Send your man to tell them to get ready."

Beatrice saw the anger cross her husband's face as he passed the window. "Do it," she whispered. "Play along, at least."

He gave his head a sharp shake and turned. "You have disrespected my wife, bursting in like this. I will get dressed, and we'll discuss this in the hall. Erol, take him...."

Simon launched himself across the room, Erol scrambling on his heels. Beatrice screamed, throwing herself between them, even as Kenrick shoved her back. Simon's blade sank into his stomach, twisted, and jarred upward.

"No!" Beatrice screamed.

Kenrick drew several sharp breaths, doubling over her cousin's fist, his eyes wide as he gripped Simon's hand with the knife.

"No!" She sank to her knees, clutching Kenrick's hand, reaching for his face. "No!" She turned anguished eyes on Simon, as her husband's knees buckled, and he sagged against her, gasping for air.

"Now." Simon gripped her jaw in his blood-stained hand. Kenrick collapsed on the floor, gasping. Simon's eyes bored into hers. "Send Erol to do my bidding. You will want to know that your son is even now with my men. A bonny lad, is he not?"

She nodded, numb but for her pounding heart, and turned her head to meet Erol's eyes. She gave a brief, slow nod, trusting he had the wits to know what she wanted from him.

Erol turned, running through the door, shouting.

Simon's hand fell away, leaving her jaw sore, and stained with her husband's blood. "Your son will accompany me." He stooped, yanking the knife from Kenrick's body and stalked out.

She sagged over Kenrick. "Don't die," she said fiercely. She pressed her night shift to the wound, where blood stained his shirt.

"Stop him," Kenrick whispered. "Do what...it...." He drew a long, painful breath, and sagged in her arms. She laid him down gently, before lifting her shift and racing from the room. If she could but get a word with her son....

Midwest America, Present

The moon rose high in the tall windows on either side of the fireplace. Shawn leaned back in his favorite leather chair, gazing at the crackling fire, a cool, sweating glass of his favorite bourbon in his hand. Two photographs lay on the coffee table. Amy with James. His mother.

Rachmaninoff pulsed. *Opus 16 number 3.* A funeral march. A song without words. Yes, that worked for him, right here, right now, on many levels. It was a funeral march. A review of a life.

He'd spent the past months memorizing music, listening to music, absorbing it into his soul forever, writing music to put back in the world—

music that had disappeared because of his and Niall's actions at Bannockburn.

He'd spent months honing his swimming, archery, and fighting skills, and somehow managed to perform half a dozen concerts and have another three books of arrangements ready for the publisher, dropped off this morning. Every one of them would help support James and Amy, and make sure Angus had every resource he could possibly need—in case.

None of it had left him with the exhaustion that had subsumed him in the wake of signing the papers, an emotional drain that came only with the most intense of concerts. And really, he corrected himself, not even then. Nothing in his life came close to the utter exhaustion he'd felt on leaving Charles's office, the words *Don't hurt your mother again* pounding in his ears, as Rachmaninoff pounded the bass notes.

He picked up the picture of his mother. A steel trap clenched his heart, crushing it, till he struggled for breath. How could he do this to her? His eyes strayed to Amy and James. The other option was to harm his own son, to abandon Niall. To risk Simon succeeding. And then they would all be hurt, anyway.

He looked up, drawn by a sense of someone in the room. Clarence stood at the French doors, coming up from the solarium. He pushed his hand through his wet hair.

Setting the photograph down, Shawn gestured at the empty glass on the table, and the bottle beside it. "Might not have access to the good stuff much longer," he said with a grin.

Clarence dropped to the couch, reaching for the bourbon. "Stayed under for two minutes and ten seconds." He poured a smooth flow of amber into the glass and lifted it. "Hit the bull's eye twenty-five times in a row, three-hundred-sixty-seven yards. Seven times at four hundred."

Managing a smile, Shawn leaned forward, clinking the rim of his glass to Clarence's, and they both drank. Shawn set his glass on the table. "We have three days. That gives me time to try to warn him once or twice more. We do a little more swimming and shooting. You booked the pool and the range?"

"Yeah, I booked them."

Shawn lifted his eyes to the two story windows and the stars winking through them. He loved this place. He wanted to be here forever.

Clarence glanced at the photograph of Carol. "You okay?"

Shawn stared at the flames flickering in the hearth. The deep bass of Rachmaninoff pulsed. "I signed the papers this morning," he finally said.

"It got real, huh?" Clarence propped his feet up on the coffee table, sipping the bourbon. "You know nothing's a done deal yet."

"As long as those scrolls don't change," Shawn said, "It's a done deal." He took another drink of his liquor. "At least for me it is."

"I said I would do it," Clarence reminded him. "I haven't been doing all this work to back out."

"Technically, you've been doing all this work for a paycheck."

Clarence's eyebrows drew together. "To begin with, sure. But these last weeks, I've been doing it for exactly the reasons we talked about, behind Glenmirril."

Shawn stared at the pictures of Carol, James, and Amy. "You're giving up everything, your whole life here, if you follow through with this."

"As you pointed out not so long ago, that's no big loss."

"It's still everything you know." Shawn abruptly dropped his feet, and leaned for the bottle, filling his glass. "You have to have a choice in this."

"I do. You said the money stays in my account no matter what. And it's added up."

"Yeah. It has." Shawn grinned. "Those bull's eyes are costing me all right." He sipped his drink and set it down. "But it's not enough to get you through another sixty years, you know, if the job situation still didn't pan out and all."

Clarence waited. Bass tones pulsed.

"I drew up the papers such that if you show up by August 1, you inherit a quarter of everything."

Clarence stared. His Adam's apple bobbled. "You didn't have to do that, Shawn," he said.

"That includes the business itself. You, Amy, Angus, and my mother would run it. You'd have your rooms down there. You'd have a job for life." He laughed sharply. "Hell, you'd have an *empire!* Or at least twenty-five percent of an empire. Which is still pretty good. You'd live in a mansion with a pool. You'd never go hungry."

Clarence shook his head. "Don't do this, Shawn."

"Don't do what?"

Clarence's fingers tightened around his glass. He stared into its amber depths. "Don't treat me like someone who would take your money and back out on you like that."

Shawn picked up the picture of Amy and James. He blinked, fast, drew a quick breath. "If I've really forgiven, I can't use guilt to force you into what I need. It has to be a choice, and it's only a choice if there's something real to come back to, isn't it?"

The bass tones of Rachmaninoff pulsed.

One fear pulsed through Shawn's heart: What if he couldn't stop Simon, and James still ended up having to come back? He thought of the letter. There was one more thing he had to do—one more person he had to see, when he reached Scotland. Just in case.

Clarence drank down his bourbon and rose. "I'm packed, Shawn. The flight's early."

Shawn nodded. "We leave before dawn."

Berwick, on the Border, 1318

Keith and Bruce went back and forth for a full hour, arguing the merits of Peter Spalding's offer.

"It could be a trap," Keith said. "How does a man get in and out without being seen?"

"Tunnels, secret entrances and exits," Bruce countered. "Every city has them. If he's a guard on the walls, he knows them. Do we know aught of any Peter Spalding?"

At that moment, a cough sounded outside the tent, and Bruce called, "Come!"

Lennox pushed through the flap. "I've asked about," he informed Bruce with a bow. "Peter Spalding is indeed married to Sir Robert's kin."

"What of the captured English soldier?" Niall asked.

"I've spoken with him. He knows Spalding, and says the governor has shamed his wife at every turn, for months now, as Spalding says."

"Does it seem enough for a man to betray his own people and turn over his town?" Bruce asked.

"He says the governor set the example, such that soldiers harass her, and shopkeepers mock her when she goes to market. It sounds to have become a very difficult situation."

"For love or hate," Niall said, "a man will do almost anything."

Bruce regarded him briefly before saying, "Tell us again how you met him."

Niall told again of the man coming out of the dark, the direction from which he'd come, the distance from the castle.

"It's been years," Bruce mused, "but I recall tales, from my time in Berwick, of a tunnel that would match where you saw him appear."

They discussed it back and forth for some time more before Bruce asked, "What say you, Robert?"

"Of all I recall of my kin's husband, I think we wager well to trust him."

"If it's a trap?" Niall asked.

"Then we fight," said Randolph. "If not, we have passed our best chance."

"Perhaps our only chance," said Keith. "Berwick is well built."

"We'll do this," said Bruce. "Sir Robert, on the appointed night, you will assemble your men at Duns Park. I'll send Douglas and Randolph to you there, with a few hand-picked men. Only then will you tell them the full truth. They are to secure that section of the walls, and send men through the town to cause panic. I will be there in the morning with our full troops. Have the gates open." To Niall and Keith, he gave one last warning. "None are to know of this."

Keith and Niall bowed to their king, giving their assent.

And in his heart, Niall heard the soft whispered words: *It's a trap! Don't do it Niall, it's a trap!*

Midwest America, Present

Shawn spent an hour after Clarence went to bed writing out four more of the *Lost Songs*, before laying down the pencil and trying to rest in his king sized bed under the forest green covers, in the spacious bedroom he loved. It was useless. He was soon up again, wandering through his mansion, remembering his first look at the place and his thought of how his father would have loved it.

He ran his hands over the gleaming granite counters in the kitchen. He opened the cupboards, admiring the rows and rows of spices. Angus would love them. He might like a few of the cookbooks Shawn had left on the counter, in which he'd printed up some of his favorite recipes—and Amy's.

Shawn wandered next to the dining area, running his hand the length of the long mahogany table. He and Amy had sat here, the first night she'd come to his home, arranging music.

He climbed the stairs to the room Amy sometimes used, and James's nursery. He touched the crib, and held a blanket to his face, smelling his son's scent on it. He looked at the books and toys, picking up several, before moving to the largest room. Shelves lined its walls. A long study desk ran the length of the room. Three boxes of books, still unopened, sat there. He trailed the shelves, studying the spines.

Clarence had used his free time, since returning from Glenmirril, hunting down the titles, and arranging them. There was a full shelf each of medieval French, English, Latin and Gaelic—along with the recordings of Shawn's lessons with Clarence. It wasn't enough, Shawn thought.

There were eight shelves of political history, two devoted to daily life in medieval times, four on the faith and philosophy a traveler to the fourteenth century would be expected to know, four devoted to biographies of the Bruce, James Douglas, William Wallace, King Edward, Mortimer, Isabella, and assorted nobles of England and Scotland, and finally—Shawn crossed the room —an entire bookcase, floor to ceiling, book after book, of medieval warfare, medieval fighting skills, hand to hand combat, weaponry, archery.

He hoped none of it would be needed. He tore open the boxes on the table and put the last books on their proper shelves. *The Butcher of Berwick,* read the title of a slim volume. Shawn studied the face on the cover. It was surprisingly like Simon. He put it on the biography shelf, and sat down to write letters—real letters, in long hand, with a pen. He addressed the first to Angus, the second to his mother, the third to James, the last to Amy, wiping at his nose, blinking his eyes as hot tears stung them.

As the sky lightened to gray, he folded them, tucked them in his shirt, and made a quick circuit of the rest of his house, his bedroom, the solarium. He walked the grounds outside one last time, along the stream and the small waterfall Amy loved, to the clearing where he'd given her the black and white kittens; finally returning as pink rays streaked the horizon.

Clarence stood in the kitchen, their suitcases at his feet and the keys to the Jag in his hand. "You ready?"

Shawn laughed, though it was not a laugh of joy. "I was born ready." He took the keys and as if choreographed, they each leaned to hoist a suitcase, and headed out to the garage.

Northern England, 1318

Fury burned in Simon's gut as he spurred his horse hard into the morning sun, riding for Lindisfarne. One of his men rode alongside each of Kenrick's. He kept the boy, Beatrice's boy, Audric, beside him, a delicate youth of twelve or thirteen, the spitting image of his mother, who had sniveled in Beatrice's arms, clinging to her in her night shift and crying shamelessly as they tore him away.

The man, Erol, had insisted on accompanying them. Simon glanced at him, riding on the boy's other side. He'd thought about killing Erol, too. He didn't trust him. But his men barely outnumbered Kenrick's as it was. He hadn't had time to risk provoking a rebellion before they'd even left Wendale, where more of her people might rise against him. Besides, Erol was harmless—a thin man who barely managed to avoid an outright stoop, and looked as if a breeze might frighten him to his knees.

As the sun reached its zenith, they stopped outside Elsdon, waiting under the edge of the forest, while he sent two of his men into town to gather food. Kenrick's men sat in a circle, waiting, stripped of their weapons. Erol, the little man, whispered fearfully to the biggest among them.

"Stop talking," Simon barked. He looked for Beatrice's boy. He found him quickly, standing fearfully by his horse, feeding it soft spring grasses. A leather thong held his soft brown hair back. It curled gently.

"You!" Simon snapped at him.

"My Lord?" Audric whispered.

"Feed the rest of them," Simon said. "There are oats in bags under the saddles and you can bring grass from the field." He watched as the boy fumbled awkwardly at one saddle. "Don't get out of my sight," Simon warned, "or one of your father's men dies."

"Yes, my Lord." The boy lowered his eyes.

Simon smiled. The boy would give him no trouble. He turned back to see two more of Kenrick's men murmuring to one another. He marched up close, pushing the tip of his sword under the man's chin. "I hear this night's sunset will be glorious," he said softly. "Do you wish to see it?"

The man nodded, gulping sharply with his head forced up at the point of the sword.

"Very good." Simon turned to his own men, who stood like fools around the others. "Why are you letting them whisper among themselves?" he asked.

"My Lord." Herbard stepped forward. "They are disarmed. They asked only when they might relieve themselves and were discussing who...."

"Such a discussion requires whispering?" Simon narrowed his eyes. "Let us not be fools, Herbard. My own men are not exempt from my displeasure."

Herbard nodded hastily, his eyes averted. "Yes, my lord."

The men fell silent, Kenrick's and his own. Simon paced between them and the horses. The boy looked up at him fearfully, standing beside a horse with his hand under its nose, letting it nuzzle his palm.

Simon turned his eyes back to the men. They sat silently under their trees. He paced back and forth, thinking through the coming days and months. Meet MacDougall, escort Niall and Shawn to London. With any luck, James Douglas would race to Niall's rescue and be captured, too. He would prove his loyalty to Edward, get his reward, subdue the Scots, and move on to greater things.

If Edward wished to be obstinate, he'd already opened communication with Lancaster.

His men returned, finally, with the food.

"Eat quickly," he ordered, with a glance at the sun. "We don't stop until we reach Alnham."

CHAPTER TWENTY-NINE

Berwick, 1318

Beside the Earl of Dunbar and Hugh in the dark night, Niall looped his pony's reins in a bush. His heart pulsed to the beat of coming battle. He and Hugh would be the first in. If it was a trap, they'd be the first to die. Under the shadowy trees of the duns, they studied one another. Moonlight scooted from behind thick clouds only long enough to flash from Hugh's eyes. He clapped his hand on Niall's shoulder. "You'll see your sons," he said.

Niall wanted to ask, *how can you be so sure?* But there was no room for such thinking.

"You'll hold Alexander at his christening," Hugh added.

"Let's go," whispered Dunbar.

Niall's heart pulsed.

They slipped out of Duns Park, jogging toward Berwick, silent as the mist through which they ran, and grateful for the thick smear of clouds in the dark sky that covered their run. When they reached Spalding's post, Douglas and Randolph were there waiting. They had not managed to take Berwick from the English for years, and the solid, towering wall at Niall's side told why.

"Just hold it till morning," Douglas reminded them. "Sir Niall, you and your men go through the town causing chaos and confusion." He studied the faces of the men in his group, some of them still in their teens. "We are here to take Berwick. No pillaging, no looting, no harming the citizenry. We wait for the Bruce to come in the morning." He pulled several of his inventive rope ladders with the grappling hooks from the bag on his back.

Niall cast a glance upward. Only one face looked down. It was impossible to tell if it was Peter Spalding waving down at them, hissing, "Now!"

Douglas, Randolph, and two other men heaved ladders upward. The man at the top caught one, hooking it firmly in place, before going down the line securing two more. He missed the fourth, which had to be tossed again. It clanged against the wall, shattering the night silence.

Niall reached for the crucifix at his throat.

Don't go, Niall! It's a trap!

His hand sprang away from his throat. It was bare.

"You've not had it in years," Hugh said softly.

"I forgot." Niall dropped his hand, embarrassed. "Instinct." He couldn't tell Hugh he'd just heard Shawn's voice. Or was it his own intuition telling him something he couldn't yet see on his own?

"Go!" hissed Douglas.

It's a trap!

Niall gave his head a sharp shake. He *couldn't* tell his commander no. He grabbed a step high above his head, the wood rough under his fingers, and hoisted himself up. He couldn't tell his commander a disembodied voice had warned him. Hand over hand he climbed, wondering if the man above, urging him forward, would stab him in the back the moment he turned to help the next man over. Or would the entire English army be waiting below the ramparts for the Scots to scale the walls?

His heart hammered, but the adrenaline pushed him forward. Hugh's words strummed through his head. *You'll see your sons again.* He clung to the promise as he threw a leg over the crenelations, and dropped on soundless leather boots in a crouch, scanning the area for betrayal. Down the length of the town walls, Hugh, Lachlan, and Patrick Dunbar landed, crouching low like mountain cats.

The clouds sailed off the moon, leaving it full and silvery, shining down. "Haste," whispered the man at the top, gesturing frantically with his hand. "Down these stairs. Out of sight."

Niall nodded, but the four of them waited until James Douglas's great black head of hair appeared over the walls, along with three other men. Together, they drew knives and swords, spread along the ramparts, and descended into the dark town. Nerves shot up Niall's arms, heightening every sensation. He might find the English army waiting at the bottom. Hugh, at his back, gave him some reassurance. His foot hit the town street. He paused, peering around the corner. Nothing moved.

He beckoned the men behind him, not daring to speak, and they moved like ghosts into the dark town, with the houses leaning in over the streets, turning them into deep ravines with only a ribbon of black clouds showing high above. "Havoc," Niall murmured. "They want havoc and panic in the town."

With Shawn's warning echoing in his bones, he led his men, more cautiously than he might normally have, through the sleeping streets. His heightened senses picked up the sound of approaching guards, and he pulled them back around a corner. The guards moved on, making bawdy cracks about their hopes for the next night.

Niall inched around the corner. "Lachlan," he whispered, "set fires. Owen, release horses from stables. Hugh, you and I distract the guards at the other gate. We want chaos, not to be caught. We must not be seen or heard."

The night wore away swiftly with the citizens of Bewick waking in bewilderment to smoke and braying donkeys and runaway horses. Nowhere did Niall see any sign that Peter Spalding had betrayed them. It wasn't a trap. So why had he heard Shawn's voice so clearly saying it was?

Berwick, England

He pressed himself against the wall of the inner arch of the Cow Gate, hoping no one would come along, and find him here, in the middle of the night, in chain mail. *The middle of the knight,* he thought sardonically. *Night knight.* Below the helmet, his smile showed. *Like father, like son,* they would say—crazy—anyone who remembered. And there were those who did. There had been a classmate or two—or a dozen—throughout the years who had taunted him about his crazy father. He'd smiled then, too, and gone about his business —the business of an undefeated record in wrestling, football scholarships, an offer from the NFL, and plenty of interest from the very girls his taunters would have loved to have look their way. *Like father like son.*

But Beatrice waited. Knowing that, he'd taken no interest in any of them.

He clutched his sword more tightly. It was the night of April 1. He smiled, waiting for the men to come over the wall, as he knew they would. There was work for this knight...this night.

Inverness, Scotland, Present

An hour after the plane touched down, Shawn exited the elevator to a hall dimmed for the night. Jenny looked up from the desk. Soft prayers whispered through his mind. *Please, God. Please.*

Her eyes lit up. "You're back!"

Shawn gave a weak smile, as he reached for her hand over the desk. "I'm leaving tomorrow morning." His smile slipped. "I need a favor. A big one."

"What would that be?" She tilted her head, curious.

"I wouldn't ask if it weren't dead serious." He glanced down the hall and seeing it clear, leaned forward, saying softly, "I need penicillin."

Doubt flashed across Jenny's face. "Can you not get a prescription?"

"The infection hasn't started yet. There's nothing to diagnose."

"Then how do...?"

"Jenny." He dropped his voice even further. "My son's life depends on it or I would never *ever* ask this of you."

"But you said there's no infection."

Shawn picked up a pad of paper from her desk and jotted a note. He handed it to her. She read it and lifted her eyes to his. "You're serious?"

"Deadly."

"Because I *could* lose my job over it."

Shawn gave a nod to the note. "If you do, you'll be better off than before. If there's a problem, show the note to Amy. She'll take care of it."

"Why not you?"

Shawn closed his eyes for just a second. Her questions were reasonable—entirely justified, in fact—but he wished she'd quit asking them. "I leave early tomorrow morning. I won't be here."

"'Tis not just money," Jenny said softly. "Or my job. 'Tis my *name*."

"What if you left it out somewhere?" Shawn leaned forward. "Please. I have to do this for James. He's going to need this. To *live*. Please."

She bowed her head a moment.

"You don't think I'm lying, do you?" he asked.

She looked up. "No. Your story is mad, I know you'd a reputation for lying, once. And yet—that's not the man who took me to dinner. 'Tis not the man I see in front of me. I believe you." She slid his note into her pocket, picked up her phone, and punched at its buttons. "There are cameras," she murmured, her eyes locked on her phone. His own phone beeped as she lifted her gaze to him. Anger flashed suddenly across her face, and she shouted, "You played me for a fool! Get out of here!"

He threw up his hands, shook his head in mock anger, and as he turned for the elevator, flung over his shoulder, "You knew it was never going to be more than a few dates!" He checked his phone as soon as the elevator doors closed. *It will be on the back counter. I have rounds in five minutes. Be fast.*

Thank you, God, he whispered. *Thank you.*

Northern England, 1318

Erol settled himself by Audric under the trees and the watchful eyes of Claverock's guards. Beside him, the fire crackled softly. He wondered that Simon believed they would fight for him, under such duress. They had heard tales of Lady Beatrice's cousin returning, and his rise to power under Edward. He supposed power brought vanity and foolishness. In the dim moonlight shining down through the trees, he met Audric's eyes, lifting his eyebrows in silent question.

Audric gave the slightest nod of his head. His lips curved in the smallest of smiles. His lips moved soundlessly.

Hemlock, Erol read in the motion. He held back his own grin. Young Audric bore his mother's delicate features. It served them well, as Claverock failed to consider that the boy also bore his father's stamp. He had a quick wit, knowing among other things, the herbs of the forest. He was the most skilled at arms of all the boys in the castle, with far greater strength than his slight frame suggested. Like his father, he had nerves that could not be rattled.

By the time Erol had gathered his things and joined the men, back at Wendale, Audric had stood trembling by Claverock's side, playing to perfection the role of foolish, scared boy. Erol had glanced from him to Beatrice, still in her night shift, held back by one of Claverock's men, and taken his cue.

Erol rolled over, meeting the eyes of the man on his other side, and gave the smallest nod of his head. Audric would likewise be passing the silent message to the man on his other side. It was several minutes before the man beside Erol rolled over—passing the message with a look and a nod.

The moon rose to its crest before Erol heard Audric rustle. From the man on Audric's other side, came a soft grunt, and a sigh. Erol felt himself nod in satisfaction. Audric had slipped his knife into the sleeping man's ribs. He would slide straight into heaven or hell, hardly aware he'd woken on the wrong side of the veil. Kenrick would be proud of his son. The lad had nerves rarely seen even in the most seasoned warriors.

Erol wondered if he imagined the tension heighten among their own men. He lay quietly, waiting, till they heard the rustle of the guard slipping into the wood to relieve himself; till he felt Audric nudge his toe with his own. He nudged the man beside him. Then...*one* soft thump of Audric's booted foot on the forest ground. *Two. Three.*

And the boy rolled to his feet, swift and silent, and Erol and all his men with him, each falling upon the nearest enemy, a swift jab of the knives Audric had spirited from the saddle bags, and those lacking weapons, a sharp fist to throat or gut. In the wood, the guard shouted. And they were bolting, slicing reins tied to trees, as Simon Beaumont jolted from sleep, shouting, as his men scrambled to their feet in confusion. The men of Kenrick rode hard.

"On them!" Claverock roared in their wake.

"Swiftly to Wendale!" Audric called, as they galloped down the moon-dappled forest path.

"We've almost no weapons," one of the men gasped.

"Their horses will not carry them far," Audric replied. "I fear they are not well."

Bannockburn, Scotland, Present

Shawn lifted his glass in a toast. "To the best mother anyone ever had." He felt good. The flight had been smooth, he had the precious drugs, and the restaurant his mother had chosen lived up to its reputation as the best in Bannockburn.

Clarence and Carol clinked the rims of their wine glasses to his. The crystal chimed in happiness.

A smile hovered on Carol's lips. "You still haven't told me what the occasion is."

"Do I need an occasion?" Shawn almost boomed the words. "Eat, drink! Be merry! Live each day like it's your last! This is going to be the best meal of your life, to launch the best...." He glanced at Clarence.

Clarence finished the sentence. "The best month on the isle of Skye!"

"I had your rental car dropped off at my house," Shawn said.

Clarence stared at his white china plate.

Carol turned to him. "You don't look as cheerful as he does. Did he strong-arm you into this?"

"Absolutely not." Clarence smiled. "He invited me, and I wanted to." He reached across the table and grasped her hand. "I owe you so many thanks in so many ways. I can never repay all you've done for me."

Carol tilted her head. "Is that what this is about?"

"Mom," Shawn objected, "you're searching for meaning where there is none. We just want to spend time with you and give you the wonderful gift of our presence." He spread his arms with a laugh. "Best of Scotland, yeah? I want to hear your favorite memories over the years."

"All of them?" she asked.

"Every one!" Shawn beamed.

"Okay, then, our movie nights."

"They were the best!" Clarence turned to Shawn, his eyes lighting up. "Remember the *Lord of the Rings* marathon with fifteen buckets of popcorn?"

"What about the time we went canoeing up north?" Shawn asked.

The food came as they reminisced the best times of their lives. Two hours ticked by, with more appetizers and a double tall mocha. Carol laughed till she cried, remembering Shawn's childhood and the years with Clarence, till Shawn looked at his watch. "Mom," he finally said, "I need to get going."

"You have to be somewhere?" she asked.

He and Clarence glanced at one another, and Shawn said, "I need to see Amy. About that spot she's filming. Let me give you a ride back to my house."

The drive was quiet. She touched his arm once. "Shawn, why do I feel like there's something you're not telling me?"

"Because you have a suspicious mind." He grinned. "Life is good, Mom. No worries."

She turned to look at Clarence in the back. Shawn raised his eyes to the rear view mirror, to see Clarence guiltily avoiding Carol's eyes. "Clarence," she said sternly, "you were never a good liar. Tell me what's going on."

"Nothing," he mumbled. "Just a nice brunch."

"No, you two are up to something," she insisted.

"This was it," Shawn said. "A surprise brunch, a month on Skye. You were surprised, weren't you? Best restaurant in Bannockburn, best meal you've ever had, am I right?"

"Oh, definitely," Carol assured him.

Shawn pulled up to the house he still paid rent on. "Here you go. I really have to hit the road to get to Inverness." He climbed out and circled the car to open her door.

Clarence joined them, pulling her into a hug. "Thank you, Carol. For all you've done for me."

Shawn placed his hands on her shoulders, studying her face. Finally, he kissed her on each cheek, and pulled her into a long, tight hug. He stepped back, touched her cheek, and said, "I love you, Mom." He and Clarence climbed into the car and backed out quickly, waving, smiling, while Carol stood in her driveway.

"Don't say anything." Shawn fixed his eyes hard and unblinking, on the road ahead. His throat burned.

"There's nothing to say," Clarence replied. "We've been over it. You still have many hours to change your mind."

"So do you."

"I've made my decision," Clarence said. "I'm at peace with it. It's your call from here on out."

Berwick

From the streets beyond the Cow Gate, all was silent and dark. James leaned his head against the wall, peering up through the vertical slits in the helmet to the arch overhead. He wanted to clutch the crucifix. But it was under a *leine,* a padded gambeson, and chain mail. He could feel it pressed against his chest. That would have to do. That, and trust.

But if nothing happened, he thought, he would find he'd built his life on a crazy story: Everything he knew, everything he'd done for years, the wife and family he'd given up.

Except—he'd given up nothing. He'd only waited patiently. He had *seen* Beatrice's journal. He'd found it when he was sixteen, and read three pages, before his father caught him, warning him of the foolishness of reading it—or his own, which had been wrapped in the same velvet cloth with hers.

From the street, just outside the thick wall against which he pressed, came a voice—a quick, whispered sound. He could make out no words. Feet brushed on stone and a dark shape appeared.

He kept his breathing slow and silent, wary of his breath rasping against the helmet and giving him away. His hand tightened on his sword. A man raced by, too fast to be seen in the crushing dark of the night. His breathing slowed.

It might not be too late to back out, he thought. He took a careful step to the edge of the thick wall to peer into the street. Excitement and fear warred in him. But he'd read the letters—those his father had allowed him to read—a thousand times. There *was* no going back. There was only the fear that nothing would happen, that he would step out in the morning, in chain mail, to the curious stares of passers-by; the fear he had miscalculated and picked the wrong night.

The clang of iron against stone sounded above him. The rope ladders? Had it already happened? He racked his mind. He could think of nothing other than the grappling hooks that would make that sound. His heart pulsed more quickly.

He had a plan. If he could just find Niall, warn him....

"Quickly!" The voice spoke clearly on the wall above him. "Down these stairs. Out of sight." Half a dozen feet brushed against the stairs. Two men, in leather armor and helmets darted past the cow gate before James could react. Then three more. He realized he had no way of finding Niall in the dark. He slipped into the street with the next group, unsure how he'd even tell English from Scots.

Inverness, Scotland, Present

"I need your help." Shawn found Angus emerging from the pool.

"You again." Angus climbed slowly up the pool stairs, water streaming down his chest. Sunlight shimmered through the tall windows, and the smell of chlorine filled the air. "Make yourself useful at least, if I have to put up with you." He indicated the blue towel on a nearby chair.

Shawn tossed the towel wide.

Angus caught it, holding the rail with one hand, and rubbed his chest. "How am I supposed to help you when I can barely walk?" He rubbed the towel vigorously in his hair. "Fight your fan girls off with my cane, aye?"

Shawn cocked an eyebrow. "Seems to me you're using the cane to fight off *your* fan girl."

"That's not up for discussion."

"Not even in exchange for your cane?"

Angus rolled his eyes. "I know the fan girls can't live without you, but I've managed. Even in my diminished state."

Shawn handed him the cane nonetheless. "Great! So you're in good shape to take James for an hour or two!"

Angus's mouth tightened for just a second, as he steadied himself between the pool rail and the cane and flipped the towel over his shoulder. But his shoulders stiffened as he said, "He's your son, and he needs you. What's this about? Where's Amy?"

"I need to take her somewhere."

"Oh, aye, that's grand." Angus took a cautious step away from the rail. "Didn't you just say she's my fan girl, but I'm supposed to mind him while you two go out?"

"It's not like that." Shawn fell into slow step beside Angus as he made his way across the wet deck.

"Carol?" Angus asked.

"Apart from being down in Bannockburn, she's busy," Shawn said. "You know how it is. She makes friends wherever she goes, has a social life."

"So do I," Angus said as he entered the locker room. "Marching with the police band tonight. Meeting some friends to go hiking." He waved to two men heading out to the pool, and stopped at a shower. "Nothing personal, but here's where we part company. I'm sure you can find someone for James, with all your money."

"Yeah, I can." Shawn placed his hand on the side of the shower, making no move to go. "I can hire a squad of nannies round the clock. But I can't hire love. He loves *you*. And you love him. And you're better than this."

"Than what?" Angus swatted Shawn's hand away, and stepped into the shower, snapping the curtain closed between them. The sound of spray erupted.

"Than to take it out on a child!" Shawn lifted his voice over the flow of water.

"Leave him out of it." Angus's voice came wearily from the shower.

"Yeah, let's leave him," Shawn retorted. "Let's leave him with a stranger because you're standing on your pride."

"Funny thing to say to a lame man." Water spattered against the curtain. "Go away!"

"She dumped me," Shawn reminded him. "Did you forget the part where she dumped me, the ridiculously good-looking multi-millionaire and picked you?"

Two men came in from the pool, glancing curiously at Shawn in his breeks and linen shirt, raising his voice at the shower, from which Angus replied, "She's living in a fantasy world."

"One she'd very much like to share with you."

"You've not told Amy, have you?" Angus asked. "If you'd told her...."

"I'd rather leave James with you," Shawn interrupted, ignoring his words, "even if you do have your head up your ass."

The water stopped abruptly, the last few drops pattering to the tiled floor. "I'm not exactly that flexible anymore, but thank you for your confidence." Angus reached out for a towel and a moment later snapped the curtain open, snapped in his towel. "What is this about? Why are you dragging me into this? James needs you."

"But he misses *you*." Shawn handed him the cane.

"How would we know that?" Gripping the cane in one hand and the corners of the towel, wrapped around his waist, in the other, Angus moved carefully to a bench and lowered himself.

"He says your name," Shawn said. "He loves the books you gave him, and the rabbit, and asks for you."

Angus turned his head aside, his hands braced on the edge of the bench.

One of the other men looked over from his locker. "You want me to get this eejit out of here for you, Angus?"

"He can take care of himself," Shawn said irritably, barely sparing them a look.

"What're these pants you're wearin'?" the other man asked, laughing.

Shawn's mouth tightened.

"Knock it off, Davey." Angus grabbed his shirt from his locker.

"Whatsa matter, Angus," the other man asked. "You know we won't really hurt him. Even if he's wearin' a girly shirt!" He and Davey chortled.

Angus pushed himself up on his cane. "Davey, don't."

"What's he gonna do?" Davey flicked the bell sleeve of Shawn's shirt.

"Davey!"

Shawn spun, hand on Davey's throat, grabbed the other man, and slammed them both against the wall. He kneed Davey's groin, and the man dropped to the floor, groaning. Shawn's fist came up. MacDougall's face swam before his eyes.

"Shawn!"

His name echoed through the roar of battle. *Shawn*, not Niall. He blinked. A balding middle-aged man cringed, pinned to the wall, hands shielding his face.

"Shawn." At his side, Angus spoke softly. "This is not 1314."

"What is it?" Shawn asked. "A time when sniveling cowards bully people because nobody's going to do anything? Because the law protects them?" His hand quivered, ready to strike.

"You'd be missing some teeth if it weren't," Angus said.

Shawn stared at the cowering man trembling in his grip, looking out from behind spread fingers.

"Maybe I would have learned and not hurt Amy."

"And Celine. And Dana," Angus said. "It dinna matter. This is the way things are here and now."

Shawn's eyes narrowed, staring at the man. "Who's he going after next? Someone who *can't* defend himself?"

"Put your fist down," Angus said. "This is not 1314."

"No." Shawn dropped his fist. "But I am."

He brought his knee up into the man's gut, watching dispassionately as he sagged to the floor, gasping for breath. "Next time," he told him, "treat people the way you want to be treated." He turned to Angus. "I'll be in the car. Hurry up."

Berwick, 1318

With gray dawn creeping up the town walls, the skirling of Randolph's men erupted over the bewildered citizens. The English guards, scattered across town dealing with fires and loose animals and fights among the townsmen, ran some for one gate and some for another, shouting in confusion as Randoph and Douglas's men exploded from ramparts and alleys. Niall and Hugh charged from the gate they'd just taken. English knights in long tabards broke ranks, one man dropping his spear, as they turned and ran, shouting, "Retreat! They're on us!"

Men in nightshirts bolted from their homes, shouting for wives and children. Flames leapt high in the air, casting orange shadows on the graying sky. The smell of burning straw and oil filled the air.

"Thousands more, coming over the walls!" Niall shouted.

Women screamed and ran, clutching children in their arms, dogs barked, the whole town filling the streets and running for the castle.

"On them!" Niall shouted, but he and Hugh only jogged, shaking their swords. They'd be dead if anyone turned and saw there were only two of them. The crowd jostled at the bridge leading to the keep, thinning and squeezing across as fast as they could. Niall and Hugh reached Douglas. From an alley across the street, Lachlan appeared. "Every horse and pig in town set free." He grinned.

Among Douglas's men, cries of glee rang out, and one young soldier dashed for an empty house. "Get back!" Douglas roared.

From the house, a woman screamed. Douglas's man burst back out, waving a pair of candlesticks. "Gold!" he shouted, and half a dozen men broke ranks, heedless of charging through the town.

"Bruce wants our town, not the people's gold!" Douglas bellowed.

But another young knight raced off, and soon, Scots ran every which way, heedless of his and Randolph's orders.

"The keep!" Niall shouted.

Douglas spun from shouting at the rampaging Scots. "Dear God," he whispered. The English were charging across the bridge. With the men scattered, there remained only a handful of Scots against the on-coming tide. "Brace yourselves," he murmured.

Eden Court Theater, Inverness, Present

As silently as they'd ridden in the car, they walked into the theater, Shawn slowing his stride to Angus's. Shawn hadn't asked what had happened to the men. He didn't care. He hadn't ask why Angus had chosen to come with him. He'd never doubted he would.

He waved to the woman at the desk as they passed. The sound of a violin floated down the hall, a series of bouncing staccatos—*Der Erlkonig.*

Who rides there so late through the night dark and drear?
The father it is, with his infant so dear.

There was no changing his mind.
Apart from Niall, there was Beaumont.

He holdeth the boy tightly clasp'd in his arm
He holdeth him safely, he keepeth him warm.

"Amy knows I'm coming?" Angus asked.

The father shudders, he quickens his pace
He holds in his arms the groaning child.
He reaches home with haste and dread
In his arms, the child is dead.

No, James would be saved. "I think I told her." Shawn entered the backstage and they stopped by the heavy black curtains. Amy sat alone on the stage, but for the conductor before her, who bowed his head listening as she leaned into the music.

Memories flooded Shawn: the day of his audition when he first saw her; the first day she came to his home; ice skating; writing music with her; the first time he'd kissed her. The longing for her, back in the 1300s, the need to get back to her.

He saw a future he craved: Angus would continue to reject her. He would win her over—completely. He would prevent James going back—somehow. He had all the money in the world. Surely he could save his own son.

The conductor waved a hand. Amy stopped, lowering the violin to her knee. The conductor lifted his head, looking at Angus. Shawn lifted his camera.

Amy turned. Her eyes fell on Angus. Her face lit. Shawn clicked.

"Angus!" Joy filled her voice. She turned to the conductor. "Are we done?"

He shielded his eyes, calling to the cameraman out in the theater, then gave her a pat on the shoulder, a few words, and with a wave to Angus, headed off the other side of the stage.

Amy came swiftly to Angus, her long skirt lifted in one hand. "What are you doing here?"

He looked at Shawn. "You said she knew."

Amy saw Shawn then, and startled. "Shawn! Knew what?"

"I thought I mentioned it." Shawn tapped at his phone, sending the picture to a print shop, and lifted his gaze to her, beaming benignly.

She narrowed her eyes. "Mentioned what?"

"We came to get you." He slid the phone into his pocket.

"Yes, I can see that." She marched to her case, laying open on a table in the wings, and began polishing the instrument to a high sheen. "But what did you supposedly mention to me?"

"He's taking James for an hour or two."

"Really?" She dropped the polishing cloth in the case and laid the violin in. "Why? I know you Shawn. What are you up to?"

"I have something to show you."

"Mm, yeah, you always did." She laid the velvet cover over the violin.

Shawn grinned, unfazed at her sarcasm. "And it was always worth it, wasn't it."

A warm blush crept up her cheeks.

Shawn's grin grew, remembering the first time he'd seen her blush. He didn't have to do this. He could stay.

She met his eyes. "It's just usually you had more class than to have Angus babysit while you tried." She swung her violin case off the table. "Angus, you don't have to babysit James while...."

Shawn took a quick step forward, touching her shoulder. There was James. There was Niall. What he wanted didn't matter. There was Simon. All humor fell from his voice. "Amy, please. It's important."

She looked from Shawn to Angus, who nodded. "You know what this is about?" she asked.

"No." Angus stared at Shawn a long moment and said, "But I think you ought go."

Berwick, Scottish Borders, 1318

With the English charging, the Scots looting, and only the few remaining, James spun to do what no one was left to do. Grabbing the reins of a pony deserted by one of the looters—riding lessons since age five—he sprang onto its back, driving it to Berwick's main gate. An English soldier shouted, brandishing his sword. James leaped from his mount, slamming into the man. Years of American football. *Knock him down, open the gate!* They crashed together to the cobblestones under the arch. In a flash, the Englishman was on top, drawing his sword back.

James's heart pounded. A lifetime of training was not the real thing. Angus had warned him. A night among the Scots, raising chaos as he searched for Niall did not prepare him: *This man would kill him.*

He arched, rolled—American wrestling—sent him flying. He scrambled to his feet, snatching the knife from his boot—a move taught by Angus—and dove in, driving the blade up.

He heard the man's grunt. Saw his eyes widen. Felt warm blood trickle from his mouth. Nothing Angus had taught him had prepared him for this. *Kill or die.* He yanked the knife from the body, shoved it in his boot and threw open the great wooden doors.

He stared at the dead man. His father had described vomiting after Coldstream. He understood now. He tore his eyes away, looking back into the town. He could help them. He could still find Niall. Warn him. Spare everyone the awful event to come. But then he might be caught—commanders questioning who he was—unable to do his real job.

He grabbed the pony's reins, and leapt on. Wendale and the Lady Beatrice. *Bailedan* weren't far—he had plenty of time. And he desperately wanted to see her—finally—after a lifetime of waiting.

He spurred the garron, and rode out, racing across the field toward Bruce's army, waving his arms and shouting. "The gates are open!"

Glenmirril, Present

They stood on the hillside high above Glenmirril, where they'd stood so long ago, when they'd hiked above the castle. A gigantic fir tree spread its limbs like a circus tent.

"In here." Shawn stooped, lifted a bough, and climbed under. "I want to show you something."

"What?" Amy asked. "You haven't answered my questions."

"I'll answer one of them if you come in here," he called from inside the cover of pine branches.

She stooped and wiggled through, letting the limbs fall back in place. A dim, dark-green world closed around them, heavy with the scent of pine.

"The west side of the tree is here." Shawn patted the trunk as he might pat Niall's shoulder. "The side away from Glenmirril."

"Okay, so? Why did you drag Angus into this? It wasn't...."

"So, did you watch closely how to find this tree?"

"It's just up the slope from Glenmirril. It's pretty distinctive. Why....?"

"I want you to come here tomorrow with a shovel and dig on the west side. Dig deep."

"What kind of game is this?" she asked. "You haven't answered anything."

"I've answered where we're going and why." Shawn grinned. "I think you'll like what you find."

"Why not now?" she asked.

Shawn lifted hands in apology. "I didn't bring a shovel. Can I see that crucifix for a minute?"

"Why?" she asked again.

"Just curious. Niall talked about it. His mother was quite upset he'd lost it, but for all the hullabaloo, I've never really seen it close up."

Amy lifted it over her head, slid it down the length of her hair, and handed it to him. He dropped down on the carpet of pine needles, inside the massive veil of branches, studying it.

Amy sank to the ground, wrapping her arms around her knees, watching him turn the crucifix this way and that. "You want to tell me about Clarence?" she asked.

Shawn lowered the crucifix to his lap. "Not really."

"You know your mother talks to me."

"Yeah, you two are close." He lifted his eyes to her. "Would it stay that way—if you were with Angus?"

"I would hope so," Amy said. "Not that he shows any sign of giving in."

Shawn watched her, holding his tongue.

"Would that make you uncomfortable?" she asked. "If we weren't together?"

"No, I want you to stay close to her." He leaned back against the tree,

gazing up into the leafy canopy. "Something very interesting happened here."
"To you?"
"No. Christina had a baby."
Amy felt as if she'd been hit in the chest. But she wouldn't ask.
"Not mine!" Shawn said quickly. "Why does everyone assume?"
Amy raised her eyebrows high. "Huh. No clue."
Shawn scowled at her.
"Who else assumed?" she asked.
"The Laird. He's got his medieval panties all in a medieval bunch. Had."
"Your reputation preceded you," Amy said. "By seven hundred years. Very impressive, really." She couldn't resist laughing as his glare deepened. "Really," she said. "What did you expect? So if it wasn't yours, whose was it? Honora Stewart—that's the old lady who has the history—said it was a big question who the father was."

"Still a mystery centuries later," Shawn said. "I don't know, either. He was born in March. That means he was conceived about the time I left. I killed her loser ex-husband, but I don't know when he got into Glenmirril." He turned his head, revolted by the unspoken words. *And if he found her before I found him. Or if it was the elder MacDougall.* He felt sick at having left her to her fate.

When he returned, she would be married to Hugh. He would be alone—with the added torture of being so near her. It didn't matter, if it meant saving Niall, stopping Simon, and saving James from having to go back.

Amy watched him, sitting against the tree, his arms looped over his knees and hands clasped. He stared at the ground. "Why won't you admit you loved her?" she asked. "I can see in your face that it hurts thinking about it."

Shawn's lips tightened.

"What was she like?" Amy asked.

He stared into the pine needle carpet as he spoke. "Quiet. Regal. Peaceful."

Amy had no desire, this time, to poke fun at him. The old Shawn never would have been drawn to such things.

He lifted his eyes. "I compared the two of you a lot."

Amy bit her lip. "And?"

"And at first, I found you lacking. And then I realized that everything I admired in her was exactly what I'd admired in you." He dropped his gaze again. "Until I bullied it out of you."

Amy's breath caught in her throat. There were levels of remorse, she realized, and Shawn's had just gone deeper. She wanted to reach across the space inside the pine tent and hold his hand, fall into his arms.

He lifted his eyes. They were bright, even in the dim light. "I'm sorry," he said, and his voice cracked. "If I had it to do over, I would. I'm sorry for everything I put you through."

"Did you do the same to Christina?" she asked.

He shook his head, once again studying the toes of his leather boots.

"That hurts," Amy said. "Even if it was as strictly platonic as you said...."

"It was," Shawn said.

"It still hurts that you would treat her better than you treated me."

This time, Shawn met her eyes and held them. He reached for her hand. Warmth shot through her at his touch. "It's not about you or her," he said. "It's about what everyone here accepted and allowed. A lot of people even applauded me for it—admired it. *There*, someone would have done me serious bodily harm if I'd mistreated her. You had no one backing you up."

"No," she agreed. "I had you and Dana and the whole world pressuring me."

"And I'm sorry I *wasn't* a better man for you."

She squeezed his hand. "Do you ever think—James came from it. And that's good."

Shawn gave a short laugh. "And saving Scotland."

"And meeting Niall."

"And starting a new fashion trend." Shawn gestured at his breeks and leather boots.

They laughed.

"I'm not sorry anymore," she said. "When I first knew, I wanted to kill you. But there were so many good times, too. I found a talent for arranging I didn't know I had, with you."

"You would have found it sooner or later on your own," Shawn said.

She slipped her other hand into his. "Thank you anyway, even if that's true. Thank you for James. Thank you for setting me on a road where I'm being asked to do arrangements for other orchestras."

"You are?"

She smiled. "San Francisco called this morning. They want an entire program of Scottish music, including my solo on *Wild Mountain Thyme*."

"Congratulations." His face lit up. "You're going to be great. When?"

"They want it in two months. Then I'll fly over for rehearsals. The concert is this summer."

"That's great." His eyes burned into hers. He pulled her into a hug. He made no move to do more. "I wish I could hear it."

"There's no reason you can't."

He sat back, releasing her hands. "Well...you know. Ben's keeping me busy. We'll see." He leaned back against the tree, with a great sigh.

The green filter of light dimmed overhead. "Maybe we should be getting back," Amy said. "It must be late."

"Yeah, we'll go," Shawn said. "But you'll dig here tomorrow?"

"Yeah, okay, but can't you just tell me?"

His eyes twinkled. "It's a surprise. You know how much I like surprises!"

She glanced around the tree, remembering his words. *Something happened here. Christina had a baby.* "James Angus," she said.

"What?" Shawn's eyebrows furrowed.

"Christina's son was James Angus," Amy said. "In Honora's family tree, it seems he's Angus's ancestor."

Shawn stared at her, seeing the future he envisioned, and said softly, "The Laird was wrong. Something good came from a MacDougall after all."

"Yes." She smiled, just a moment, before saying, "You started telling me something happened under this tree, and it started with Christina had a baby."

"Oh. Yeah." Shawn looked around the area. "They hid her. You know, not so accepted in that time to be a little bit pregnant with no apparent father."

"Because Duncan had divorced her by then."

"*She* divorced *Duncan.* Anyway, yeah, when James Angus was born, they hid him under here in a box." He told the story of the purported hunt that drew Lachlan to the area, and the discovery of the child.

"Like Moses," Amy said. The tree suddenly seemed as full of meaning and history as the great cathedrals. It was all too easy to imagine Niall crouched under the boughs, listening to the snuffling of the boar.

"Let's hope he also did great things for his people." Shawn climbed to his feet. "It's going to get dark," he said. "Let's get you back to Angus. I think you should just stay there for the night."

"What?" She rose, too, laying a hand on his arm. "Where are you going? Won't we get James and go to the hotel?"

"I messed up the reservations. There's no room." Shawn's jaw tightened. His breathing became shallow and quick. "I'm not father material."

"What are you talking about?" she asked. "You walked on the shore with Niall's son and took him to see horses. My nephews love you."

Shawn shook his head and spoke sharply. "Amy, you've always been naïve. You've always seen what you wanted to, especially about me. Except for giving you James, I was never any good for you, and I'm no good for him." He pushed through the boughs, half-ducking to wiggle out, and by the time she followed, he was marching like a warrior to battle down the path to Glenmirril.

She jogged a few steps to catch up. "Why are you saying this, Shawn? What does this have to do with the tree and tomorrow?"

When he finally spoke, another twenty feet down the path, his voice was *pianissimo.* "You have a great future, Amy, with someone who loves you better than I ever did. If he had any sense, he'd be calling to find out what we're up to." Shawn crossed onto the parking lot, aimed his fob at his car, and gave it a hard snap.

She climbed in, and they drove in silence to Angus's home. There, he pulled a bottle of wine from the backseat. "A celebration," he said, and suddenly snapped his fingers. "A corkscrew! No, go on in, and I'll run and get one." He pressed the bottle of wine into her hands.

"I'm sure Angus has a corkscrew," she said.

"No, let's not bother him. I'll just go grab one."

"Well, at least come in and ask—and James will be asleep soon. You need to see him before...."

His jaw tightened. He swallowed hard. "Don't make this hard, Amy."

"Make what hard?" she asked in confusion. "Look, just come in and find out if Angus has a corkscrew and if not, at least see James before he falls asleep."

Shawn threw the door open abruptly, and almost stormed into Angus's house, with only a perfunctory knock on the door. By the time Amy followed, he stood in the front room, clutching the sleeping James tight against his chest. In his chair, Angus looked on in surprise. Shawn's head bowed over his son's peaceful face. He kissed his forehead, several times, and looked up at Amy. Tears streamed down his face. He kissed James again, held him close for a long time, and finally handed him down to Angus.

"What's going on, Shawn?" Angus asked, as he accepted the child.

In lieu of an answer, Shawn reached to shake Angus's hand, then turned to hug Amy, his face buried in her hair. "I loved you," he said. "The west side tomorrow. Dig deep."

He released her and almost ran from the house.

Amy's heart pounded. "Angus, what did that look like to you?"

Outside, they heard his car start.

"A man saying good-bye," Angus said. "Amy, he's going to try to go back and save Niall."

"He can't," she said. "He doesn't have the crucifix." Her eyes opened wide with the sudden realization. "Angus! He asked to see it. He never gave it back!"

Berwick, Scottish Borders, 1318

The English charged from Berwick Castle, a treacherous wave rolling toward the few Scots.

The words *It's a trap* shot through Niall's mind. But Peter Spalding had not set a trap. He'd let them in, given them free rein to spread chaos. It wasn't his doing the Scots had broken ranks or that the English had rallied and charged before Bruce's arrival.

From the cluster of Scottish knights, William Keith broke, running straight at the charging knights.

"What in God's name is he doing!" Douglas demanded.

"The horses!" Niall saw what he was doing. Those that had been scattered were in the alleys of the town. Keith was running for them, grappling at reins, throwing himself on one, and dragging another back. Niall was on his heels, racing down another alley, leaping onto the nearest beast, grabbing another, and driving it back to the Scots. They had a better chance on horseback.

The horses flew before him. Galloping back onto the main road, he saw Keith wheeling, plowing down another street, driving more horses back to the

Scots. Douglas, Hugh, and Randolph had managed to mount. They met the oncoming foe with a crash, as Keith and Niall burst from side streets, driving more horses toward their small band, and spinning to fight.

"The gates!" Niall shouted. "They have to be open for Bruce!" He wheeled his horse, to find his way blocked by three English knights. The Scots were surrounded! Allene, his sons, Alexander whom he'd never even seen, flashed through his mind, freezing him. His heart pounded. He didn't want to die!

Keith spun his horse, rose in his stirrups screaming *hiya!* and charged into their midst, swinging. Niall broke from his paralysis as Keith crashed, one against three, into the English. He charged after him, leaning low on his horse, driving it mercilessly, and plowed in behind, scattering the small group.

"Get the gate!" Keith shouted. "I'll hold them."

Niall raised his shield against an attacker and drove his sword at one of the English on Keith.

"The gate!" Keith shouted. "Open the gate!"

His heart sinking, fearing what he'd come back to, Niall deserted Keith. He skidded before the arch, confused. The gates were flung wide, and a rider raced away, waving wildly at the army beyond the city walls. His heart thrilled to the sight of Bruce's force thundering across the field. He launched himself back on his horse, charging into the fray, shouting, "Bruce is coming! Bruce is coming!"

Action froze. Panic flitted across the faces of the English, beneath their helmets, and the energy of battle shifted, as they wheeled their horses, shouting, and bolted for the safety of their castle.

"Hiya!" Niall roared, brandishing his sword, driving, with the Scots, after them, into a melee of horseflesh, armor, swords, and pounding hooves.

Inverness, Scotland, Present

The phone trilled Carol's ringtone. Amy's instinct leapt into high gear. She knew she was right. She heard Carol yelling before she finished saying hello.

"Amy! Has Shawn been there? What's he planning? He left me papers, his bank account numbers, signed everything he owns over to you, me, Angus, and a trust for James. He left his phone behind. What's he doing?"

Amy's knees turned watery. She sank onto the couch, Angus reaching for her. "Carol, he's going back."

"Back? Back where?"

"Are you okay, Amy?" Angus asked.

She nodded, but her voice shook as she spoke. "Back to save Niall. He wouldn't tell me what, but something bad is going to happen to Niall."

"Niall who played the harp?" Carol sounded as panicked as Amy felt. "Where is he? Why is Shawn signing everything over to me? Is he upset about something? He's not going to commit suicide, is he?"

He may as well. Amy held the words for Carol's sake. Her heart hammered in her chest. Her arms tingled with agitation and nerves. "No, but Carol, we have to stop him all the same. He won't be coming back if he does this." To Angus, she said, "He signed everything over to you, me, and Carol. Do you know where Niall died?"

"What do you mean where Niall died?" Carol cried over the phone. "How's he saving him if he's already dead?"

"He's been at Berwick, Lochmaben, Glendochart, and Glenmirril." Angus strained forward in his chair, holding James. "Tell her to look for the scrolls. Maybe he left them."

At the same time, Carol asked, "What's happening? What's he doing?"

Amy took a deep breath. "Carol, I need you to search his room for some scrolls. They're in an oilskin bag, dark brown...."

"I know them," Carol said. "I'm on my way up the stairs. Any idea where he would have put it?"

"The closet. Under his bed. In his desk. I don't know."

"Okay, I'm looking," came Carol's voice over the phone, and the faint sounds of rustling.

"Bring your laptop." Angus said. "Bring me my phone. I'll see if Charlie knows anything, but my guess is Berwick."

"Why was he at Lochmaben?" Amy asked, even as she ran down the hall to the kitchen. She came back with Angus's phone, almost dropping it in his waiting hands as she spun on her heel and flew up the stairs to get her laptop. "Have you found it?" she asked Carol.

"It's not under the bed. I'm trying the front closet."

By the time she got back down, Angus had Charlie on the phone. "Charlie, it doesn't matter why," he said insistently. "I need to know anything, anything at all, about where Niall Campbell died. Do you mean to tell me in all these years, you've never come across anything about his death? There was a trap."

Amy could hear the squawk of Charlie answering as she logged on, clutching her own phone to her ear. "Where are you, Carol?" she asked.

"Going through his desk. It's not here. I couldn't miss something that size."

"Try his bedroom closet." The internet sprang to life.

"Can you go to the archives right now?" Angus demanded. "You know I wouldn't ask if it weren't important."

"I'm in his closet," Carol reported.

"How long will it take you?" Angus asked.

"He can't do this!" Amy pecked with one finger in the search box, while holding the phone.

"Good, very good!" Angus sounded relieved. "Search everything. We know there was some sort of a trap, I'm guessing at Lochmaben or Berwick. I'm guessing the date is April 2 or 3, 1318, but I could be wrong on the year."

"I found it!" Carol shouted in triumph.

"I'm sure about the date," Angus said.

"Thank God!" Amy breathed. "Open it. Look at the dates."

"I'm sure we're looking for something that happened this exact day," Angus insisted into his phone.

"Find the latest ones, Carol." Amy stabbed enter, and the search results flashed up on her screen. "That's where we'll find out where he's gone."

"Call me as soon as you find anything," Angus said. "Anything at all."

"Amy," came Carol's shocked voice, "this is all in—what is this? Medieval English?"

"Yes, it must be." Amy hit the first link promising information on *Niall Campbell Berwick.*

"As fast as you can," Angus said. "Please! Aye, I'll tell you later, but right now I need anything, as fast as you can find it." Angus snapped his phone shut.

"Angus," Amy said, "it's in medieval English."

"She'll have to wade through and see what she can find, or start driving up here." He punched a number as he spoke. James stirred in his arms. "Marjory," he said. "I need your help. Right now. Anything you can find from anywhere—on the internet, anything you've ever read or heard—on where Niall Campbell died."

"I can make out some of it," Carol sounded uncertain. "Is there anything in particular I'm looking for?"

"Is there anyone you can call?" Angus insisted. "Helen O'Malley? Do you know her?"

"Anything about Berwick or Lochmaben, I think." Amy scanned the page as she spoke to Carol. "Otherwise, Angus wants to know if you can get up here as fast as you can. He can read it." A thought occurred to her. "Angus, he'll have left all of this till the last minute. He's not going to have left us time to spare for her to drive here and us to drive wherever he might be."

"What do you mean he won't have left us time?" Carol wanted to know.

"Let me get you her number." Angus touched Amy's sleeve. "Where's Helen O'Malley's number? Marjory will call her and see what she might be able to find out."

"Just a minute," Amy said to Carol. "I need to get Helen's phone number. See if you can find anything about Berwick or Lochmaben, and I'll call you back." She ended the call and scrolled through her contacts. "I'm scaring her," she said to Angus. "What in the world am I doing, scaring his mother half to death? How could he do this to her? To James?"

"Do you have the number?"

"Right here." Amy read it off.

Angus repeated it into the phone. "Thanks, Marjory, and give her my apologies about the time, but 'tis vital we find out now. Morning will be too late."

He ended the call. They stared at each other.

"Let's stop and breathe," Angus said. "I'm thinking tea."

"We don't have time for tea," Amy said. "I know him. Whatever's he's doing, he's left himself exactly enough time to get there to make sure we don't stop him."

"We've no time to panic," Angus argued. "We must think this through. We know he was at Lochmaben and Berwick."

"And Glenmirril. And Glendochart."

"Why four places?" Angus asked. "Were their letters unclear? Do they themselves not know where he was killed?" James stirred in his arms.

"What about James?" Amy asked. "We can't take him with us. What if he disappears? Because if it has to do with that blessing and prophecy, he's Niall's descendant, too."

Angus stabbed at his phone and put it to his ear. "Mairi. Can you come over? Right now? Thanks."

"Just like that?" Amy checked the wall clock. "It's past eleven at night."

"She knows I'd not ask if it weren't important. Now why four places he went shouting?"

"Okay, he knows you were able to get a message to him at Glenmirril," Amy said. "So he must have figured he could get a message to Niall there, before it happens. Can we rule out Glenmirril?"

"As long as Charlie's heading there," Angus said, "let's rule out nothing." He placed another call. "Charlie. I need you to look around Glenmirril, as long as you're going. See if Shawn Kleiner is there."

"We still can't get to three other places." Amy's phone rang as she spoke. She hit *talk*.

"Amy," came Carol's voice, "I'm starting at the end. I've found the words Lindisfarne—I think. I'm sure. And Berwick. But I'm having a hard time making heads or tails what happened there. And these are dated 1318. What does it have to do with Shawn?"

"Carol." Amy almost breathed her name in compassion. "Please. Just tell me what it says. Is there a way you can scan it and send it to me?"

"I don't have a scanner. I don't know enough people to know who would. Especially at this time of night!"

Amy turned to Angus. "Should she drive up here?"

Angus reached for her phone. "Carol, read it out loud, as best you can. If need be, spell every word." He paused, before saying, "Aye, he could. Shawn was fluent in medieval English, French, Gaelic, and Latin." There was another pause before he said, "No, 'twas no trick he played on that talk show. He spoke all four fluently. Please, Carol, later. Read the scroll to me. Letter by letter if you must." He lifted his eyes to Amy. "Can you take James? And get me pencil and paper?"

She nodded, swallowing hard. James squirmed, threatening to wake up, as she took him. But he settled on her shoulder as she hurried up the stairs and laid him in his crib. As quickly, she raced back to the front room with paper and pencil, and laid them on the coffee table for Angus.

His phone rang. She picked it up and hit talk. "Charlie? Yes, Angus is on my phone with Shawn's mother." She noticed the page still up on her computer, and scanned it as she listened to Charlie.

"I've searched all of Glenmirril. I've shouted his name a hundred times. If he's there again, he dinna wish to be found."

"Okay, thanks, Charlie. Anything in the archives?" She hit the back button.

"Can you spell that word?" Angus asked.

"I'm just after going there now," said Charlie. "I'll call you if I find anything, but 'tis unlikely."

"Okay." She hit the second page promising information on Niall Campbell and Berwick. There were 656,000 results. And that was for only one spelling of his name. And the information might be found by searching Holy Island or Lochmaben. She bit her lip. It was useless. "Thanks, Charlie, for trying. Please, if there's anything at all, any chance, we need it."

She thought of James, asleep in his bed upstairs, unaware his father was even now planning on walking out of his life forever. Anger washed over her. How could he be so selfish! At the same time, she thought of Niall. What had driven Shawn to do such a thing? She couldn't bear the thought of something awful happening to Niall, either. "Bye, Charlie." She hung up the phone, and scanned the second page. Angus's pen scratched. She glanced up, assured there was nothing else she could do to help there, and went back to the search results for the third page.

Angus's phone trilled. Helen O'Malley's name popped up on the screen. "Amy," she said, before Amy finished saying hello. "I've just had a call from a Marjory, at the Glenmirril archives. She said you need to know where Niall Campbell died."

"Right," Amy breathed. "I'm sorry for disturbing you again, especially...."

"Oh, 'tis great craic!" Helen said. "And wouldn't you know now, I put in a few calls, checked a few sources, and I found a paper about Edward II that mentions the hanging of Nel Cambel, Highlander, in London, with one Alexander MacDougall receiving payment for bringing him in. It has to be him!" She sounded excited, as if finding information on a treasure hunt.

Amy felt sick. But London was what they needed to know. "Thank you," she murmured, and turned to Angus. "Angus, she says he died in *London*. There's no way we'll find Shawn if he goes to London."

The doorbell rang, and Mairi flew in before they could rise from their seats. "Is everything all right?" she asked in a rush. "Angus, you're all right?"

"Fine," Angus said tersely, and scratched more of Carol's words on the paper. "Skip to the paragraph above Berwick," he said. Amy and Mairi stared as the sound of Carol's voice came over the phone. "We'll explain later," Angus said. "Please, we just need to know the details."

Mairi turned pale. "Niall?" She looked to Amy for an answer as Angus leaned in, scribbling.

"Lindisfarne?" he asked. "The causeway? Can you spell the word? Amy, get anything we need for a long drive."

"Where's Lindisfarne?" Amy asked.

"By Berwick," Mairi told her, and to Angus, "Why are you going to Holy Island in the middle of the night? Sure they've not called you in for a rescue!"

"Mairi." Amy rose from the couch. "Help me. What do we need? Food? James is asleep upstairs. We can't take him."

"You're going? On a rescue?" She spun. "Angus, they canna call you in!"

"Please, Mairi," he said, "stay with James." He pushed himself up from the chair, reaching for his cane.

"Food." Amy hurried to the kitchen, gathering bread and apples. She filled his thermoses with water, and pushed it all into a bag.

"Warm clothes!" Angus shouted down the hall. "Extra everything."

"What happened to Niall?" Amy called, frantic as she dashed up the stairs. She rummaged in his drawers, shoving clothes and extra socks into a backpack.

"He's gone to the car with the food." Mairi appeared at the top of the stairs. "Can I help you?"

"I think I have everything." Amy gave Mairi a quick hug. "His formula is in the diaper bag. Diapers and clothes, too. Thank you." She hurried down the stairs and out the front door.

Angus waited in the car. As Amy slid into her seat, he said, "The trap was on the causeway going to Lindisfarne—early tomorrow morning in 1318. It's where he's going." He handed her an envelope. "It looks like Shawn left you something on his way out."

"We've lost an hour," she said, staring at the envelope.

"'Tis a miracle we were able to find an answer that quickly," Angus said grimly. "We'll count on that, aye?"

Road to Berwick, 1318

Allene rode endlessly, blind-folded, through light and dark, light and dark, struggling to stay alert, to note everything around her and keep count of the days. It was difficult, between the mind-numbing boredom of riding hour after hour in the sightless world behind the blindfold, the terror, the constant pain in her wrists from the binding ropes, and the ache of muscles unaccustomed to such long rides.

They slept in the open, forcing her to eat with the blindfold on, and untying her wrists several times each day to let the blood flow.

The sun bore down on her left each morning, telling her they traveled south. She strained to hear what she could of the deep voices that muttered around her now and again. She recognized none of them, but they must be MacDougalls. There was no one else who would want to do such a thing, and Christina, who knew MacDougall best of all, had warned her.

After many days' travel—five, four, six?—her garron's gait changed as the world beyond the blindfold grew darker, as night fell again. They rode over soft, squelching ground with the smell of the sea and the sound of waves all around. Days of riding...south...soft ground...the smell of the sea. They rode across the causeway to Holy Island, she guessed. Her alarm spiked. It was not far from Berwick. And MacDougalls—if such they were—taking her anywhere near Berwick meant they were after Niall.

Road to Holy Island, Present

Shawn knelt in the dark church, Clarence at his side, in matching medieval breeks, boots, and shirt. Shawn's mind refused to settle on God, but stayed stuck on Niall. The church had been here for centuries. He hoped Niall had visited it in his own time, in his days besieging Berwick. Maybe the times would simply come together here and now and somehow he could be spared this choice. He could stay with his son, his mother, even with Amy. He could hear SFO play her pieces.

But the church remained dark, still, and very much the desolate ruin of the twenty-first century.

"Shawn." At his side, Clarence whispered. "She's not stupid. She's going to figure it out and come after us."

"They don't know where it happens." Shawn closed his eyes, trying to pray, as Niall would. But no words came, no sense of God anywhere near, and he reluctantly rose. They got back into the car, Clarence taking over the driving, and pushed on past Berwick, taking the road to Lindisfarne.

"You checked when the tide is out?" Clarence asked.

"I checked everything and double-checked." Shawn pulled a leather packet from inside his shirt, protected inside two layers of plastic Ziploc bags. From it, he took a handful of pictures: The picture he'd taken just hours ago of Amy in her concert black with her violin on her knee: the Glenmirril Lady. James on his blanket, laughing, his rabbit's ear clutched in one hand. His mother smiling by the sea, reminding him of so many family trips to lakes.

He ran a finger over the small leather Bible tucked in with the pictures. Christina would treasure it. His heart hurt, thinking of her married to Hugh. So near him and yet unattainable, and Amy, too, beyond his reach.

It didn't matter. Christina would treasure it.

Lastly, he drew the small pair of vials, taken from Jenny's desk, from his pocket, and added it to the rest. He would save Niall and he would save James, no matter the cost.

He put everything inside the leather wrap, re-sealed them inside the bags, and tucked them inside his shirt, inside the wide ace bandage strapped tightly to his body. Then he knotted the belt of his trews more securely, holding the shirt in place. They would stay there, safe and dry.

Berwick, Scottish Borders, 1318

A hand grasped Niall's reigns, bringing his horse skidding to a stop at the edge of the bridge that crossed over a ravine to Berwick Castle. At the far end of the bridge, the portcullis slammed down with a sharp iron *clang!*

Around him, dozens of other horses and ponies slammed to abrupt stops. At their head, Douglas glared at the castle, shouting and shaking his sword.

"No point crossing!" Hugh lifted his voice to be heard. "They'll run out of food by and by."

Men crowded around the bridge, while faces appeared, across the ravine, on the ramparts. Bruce broke through the crowd, astride his sturdy garron, and gazed up at the men now looking down from the safety of the castle ramparts.

"Edward will send ships," Niall said.

"We've our own ships," spoke Conal.

"Aye," agreed the Bruce, "their ships will not get through."

As the melee settled, men looked to their king. "Start the watch," Bruce called. "And a feast. We've taken Berwick!"

Cheers filled the air, the grim aspect of the previous months' siege melting under the noon sun, to be replaced with the air of a market day. "Sir William!"

The young knight who had saved the day rode forward.

"Pick a score of men and bring us deer and boar!"

"James! Thomas!" Bruce called to Douglas and Randolph, wheeling his horse back into the cobbled streets.

"Find yourselves places to stay until we take the castle," Douglas called back over his shoulder.

The men dispersed.

"Another siege," Owen groused.

"Mayhap the Bruce will have work for us again," Conal said.

Niall's mind turned to Allene and his new son. The Bruce had already sent him home once. He couldn't possibly hope for a second such gift from his king.

Lachlan slapped his shoulder. "Let us find a place. Surely there's a harp somewhere in all Berwick. Will you teach me to play?"

Niall smiled. "'Twill do well to pass that time."

"Sure an' I'll have you singing before the siege is done, too," Conal added. "Owen, will ye join our fine band of bards?"

As the sun burned high overhead, William Keith, grinning, shouted. "Who will ride on the hunt?"

Men shouted to go and soon a group charged after Sir William back down the cobbled town streets, voices raised in excitement. Niall and Hugh followed, more slowly, with the men of Glenmirril.

"Sure you'll soon be home with Allene," Hugh boomed.

"'Twill be but weeks," Conal added.

"Find a harp, Lachlan," Niall said. "By the time we get home, you'll be playing love songs for Margaret!" He felt better than he had in months. He hadn't felt this light since the day they'd set out from Glenmirril to take Shawn to Iona. *Why are you so happy?* Shawn had asked. *You could die any day.*

Niall remembered his answer. *Aye, I might. I haven't time to be sad.*

He set the thought aside. He would see Allene soon.

CHAPTER THIRTY

Road to Holy Island, Present

"Call Carol," Angus advised, as he sped through the dark night. "Then get some sleep if you can."

"I can't push this insane story on her." Amy reluctantly placed the call.

"'Twould be crueler to leave her hanging."

"I know," Amy murmured as the phone rang. "But I still hate it."

"Did you find him?" Carol burst breathlessly onto the phone. "Amy, what's he doing?"

Amy took a calming breath. "Carol, I told you before, and I know it's hard to believe. But you have to look at all of it: his scars, the ring, the chain mail he showed up in, the way he keeps thinking he's been gone two years and thinking he's a year older than he is. He's fluent in four medieval languages. It's because he was living as Niall Campbell, who was expected to be fluent."

"Amy, please...I can't...this can't be...."

"Carol." Amy clutched the phone, trying to steady her voice. "You *have* to believe it. You saw the letters they left. You can see the dates on them. Niall is in danger. He's going to die a horrible death, and Shawn is going to try to go back in time to save him."

"This can't be true," Carol protested weakly.

"It is true," Amy said.

"He's taking Clarence," Carol said. "I can't believe he'd do this. He wouldn't really just leave us, would he?"

Amy took another, steadying breath, seeing the truth clearly. "Carol, to save Niall, yes he would. He has. He *is*." A hot lump formed in her throat. "I'll call you back when I know more, but we're still a long way from Lindisfarne."

They talked another few minutes, Amy with her eyes closed, answering questions she'd answered often enough to herself, to Angus, to Clarence, the very questions she'd asked Niall in his brief time here, and hating every minute of it; hating telling this story to a mother who had been through enough.

"You're telling me," Carol said, "if this works, he's as good as dead. He's never coming back?"

"That's what I'm saying," Amy confirmed. "He clearly doesn't expect to, anyway." Her voice caught. Maybe he *wanted* to stay. A blade of glass sliced through her heart. She glanced guiltily at Angus. She didn't want him gone.

"I'm looking up directions and driving out," Carol said.

"Okay." Amy wanted to reach across the space and hold Carol's hand. "But Carol, check the tide tables. It's dangerous to go out once the tide's coming in. Don't try it." She ended the call, and leaned against the headrest with a sigh.

Angus reached across the gear stick for her hand. "It'll only add to the confusion."

"What do I tell a mother who's facing losing her only son—again?" Amy asked.

"He left you a letter," Angus said softly. "You've time to read it now."

Outskirts of Berwick, 1318

MacDougall held up his hand as they reached the outskirts of Berwick. Behind him, his men came to a stop with a jingle of tack and creak of leather. He glanced at the girl, Allene. She sat atop her horse, blindfolded, gripping the pommel. Niall's wife. MacDougall smiled, anticipating, and turned his eyes to the land before them. It was devoid of Scottish troops. Below, men walked the town's ramparts. And against the sunset, Bruce's flag flapped in the breeze.

"They've taken Berwick," Roger said.

They stared at the town silently for a moment, before MacDougall spoke. "Seek the boy Taran." He smiled in the dusk. "You know what to do."

Roger smiled, too. "'Twill be done, my lord."

"Make sure he believes I'm traveling alone." MacDougall lifted his hand again, beckoning. His men fell into line as he led them in a wide berth around Berwick, heading south toward Lindisfarne. A day—two at most—and Niall would be dead. He would join with Claverock and rise to greater power and wealth than he'd ever had in Scotland. He glanced again at Allene, and his smile grew. Yes, Duncan's vengeance was here. And it felt good.

Road to Lindisfarne, Present

Where do I start? Shawn's letter begins.

We've been through so much. Or rather—I've put you through so much. Yes, that's more accurate. As I write, I see how true it is. Except there were so many good times, too, and I'll always treasure those. There's a lot to say and not much time to say it. I put off writing this letter because, well, because it's hard. I don't want to.

I'll start at the beginning. I saw you the first day I went to audition. You were walking down the hall with Patrick, laughing at something he said. I fell in love then and there. You were the prettiest girl I'd ever seen, your hair, your face, your eyes, everything. I didn't tell you that often enough.

Looks are hardly a reason to fall in love, though, and beyond that, I never understood what attracted me to you. You know you weren't my type. Take that as a compliment. Dana (I'm sorry for bringing her up) once said...you were what I needed. I had a lot of time to think, back in the 1300's, without my usual busy-ness, without parties and drinking. She was right.

I was a better man with you.

I know that must be hard to believe, but I think my descent was slower because I tried to live up to the good you saw in me. With someone else, I would have been worse. I'm sure that's hardly consolation, and yet—I hope it means something to you.

The way I treated you was never about you, but about the brokenness in me. I saw that in my two years away. I'm sorry. I know I've said it. I want to say it again. I want to say it for all time. Some days, I admit, I'm sorry for myself. I blew it. The best thing I ever had. But I'm sorry, too, for the pain I caused you. I'm sorry for the pain that will come to James because of the things I've done.

I still love you. I love you in a better way than I did. I learned a lot from Niall and Christina and all of them. I wish I had learned those things before I met you, and maybe we would have had something really great that could have lasted forever.

If you haven't figured it out already, I'm going back. I know I should say I'm going to try, but, well, you know how I feel about the word try. I owe you explanations. So....

In a small way, it's about you. Angus and I have had some good talks. You're torn, and always will be as long as he and I are both here. I hated Angus at first for taking you from me. I guess I don't need to tell you that. But I know I did it myself. If I really love you, I need to put you first. When you thought I was gone forever, you were free to love him completely. If you had chosen me, you would have been just as torn.

But it's not just about you and me. Clarence asked if I'm running away from problems. I'm not. I could have dealt with people thinking I'm crazy. What the hell—they're buying more albums, aren't they?

I bite my lip, laughing a little. It's so like him.

I could have dealt, somehow, with being a man who has killed, living in a civilized world. I could have dealt with having to explain this disappearance and I would have eventually remembered what age I'm supposed to be.

But what I can't deal with is leaving Niall and all the people I loved that night. Even more, I can't deal with abandoning Niall to a brutal death. I have to try. No, you know I don't try. I do.

It has torn me in two, knowing I have to abandon my son, my mother, you.

Or abandon my brother. I know how much it means to you for James to have his father. Thank you for that. For not hating me or wanting to shut me out.

I know I'm abandoning you, too, and the dreams you had for James. I'm causing you pain by leaving him. I'm begging for your understanding and forgiveness for doing this to him. You know in your mind what it means to be hanged, drawn, and quartered. But can you grasp the full horror of it? This is what is going to happen to Niall. I left the scrolls in my closet. You can read them. Please understand. I am forced to choose between my son having a better life and a father, or my brother being tortured in the most awful way known to mankind and killed. I wrestled with this decision, yet I think I knew from the start that I couldn't, wouldn't, abandon Niall.

One thing you should know. You deserve to know. Christina has married Hugh. I am not abandoning you or James for her. I am going for Niall.

Eamonn says I can't go back and forth forever. Even if I thought I could, I don't think I should. I can't yank the people in my life back and forth, here or there. I can't keep making up stories for why I keep disappearing. I can't keep hurting my mother, or sending you and Angus on wild goose chases.

Please forgive me, Amy. If I could save Niall and be there for James, I would do both. I can't. I see the love given to him by you, my mother, Angus, Angus's family. Once more, I'm humbled. I owe you and Angus a great debt.

Niall owes you a great debt. I know I'm leaving James in good hands, I know you and Angus will take my mother as your own and not leave her alone, which allows me to do this for Niall. You cared for him, too. I hope it is enough for you to understand why I have to do this.

Maybe I should have stayed there to begin with. But in my selfishness, I am profoundly grateful I had this time with you, that I was able to meet my son, to take back pictures of both of you, to meet and come to appreciate Angus and know he'll be here with you. I'm grateful for the chance to tell my mother good-bye and give you the full apology you deserved.

Most of all, I've done all I could to make things right with you, and I'm grateful I had that chance.

My lawyer has put my affairs in order for you, my mother, and James. I've left instructions with Ben and an account which will be funded from the royalties, to continue promoting the albums. I've paid off the house and signed it over to you. Sell it or keep it for yourself and Angus or James. By the time you read this, I'll have taken you to the tree to show you where I'll bury coins, jewels, gold, whatever. Mint condition medieval coins and jewels will be worth a lot. It will help take care of you and James. I know Angus is capable, I know you have a good job and a great future and don't really need my help, but it's the only way left to tell you, and one day, James, how much I loved both of you.

Amy, I don't know what else to say. I don't want this letter to end. I don't want it to end with you, even now, and signing my name to this letter is the end, except for seeing you in person and showing you the tree.

Maybe you could pray for me and Niall. Despite Niall's best efforts, I'm not much good at it, but I think maybe it would be a good idea in this case. Please—give James a kiss every night and let him know it's from his father who loves him. I love you, and I've come to respect and like Angus.

There is no name, but the symbol, the flattened S that was also a trombone. I press the back of my hand to my mouth.

Berwick, 1318

With guard duty, hunts, and the business of reasserting Scottish rule of Berwick, Niall looked forward to time passing with a minimum of boredom, and with hope for the future; with playing harp for the men in the evenings, and teaching Lachlan.

The voice calling *It's a trap* had fallen silent. Bruce had given him, that evening at dinner, a ring and a bag of gold, for his part in their final success. He lay in the dark, sleeping—or trying to sleep—near the town gate with his men, reflecting on the past weeks.

He lacked only the promise of a date he could go home—and letters from Glenmirril—to make his contentment complete. It wasn't as much Bruce's gold or commendation as the feeling, once again, that his life mattered. As he'd taught Lachlan that afternoon, the stress had faded from Lachlan's face, and he saw once more the young and cheerful Lachlan he'd known. As he and Lachlan talked quietly in the night of Margaret and Allene, he felt his life mattered. And, he thought, if a king saw such value in his music, he ought to, as well.

His head cushioned in his hands, and men snoring around him, he stared up at the stars twinkling above the town walls. The Laird had promised him they were there. And he saw them now. After three failed attempts, they had finally taken Berwick. And he had helped. He began to believe Douglas and Hugh. God had called him here and now for His own purposes. He must stop envying Shawn. God had his reasons for granting Shawn the life he did, but he, Niall, was called to something different.

And yet—he couldn't sleep. He sat up, looking around at his men sleeping near the town gate. Restless gnawed at him. He rose, touching Lachlan's shoulder. Lachlan stirred and looked up. "I'm going to St. Andrew's kirk," Niall said. It had been good to spend time alone with God at Glendochart. Maybe it would settle his mind further.

Taking his helmet and sword, he walked through Berwick's dark streets, greeting now and again one of the Scots patrolling the town. The April night was warm after the winter, but still not what Shawn would call balmy. Funny word. *Balmy, balmy.* Letting himself into the dim chapel, he knelt in the first pew and bowed before the cross. *Life was good.* His heart settled to a peace he hadn't known since Shawn had left.

A noise at the door caught his ear. He turned, his hand sliding to his sword. "Niall."

"Taran?" Niall's hand relaxed on his sword.

"MacDougall has Allene," Taran gasped. "He's going to kill her."

"Where?" Niall was already on his feet.

"He's riding with her even now to Lindisfarne."

"'Tis a hard night's ride!" Niall's heart kicked into a sharp staccato as he strode for the door. "How do you know?"

"One of his men sought me out, telling me to warn ye."

Niall's brows furrowed. There was a question he should ask, but one thought blazed through his mind. "Let me get my men."

"There's no time," Taran insisted. "We've lost an hour already, in my search for ye." His voice rose in excitement. "'Tis but MacDougall. His man mocked him. They've deserted him, Sir Niall. Come away now, if we're to catch him up."

Niall checked that his father's sword was secure on his back, his dirks in his boots and belt, and scooped up his tartan. He and Taran hurried through the dark streets, while he waged an internal battle between taking the time to get his men, and falling too far behind MacDougall. But if MacDougall was alone, against himself, Taran, and Allene, there should be no problem.

"The horses, my lord." Taran stopped at the stables. Mist crawled along the ground, swirling cold around their ankles. Two ponies snuffled there. Shawn's horse lifted its head in its stall; its ears perked and swiveled, as its large liquid eyes watched Niall. "I've no apple tonight," he whispered, and vaulted onto its back. "Take me there, Taran," he said.

Taran scrambled onto his own beast, with a quick, fearful look over his shoulder, and kicked his pony's sides, ducking low. With a word to Randolph's man at the gate, they were out into the countryside. Niall spurred his animal on, catching up to Taran in a dark scattering of trees.

Their horses pounded through the thickening wood. Something sat wrong with the story. But though Niall searched the details, he couldn't pinpoint it. The problem, he decided, was the nerves shooting up and down his spine screaming that MacDougall had Allene. He could think of nothing else. He gripped the pony's mane, leaning close over its neck, as its feet shot in and out, carrying him closer to the thrice shilled trumper, the skalded skaitbird! He would kill him!

Holy Island, 1318

Allene woke in the dark, her head throbbing. The familiar terror climbed high in her chest. Only after the long ride across the soft damp trail, across the firm land she was sure must be Lindisfarne, only after escorting her into shelter had the blindfold come off. She had fallen immediately into a deep sleep on a pallet of hay.

Now, on waking, she lay still, studying the dark, cavernous space around her. A group of men sat around a crackling fire in the center of a stone floor, cooking meals. Beyond them, through stone columns, she saw stained glass windows. She was in an old, abandoned kirk. It had to be Lindisfarne.

She turned her eyes back to the men around the fire: Large men with black and red beards. One old and gaunt man. The flames sent shadows dancing on their faces. She recognized none of them.

The largest of them turned to her. He studied her a long moment before rising. She pushed herself up hastily, pulling back into the corner. Pain shot through her wrists. But they were unbound. Her heart pounded as he approached. She had no knife to fight him.

He tossed a kerchief at her, wordlessly. Bannocks spilled out across her skirt. The rest of the men stared at her, saying nothing. She glanced down at the bannocks and back to them. The man who had tossed them stood, watching her. The old man stared at her steadily. Her stomach rumbled.

She ate the first bannock slowly, trying to stay calm, trying to see a means of escape. She couldn't fight them. The smell of the sea was strong. If she was indeed on Lindisfarne, she was trapped here until the tide went out, even if she escaped the kirk.

She was sure she was right in guessing MacDougall intended to use her to get at Niall. And she saw naught she could do with the information. She prayed as she ate the bannocks that sank in her stomach—*My Lord, I beg for a miracle. Are Ye there? I need help!*—watching the men as they watched her.

The man spoke. "The lad, Taran, is at Berwick, is he not?"

Her heart pounded. Cold shot up her arms. "Taran has done naught to you." Her voice shook. If these were MacDougalls, they may treat her as Duncan had treated Christina. She was foolish to speak boldly. "Leave him be!"

He regarded her steadily, his black eyes glittering. She trembled, waiting for him to strike, bracing herself. But he smiled. "Aye, he's at Berwick." He turned, saying, "I think we'll have a grand morning." He grasped her arm, yanking her to her feet, and pulled her, resisting, toward the door.

There, Allene saw the man to whom he had spoken, blending into the shadows in his dark cloak, his face shielded by the hood. "Time to go," he said, taking her arm, and to his men, "You two—take up your posts on the dune. The rest, wait here till I bring him in."

"No!" Allene shook her head, pulling back.

Two men lifted bows and headed out the kirk door.

The cloaked man dragged her out into the misty kirkyard, just beginning to lighten from black to gray. She yanked against him, but his fingers bit into her arm as he dragged her down the moonlit path between leaning tombstones, and out through the kirk yard gate.

"I've a knife," he said. "You can die here if you get too difficult." He pulled her far out onto the damp seabed exposed by low tide, to a currach waiting where the water began. "Get in." He slid the knife from his belt. "Or you die here."

She climbed in slowly, watching the blade. James, William, and wee Alexander filled her mind. Even if she could fight him, there were more men in the kirk. She had nowhere to flee to. *God, my father, help me.*

"Move." He pressed the knife to her back.

Heart pounding, she lifted her skirt, the cold water soaking one boot as she settled into the boat. *Jesus, send your angels!*

He pushed the boat deeper into the water and climbed in after her, the hood falling back.

She stared at MacDougall, The moon faded above as he began to row, taking them deeper. She thought of throwing herself over, running for dry land.

He smiled. "You've nowhere to go if you do reach the island."

She stared at him, prayer for Niall pounding in her mind. And a sense of peace came over her.

"He shouldn't have killed Duncan," MacDougall said.

"You attacked our home." Her voice trembled. "And 'twas not Niall who killed him."

"Aye, 'twas Shawn." He smiled. "I believe he'll come for Niall, just as Niall will come for you."

"Then ye're a greater fool than ever we'd thought," she said. "Shawn cannot come. Not if he wanted to."

"Then Niall shall die in his stead."

"When does your hatred stop?" she whispered.

"When Niall is dead. Better yet, when your sons die as mine did."

The peace grew, and Allene shook her head. Her words came out strong. "Nay. Ye'll find, Alexander, that when one has spent a lifetime hating, one no longer knows how to live without it. You'll find another reason, for you are far more trapped in your hatred than I am in your wee boat."

He glared at her.

Road to Lindisfarne, Present

"Two a.m. Two hours until we get there." Angus breaks the silence in the dark car, for the first time in half an hour.

"Angus?" In the dark, I study his profile as he drives.

He reaches for my hand. "Is it that bad? You sound like you want to cry."

"Yes," I say. "How much did you understand from what Carol read?"

"It happens at the causeway." Angus stares straight ahead.

"Come on, Angus!" I speak more harshly, in my agitation, than I intend. "I'm not a child to be protected. He's captured at the causeway, but he's going to be hanged, drawn, and quartered in London."

"He told you in the letter?"

"Yes. He told me—what's serious enough to leave James and his mother."

We drive in silence for a time, along the dark highway. Occasional passing headlights flash light across our faces. Neither of us makes a move to bring the stereo to life. "We both met Niall," I finally say.

"Aye. And I wish I'd known who he was when I did."

"It's one thing," I say, "to read that someone was hanged, drawn, and quartered. It's another to know it's going to happen to someone you know."

His hand leaves the wheel and wraps around mine, but he doesn't answer.

"If we stop him," I say, "we condemn Niall."

"It's unthinkable to let him do this," Angus counters. "He's throwing his life away, walking away from everything, deserting his son and mother."

"To save a life. To save Niall from a horrible death. We know he dies eventually, but to die like that?"

"He can't just walk away from his mother, his son. 'Tis hardly different from suicide."

"I know." I lean back against the seat as Angus guides the mini at high speeds along the highway.

Ten minutes pass and he asks, "Is that what he talked about in the letter?"

I nod. "The dilemma—James or Niall." I stare out the window at dark fields passing, and the faint glow of stars overhead. "How can I stop him? And how can I let him go?"

"Are you still in love with him?" Angus asks.

I squeeze his hand. "Why are you asking, Angus? Haven't I shown you over and over I chose you?"

"But you're still in love with him."

"Yes." Irritation rises. "The irony is, he wrote in his letter that if I chose him, I'd still be in love with you. He feels exactly as you do."

Angus says nothing.

"I'm condemning Niall myself if I stop him," I say

"We can't turn back," Angus argues. "How do we tell Carol we didn't try? And neither of us is going to tell her we tried and failed, if we didn't."

"No, I won't do that." I twist in the seat, reaching for the bag in the back. I hand Angus a water bottle. "He didn't say why he took Clarence."

"Clarence has likely had a hard time of it, coming out of prison, and Shawn saw this as a fresh start for him. Or maybe he knows more about the trap and needed Clarence's help."

"What if it goes wrong?" I ask. "What if they're both captured and executed in London?"

"We daren't think on it," Angus says. "We do what we can and pray God's will be done, because these problems are too much for me to know any more what's right and wrong."

I reach into the bag for my own water and two of the roast beef sandwiches Mairi made. I hand one to Angus.

He eats the whole thing before speaking again. *"Would you ha' come back if I'd not had the accident?"*

I finish my own sandwich. *"You've convinced yourself I came out of some Florence Nightingale syndrome. What you've never quite believed is that I came because when I faced the very real possibility of losing you—I knew I would do anything not to."*

"And yet you're torn."

"Please." I turn, staring out the window. *"I told you once if I could change every memory and emotion for you, I would. I can't. But I can put him in the past."* I shake my head. *"Bad choice of words."*

"Where did he take you last night?"

"Last night?" It seems a million years away. *"There's a tree above Glenmirril—two trees grown together."*

"I know the one," Angus says. *"Couples like to go under it."* He turns to me, with only the stars illuminating his face, a silent question in his eyes.

"No," I say.

"Then what did he want?"

"He told me to dig on the west side of the tree, under the boughs, this morning. He explained in the letter he's going to leave coins and whatever else buried there, things that will have value today, to take care of James."

CHAPTER THIRTY-ONE

Road to Lindisfarne, 1318

They burst from the forest onto the coast, and veered right, giving the horses their heads in a wild gallop. Niall forced his mind from anger to prayer, drawing in *Aves* and sighing out *Paters* with the rhythm of hooves rolling across the new grasses bursting from winter's grasp. Waves undulated softly on their left; moonlight glittered on the waters.

Niall broke the silence at last, asking, "How does he plan on getting to Lindisfarne?" The familiar ghost of his brother rode along him. He'd gotten to Ireland and back, he reminded himself. "Is there a boat?"

"We can cross the causeway at low tide," Taran answered.

They pounded along beside the North Sea for some time before Niall asked, "When *is* low tide?"

"I don't know, my Lord. But it comes every six hours. And mayhap we'll catch him up before he reaches it."

Niall sighed an *Ave*. "What if she *hasn't* six hours? Did they say aught of his plans?"

"Only what I told you, my Lord."

They drove their garrons hard. Niall's mind filled with every memory of Allene—her red gold curls he'd first loved, the day he'd kissed her behind the oven, the night of their marriage. He savored—*treasured!*—each moment, each touch. Lying beside her in bed with wee James and the joy on her face! *He couldn't lose her!* He leaned into his horse's mane, praying for its strength.

Far over the water, the black sky lightened, the rich velvet blue unfolding on their left, inching up the arc of the heavens, followed by a pearl-gray on the eastern horizon, chased in turn by streaks of pink.

"Ahead," Niall breathed, and Taran nodded. The low bulk of the island showed black against the shell-pink dawn. "The tide is out."

They pulled up their horses at the muddy edge of the causeway. MacDougall had beat them across the stretch of damp sand and somewhere across it, he held Allene. Perhaps even now he stood on the shore with an arrow nocked, watching and waiting. The ponies danced under them, panting from the long ride. Niall's animal whickered and shook its head hard, letting out a snort to tell Niall what it thought of his demands. He patted its neck. "Just a wee bit more," he soothed it. "We'll kill Darth Vader and go home, aye?"

"Who?" Taran asked. "'Tis but MacDougall."

Niall smiled grimly. His nerves stretched taut, studying the long narrow path crawling between the sea's cold fingers, to the island. "There can be quicksand." He trotted the pony out carefully, trying to judge if the tide was coming or going. His horse lifted its hooves daintily, disturbed at the change of ground under its feet. He urged it forward, step by step, watching the water far out. It crept closer.

Halfway across, he was sure the tide was coming in. He pushed the pony to move faster, wary of MacDougall watching from somewhere, waiting with an arrow ready to fly. He pulled his targe from his back, tugged his helmet firmly over his brow, and nudged the pony forward. Water lapped against its hooves. It splashed forward, giving an angry shake of its head.

Causeway to Holy Island, Present

"Now what?" Clarence slid the car into a parking spot at the far end of the causeway, on the island, and killed the engine.

"You pray, because apparently you're better at that sort of thing than I am." Shawn touched a hand to the pictures and Bible, assuring himself they were snug against his abdomen. "We watch for him and hope this is really a magic crucifix. And if not, that whatever *did* cause it will happen again." It had happened, the last times, on the feast of St. Columba. He didn't want to think about that. *Please, God. I know I haven't been that great, but I'm doing this for Niall, so for his sake...?*

He glanced at his watch before removing it and dropping it in the cup holder. "We're in plenty of time. It happens at dawn." The sky was still dark. "You've taken off watches, anything that might raise questions there?"

Clarence climbed out of the car, looking up and down his trews and linen shirt. He patted his waist and boots, where three knives were stowed. He checked his wrist. "I'm good," he said. "Even now, a part of me is wondering if you're pulling an elaborate trick on me."

"No." Shawn spoke grimly. "There are no cameras waiting to jump out. Believe me, I wish this weren't real." He threw open the back door, and began pulling out the chain mail, the hauberk and coif, in which he'd arrived back in the 21st century. He laid them on the ground with the sword and his tunic, sat down beside them, and began unlacing his boots. "I can't swim in them. Honestly, Clarence, I'm thinking if this switch has anything to do with the blessing and prophecy, you're not his descendant, so maybe you can't make the crossing, anyway. Maybe I've dragged you on a hopeless trip. Maybe I got your hopes up for nothing."

"We'll see." Clarence dropped onto the ground beside Shawn, looping his arms around his knees.

"If you don't, you'll take care of my mother, right? Like she's your own?" He tugged the first boot off.

Clarence slapped him on the back. "Have you forgotten who you're talking to? I'd take much better care of her than of my own mother."

"Hey, you're the big Christian between the two of us," Shawn returned. "You wouldn't take care of your own mother?"

"Yeah, I would," Clarence said. "But taking care of someone with real love —that would be Carol." He blinked. His lip tightened.

Shawn worked at the lace of the second boot, the one the cordwainer had made in a rush in October of 1314, grousing about being hurried, while the Laird pushed him to reproduce Niall's boots exactly, and haste, so Niall could get to Bruce's parliament at Cambuskenneth. "As long as I know someone's there for her," he said.

"Amy and Angus will be."

"Yeah. They will." Shawn yanked off the second boot, then tugged his period woolen socks off, preparing to go in the water. "You remember all I showed you about fighting with a sword? Niall rides out on the causeway. The archers come out on the sand, two of them. You pick them off with the bow and arrow from behind. You have three knives if you need them. Kill them, then swim out to help me. You take Allene. I take Niall."

"You're sure you can tread water long enough?"

"I'll stay on shore till we see him coming," Shawn said. "Then I swim out under water and wait behind the boat till you take out the archers."

"What happens if I don't make the crossing?" Clarence asked.

Shawn shrugged. "Then I get two arrows in the back. If they're good shots. I'll rock the boat, make a lot of commotion, hope they miss."

"There's no reason to think I'll go with you," Clarence said.

"Sheer force of will on my part." Shawn sat in the scrubby grass, casting an eye to the sky. It had lightened to charcoal gray, the stars faint smudges of light behind wisps of cloud. He didn't want to think about Clarence's question. He didn't want to think about why he'd been so foolish as to fail to consider this very important point, when he'd planned everything else to the last detail. In the end, it changed nothing. He'd do whatever he could to prevent Niall's capture.

They sat on the shore, waiting, their eyes darting between the causeway and the long road to the old priory, down which MacDougall's men would come.

"You think the car will disappear instantly or fade away?" Clarence asked.

Shawn wanted to laugh. It was only tension, making the idea of his car appearing to medieval archers seem funny. "Last time, at the battle, I could see it all, past and present. So maybe the archers will see it. Maybe it'll help us.

Shock them, you know. You got plenty of arrows? You checked your knives?"

Clarence re-counted the arrows, patted his boots, and touched the knife at his belt. "All here."

"Aim small," Shawn reminded him.

"Twenty-five in a row, remember? I can do this."

"Don't shoot me by accident, okay? Shoot the ugly guys with the bows."

Clarence clasped his hands and bowed his head low over them.

"What are you doing?" Shawn asked.

"Praying."

"About shooting? You're the master."

"About everything. Me making the crossing at all is a long shot. Forgive the pun."

"I'll forgive anything as long as you take out those archers."

"Shut up," Clarence said. "I'm working."

Causeway to Lindisfarne, 1318

"Niall!" The voice sounded on the wind, a plaintive cry floating like a ribbon in a storm, as Niall rode out onto the soft mud of the causeway.

Niall spun. His heart jolted. Far out in the bay nestled in the crook of the island's curve, a currach bobbed against the blazing orange of the rising sun. It rocked with the animated motions of the two people in it. Niall squinted, trying to see. The red hair jumped out at him, and the jet black hair of MacDougall. Clasped in his grip, Allene struggled. Her face came into focus, stricken with terror.

"Come out and get her," MacDougall shouted. "Which one are you? The one who's afraid of water, or the one who fought like Douglas himself on the sea of Jura?"

Niall eyed the expanse of damp sand and water between them. The water was rising as the tide came in. Taran's pony edged up behind him. The lad looked more scared even than Niall felt. Niall scanned the shore, searching for anything he might use to float. There was nothing. *When you pass through waters....*

He would have to ride into it. A scant few furlongs. MacDougall had no bow, and his hands were full holding tight to Allene, one hand clamped over her mouth. Niall touched the dirk at his belt, felt the weight of his sword.

Across the water, MacDougall threw his head back and laughed. "What are you going to do, *Sir* Niall? Throw it at me? Was it you who earned the knighthood? Or Shawn?" MacDougall taunted from the currach.

Niall's heart slammed in his ribs. *He knew!*

And guilt swarmed him. He had not earned knighthood. Shawn had.

"You did grand on Loch Tarbert," MacDougall shouted.

Niall yanked the reins, leading the horse into the water. It balked once, then plunged forward. The sea swirled quickly up to its fetlocks.

"You've no choice, Niall," MacDougall called. "They say you jumped in the water at Jura and pulled a man to shore. Come out and I'll hand her over. I'll never bother you again if you'll just swim out here."

A scream erupted on the water. Allene wrestled MacDougall, her head down over his arm, one hand flailing. He yanked his hand away, streaming with blood and swearing, as she shouted, "Niall, get back, there are...!"

MacDougall struck her head. She sagged in his arm. The boat rocked perilously. "Can she swim, Campbell?" he shouted. "Will you come and drag her to shore, too?" With Allene limp, he tugged at the oars, pulling the boat farther from Niall, closer to Lindisfarne's shore, with the Priory rearing up behind him, black against the dawn.

The sea rose, slowly, as the tide came in.

Niall charged in. The water's icy shock soaked his boots, and cut into his ankles. He fought the horse, whinnying and stamping its feet high over the frothing tide; he fought his own rearing panic.

Allene stirred; struggled. MacDougall gripped her hair, yanking her head back. The water rose to Niall's knees, the cold biting in. He tried to remember what to do when it rose past his shoulders. *When you pass through waters, they shall not sweep....*

"Are you coming?" MacDougall called. "She may fall in!"

The water rose to Niall's thighs; he tried to force himself to calm, to push out his brother's bloated face, and remember all Shawn had tried to teach him about swimming. *Think of happy things.* Icy water bit into his waist; he thought of Shawn playing harp on the shore; James laughing and reaching for the waves, as he plowed ahead.

When you pass through the waters, they shall not sweep over you!

Shawn seemed to shout in his ear. *Sing into the water!* Alexander's face stared up, bone white, bubbling up from below the water, eyes black and unseeing. Niall closed his eyes and opened his mouth. *"The minstrel boy to war is gone,"* he sang softly. Alexander's face receded. He snatched the sword off his back and threw it to Taran. It landed with a splash. *"In the ranks of death you shall find him!"*

MacDougall laughed. "Going to sing me into submission, Niall? There's a new battle plan if ever I heard one."

"His father's sword he has girded on!" Niall threw himself into the sea, singing louder. Water hit his chest, a sluice of cold shock, slapping words from his mouth. He gasped in shock, drawing in water, and spit it out, coughing. Alexander's pale white hands reached.

When you pass through the waters, they shall not sweep over you!

In the currach, Allene shrieked. "Help him!" Niall coughed, his lungs burning, focusing on the sound, praying, seeing in his mind how Shawn had moved his arms. He swung one over his head. *Legs straight,* Shawn had shouted at Hugh over and over. *Two blades slicing past each other.* He breathed out into the water again. *"With his wild harp slung behind him."*

♫

Shawn tried to settle his own thoughts to praying. But he'd never been one for sitting still. The best he could manage was, *Come on, God. There had to be a reason for this. Come on, help me out. Don't let this happen to Niall.*

Niall would have found such prayer irreverent. Overly familiar. Shawn's mother would say it was trusting God with child-like faith. Shawn hoped that in this instance, God favored Carol's view, as he couldn't do any better. He raised his eyes to see Clarence bowed over his clasped hands, his lips moving. It wasn't much to hang his hopes on, Shawn thought, this invisible and silent God people claimed was out there somewhere. But then, he had nothing else. In the east, the sky turned silver gray, with the thinnest line of pink showing under it.

"Come on, Clarence." Shawn rose. "Show time. Get behind the dunes."

His car still sat in its spot. He felt like a fool, sending Clarence to hide from archers who had walked the shore centuries ago, with his car right there. No wonder people thought he was crazy. Still, he couldn't sit around and have the time switch happen with himself and Clarence in plain sight. Clarence took up his bow while Shawn gathered his boots and mail and weapons, and they climbed the slight rise. Scrubby grass bit into his feet.

The sky lightened high up. Fiery orange blazed across the eastern horizon, showing the black silhouette of the castle and the small town on Holy Island "Watch the the castle," Shawn reminded him. I'm going in." The sea had risen with the incoming tide.

Something glinted far across the water. He straightened. It was a car on the mainland. His mouth tightened as if he was going for a high F—the really high F—trying to guess how this would affect the time switch. It had happened before with thousands of spectators, he reassured himself. A single early morning driver changed nothing. He settled back, searching the road, and immediately tensed at a second flash of sunlight.

"Clarence!" He yanked his arm, pointing. "Do you see it?"

Clarence turned, lifting a hand to scan the shore. "The car?" he asked.

"No, someone coming out on the causeway."

Clarence squinted, leaned forward, and finally shook his head. "No, I only see the car."

Shawn muttered an oath, and spun. Far out, a boat bobbed on the water, a man gripping a woman tightly. Shawn dropped back behind the dunes, watching. She struggled, silent at this distance. Shawn grabbed Clarence's sleeve, pointing. "Do you see them?"

"I see an old man tossing nets out," Clarence said.

Shawn saw the old man, and shook his head. "No, a middle-aged man with a woman. She's in a surcoat, fighting him."

Clarence scanned the shore.

"The switch is happening!" Prayers screamed through Shawn's head. *Please, God. Please, God. Please. Let me save Niall. I need Clarence's help.* He heard no answering voice. He felt no inner urge. He had no peace, only the familiar adrenaline rushing to every part of his body.

"Campbell!" MacDougall's shout sounded across the water. The currach rocked as Allene struggled. Shawn scanned the area, and his heart pounded his sternum. *Niall! He saw Niall!*

"Come out and get them," MacDougall shouted. "Which one are you? The one who's afraid of water, or the one who fought like Douglas himself on the sea of Jura?"

Niall's hand went to his sword.

Shawn's hand went to the crucifix. He didn't believe it for a second. It was ridiculous. But what if he took it off and gave it to Clarence? Would Niall fade away into the past? Would Clarence see them, instead? He gritted his teeth. He didn't believe in magic.

MacDougall threw his head back and laughed. "What are you going to do, Niall? Throw it at me?"

Niall stared helplessly from the shore.

"Come on, Niall," Shawn muttered, as if watching a play. "Think of happy things!" He looked to the ridge, and saw the two archers coming. He couldn't shoot them from here. He couldn't wait any longer to start swimming, or Niall and Allene would drown. He couldn't leave the archers here on the shore to shoot him in the back. "Clarence, you don't see them? Coming down from the castle, there, heading to shore."

"I see an old man fishing!" Clarence hissed back. But he lifted his bow, and slid an arrow from his quiver.

Shawn lifted the crucifix over his head, gripping it in his hand even as he inched higher up the ridge, trying to keep the whole scene in view. The car roared across the causeway, waves washing up at its wheels.

"You're a strong swimmer," MacDougall shouted. "Come on out and I'll hand her over."

Niall scanned the shore.

"Come *on,* Niall," Shawn muttered. "Can't you remember anything I taught you? Sing into the water!"

The archers crossed in front of the dune, and raised their bows on the shore. "You don't see them?" Shawn whispered frantically.

Clarence shook his head. "Can you still see our car?"

"Yeah, I see it. But I see them, too. MacDougall's out on the water with her. Just like the letters said."

A scream erupted across the sea, jolting Shawn's attention. MacDougall yanked his hand from Allene's mouth, blood streaming down his wrist. He smashed his fist into her temple, and she sagged.

Niall drove the pony deeper into the water.

"A shame if she falls in the sea," MacDougall shouted.

"Take the crucifix!" Shawn thrust it at Clarence. He had to take a chance, one way or the other. He could see Niall. He prayed the past wouldn't fade away. With a last glance at the archers, taking up their positions, and the car speeding across the flooding road, he crept across the dune behind their backs, and with their attention focused on the boat in the water, raced across the last muddy stretch of exposed seabed, praying Clarence would make the switch.

Causeway to Holy Island

"There's his car," I breathe. "Angus, the tide is coming in."

"We'll be fine," he says.

"This isn't the rescue raft. It can't float."

"It can drive very fast." But Angus doesn't smile as he pushes his foot down hard on the gas. The waves creep closer. I close my eyes, praying. "Do you see him?" I ask.

"Aye, he's just gone over the dune with Clarence. I think it's them."

"Why would he go over the dune?" I open my eyes. The sun glares a fiery half-circle of orange on the eastern horizon, and the castle stark black before it. Streaks of pink spread across the horizon, with wisps of cloud floating in front of them. Water sprays up from the front wheels in great sheets. On the sea, an old man throws fishing nets into the water. A half mile away, Shawn comes over the dune, inching, crouching. "What's he doing?" I ask. "He's acting like he's afraid this old guy is going to see him."

"He's going in the water," Angus says.

"What if he's already making the switch?" I ask. "What if he's already seeing Niall?" Shawn is moving, crouched, across the damp sand.

Angus glances at me. He takes hand. "Letting him walk out of his son's and mother's life is like watching him commit suicide."

"Stopping him is like killing Niall."

Shawn is wading into the water, trying to keep his body low and flat. The water sprays as high as the car. I squeeze my eyes shut, unable to look.

Causeway to Holy Island

The sea closed, chilly, around him, as Shawn flattened himself into the water. MacDougall's attention stayed on Niall. Cold bit through his breeks and linen shirt. The Bible, small as it was, weighed him down. But he was a strong swimmer. With a breath deep enough to play *Greensleeves* largo, he glided under the water, and swam, covered by the waves, praying the archers wouldn't notice him when he had to catch more air, praying he would come up where

MacDougall wouldn't see him. *Praying,* with a frantic *Come on, God!*

He slid under the water and swam, half blind with the murky sand swirling so close below his body, until the sea grew deeper. He pulled himself through dark waters with powerful strokes, searching for the bottom of MacDougall's boat, praying, *praying,* Clarence would make the switch.

Finally, Shawn raised his head carefully above the water. He'd judged right. He'd come up behind MacDougall's boat. *Thank* you, he thought. He swirled his hands, treading. Niall flailed in the sea. But he was swinging his arms! He gasped and sank and rose, but he was nearing the boat!

The archers stood on the shore. The car crossing the causeway threw up sprays of water high enough to shield it from view. Shawn hoped they made it. He considered swimming under the boat and pulling Niall from the water. But the archers waited on the shore to kill them both if he let himself be seen.

He considered upending the boat. But then he'd have both Niall and Allene in the water, and *still* two archers on the shore. *Please, God,* he muttered. *If you won't listen for my sake, will you at least listen for Niall's?*

Across the water, the car roared off the causeway onto the island. Shawn felt a rush of relief for the unknown occupants. It skidded into the parking lot, and people tumbled from either door. A shock colder than the water hit him as he recognized Amy and Angus. Amy ran to the shore, right beside the archers, Angus hobbling behind, waving, shouting his name. "Oh, damn," he muttered, "go home! Don't mess this up or Niall's dead."

♫

It seemed only moments before Niall opened his eyes, gasping as sea water plunged down his throat. But he was at the boat, grasping it with one hand, while MacDougall, above him, laughed. He stooped, reaching down. Niall grabbed his hand, wrist to wrist, instinctively, and MacDougall hauled him, spitting and coughing, doubled over on his knees, into the boat.

Niall's mind scrambled, screaming to get up before MacDougall planted a knife in his back. Allene would delay it, at least. He shoved his hands against his knees, yanking the dirk from his boot. MacDougall laughed. Allene gasped, shouted, "Niall, look!"

Niall's head swung in confusion. Then his heart sank. Two archers stood on shore, each with an arrow ready to fly. "They risk hitting you," he told MacDougall.

In his moment's distraction, MacDougall had Allene in his grip, his knife pressed to her neck. "After the arrow goes through her," he said. "'Tis a chance I'm willing to take." He smiled, showing the yellow teeth. "Are you? Row."

Niall turned to him slowly, studying the knife.

"Don't, Niall," Allene whispered. "He'll kill you if you go there."

MacDougall jolted the knife tight, forcing her head upward. A thin line of blood appeared at her jaw.

♫

Treading water behind the boat, Shawn saw MacDougall's knife jerk upward, and heard Allene's sharp intake of breath.

Amy waved her arms on the shore, shouting. Angus following more slowly. Shawn prayed she couldn't see him in the water. They couldn't stop him, he decided. He was already half in 1318.

"Shawn, you have to come back!" Amy shouted.

Shawn tread water, ignoring her, still deciding whether to upend the boat. The arrows would hopefully fly safely over the water—because he couldn't see Clarence. It looked like he was on his own, to get both Niall and Allene to shore, out of the archer's reach. It was impossible. One of them would drown.

Come on, God, he muttered. *You've got to help me! I know I'm not worth it, but Niall and Allene are. They've been faithful to you! Please!*

♫

Clarence watched the couple jump from the car. *Amy! Angus! They had to leave!* He almost rose, but something, an invisible hand on his shoulder, pushed him back to his knee behind the dune, waiting, watching.

"Shawn!" Amy shouted. Her black hair swung against her back. She waved her arms. The old fisherman on the sea looked up in irritation. Angus hobbled up beside her with his cane, leaned forward, bellowing, "Shawn, you can't do this to James! Come back!"

Clarence tried to rise, to tell them Shawn had made his decision. But the force kept him on one knee. His eyes narrowed, his breath became still, wondering what it was. He closed his eyes a moment, two moments, in silent prayer, opened them again, and saw four people side by side on the shore—Amy, Angus, and two men wearing long medieval tabards and helmets.

His heart stopped for a split second then rushed to catch up. He swallowed convulsively, then looked around quickly, searching for a movie camera, actors, anything to indicate it wasn't a joke, a misunderstanding. But they stood each with a foot forward, bows raised.

This was his part!

His hands shook as he rose from behind the dune, nocking an arrow to his own bow. He drew a deep breath. What if they weren't medieval archers, but actors out on the shore for some reason Shawn didn't know about? Filming a commercial?

"Shawn! Come back!" Amy shouted.

"Stop yellin'!" the fisherman hollered at her. "Ye're scarin' me fish away!"

It dawned on Clarence...the archers didn't react to Amy and Angus, though they stood side by side. He inched forward, afraid to kill, lest they be modern men. Shawn couldn't have set him up. *He wouldn't!* The bow trembled in his hands, fearing a murder charge.

He inched closer.

"Is he mad?" one of the archers asked. "How does MacDougall expect us to shoot Campbell with him right there in the boat beside him?"

He spoke medieval English. Amy and Angus seemed oblivious to two medieval archers beside them. MacDougall and Campbell were the right names. Clarence lifted his bow.

♫

Pain flashed across Allene's throat, hot, razor-thin pain. She held her breath, pressing back against MacDougall, trying to keep the pressure of the knife off her throat. On the shore, the archers raised their bows. James, William, and wee Alexander flashed through her mind, left at home, without parents. Hugh and Christina would care for them, she reassured herself.

"Pick up the oars, Niall," MacDougall spoke softly in her ear, the voice of one willing, even eager, to kill.

Beside her in the boat, Niall hesitated.

The knife jerked upward, lifting her head till her neck ached, trying to keep the blade from digging in.

"D' ye want to see her die before your eyes?" MacDougall asked. "Row!"

"Please, Niall, dinna do it," Allene whimpered.

MacDougall jerked her head upward with the hilt of the knife, pressing into her windpipe. She struggled on tiptoe, for breath.

Niall stooped, reaching for an oar.

♫

The scene wavered before Clarence's eyes. Something shimmered on the water. For just a second, he thought he saw a second boat, people fighting in it. Then it was just the rising sun, glancing off the water.

"Clarence!" Shawn shouted from the water. "Shoot them!"

Amy spun toward his voice. Her black hair whirled in a wide circle, brushing right through the archers. They didn't react. Her eyes widened at the sight of Clarence on the dune, his arrow pointed at her.

"Are you mad!" Angus shouted, throwing Amy to the ground, covering her with his body.

"The archers, not you!" Clarence yelled.

The archers turned. Through the vertical slits in the helmet, a pair of brown eyes met Clarence's. His insides sank, fearing a mistake, terrified of killing two innocent men. He loosed the arrow.

♫

"Shoot!" MacDougall roared in her ear. "Shoot her!"

His words echoed across the water. Allene's breath came in short, shallow pants of fear.

"See how it feels, Campbell, to watch someone you love die!" To the archers, he shouted, "Shoot, damn you!"

Allene watched them, eyes wide. Her lip trembled, wondering how badly it would hurt. She'd seen arrow wounds, heard men moan in pain, watched their faces go gray-white in agony.

The archer raised his elbow, drew, and collapsed on shore.

MacDougall gaped. His knife dropped half an inch.

Niall lunged.

Allene threw up an arm, dislodging the weapon, and dropped to the floor of the boat. It rocked. MacDougall tumbled backward, splashing into the water. She grappled at the edge, Niall reaching for her hand, swaying.

"Allene!" A deep voice shouted. "*Allene!*"

She toppled into the water, the boat flipping over on her. The water soaked her surcoat and underdress, weighing her down, dragging her beneath the sea. She tried to swing her arms as Shawn had taught Niall.

"Angus, help me!" the same voice shouted.

She sank, murky water closing over her head.

♫

Angus swore under his breath. The man's arrow shot through the air, brushing the space next to Amy, and disappeared. Suddenly there was a medieval archer, collapsed on the shore, and another beside him, spinning to confront his attacker.

Shawn thrashed in the water, shouting. "Angus, help me!"

The archer charged Clarence before he could get an arrow nocked. Angus stood, torn between Clarence and Shawn.

"Help Shawn!" Clarence shouted. His face was tight. He ducked, snatching knives from each boot, and launched himself at his attacker.

"Niall!" Amy scrambled up from the shore. "Angus, help him! He can't swim!" She ran, yanking her long-sleeved shirt over her head as she splashed into the water.

Angus saw them, the boat tipping, the woman, and the man who looked like Shawn, tumbling, falling, Shawn reaching for the woman. Angus tore at the laces of his boots, fumbling, tearing, ripped his sweatshirt and jeans off, and launched himself into the water, praying, praying he'd be on time.

♫

The man came at him with more force than any prison inmate ever had. Clarence glanced to the fight in the water, to the man with the black hair hanging to the side of the craft, striking out at Shawn, at the woman sinking. In the moment's distraction, the man was on him.

Clarence's arm flashed up, blocking. Sunlight flashed off the blade as it glanced past his temple. He jabbed upward with his left hand. The man sucked in his gut, dodging backward, and came at him again, his knife low, ready to swipe up into soft tissue. Clarence dodged, circled, trying to ignore the screams from the water, Amy shouting on the shore, wading farther in, Shawn bellowing.

His focus cost him again, as a blade sliced through his sleeve. Red welled through the white linen. His heart raced. This man had been dead for hundreds of years! It wasn't possible. The impossibility distracted him, the man threw himself on him with a snarl, driving the knife.

Instinct reared. Clarence side-stepped, snapping back to every prison fight, driving the impossible scenario from his mind. Just another inmate! He plowed in, knives flashing, slashing, fury at his mother, at that last boyfriend drive him; dodging, attacking, panting, and the man was backing off, wary, circling.

Angus threw himself into the water.

Clarence's heart pounded like a jack hammer, adrenaline rushed, his hands flashed in and out, stabbing, driving all the anger of the last months, last years, last lifetime, into it. Blades scraped chain mail, sank in flesh, in and out, and the man collapsed on his knees, face going gray.

Clarence fought with his laces, tore at his boots and shirt, and splashed into the water.

♫

"Get Allene," Niall gasped, trying to grab for the boat. But it drifted with the tide, slipping from his reach.

"You're drown...."

"Save Allene!"

Near the boat, Shawn tried repeatedly to escape MacDougall's grasp, but the man lunged at him over and over, punching, grabbing, pulling.

The last of Allene's red hair floated on the surface, and disappeared. Angus gulped air and dove, powering down with arms still strong, peering through water turned murky as the fight above churned up mud from below.

He saw her, a dark shape floating in the dark, churning silt. Her hair floated, her dress puffed as she sank. He pulled himself through the water, cursing his weakness, and grasped her around the waist, heavy in the woolen surcoat, kicking up with weak legs, pulling with his free arm, praying she could be saved. He moved as in a dream, pulling, kicking, barely moving. His lungs burned. She was too heavy.

♫

Shawn saw Angus disappear under the waves, a flash of bare feet, and he was gone. Niall flailed, his head slipping under water, and burst back up, gasping, spitting. Shawn tried to flip under the water, but MacDougall grabbed his hair, clinging to the upturned boat with one hand.

"I knew you'd come," he gloated. "Who are you?"

"The one who's not afraid of water!" Shawn snarled. He swung at

MacDougall's hand, wrenched his head. The man held on tight, his yellow teeth bared, his black hair slicked back from his high forehead. Shawn punched, but water slowed his movement. MacDougall kicked. Niall slipped under the sea again, shouting. Angus had been gone a long time.

Don't panic, Shawn warned himself. *Don't panic.*

He twisted, diving deep, hoping to drag MacDougall down with him. But the man held tight, trying as they both went under to grab him around the neck. Shawn twisted, kicking, caught suddenly in powerful arms. MacDougall would have to let go or drown, Shawn reminded himself. *Stay calm!* He wiggled and slipped through, as he'd learned to do when rescuing a drowning victim. A hand caught his ankle as he tried to reach air. He kicked, and suddenly, broke the surface, drawing breath, gasping, shaking water from his eyes, flinging drops across the surface.

No Allene! No Niall! No Angus! Amy waded farther in. She wasn't much of a swimmer. She'd drown trying to help. "Get out!" he shouted.

Then Clarence was running behind her, splashing, shouting, "Get back," legs lifting high, and dove, hands swinging overhead.

It gave Shawn hope, energy. MacDougall broke the surface, eyes wild, flinging water from his hair, searching for Shawn in his moment of blindness.

"Be careful what you wish for!" Shawn said, and punched him, hard, in the face, punched him again, punched a third time, his knuckles connecting with MacDougall's larynx. A crunch, cartilage crushed. MacDougall sagged, sank.

Angus broke the surface, gasping. "Get Niall!" he yelled at Shawn.

Shawn gulped air and dove, searching the muddy water, fighting panic. He couldn't, wouldn't come this far only to watch Niall die. He pulled himself through the water, muscles flexing in his arms, legs driving him, searching, searching, and there he was, thrashing, panicking, his hair swirling in clouds around his face. He saw Shawn, strained toward him.

Shawn ducked, flipped in the water, coming in behind him. Niall twisted, eyes panicked, reaching for him. Bubbles fluttered from his mouth.

Shawn kicked again, circling behind, and grabbed fast. Niall twisted in his grasp. Shawn hung on tight, hoping—*praying*—that his months of training had paid off. Niall was heavy, weighed down in his gambeson and breeks and boots. Shawn kicked his legs, strained upward with one arm, grappling to hang onto Niall as he struggled, panicked. The weight dragged at him, pulled on his legs, his free arm. Niall twisted and squirmed, bubbles flying like a flurry of thirty-second notes from his mouth. Shawn pushed upward through swirling mud, his lungs starting the slow burn that warned him he had only seconds left.

♫

I tense on the shore, furious with myself, knowing I can do nothing to help. I pace, fists clenched, march farther in, the tide rushing to my knees, wanting to help, back up, knowing I can't, knowing Shawn is right and I'll only make it worse for him and Angus; for Clarence, cutting through the water toward them.

Angus bursts up from the water, struggling, gasping, shaking water from his eyes. A cloud of red hair swirls around his head, floating in front of him. He struggles to keep her above the waves. He's in no shape for rescues after his long convalescence. I wade in up to my waist. The current tugs at my legs. I strain to stay on my feet, take a step back.

Shawn and Niall are nowhere to be seen. Clarence raises his head in the water, and suddenly dives, near where Shawn went under.

The muscles in my back scream with tension, straining forward.

"Air ye mad?" the old fisherman yells. "'Tis dangerous! Get out!"

"Don't you see them!" I shout back. I clutch my arms around myself, my lips tight. I can see Angus struggling under Allene's weight, without the full use of his legs. I take another step, farther in. I'm not a strong swimmer. I know I shouldn't. But I can't stand on shore doing nothing.

♫

A dark shape sliced through the water. Shawn's lungs burned. He fought the urge to draw in breath, hoping Niall wouldn't either. Clarence slammed into them, swinging at Niall's temple. Niall sagged. Clarence grabbed one arm and kicked for the surface, motioning for Shawn to follow. Shawn tugged Niall's other arm, kicking, slicing water, pulling upward. His head broke the surface. He heaved in air, gasping, flinging water from his eyes, hauling Niall up. Clarence's head burst through the waves. Shawn drew in breath, shouted, "Get him to shore!"

Niall lay limp in the water, easy to manage now.

"Where's Allene?" Clarence yelled, treading water with one hand locked across Niall's chest.

They looked around. Angus flailed, struggling to pull her toward shore. He slipped under the water. Her face, grey-white on the choppy waves, followed, only her hair floating. "Take him in and get back here," Shawn ordered, and dove after Angus. They were close to the surface. It took little to haul him up.

"Too heavy," Angus gasped. "Wet clothes. Can't get...them off."

"Can you get yourself in?" Shawn asked.

Angus nodded, and they managed to shift Allene, pale and limp, to Shawn. He clasped her in the rescuer's grip, and kicked, Angus swimming beside him.

"You've saved them," Angus said. "I can still see the cars on the shore. Come back with us."

Shawn glanced to the island, seeing the two cars side by side, and Clarence ahead, dragging Niall in. Amy stood in water to her waist, reaching for them, grabbing Niall, wading backwards to help Clarence pull him up on the sand.

"Get out," Shawn gasped. "If Amy crosses...no one left...for James. My mother. Get out." He struggled under Allene's weight. The dummy hadn't prepared him for the real thing, especially not a full length woolen surcoat.

Angus, freed from the weight, pulled an inch ahead.

"You have to come with us," he insisted.

"Go!" Shawn barked. "Can't leave James...orphan! Go!" Allene's weight grew heavier. He struggled to keep his head above the surface. Angus tread water, waiting.

"Go!" Shawn yelled, furious at the thought of James being abandoned. Heat prickled his eyes. He'd *counted* on Angus, counted on Amy, to be there for James and his mother. "Go!"

"I still have more upper body strength," Angus argued. "I'm close enough, I can do it now. Go back to Amy."

"And have you disappear in time while I stay?"

"She loves you," Angus said. "He's your son. He *needs* you!"

Shawn kicked his legs, dragging Allene another inch. On the shore, Amy leaned over Niall, who convulsed on his hands and knees, spitting out sea water. Clarence stood, hands on knees, doubled over, coughing. Shawn felt the strain in his arms. He could go back to Amy. The evening under the tree warmed his mind. If it was just the two of them, without Angus, it would be different. It could be. He kicked, scratched at water with a weakening stroke, gaining another inch.

Amy looked to him, her face etched in fear. He wanted her. He *wanted* his life with her!

"You can't give yourself up," he gasped to Angus. "Get her out of here."

"You're not going to make it."

Water lapped over Shawn's face. He swallowed sea water, choking. Angus pushed him aside. "Get yourself in. She's too heavy for you. I've had enough rest to finish." He locked his arm around her, while Shawn swam for shore, freed of her weight.

"Angus!" Amy heaved herself up off the sand and ran, wading into the water. It climbed, higher and higher, to her knees, her waist.

"Get her out," Angus shouted. "The current...."

Shawn looked from Angus, pulling Allene through the waves with renewed strength, to Amy, now up to her waist, entering dangerous undertows.

"Please," Angus said. "You can't carry Allene that far. You can get Amy out."

There was no time to argue. Shawn shot ahead, stroking as he'd once done racing his father across the lake, speeding for bragging rights, for the first bowl of ice cream, arm flashing over arm with a burst of adrenaline. She didn't know how strong the currents could be!

She came out further, the waves chopping above her waist, to her ribs. "Help him!" she shouted.

"Get back to shore," Shawn yelled back. "It's dangerous." The cars flashed sunlight on the shore. Two bodies in MacDougall's colors lay on the beach. An old fisherman shouted at him.

He could feel the current, and suddenly, Amy's arms flew up, and she disappeared. Clarence, on the shore, looked up from where he bent over Niall. Niall vomited onto the sand.

His legs sliced through gray sea, his arms windmilled over his head, he twisted his head, grabbing air, and propelled himself up into the shallows, scrambling to get his feet under him. The current swirled around his ankles, his knees, tugging. He jumped into it, grabbing, found her arm, and pulled, fighting the current, dragging her up the sand.

Tears streamed down her face. "Get Angus!" she protested. "Shawn, *help* him!"

"Get out!" Shawn yelled in her face. "Get in that car and drive before you're trapped here!"

"Angus," she yelled.

From the water, Angus shouted, "Go, Amy! Before you're caught here!"

Shawn wrestled her, dragging her by one arm to her Renault. Allene's red hair floated on the water alongside Angus's black curls. Clarence leaned over Niall, pumping his stomach, forcing sea water from his lungs. Shawn fought with the door, trying to push Amy in. "Don't you get it!" he shouted in fury. "You'll be stuck in medieval Scotland. James will be left an orphan in our own time. *Get out!*"

She shook her head, biting her lip, pushing at him. "Angus needs help."

He made a sound of anger, spun, and pelted for the beach, into the sea. Ice water shocked his ankles, his bones, his knees. He pulled with all his might through the water, desperately hoping the door would remain open long enough for Amy to get back, grabbing Allene's shoulders. He looked back. The car wavered, shimmered. Panic hit his heart hard, like Bruce's stones rolling down off Glen Trool in an avalanche.

He yanked on Allene, shouted to Angus, "The cars! They're fading! Go!"

Angus shook water from his eyes. Amy had come back to the shore shouting for him.

"Only one of us is going to make it," Shawn gasped, struggling to hold Allene's head up over the waves that fought for her. "Go. Take care of my son."

"I've never walked away from a rescue!"

"Then swim away!" Panic mounted as the cars on shore shimmered and wavered. "You have to rescue Amy! She won't leave without you. You have to teach James to fight." He pulled Allene from Angus's grip. "I have the antibiotics to save him."

Angus swallowed. "You'll break her heart."

"You'll break it more if you abandon James and leave her trapped here!"

Angus cast a glance at Allene. "Promise me you'll try to get out," he said.

"Promise. Right behind you! *Go!*"

Angus dove ahead, arm over arm, drawing himself to the shore.

♫

I meet him in water up to my waist, pressing the cane into his hand, pulling his arm, and shouting, "Hurry, Shawn. Is he coming, Angus?"

"Aye, right behind us." Angus pulls through heavy water, his legs weakening, till he all but falls on the shore beside Niall, who rises now on his knees, looking sick. He reaches a hand. "Amy."

I drop to my knees, hugging him. "Leave me messages, too, Niall, you and Allene. Thank God you're alive!"

Angus and Niall stare at one another. Angus offers his hand. "Niall?"

Niall nods. "Angus." He switches to English, an almost perfect modern American accent. "I wish we'd more time. God go with you. Haste, your car!"

Angus turns. Shawn struggles in the water.

"You haven't time," Niall snaps. "Get her to her son."

My lip trembles, looking from Niall to the amazement on Angus's face, to Shawn fighting the current.

"Come, Amy, we can't abandon James." Angus moves as quickly as he can with his cane.

The car wavers, light shooting through it. The thought of James drives me away from Niall, away from Allene, even away from Shawn struggling in the water. I push Angus into the passenger door, and fly to the driver's side. The fisherman is gone. Clarence watches from his knees at Niall's side.

Panic scratches at my heart. The fisherman is gone! It's too late!

I twist the key, relieved to hear the engine start, and throw it into reverse.

"Shawn...." I say.

"He's right behind us in his own car," Angus barks. "Amy drive away from this shore and the crucifix before James is left alone." He bows his head in his hands, looking as distressed as I feel.

He's never lied to me. Never.

I know he's lying now. Shawn is not right behind us. I have no choice, if I'm to get back to James.

With tears burning my eyes, I shift into drive, and hit the gas pedal. I can't think of it as leaving Niall and Shawn behind. I can only think of going to James and Carol.

FINALE

CHAPTER THIRTY-TWO

Lindisfarne, 1318

Shawn dragged Allene to the shallows, his eyes on the car as it sped across the half-submerged causeway, disappearing into the future. His heart screamed to go with them. Back to Amy. Back to James. *He wanted his son!* His mother. His beautiful home. Writing music. Playing on the world's great stages. He watched the car roar away, with sprays of water like great angel wings, hiding it from view.

Niall waded into the water, grasping Allene under the shoulders, helping pull her, heavy in the long skirt, to the shore that tried to creep away before them as the tide washed in. "Save her," he whispered. He dropped to his knees, tears streaming down his face. "Save her, Shawn!"

Shawn's attention snapped back as he lay her on her side on the damp shore. Her face was gray. Clarence joined him, slpping her back. Water spewed from her mouth. "More," Shawn snapped.

Clarence shook her, saying, "Wake up, come on, Allene!"

Together, he and Shawn pulled her up to her hands and knees, Shawn clenching his hands together under her stomach, pumping. "Again," he said.

They repeated the procedure, as Niall watched, hands on knees, too exhausted to protest or question, and suddenly, she arched violently, spewing sea water, and retching on the sand. Niall sprang forward.

"It's okay!" Shawn warned him back. "She needs to get it out."

She knelt on all fours, her auburn hair in wet, tangled knots, heaving, before suddenly collapsing.

"Allene," Niall whispered, dropping to his knees at her side.

She struggled up, blinking, and seeing him. "Niall?' She rasped over the sea water still in her throat, and suddenly arched and retched again. He stroked her back, her hair; he lifted her to her knees, wiping tears from her eyes.

"You're alive," she whispered, wiping at her mouth. "How?"

"Shawn is here." He wrapped her in a tight embrace. Tears stung his eyes.

"We have no time," Shawn said. "You okay, Allene?"

She coughed, but nodded, as Niall and Shawn dragged her to her feet. "We have to get off the island," she gasped. "He has men in the church."

"Clarence," Shawn said, "get the currach in."

Clarence, with a last glance at the medieval couple embracing on the shore, waded into the water, swimming out for the abandoned boat.

Shawn took both their arms, pulling them into the water, up to their knees. Both seemed too dazed to question. Clarence was swimming in, pulling the currach behind him. "Get them in," he said.

Niall and Shawn managed to tumble the still-dazed Allene into the small boat, and Shawn shoved Naill in behind her. He and Clarence climbed in from either side, careful not to tip the small craft, and rowed for the mainland as Niall held Allene, stroking her hair, his cheek to hers, oblivious to all around him.

On the far shore, Taran waded into the sea, pulling the craft up onto the rocks, shouting, "My lord! My lady! Thank our good Lord!" He took Allene's elbow and with Niall, helped her, stumbling, out to shore.

Shawn glanced between Taran's and Niall's garrons. "Can we get more horses?" He boosted Allene onto Taran's mount.

"Shawn?" Allene stared as if at a ghost.

"A farm house down the road a wee bit," Taran answered.

At that moment, a tall, thin figure emerged from the woods. Shawn stared, doubting his senses. "Brother Eamonn?" he said.

"Brother Eamonn?" Niall repeated.

The old monk strode forward, strong for his years. "Let us get her to the farmhouse. I will stay with her." He grasped the reins, looking from Shawn to Niall. "You must make haste to Monadhliath."

"I canna simply leave...." Niall started.

"I've spoken with milord Douglas," Eamonn said firmly. "He's sending men and horses. Word is already going to Beaumont that MacDougall is dead. Simon will soon be on his way. You've not a moment to waste. You *must* get to Monadhliath first."

Northumbria, 1318

"*How?*" Simon roared in Herbard's face, as they fixed oatcakes over fires in the morning. "How does a group of *unarmed men* wound and kill *half* the men of Claverock?" He grabbed Herbard's shirt, shaking him. "Alred is dead! *Dead!* So is Elard, and five more too injured to ride!"

"My Lord! I...."

His men stood in a ragged circle, barely breathing, avoiding his eyes.

"Half our horses are swaying on their feet," Simon shouted. "Two collapsed altogether!"

Herbard said, "They'll be well by...."

Simon shoved him to the ground. He crumpled, to the sound of a bone snapping. Simon spat on him. "What use are you?" Hands on hips, he glared up at the sky beyond the forest canopy, bright with the dawn colors.

Herbard whimpered on the ground, clutching his arm to his chest.

Another of his men approached—but only to a safe distance. "My Lord?" He uttered the words smartly, staring straight ahead.

"What else?" Simon growled.

The man faltered for a moment, then pulled himself straight, not meeting Simon's eyes. "They've taken ten of our swords."

Simon spun, slamming the flat of his blade against a tree. A string of oaths spewed from his mouth, as he rounded on his remaining healthy men. "*How?* They were unarmed! You pathetic oafs! Not a one among you deserves to remain with the house of Claverock!"

He paced to the edge of the clearing and back. It didn't matter, he told himself. He still had his fleet. MacDougall would have Niall and Shawn, by the time he reached Lindisfarne. They were worth a good price. He would yet sway Edward to believe in his loyalty, and barring that, Lancaster.

"Get up," he snapped at Herbard, and to another of his men, "Bind his arm that he might yet be any use."

As the man scurried to do so, a rider raced through the forest. On seeing them, his pony skidded to a halt. He slid off, as the animal drooped its head in exhaustion. "My lord Claverock! Thank God I've found you!"

"Good news?" Simon snapped sardonically.

The man shook his head, missing the sarcasm. "Nay, my lord! MacDougall is dead. They say James Douglas is even now mounting three score men to ride for Lindisfarne to retrieve Glenmirril's daughter. And the Laird of Glenmirril coming with another score!"

Simon looked around his men. Roughly thirty remained alive, armed, and in shape to fight. "Pull the dead into the forest," he said. He had one other desire to fulfill in his own time. It would take but few to accomplish that. And when he had, they would rest, regroup, and return to Edward to plead his case, to convince him he had been mistaken. "We will bide a short while with the monks of Monadhliath."

When he had carried out his vow to the future Brother Fergal, when he had killed all but those needed to prepare food for his men, their monastery would make an excellent refuge. None would think to seek him in the fastnesses of the Scottish Highlands. He pulled at his reins, turning the horse's head north.

Northern England, 1318

"The Lady Beatrice." Having ushered James across a large stone chamber, hung with colorful tapestries, past half a dozen women sewing and whispering as they glanced at him, the guard stopped before a woman, his own age, James guessed. She sat in a high-backed chair, apart from the others, gazing out an arched window to the fields and woods beyond, though a book lay open on her lap. "My Lady."

He shouldn't have come, he thought. He should have asked at Berwick, or at a town along the way, where to find the lost village. But Angus had guessed it no more than two day's ride from Wendale. It would take them far longer than that to reach Monadhliath and return so far. He had time to meet Beatrice.

And he had waited a lifetime to do so! Oh, how he had waited!

She turned, her gaze flickering over James. He felt himself rise taller, to his full six feet, and straighten his back, under her scrutiny. She was lovely! Fairer than he'd even imagined!

The guard spoke. "I present James."

"Thank you," she said, and in her voice, James heard the musical quality his father had described in Christina's. "You may go."

The guard bowed and left.

"James." She turned back to him. "Just James?" The woman arched a slender eyebrow. Her lips curved in a smile.

James couldn't help responding with one of his own. He was his father's son after all. She was fair. Soft honey blonde hair showed under her barbet, and the gauzy veil of it surrounded a heart-shaped face. Her smile waned; she offered her hand. James dropped to one knee, taking her hand and grazing her fingers with his lips. *Everything* stirred in him.

He *was* his father's son, after all.

He raised his head, meeting vivid blue eyes. His stomach lurched. He knew how it all ended. *And he couldn't wait for it to start!* But now was not the time. She was recently widowed, grieving—and he had a job to do. He bowed his head as he let go—reluctantly—of her fair fingers. "Just James," he agreed. "But I believe you'll be happy to hear my request, without knowing more."

"Rise," she said.

He did so.

"Tell me."

"I seek Simon Beaumont, Lord of Claverock."

Her eyes grew dark. "Ah, Lord Claverock. You are a friend of his? I've known him since I was a child."

"For that I offer my sympathy," James said.

"You are acquainted with his ways, then?"

"Quite well," James replied.

"His mother and mine are kin. He delighted in dropping spiders on my embroidery when I was but a child."

James lowered his eyes, thinking of the long-ago woman at Berwick, slain in the act of giving birth; of the girl Colleen killed behind the market, of Graham Dromond, Seamus P. Martin, and more, whose names he had never learned, murdered in alleys in the twenty-first century. They had walked with him through his early teen years, as he had learned what he was called to do.

"I see by your countenance," the woman said, "that you know dropping spiders on a child's embroidery is by far the *kindest* thing that man has ever done."

"Aye." James lifted his eyes back to her. Flirtation was gone.

She nodded to the women. "Ava had a sister. She threw herself into the moat after Claverock's...*attentions.* He was sixteen at the time. He's only become more vicious since then."

"And he seeks to do far worse still," James said. "I believe he was here and you might know which road he has taken."

She glanced at the women sewing at the other end of the room and down at the floor. She raised her eyes and spoke softly. "Come." She rose, a slender hand lifting the hem of her skirt, and glided to the window, putting distance between them and the women. She lowered her voice still more. "He killed my husband, here at this window. If he suspects I have done aught against him, he will kill first my son—most likely as I watch—and then me."

"The hope," James said, lowering his own voice, "the *prayer,* which I hope you and your ladies will offer until you hear news of me again, is that he will no longer be able to kill anyone."

"You seek his death," she clarified.

James gave a deep bow of his head. "I do, my Lady. For you. For your husband. For all those Claverock has hurt." He raised his eyes, meeting hers. "And even more, my Lady, for all those he plans to harm, I do."

"He will stop at Bellingham and from there take the road to Alnwick. I know not where he goes after that."

"And *Bailedàn,* my Lady—can you tell me how to find it?"

"*Bailedàn?*" She frowned. "That is in Scotland. There is no reason he would go there."

"I believe he will find himself there," James said. His father had left explicit instructions. The Laird and Douglas's men would meet Claverock near the spring just north of *Bailedàn.* No amount of research, however, had ever turned up the location of such a place. "Do you know where I may find it?"

"It is far north, on the edges of the Cairngorms, by the healing well."

His heart sank. "That is many days' ride."

"Aye," she agreed. "Nearly four, an' you ride hard and all goes well."

James drew a breath. There was no counting on all going well. He knew the land well in his own time—but there were now vast forests where there had not been.

She stared out the window, and added, "You understand he will take his vengeance on me, on my son, and on all who dwell here, if he thinks I have told you aught of where to find him."

"Thank you, my Lady," James said softly. "None will know where I learned of it. He'll not be back to harm you."

"He has my son," she said, more *pianissimo* still.

James wanted to touch her hand, there on the window sill, to reassure her. He held back, saying only, "Your son is as stalwart, brave, and clever as his father. He will be back shortly, leading the men of Wendale."

She turned to him, pain in her eyes. "I pray you speak true. But how can you know these things?"

He smiled. "I read a beautiful story." He didn't add that it included her. "I vow, my lady, I *will* return to tell you personally he is no more threat to you."

"I pray you do." She turned, moving gracefully back to her chair, and past it, past her women, saying, "I'm very sorry, James, that I am of no help."

"My Lady," he said, "we know at least that the stag *is* in the forest. What you have shown me is enough." He bowed over her fingers, brushing his lips against them. What pleasure! After a lifetime of waiting for her! He rose to see the light in her eyes.

But his journey was so much father than he'd planned. He feared he would be too late, and that history would change again. Nonetheless, he gave a promise. "I shall bring him down."

Berwick, 1318

"Hail! Well met!" One of Bruce's guards stepped from the gatehouse, laying a hand to the nose of the Laird's garron. "Glenmirril, aye?"

The Laird nodded. "My daughter has been abducted by Alexander MacDougall. We believe they're at Lindisfarne. We seek Sir Niall."

The man nodded. "Aye, we've news of that." He waved Glenmirril in.

"Good news?" asked Darnley.

"You'll want to find James Douglas. Find him at Bruce's headquarters." He pointed down the street. "The house beside the kirk."

"*Good* news?" Darnley pressed.

"Not bad," the man replied.

His tension draining just a little, the Laird led his men through Berwick's cobbled streets and a citizenry once again adjusting to new government. Women shouted from stalls about fish and wool. Sounds of construction rang from another street. At the far end of town, troops guarded the keep, waiting for the English to starve or surrender.

They quickly found the kirk and the house beside it. Inside, James Douglas leaned over a table, studying a map, with one of his men beside him. He looked up as his man announced their presence, and straightened, filling the small room with his height and great black beard. "Malcolm MacDonald of Glenmirril! 'Tis good to see you." His face was grim. "Your good-son has had quite an adventure."

"Is he *alive*?" the Laird demanded. "And my daughter?"

"Aye!" Douglas nodded vigorously. "I'm not a 'tall clear what happened out there, but he and the silent Brother Andrew arrived...."

"Brother Andrew?" MacDonald's pulse quickened. "Nay, he is long gone! 'Twas Alexander MacDougall! Did ye not stop him?"

Douglas's great black eyebrows beetled. "Well, now, Sir Niall *said* 'twas Brother Andrew and that MacDougall drowned in the tide off Lindisfarne."

MacDonald was silent a moment before asking, "*Niall* said 'twas Brother Andrew?"

Douglas nodded.

"And my daughter?" Hope surged in his heart.

"She is safe at a farm house with the old monk, Eamonn."

"Him!" MacDonald snorted.

"Your daughter is safe," Darnley said mildly. "It seems all is coming aright as our strange monk foretold."

"I've sent men to retrieve her," Douglas said. "She ought be here soon."

"And where is Sir Niall?" Darnley asked.

"That brings us to the moment." Douglas scratched his chin under the big beard. "They arrived in haste, speaking of Claverock." He coughed. "That is to say, *Sir Niall* spoke of Claverock. Brother Andrew seemed, for once, to honor his vow of silence. A peculiar monk."

"We seem awash in peculiar monks," Darnley said.

MacDonald found himself smiling. "Most peculiar, indeed," he said. "Though I find I'd grown rather accustomed his peculiar ways and perhaps had even missed them." His smile slipped. "What did they say of Claverock?"

"He was part of the plot with MacDougall to abduct your daughter and kill Sir Niall." Douglas dropped into his chair. "With MacDougall dead, he has retreated to Monadhliath—to do harm to the monks, they believed. You have perhaps heard tales of his rise in England?"

"That and wicked plots against Scotland," the Laird said.

"Aye." Douglas nodded. "That being so, I gave them permission to follow, he and his men, assuring them I would send my own close behind. The Bruce is none too happy with Claverock being on Scottish soil a 'tall and would fain see him dead or brought back in chains." He nodded at the soldier studying the maps. "Osgar is but now looking over the route. We'll be glad of your arms, as they are few enough I can spare." He glanced at the window and lowered his voice. "We hear strange tidings of Claverock and weapons he has built."

"Aye," the Laird said. "He may at that." He glanced at the door, beyond which his men waited. "We shall ride with them. Keep my daughter safe and assure her I'll soon return for her."

Douglas turned his gaze to Osgar, waiting by his desk, helmet in hand. "Brother Eamonn, when he sought me out, strongly advised me you will find them on the road just north of *Bailedàn*."

"'Tis a long way," Malcolm said.

"Aye. Ride hard an' ye wish to catch them."

Northumbria, 1318

Several hours north of Wendale, James took his rest in a small town. He had studied enough maps. He had walked the ways in his own time, six months of every year—as well as the old medieval routes *could* be walked. He knew exactly where to find the healing well in his own time.

It was nearly two hundred miles northwest—hardly the day and a half ride he had expected. He could do it in maybe four with the charger Beatrice had given him—if he could go straight there. But the landscape had changed. The roads had changed. Despite years of orienteering, he would likely have to ask along the way to confirm he was still on the right path.

With his plan set, he bought a skin of ale and a bundle of biscuits wrapped in a cloth from the innkeeper's wife—paying her with a coin left in the trunk by his father—and pushed Claverock's sword into the scabbard on his back. He mounted his pony and rode hard for the border, his jaw grim. He hoped he could be there on time—and that the one letter that never changed would not change this time.

Glenmirril, Present

We drive straight to Glenmirril, stopping only to collect a shovel and two large cups of extra strong coffee. I'm numb with shock as we crawl, Angus and I, into the green world beneath the ancient spreading limbs. "He can't really be gone." I'm aware I'm repeating myself, as Angus starts digging.

"Help me," he says, and I do. He is worn from the swim, from a long night driving. He struggles to shovel, unable to lean on his cane. We take turns, deeper and deeper till I think I must have imagined it all, or maybe Shawn died before he could ever get back to Glenmirril, or maybe he just didn't follow through. Or maybe, somehow, he didn't cross over at all.

And then...the shovel strikes metal.

It takes a call to Clive, who arrives with an assortment of ropes, before we can lift it out. It's heavy. The three of us look at one another under the tree. Clive leaves and comes back with a sturdy cart from the archives, and Angus's friends Jack and Charlie. Together, the four men lift the iron box onto the cart, and among us, we get it down the hill and into Clive's car, studiously avoiding stares from Glenmirril's visitors.

Mairi has brought Carol to Angus's home. There, we break open the chest.

A large oilskin envelope stretches out over the contents.

Carol shakes her head. "I don't believe it," she says. "Please tell me. Where have they really gone?"

"To save Niall." Angus takes her hand, as I lift the oilskin gently.

Gold winks underneath. Coins. Chains. Jewels, even. I stare in amazement. "Just like he promised."

Carol stares, too, her eyes wide. "How did he know it would be there?"

"Because he intended to leave it," Angus says.

I slide the parchments from the oilskin. The crucifix—Niall's crucifix that his father got from the monks of Monadhliath—slides out with them. I clutch it in my hand, my eyes squeezed tight. It is almost too much to comprehend that he took it from me just last night, less than twenty-four hours ago, and yet it has lain hidden in this chest, buried deep under an ancient tree, for the past seven hundred years. I drop it back over my head, tucking it inside my shirt, and turn my attention to the parchments.

The ink is remarkably intact. He has been—was—very careful with it.

"Read it," Angus says, his voice almost reverent.

Dear Amy, *I read softly.* We rode hard as soon as we reached the mainland. By the time we found Douglas and got leave, it took us four days to reach Monadhliath. We managed to snag gear for me and Clarence and even another 'Brother Andrew' robe. In the confusion before we managed to cover my face, I think people just thought Niall was everywhere, if they noticed at all.

It was our bad luck we nearly ran into Simon just an hour from Monadhliath. He saw us in the distance and gave chase. We reached it first because we were fewer in number, had the garrons, and of course know the hills far better than he does.

Monadhliath, 1318

"Open up! Brother William!" In the falling dusk, Niall shook the tall iron gates of the monastery. Shawn, in his gray robe, yanked the bell rope. "Let us in!" Hugh, Clarence, Owen, Lachlan, Conal, and Taran clustered behind them, Taran looking over his shoulder, scouring the darkening hills for their pursuers.

Brother William hastened from the buildings. His face lit at sight of Niall. "Sir Niall! Hugh!" Then he saw the robed monk. He scowled. "Did I not quite specifically ask that this monk of unknown origins...."

"Peace, Brother William," Niall said. "Lord Claverock is on his way!"

"The Butcher of Berwick." Brother William twisted his key in the lock and yanked the bolt on the gate, asking even as Niall's company pushed through, "What have we done to anger him?"

"We haven't time." Niall took the prior by the arm and led the group through the rising gloam, into the front wing, to sconce-lit passageways.

"Three to the abbey," Shawn said from under the hood, "three to the church. Go!"

Brother William puffed up like a rooster. "I'll be giving the...."

"Haste, men!" Hugh beckoned, charging down the torch-lit hall.

"Knives, Brother William." Niall interrupted, handing him three. "One in each boot, one in your belt. You were a great fighter, not long hence."

"I shan't let them in," Brother William said indignantly.

Niall smiled. "Brother William, let them in, or they will tear down the gates. Play the fool, deny knowledge of our presence."

As he spoke, Shawn pulled off the gray robe, revealing his face. Brother William stared in shock. "It was me who copied your music." Shawn grinned. "Let them search. All will be well."

At that moment, a shout sounded from the front gate, and the bell set to angry jangling.

Shawn's jaw tightened. "Let's go, Niall. You'll be fine, Brother William."

They turned and ran through the halls of the monastery, into the cloistered gardens, through the gate to the graveyard. Dark had come swiftly. They slipped down the path, into thickening mist. Niall stopped suddenly, staring.

A ghostly young monk stood on the cloister walk, staring, wide-eyed at him. Niall grabbed Shawn's arm and pointed. Shawn smiled. "You'll be fine," he told the wraith-like monk. He and Niall ducked into the graveyard.

♫

"Open, you fool!" Simon shook the monastery gates, glaring through them into the mist rising over the dark garden. The moon hovered low behind the kirk's steeple. Hatred of the future Brother Fergal, standing with his fire stick at this same gate, reared in him. It swirled in a black brew with anger at Audric, at Erol. And Wendale—he would deal with them after he regrouped. He would rebuild his weapons here. "Open to the Lord of Claverock!" He rubbed at his right arm. The wound ached again, today. Though it hardly mattered against monks and the small group he'd spotted in the forest, it was yet another irritant.

A strongly built monk appeared, a dark shape in the dark doorway of the front wing, hands in sleeves.

"'Tis the prior, Brother William," Herbard said.

"Haste!" Simon roared. "You harbor my enemies!"

The man passed the garden, ambling toward the gate with no hint of haste. "We've no enemies here," he said. "We are men of peace."

"*Open* the *gate!*" Simon's hand shot through the bars, grasping the monk's robe. "Open it or my men will run you through here and now."

"Peace." William felt for his key at his belt. "You are welcome to our hospitality."

Simon let go of his robe as the key clicked in the lock. One of his men shoved the gate open, nearly unbalancing the prior.

"Where are they?" Simon grabbed his arm.

"You are welcome to search for any you believe to be here."

Simon glanced around the black stretch of lawn and gardens, and shoved the monk down the path, followed by a dozen of his men. "It will go better for you if you tell me where you've hidden them." Inside the monastery, he looked up and down the hall, and chose the right, dragging the monk along.

"I know not who you think is here."

Simon kicked open the door to the cloister. He shoved the prior out onto the covered walkway, shouting, "Where *is* he, Brother William?"

Brother William stumbled, falling to his knees in the garden, as Simon's men, in their black tunics, poured into the courtyard, boots thumping, swords clanging as they swarmed the abbey, the church, the graveyard.

Shawn melted behind a high sepulchre that seemed to float above the mist. Its white marble gleamed in the moonlight, etching light and shadow over the face of the knight atop it, clutching the hilt of his sword for all eternity. Across the cemetery, he could just barely see Niall, flattening himself against the stone wall that enclosed the place on three sides.

Four men burst in through the iron gate. "Search behind every stone," one of them said. It was only moments before the first approached Shawn's hiding place. Shawn reached out, his hand sliding over the man's mouth, knife across throat. He dropped him, lunged at the second. Across the tombs, he saw Niall do likewise. Two down.

"Where is Campbell?"

Brother William's eyes darkened. "Find him yourself, Beaumont!" He lifted his head, shouting after the soldiers, "You'll not find Niall!"

Simon swung, his fist crunching against the monk's head. The man fell, palms in the cold dirt. Blood streamed from his temple. He shook his head, dazed. A motion in the dark caught Simon's eye. He turned to see a hazy monk, tall and slender, with wispy, thin hair even in his youth. *Brother Eamonn!* Simon's lips curved into a smile as their eyes met. The young Eamonn would see his power! He would fear him! And all would change!

In the graveyard, Niall heard Brother William's defiant shout. He shoved his opponent away and bolted through the white tombstones.

Another motion—and Simon spun as the prior lunged, a knife glinting in the moonlight. Simon swung his sword up. William slipped under it, and as Simon twisted after him, Shawn—or Niall—burst from the dark, misty graveyard, weapon swinging. William's knife flashed. Simon danced backward, his sword up, slashing, dodging, as the man drove in; ducking backward, sideways, and swinging again as the prior lunged with his knife.

Two men closed in on Shawn. He waited in a half crouch, sword in one hand, knife in the other.

"Surrender, Campbell," one of them barked. "Ye can't win. We're all over the abbey."

Shawn bared his teeth, letting blood lust wash over him. They would kill Brother William. They would kill Niall, Allene. They would kill his own son. He lunged in the space between them, swinging his sword with a powerful backhand, thrusting the knife, spinning.

"Anselm!" Simon shouted. "Bernar! Get back!"

Niall was on him, driving in hard, his jaw fierce in concentration.

Shawn burst from the kirkyard, shouting, "Your men are dead, Beaumont!"

Simon stared at him a split second. Blood streaked his face. Suddenly, Niall's men poured from the abbey, from the kirk, all staring silently at him.

Simon spun and ran.

♫

He charged down the hill in the dark, the path lit by moonlight. How could it have gone so wrong! Herbard waited there with the remaining men—not even two score now. "My Lord?" He looked up the hill. "Where are the others?"

"Mount!" Simon snapped. "We ride!"

"To where, my Lord?" Herbard was already struggling up onto his horse, assisted by one of the men.

Simon glanced back over his shoulder. There was no sign yet of pursuit. It would take them some little time to saddle their horses, clean their swords and be after him. But it wouldn't take long. "Away."

As he stepped into his stirrup and threw his leg over his horse, he looked around the hills for the best cover. "South." Wendale, perhaps? It was many days' ride to Claverock. His mind churned, trying to think where he might take respite. Pain surged through his arm. *How had everything gone so wrong?*

His men fell into a quick trot behind him, as fast as they dared go in the dark, dozens of hooves raining thunder on the dirt path, splashing through patches of moonlight shining down through the trees.

Under Simon's armor, the sweat gathered, turning him cold in the night air. He wanted, suddenly, to be home with Alice. *Give yourself up to the love you feel.* He grit his teeth, rage and anger warring with the desire that burned as deeply as the pain surging through his arm, to be home with Alice, to lie with her in bed, talking in the soft glow of the fire in the hearth, or walking in the spring gardens with her. He wanted to see her smile, see her eyes light up, to revel in the love she seemed, inexplicably, to feel for him.

They broke out from the trees, onto bare hillside, still with no sign of pursuit. He glanced back once, and turned his horse west. "To Claverock," he announced. With the moon lighting the ground, he kicked his horse into a gallop.

Lancaster was glad to work with him. The weapons were not yet complete and none but him knew the final step. Lancaster and the barons of the north were eager to ally with him. He smiled. Yes, his plans would proceed. He needed only a brief time to regroup. He still held the last secret to making the *guns* work. Edward could do nothing without him.

A shallow stream shimmered ahead in the moonlight. He led his men into it, ending the trail of hoof prints that gave them away, and they splashed their way west.

Near the Healing Spring of Bailedàn, Scotland, 1318

"We know not what we'll be facing," the Laird muttered to Darnley as they rode the forest path. He didn't care to have the men—especially the boys, young Wat and Jamie MacPherson—hear his concern.

"Aye," replied Darnley, "and we have stayed too long behind our walls, becoming old men as we depend on Niall, Hugh, and Conal."

The Laird's mouth tightened grimly, acknowledging.

"We're coming to the road of which Brother Eamonn spoke," Osgar said. He turned to scan the line of riders, several in their teens. "Weapons to hand," he murmured. "Softly now."

Moments later they emerged from the forest, into bright sunlight. Behind them, leather and chain rustled as men adjusted helmets and hefted swords. Ahead lay the small hamlet of *Bailedàn*.

"Village of Fate," the Laird muttered. "We shall see what our fate is, will we not?"

Osgar nodded grimly. They guided their horses through the hamlet and up over a ridge.

"Ahead." Darnley pointed. On a small hill north of them, sunlight flashed off the armor of several dozen men.

"God willing they've only the weapons we know," the Laird said, "for it looks from here that we are a wee bit outnumbered." He glanced back, his eyes falling in particular on the boys, well trained, but untested in battle.

"Are we certain 'tis Claverock?" Darnley asked. "They fly no colors."

"Clothed in black has been his way, as 'twas his father's," the Laird said.

"Prepare yourselves," Osgar called over his shoulder. As he did, a shout rose from the men across the small vale, and they charged.

"Stay with young Wat," the Laird called to Darnley. He clapped his helmet down tight, swung his sword once and spurred his garron, racing into battle.

Midwest America, Present

We arrive at Shawn's house still in shock—stumbling in after a long flight into MSP, fending off reporters asking about Shawn's disappearance while James buried his head in my shoulder in distress at their shouting and pushing and the lights flashing; after retrieving Carol's Great Dane from the sitter, and the drive down to this place I have long loved.

Angus sets his suitcase just inside the music room, staring up at the two story foyer, and into the great room. He moves slowly on his cane. James wiggles from my arms, sliding to the floor, and is immediately up again on his feet, moving with his hands along the walls, crying, "Ang! Angs!"

I let go of Cain's collar. He bounds through Shawn's house, sniffing every corner.

Angus and I move silently through the house together. He looks at the kitchen, at a note Shawn left him, two cookbooks. He opens the cupboard to shelves laden with a dizzying array of premium and exotic spices. He hands the note to me.

Tears fill my eyes as I scan Shawn's message to Angus, his wish for him to be happy here. He has been planning this for a long time.

I'm touched at his thoughtfulness even as it hurts that he lied to me, even as anger at his deliberate abandonment rears, even as I'm grateful he saved Niall from that horrible fate. Even through the guilt at my own refusal to leave that shore without Angus. I condemned Shawn by that refusal.

I open the refrigerator. A bottle of fine wine (what other kind would Shawn get?) stands prominently on a shelf with a gift tag tied with curling green ribbon around the neck. "Welcome home, Amy and Angus," it says. "I want you and James to be happy here." The flattened S, the trombone, is sketched underneath the words.

Blinking back tears, I pour two glasses. Angus leans on my arm, forgoing his cane, and James toddles beside us, clutching Angus's free hand as we head into the great room to the white leather couch and chair in front of the floor to ceiling stone hearth Shawn loved. It is laid with kindling, ready for a welcome-home fire. He's thought of everything. Guilt and grief consume me.

Angus lights the fire, poking twigs around, and joins me on the couch. I curl into the curve of his arm as James crawls into his lap.

"Should we have agreed to let Carol stay?" I ask.

Carol wavered a bit before choosing to spend that month at the cottage Shawn and Clarence got her. We wavered a bit about leaving her. But in the end, she insisted she needed time alone to grasp this horrible truth. And she had to be there, she said. She had to be near them, even if in a different century. I understand. I did the same, the first time he disappeared.

"It had to be her decision," Angus says.

Cain returns from exploring the breakfast nook, and plunks down on the fur rug in front of the hearth. He drops his head on his paws and lets out a sigh as big as he is.

"Are you ready to read the next letter?" Angus asks.

"In a few minutes." I finish the wine slowly, watching the fire crackle, before rising and going for the parchments. Years worth of them, by the look. That's good. I take out the next, in which Shawn tells of the battle at Bailedàn.

Southern edge of the Cairngorms, 1318

"Kill or die," Simon roared at his men.

Give yourself up to the love you feel.

No! Not until he had power! "No quarter! No mercy!" he roared, wheeling his sword over his head and driving down the hill.

They charged, dust swirling up around spinning fetlocks. Sweat tickled Simon's back under his gambeson; his right arm stung from the arrow injury. He hardly noticed in the consuming pyre of battle lust. He peered through the slits in his helmet, across the field as his horse galloped, already picking out the slight rider—a boy. He zeroed in on him, his sword slamming hard into the older rider at his side; teeth bared, a growl rising in his throat, he drove at the boy, sword flashing, wheeled, hacked.

They were down, falling one by one! A youth stumbled from the field, clutching his arm as he fell to his knees. Another wheeled his horse.

"Retreat!" one of them shouted. "Retreat!"

Give yourself up to the love you feel! Simon laughed, watching the enemy disperse. Yes, as soon as he had power, he would!

And above the fray, a short, staccato shout...*Hiya! Hi! Hi! Hiya!*

Claverock, 1318

Alice sat in the window sill, smiling as she looked down into the gardens, just breaking into bloom, such beautiful sprays of color amidst the greens of the herb gardens. She would walk there with her child. And Simon, when he came home from all his meetings with Edward and Rolf, from studying his ships and supervising the replenishing of Edward's army—Simon would walk with them.

She'd heard tales, whispered stories. They were hard to believe. He had been kind to her. Attentive. He laughed at her little jokes and marveled at the new windows and the flowers on the sill. His eyes lit when he looked at her— yes, they lit and his face softened!—and he told her tales, at night in the quiet of their bed, after he had made love to her, of the places he'd seen, of the great Longshanks, of a world far beyond the sea with streets of gold and mythical beasts he insisted were real.

She missed him.

Yes, when he returned, they would walk in the gardens, hand in hand.

"My Lady."

Alice turned at the sound of the healer's voice. She rose from the window seat and from her daydreams, to see the old woman she had known since she was a child, holding a small glass potion bottle of wine—lighter by several shades than it had been.

"You are with child."

Alice's smile grew. She barely noted that Godelva's did not.

"My Lady." The woman set the bottle on the table, "I wonder that you appear happy."

"Godelva!" Alice exclaimed. "Why should you think otherwise! But of course I am overjoyed!" She stopped at sight of the midwife's face. "Sure the wine shows I am with child? I've not miscarried?"

"My Lady, you are." Godelva indicated the wine. "You can see it has changed color." She joined the girl at the window. She touched her hair, and said softly, "You know the reputation of the Lord of Claverock."

Alice turned away, staring down to the gardens, her mouth pursed. "Ugly things happen in war."

"Indeed, my lady, they do. But some men are uglier than others."

Alice drew in a quick breath. "I hear the whispers. But they are wrong. That is not the man *I* have seen, these past months."

Godelva took her hand. "My Lady, I've known ye since a wean. Do ye trust me?"

Alice nodded, her face turning pale.

"And can I trust *you*?"

Alice nodded more fervently still.

"My dear one, he has ordered me to make sure you do not give birth."

"No." Alice shook her head. "No, Godelva, he'd not do that to me. Surely you lie."

The old woman patted the back of her hand. "My lady, no harm shall come to you. But let us pray, aye? For this child of yours will be father to one who will save many people and many lands."

Midwest America, Present

We wake up, blinking in the morning sun that pours through the tall windows on either side of the hearth. James is asleep on the floor with Cain. His favorite book, about the knight, is clutched under one arm.

Angus makes breakfast. I see his delight with the bounty of spices. It vies with his discomfort at taking over Shawn's kitchen this way.

Finally, we explore the house. He tries to hide it—but he gawks at the solarium exactly as I did the first time I saw it, at the sparkling blue salt waters and stone walls and glass ceiling. We find Clarence's suite—unlocked now—and the bows and arrows and targets stored there. It hits me again: He's been planning this for a long time.

It hurts—he once again lied to me—even as I wonder:

What I would have done, had he told me.

Angus is tense in the solarium. He won't tell me why.

He won't enter the master bedroom at all and I begin to understand. This is a home he did not earn. It is not in Angus to take—to accept—what he himself did not earn.

"Why did you agree to come, then?" I ask.

He shrugs. "My gut told me 'twas the right thing to do. Maybe I was wrong this once?"

"It's in his will," I say. "A third of all of this is yours." I smile. "Whether you like it or not."

A quarter if Clarence returns. But we know he's not returning.

Angus says nothing. We climb the stairs together—moving slowly behind James who holds the rail, two feet to each stair—to the four rooms up here. One of them has been turned into a library. The shelves are carefully organized. Autobiographies. Fighting skills. Politics. Medieval faith and philosophy. Languages. Three shelves hold a small armory: bows, arrows, wooden swords and knives.

James gives a short shout of glee. His face lights up, and, holding the library shelves, he scoots as quickly as he can to them. He drags a heavy wooden sword off the shelf, shouting, "Swor! Ang, swor!" It crashes to the floor, its weight too great for him. He beams as if it's Christmas morning.

My blood chills as if I hear a funeral dirge.

The words of Der Erlkonig spin through my mind.

He holdeth the boy tightly clasp'd in his arm
He holdeth him safely, he keepeth him warm.

Southern edge of the Cairngorms, 1318

James charged over the ridge into the midst of battle, slashing with Claverock's great sword. One black-clad enemy fell. James twisted in the saddle, swinging the sword back to smash another with the hilt, knocking him from his horse; thrust into an abdomen; crunching of chain. Something hit him hard on the back, knocking him from his mount. His helmet flew, lost among hooves and men.

He rolled, sprang to his feet. Claverock in his sights! He grinned. "I'm here for you, Beaumont!" he shouted. "Look at my face. You know who I am!"

"And I know history can change!" Beaumont laughed, then suddenly let out a growl like a wolf, and lunged. James swung, forcing back the first attack as their blades clanged and slid apart.

Beaumont thrust. James dodged. They parried, circled, swinging, blades clashing. A blow glanced off James's leg. He limped backward, the breath nearly knocked from him in the wave of pain.

The Scots had rallied. He grit his teeth and forced himself closer. "Village of Fate, Simon," he hissed. "With your own sword you left in the future."

Simon swung, catching James in the shoulder with a blazing fire of pain. James reeled, stumbling backward before catching himself on his good leg. He knew sword fighting well. Simon's swing had been weak and missed its mark.

Still, Simon laughed. "Perhaps it is not the weapon, so much as the warrior, that counts." He raised his blade, moving slowly around James. "Your time is no match for the Lord of Claverock."

James watched, his own weapon raised. Fire burned up his calf. His leg shook. And suddenly, it buckled, dropping him to one knee. As if he'd watched it a thousand times in a dream, he felt himself above the battlefield, seeing men stumble out of battle madness and back into reality, looking around the field, assessing. The fighting was over, and looming over him in black leather, Simon laughed. "History will change!" With a gleeful shout, he dove.

James watched, the world sliding in slow motion as Simon's sword rose, weaker than it should have been, rising to strike at his neck.

Pain consumed him, fire climbing up his leg, through his arm.

He'd practiced it a thousand times with Angus. He'd practiced it with one hand, with two. He'd practiced it blindfolded, practiced it woken from deep sleep. He'd practiced it with weights on his arms, growing so strong his school tested him for steroids.

With a smile, James drove his own sword up; up through mail that melted before the blade, up into Simon's gut, up to the blue sky beyond. Simon's eyes widened. Blood spewed from his mouth. His stomach muscles heaved against the blade, fighting it.

"One letter never changed," James whispered. And he sank into darkness, into voices, into hands pulling at him. His arm burned! Pain unlike anything he'd ever felt. He groaned as they pulled the mail off. The world wavered around him. He opened his eyes as they pulled off his shirt.

"Stay with us now! Stay with us!" A big hand patted his cheek. Cool air touched his back, his chest. They were propping him up on a rock. "Stay with us! You've lost blood. We're binding the arm."

...an old man...bushy beard...blazing red and gray eyebrows. James smiled weakly. "The Laird."

"Aye, an' you the spittin' image of our Niall. Who are ye?"

"James," he whispered.

"James who?" demanded another.

"Just James." Everything wavered. He swayed.

"Stay with us!" the old man shouted.

"It's okay," he whispered. "My father is coming." And he had a thick journal yet to write. He smiled, thinking of Beatrice. Darkness swallowed him.

Midwest America, Present

I didn't earn this. The thought plagued Angus, as they stood in the library, James desperately trying to lift the wooden sword off the floor. It had plagued him through the previous night in Shawn's kitchen, cooking a late dinner for himself and Amy and James with Shawn's spices; through sitting with her on Shawn's leather couch in front of Shawn's hearth. He was grateful they'd fallen asleep there—avoiding the discomfort of going into Shawn's huge master bedroom and sleeping in his king-sized bed under his forest green comforter.

He couldn't bring himself to look at the solarium again. Everyone expected him to live here. As foreign as this huge house in this huge land was to him, he thought he could even grow accustomed to it except for the one problem: he had earned *none* of it.

But now, as he studied the titles on the shelf, as his eyes skimmed the weapons, he understood. *Shawn knew.* He knew James's destiny. A large manila envelope lay on the front edge of the big mahogany table, his name scrawled on it in Shawn's strong hand.

He glanced at Amy. She picked up James, who protested, reaching for the weapon on the floor. "I'll get some eggs going." She carried the protesting James toward the door, finally relenting and letting him take the smallest of the swords along.

Angus sat down at the long table. He picked up the envelope, staring at it; tapping its edge several times on the polished wood. He knew what was in it. His gut told him. He didn't want to open it. He wanted to live, for another few moments—for another fifty years—in a world where the truth inside that envelope didn't smudge the beautiful colors with black and gray.

But it was going to happen. He had to face it.

He opened the flap and slid out a thick sheaf of papers. He flipped through. Apart from the top sheet, they were all printed—a lengthy set of instructions covering the next twenty some years. And on the top sheet, in Shawn's bold handwriting, as if he'd read Angus's mind:

Don't worry, you'll be earning it. I'm bringing back penicillin to make sure he survives the injury. Your job is to have him as well prepared as he can be, so it's just an injury.

Just was underlined. Three times.

Your job is to be there for Amy. I was a coward. Once again, I tried not to hurt her. I didn't tell her. I left the dirty work to you, and I'm sorry. You are the better man. You always have been. And I am honored if I can call you friend and grateful I can count on you to do all that needs to be done there—for my mother, for Amy, for James. Thank you, Angus. Once again, thank you.

Southern edge of the Cairngorms, 1318

Niall, Shawn, and their men thundered up over the ridge as the sun sank. They skidded to a halt, scanning the small vale below. Dead and wounded scattered the valley, drenched in a bloody sunset. Below, they could see the Laird sitting on a rock, the half-naked body of a young man sagging in his arms like the *Pieta*. He looked up, seeing them, and shouted unintelligibly.

"James!" Shawn kicked his horse, urging it onward, panic burning his heart, and in moments was sliding to his feet. A ragged bandage, bright with blood, covered his left arm, shoulder to elbow. Another bound his left leg.

Shawn fumbled with the leather bag on his belt. The case he'd taken from

Jenny's desk fell to the dirt. "Hold him steady," he said. Clarence and Niall were at his side, helping the Laird prop James up. "I need his arm. Fever?"

"Aye," the Laird said.

"'Tis a vicious wound." Darnley looked up from where he wrapped a sheet round the body of a youth.

"How long has it been?"

"We met them at noon."

Shawn's hands shook as he tried to open the case. He kept his eyes off James's face. He'd been a baby—just days ago. And now he was dying in front of his eyes and he couldn't get into the damn case! It sprang open suddenly.

Clarence caught the precious vials before they could hit the dirt. "Steady, Shawn," he said calmly. As the Laird again supported James's weight, Clarence knelt beside him. "Get them away from seeing this," he said softly to the Laird.

"Back off, ye eejits!" the Laird barked. Men scurried.

"The syringe," Clarence said.

Shawn's hands shook as he handed it to Clarence.

Clarence slipped the plastic cover off and slid the needle into the first vial. "Hold him steady." He pulled the plunger up, drawing life-saving elixir into the syringe, pressed the needle to the upper arm, and thrust. "Let's get his wounds clean." He looked around, catching Lord Morrison's eyes. "Find water in the village. Get it boiling."

Morrison drew back in affront.

"Peace," the Laird said to him. "Let us save the lad's life, aye?"

Morrison hurried off.

Shawn's heart slowed. He let his eyes, finally, go to his son's face. Just days ago, he had kissed him good-bye, a baby with round cheeks. Now he lay, a strong, wiry man in his twenties, sharp planes to his jaw—his breathing shallow. "Don't die," he whispered. He bowed his head in prayer.

They waited, as men buried the dead, as the sun sank behind the western hills, through the night by a campfire, and a second shot of Jenny's penicillin. As the sky lightened to gray James's breathing evened out. Shawn touched his brow. "You saved his life," he whispered to Clarence. "I couldn't do it."

Clarence grasped his hand, his head bowed. "It's the least I owe you."

James's eyes fluttered open as the sun stretched golden fingers over the eastern hills. His gaze flickered around the men as they gathered back around him—the Laird, Hugh, Niall, Shawn—slowly focusing. "Dad? Dad!" He squeezed Shawn's hand. Then he grinned. "Angus taught me well. It's just an injury."

FINE

Glenmirril, 1318

The Laird's party rode through the night, approaching Glenmirril under the darkened forest canopy, passing through a forest exploding with the dark foliage of new growth and the sound of birdsong already beginning in the dim hour before dawn. It was good to ride with Niall again. It was good to have his son ride beside him. It was good, even from under the hood of Brother Andrew, to be once more with Conal, Lachlan, and Owen, to hear their cheerful talk around him under the stars shining above the great oaks and firs.

Berwick Castle had surrendered.

They were going home.

As they drew closer, the Laird called back, "Niall! Brother Andrew! Ride ahead. Tell them we're home!"

When they had put a few furlongs between themselves and the group, Niall spoke. "I tried awfully hard to be rid of you,"

Under the hood, Shawn grinned. "Likewise. But it seems once again, I had to come and rescue your medieval heiny."

"Ah, and the price I'll pay for it," Niall sighed. "I suppose I'm stuck with your irreverence and peculiar notions for many years to come."

"You can only hope. In the twenty-first century, they're still paying good money for just a picture of me. *Best of Scotland,* remember? That's me!"

Niall laughed. But his next question was sober. "Is it a shock to see your James a man? I canna think that I should cross the bridge into Glenmirril to find my sons grown."

Shawn was silent for a full minute before answering. "I missed his entire childhood. I'll never get that back. But I was only a few days without him. Amy has lost him for the rest of her life."

"It seems, from Brother Eamonn's experiences, that it could be no other way. You saved his life."

"No," Shawn said. "Clarence saved his life." His father washed through his mind, his heart, and he felt, for a moment, that his father, not yet born, smiled down on him from some vast incomprehensible heaven where all times were one. *Your forgiveness saved your son's life,* he seemed to say. *May it be well with my sons.*

"You both saved his life," Niall countered.

They rode up over the last hill and stopped, their ponies side by side, just under the edge of the forest, seeing Glenmirril below. Its walls dark against a sky as rich and cobalt blue as Amy's eyes. Stars danced above it, sparkling silver flashes off the softly undulating waters of the loch behind it. Watch fires burned at intervals on its ramparts.

Emotion swelled in Shawn's heart. He missed Amy. It hurt to think he would never see her again, never watch her play violin again, never hold her in his arms again. He hated the thought of her left bereaved. Of his mother left alone.

It was distressing to know that, bit by bit, his son would tell him more than twenty years of their lives. He wanted to hear a story he was sure was beautiful —and he dreaded it, dreaded seeing a world he could never more be part of, and dreaded watching the young woman he'd loved, the mother of his child, age in a matter of weeks. And his mother—James knew an Amy approaching fifty, and a Carol of nearly seventy. It hurt. It ached with an unimaginable pain. He *wanted* them.

And as happy as he was to be with Niall and Hugh and all of them, he was now also riding toward the pain of a lifetime of Christina forever in his sight, and forever out of reach.

Beyond the castle, the night sky faded to charcoal gray. Shawn cleared his throat. "There's a name we haven't mentioned."

"Christina," Niall said.

At the sound of hoof beats, Shawn turned. Hugh rode up beside them. His teeth and eyes and the first streak of gray in his beard all flashed white in the moonlight.

"Is *congratulations* what's said in this time?" Shawn asked. He liked Hugh. Loved him, even. He was brother and uncle. He would be the kindest and best of husbands to Christina. Such knowledge did nothing to stop the sharp ache that left him feeling he must gasp for breath.

Hugh looked blank. "Congratulations?"

"Christina," Shawn said. "She said she'd agreed to marry you."

A big grin broke out under Hugh's black beard. "Aye, well she did do to that, now." He beamed.

"Hugh," Niall said with a grin, "Perhaps you ought tell Shawn all about it."

Shawn's jaw tightened. "No, really, I don't need details." It *hurt.* Half a dozen stars winked away in the night sky.

"'Twas when we met her as Lochmaben's men escorted her home. However, I'd work for the Bruce. I've not been home yet to follow through."

The tension sagged out of Shawn. Hope flickered in his heart. "You...haven't married her?"

Hugh shook his big head, the grin growing. "And I believe I shan't."

"But she agreed. She promised. I can't...."

Hugh's face became sober. "I'd ha' been joyed to wed Christina. We'd have been good and faithful companions. But none with a heart would think to stand between two who love as you and she. I'd no think of marrying her, now you're here."

Elation filled Shawn's heart, a slow bloom of joy, of gratitude.

"Niall," Hugh said, "In faith, you ought ride ahead and have Christina wait in her rooms, else we'll have a scene to explain in the courtyard." He laid a big hand on Shawn's gray-robed shoulder, and they watched as Niall rode out under a pearl-gray sky, waving to the guards on the parapets. The drawbridge lowered quickly.

Midwest America, Present

I stand at the stove, watching the omelettes, half unseeing. They are rich with the onions, peppers, mushrooms, three cheeses, and spices Shawn would have used. The soft pearl gray of dawn shows through the floor to ceiling windows beside Shawn's fireplace, and through the glass ceiling of the solarium, that I can see through his breakfast nook. James stands beside me, clutching my jeans and waving his wooden sword. Tears sting my eyes. It's easy enough to see what the array of books and child-size weapons mean.

I think back to that night Shawn and I first climbed the tower stairs. He was so arrogant. I was so weak—so much more so even than I knew at the time. How much we've both changed! The man he was then wouldn't give up a party for the life of his own child. No, I haven't forgotten that first child.

And now he's given up everything to save a life.

And this time I will have no choice but to give up my son.

"Da?" James asks.

I try to smile. "He's not here," I say. How do I explain disappearance to a child his age?

And how do I cope, myself, with knowing that one day, James is going to follow? I guess I could convince myself it's only what Shawn believes—that it's not necessarily going to happen. I slide the first omelette from the pan. I don't want to live in blind denial. I mean—of course I do. It would feel so much better to believe my son is going to be with me forever.

...until the moment he left.

I would still have to deal with it then.

The first washes of pink and orange rise in the windows beside the fireplace.

I slide the second and third omelettes onto plates. I know Simon. I know I have to let James—encourage him—even help him—leave me forever.

"Da?" James asks again. I press the heel of my palm to my eye, pushing

back tears, and lift him up to my hip. "Later," I say. Much later. Pain stabs through my heart. It's hard to believe I'll never see Shawn again. That he's just gone. It's hard to wrestle with the guilt of refusing to leave without Angus—I made Shawn stay.

And mixed with the pain is the selfish question of why, if he was going anyway, he can't deal with Simon. Let me keep my dear, precious son.

I believe Brother Eamonn would say it is not Shawn's destiny to do that.

Tears slide down my cheeks.

The windows beside the fireplace, and the glass ceiling of the solarium explode with pink, orange, red, with the blazing glory of dawn.

I hear Angus's footsteps heavy on the stair, a waltz rhythm with the downbeat of his cane. Cane-step-step on the carpeted stairs. Cane-step-step. I try to brush away the tears, to hide from him that my heart is half with Shawn; that it always will be. But the tears are flowing. I turn away, trying to hide. Cane-step-step on the granite floor. Cane-step-step.

He turns me around, his arms circling me and James both.

"I'm sorry," I say. Sniffles garble the last syllable.

He strokes my hair. "I've stood on pride," he says. "'Tis alright to love him."

He rocks me and James, standing by the stove, and when my tears stop, he pulls away. He clears his throat.

"You've forgotten an egg." Dawn explodes in the kitchen, shining off the granite counters.

"No," I say. "They're on the plates. They're ready."

He pushes the egg carton at me. Confused, I open it. One cup holds a small piece of purple satin, in which a ring glints in the blaze of dawn. "It seems," he says, "that I've inherited a wee castle and though I'm completely unable to kneel for my lady as I'd like, I would be honored if you would be my queen."

I bite my lip, caught between tears and laughter. "Only if you don't mind me saying that would be the corniest proposal ever, if it weren't for the circumstances."

He grins—the happiest I've seen him in months. "Aye, but there are those rather peculiar circumstances, for we've a knight to raise, have we not?"

"Knight! Knight!" James waves his small sword. "Angs!" He plants a kiss on Angus's cheek.

"Yes," I say. "Yes, I want nothing more than to marry you."

Glenmirril, 1318

Shawn slid off his garron amidst the crowd, as the wisps of cloud turned pink in the gray sky.

Clarence stood beside his own pony, looking bewildered at the swirl of life and chorus of voices around him, at women and children rushing forth, seeking husbands and fathers. From the kitchen stairs, Bessie emerged into the courtyard, carrying a large basket brimming with bannocks, her cheeks red from the great hearths below. Dark tendrils of hair curled from under her kerchief.

Clarence straightened, his eyes on her. Bessie glanced at him, and suddenly her cheeks flushed even more. She dropped her eyes. Shawn glanced between them. And suddenly, Brother Eamonn was there, saying, "Clarence, sure you can help Bessie with her basket, aye? Bessie, are you needing more wood for the hearth?"

"I can help with that, too," Clarence said quickly. He hurried off.

Brother Eamonn winked at Shawn. "I made a promise to Bessie—and to Beaumont. I'm afraid neither of them believed me."

Before Shawn could react, the old monk faded into the crowd. At the same moment, a voice rang out, "M' lord Shawn!"

He spun to see Red gaping at him, his eyes wide. "I *told* you not to call me that!" Shawn said, but he couldn't help letting out a whoop of joy, and hugging the boy, hard, slapping his back.

"They said you were never coming back, ever!" Emotion stung the boy's voice, and Shawn thought of Red's father dying, his mother dying, the farrier who had raised him killed by soldiers.

"I'm back," he said. "And I'm not leaving again. There's someone I want you to meet." He looked forward to introducing James to the people of Glenmirril. He was walking into a beloved family who would embrace him as their own. "But first, Red, I need to see Christina."

He glanced up at the second story windows, still dim, almost expecting to see her. But they were empty. Leaving his garron with Red, he hurried across the courtyard, up the circling stairs inside the tower—the stairs he would one day, seven hundred years hence, climb with Amy, arrogant and obnoxious and cruel. How he'd changed! He climbed them in all haste, and broke into a run at the second floor, charging down the hall to Niall's chambers, the chambers that held his old room, the room that was now Christina's.

He threw the door open.

She stood at the window by her easel, looking out to to the pink and orange wisps of cloud over the loch. On her hip sat a child barely older than his own James had been, when he had left just days ago. Shawn threw back his hood, staring at him—at MacDougall's son, with the ruddy cheeks and black hair that would one day be Angus's. The boy laughed and let out a squeal of happiness.

Shawn grinned and began tugging the robe over his head.

Christina turned. Her eyes widened. A hand flew to her chest, as she drew in a long, slow breath. "Shawn?" she whispered. A tear rolled down one cheek.

He nodded. He threw the robe aside, and by the time he reached the divan, she was there, she and James Angus, her free arm clasped around his neck as he enveloped them both, burying his head in her hair. He felt the tears slide down his own cheeks.

Behind him, the door opened. He couldn't tear himself from her, from the scent of her hair, as Niall, the Laird, Hugh, Brother David, Red, and his own son, James, all crowded in.

Christina pulled back suddenly at sight of Hugh. Her face fell. "Shawn, we canna...I've given my promise...."

Hugh grinned. "Marry him, Christina, with my blessing."

The Laird spoke, with a broad grin. "I believe *this* time you'll be happy to speak your vows anon?"

"I will!" Christina's voice broke as she lifted her eyes to Shawn's.

He touched her cheek, the ivory skin; met her dark eyes. He'd given up everything, the great stages of the world, his beautiful home, Amy—to maybe die a horrible death here someday.

But Niall was alive. He'd saved his brother's life.

And Christina's eyes glowed with love.

The door creaked again, and Clarence entered, Bessie shyly at his side. Grief filled Shawn, at all he'd lost, at never seeing his mother or Amy again, as he met Clarence's eyes. "I miss them, too," Clarence said. "That's why it's called selflessness."

Christina laid her hand on his chest. "All will be well," she whispered. "We'll keep them alive in our hearts, aye?" She looked to James. "Your son?" She breathed the word in amazement.

James grinned, acknowledging, and reached out, taking James Angus from her. With Amy in half of his heart, Shawn wrapped his arms around Christina, his face pressed to hers. Gratitude to her God, and humility, filled his heart. Maybe—maybe he could at least try, for her sake, to be friends with Him. Just a little bit. Yes, selflessness hurt. And yet—he felt right and whole.

He pulled James, James Angus, and Clarence into the circle, with himself and Christina, and joy filled him. "All is well," he agreed.

The End

The Castle of Dromore

Prelude

I park in the tall waving grasses surrounding my new home, and leave the car, clutching a small key that seems anachronistic to the medieval castle before me. But it fits neatly in the large wooden door that fills the gatehouse entrance. A moment later, I'm inside. The wonder of a castle is that it's not just a home, it's a whole new world, alive with all that once was and all that might be.

I shut the door softly behind me. High stone walls mark the boundaries between my new Eden and the real world. Some of those walls are crumbling at the top. The tower and one wing, I know from my previous visits, are refurbished, with kitchen and bathroom facilities, and enough bedrooms, each with arched stone casements, for all the kids. Other wings will give them lots of twisting passages in which to run and play.

But the real treasure in this castle is its courtyard, overgrown with three large oaks spreading their summer-green limbs out over a jumble of wild grasses, out-of-control blackberry bushes, and an artist's palette of wildflowers springing up everywhere. The scent of wild roses drifts through the air, accompanied by the humming of a bee. It is here I will have my morning coffee. No day could start better. There is a wild beauty to God's planning that no master gardener can match.

A rustle catches my ear. There, half-hidden by ivy climbing over an ancient arbor, stands a woman in a long jade-green dress, with sleeves trailing to the ground. A veil barely holds her wild red hair in place. I blink, and she is gone.

Chapter One

Lisa ushered the five boys up to the castle's 'front door.' At least, she supposed it was their new front door, though like none she'd ever had before. It was really a pair of doors, standing at least ten feet high, arched at the top to fit into the curve of the stones overhead.

Ryan reached up to touch one of the rounded brass studs, dozens of which adorned the door. "What are these?" he asked.

Justin, a year older, jumped into the nano-second of silence before Lisa could draw breath and answer. He had a gift for that. "Back when castles were fortresses, enemies would attack the front door with axes, among other things. These dulled the blades of the axes."

At seven, Dylan understood enough for the words ax, enemy, and attack, to worry him. He looked up to Lisa, his eyes squinting in the sun behind her head. "Are enemies going to attack us now that we live in a castle?"

"No." She fit the large iron key into the lock, wondering that such a key and lock should have lasted so many years, and curious if something more modern could or should be installed. But she liked this key, and saw no real reason to change anything. The lack of central heating would be the more pressing expense, anyway, once winter claimed their hilltop. The boys, too, got a kick out of this being their house key, although it would be less than practical, its size and weight, in her purse.

With a twist and a stern warning to the five boys prancing at her side not to barge through, she pushed the right-hand door open. It creaked on its hinges. She examined them, to see an ancient pin mechanism. She should probably oil it, she thought.

"Wow!" exclaimed Jacob. "It sounds like a ghost!"

Again, Dylan looked up to Lisa. He took her hand. "Mom, there aren't ghosts here, are there?"

She cleared her throat. "They charge extra for the ghosts. You know I don't have a lot of money, Dylan."

He took it as a no, let go of her hand, and raced with his four brothers into the courtyard.

Watch for news of *The Castle of Dromore* and other releases at:

Gabriel's Horn's Sites:

www.gabrielshornpress.com
www.twitter.com/gabrielshornbks
www.facebook.com/gabrielshornpress

Laura's Sites:

www.bluebellschronicles.com
www.twitter.com/lauravosika
www.facebook.com/laura.vosika.author
http://bluebellstrilogy.blogspot.com